Diagrammatic Algebra

Mathematical Surveys and Monographs

Volume 264

Diagrammatic Algebra

J. Scott Carter
Seiichi Kamada

AMERICAN MATHEMATICAL SOCIETY
Providence, Rhode Island

2020 *Mathematics Subject Classification.* Primary 18M15, 18M30, 18N25, 55N99, 57K10, 57K16, 57K45, 57R42.

For additional information and updates on this book, visit
www.ams.org/bookpages/surv-264

Library of Congress Cataloging-in-Publication Data

Names: Carter, J. Scott, author. | Kamada, Seiichi, 1964– author.
Title: Diagrammatic algebra / J. Scott Carter, Seiichi Kamada.
Description: Providence, Rhode Island : American Mathematical Society, [2021] | Series: Mathematical surveys and monographs, 0076-5376 ; Volume 264 | Includes bibliographical references and index.
Identifiers: LCCN 2021035872 | ISBN 9781470466718 (paperback) | ISBN 9781470468156 (ebook)
Subjects: LCSH: Algebraic topology. | Homology theory. | Intersection homology theory. | AMS: Algebraic topology – Homology and cohomology theories [See also 57Txx] – None of the above, but in this section. | Manifolds and cell complexes {For complex manifolds, see 32Qxx} – Differential topology {For foundational questions of differentiable manifolds, see 58Axx; for infinite-dimensional manifolds, see 58B
Classification: LCC QA612 .C37 2021 | DDC 514/.2–dc23
LC record available at https://lccn.loc.gov/2021035872

∞ The paper used in this book is acid-free and falls within the guidelines established to ensure permanence and durability.
Visit the AMS home page at `https://www.ams.org/`

10 9 8 7 6 5 4 3 2 1 26 25 24 23 22 21

Contents

CHAPTER 1

Introduction

Algebra may be as old as civilization itself. Apparently, Babylonian clay tables not only contain systematic tabulations, but also contain algorithms for obtaining the tabulations. At its inception, algebra encrypted arithmetical rules via the use of variable quantities. The algebraic idea has evolved and currently encompasses virtually all of mathematics.

That is, symbols encode conceptual constructs as disparate as measure theory, topology, and indeed that which we now call modern algebra. On a day-to-day basis, mathematicians manipulate symbolic expressions according to certain rules. The expressions are combined in equations, or more generally relations, and algebraic manipulations (rules of rewriting the symbols) allow one to reinterpret relations in a more simple setting. Often algebra appears, both to a novice and to a practitioner, as a method of interchanging the relative position of linearly arranged symbols. Or more generally, one string of expressions is replaced by another such string.

In diagrammatic algebra, symbols, icons, or glyphs that represent an algebraic quantity or a topological quality are arranged in the plane. These are interconnected according to specific rules. Replacement rules allow one, usually connected, configuration of glyphs to be replaced by another. These rules reflect the algebraic or topological context that the glyphs represent.

More is true. Higher order glyphs can encode the replacement rules, and configurations of these may also be subject to replacement rules. In this way, relations among relations can also be represented in a glyphographic fashion.

Let us demonstrate an example that will be developed in some detail herein. An associative binary product "$(\cdot)$" may be given. As usual, write $a \cdot b = ab$. In a not so usual fashion, write 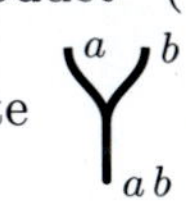

. Then the re-association $a(bc) \longleftarrow (ab)c$ may be written as

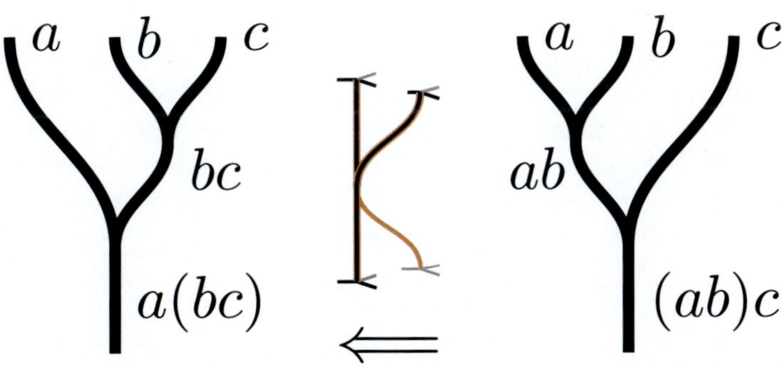

where the glyph  is named an "associator." Much later within the text (Chapter 10) a relation among associators is posited:

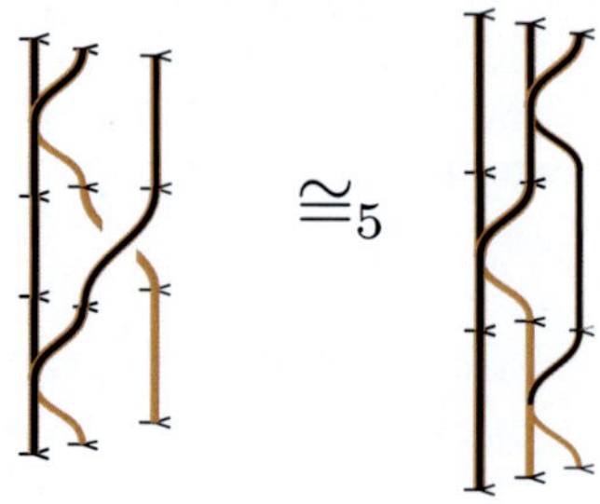

This relationship is motivated from algebraic considerations, but its representation is a distillation of the topological depiction. Using some techniques borrowed from the theory of abstract tensors, the edges of the glyphs are indexed by a module basis, and an abstract tensor relation is formulated.

$$K^{dg}_{ab}K^{ef}_{dc} = K^{hj}_{bc}K^{e\ell}_{ah}K^{fg}_{\ell j}$$

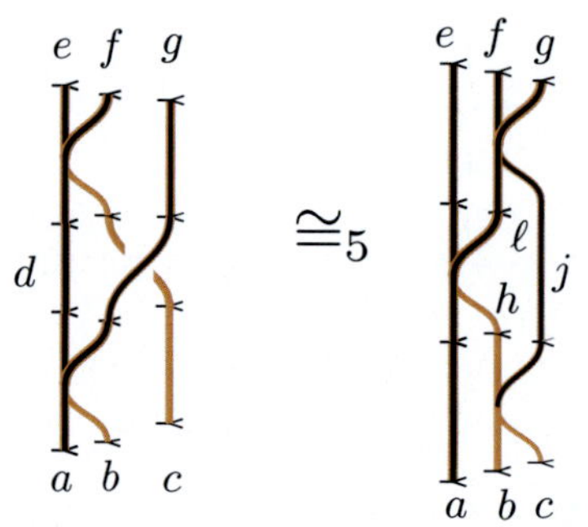

A solution to this equation (and several more) is given by means of an additional structure upon the basis of the module. That structure, a "shalgebra," is motivated by means of other diagrammatic relations that are akin to the relations between the associativity trees[1].

So one aspect of diagrammatic algebra is the use of relations between diagrams to express relations between algebraic expressions. When such diagrams are used judiciously, algebraic identities become geometrically apparent. For example, the functorial relations between 2-dimensional topological quantum field theories and Frobenius algebras is one realm for which the diagrammatic schema are facilitating. Another realm is within knot theory and its close relationship to quantum groups.

---

[1]The convention of using a Y to represent multiplication rather than a ⅄ is not consistent throughout the text. Be careful within the local context.

Within the text, we start from a very basic level, and establish some of the diagrammatical yoga within the linear algebra arena. Then we move towards the topological. Relationships in Frobenius algebras correspond to surface-like structures (foams). Diagrammatic calculi (different families of glyphs and relations among them) are established in different contexts. For example, embedded surfaces in 3-dimensional space are represented by a simple family of glyphs that represent birth, death, saddles, forks, cusps, and exchanges between fold lines. Knotted strings and knotted trivalent graphs in 3-space, as well as knotted surfaces and knotted foams in 4-space have related glyphographic calculi. In all of these cases, a limited (finite) family of glyphs represent the geometric context, and finite families of relations between compositions of glyphs are established. These diagrammatic representations are extended to study relationships among handle attachments, and codimension 2 braidings of 3- and 4-dimensional spaces.

Our methods are foundational. They allow for accurate diagrammatic depictions of higher dimensional knottings and isotopies between them. We also play back and forth between the diagrammatic and the traditional algebraic suzerains to demonstrate that many of the relations between glyphs correspond to cocycle conditions for an algebraic structure whose very definition was diagrammatically motivated.

The study of higher dimensional knotting was initiated by E. Artin nearly 100 years ago. In our previous works in this series [**15, 40**], the authors discussed the problems that are associated with diagrammatically depicting the unseen provinces of higher dimensional spaces. Both of these books discussed the problem from the point of view of surfaces that are embedded in 4-dimensional space. At the times of those writings, the diagrammatic theory of knotted surfaces was in its infancy. But these diagrammatic theories have grown in parallel with the uses of string diagrams to represent categorical compositions. The current book is a sequel to both of our previous works that can also be seen as an introduction to the topological as well as the categorical.

Two very intricate and beautiful dominions inform the narrative. First, the constructions of small categories for which the collection of arrows is also a category work well within the confines of globular 2-categories. More generally, our interests include category structures upon the collections of 2- and 3-arrows as well as higher order arrows. When the categories of concern are globular, finitely generated, and the identities are strict, then these constructions are rigorous. Strictness of the identities allows that compositions can be expanded by identities, or when identities are present these may be contracted. Exchangers (interchangers) are defined when arrows and relationships between them are finitely generated. Our assertions of rigor have the certitude that all of the constructions given within this volume have been implemented within the publicly available software GLOBULAR.SCIENCE (See Chapter 12).

Second, singularity theory informs much of the work in Chapter 6. In that chapter, the singularity theory is apparent since we write about births, deaths, saddles, forks, cusps, and relationships among them. Maps from a compact surface to the plane may be approximated locally by cubic functions, but such parametriztions are rarely, if ever, made explicit. Moreover, knot theory is informed by smooth projections to the plane. Surface knots and knotted foams are similarly projected using general position maps. Chapter 10 focuses upon eight relationships among embedded foams. Seven of these correspond to (codimension 2) singularities. While we don't write of them here, there are even higher order relations among foams that correspond to codimension 3 singularities. These also have interpretations as cocycle conditions.

The categorical and singularity underpinnings of the current work will be left to others. This work is written in a more casual, topological, and descriptive mode.

Other diagrammatic theories. Within the authors' lifetimes other calculi have developed. The Kirby calculus [**50**] allows experts to manipulate decorated knot diagrams in order to describe and relate both 3- and 4-dimensional manifolds that are thereby represented. We dub this a form of diagrammatic algebra: diagrams are drawn, and rules that allow us to manipulate the diagrams are used to determine whether two diagrams represent the same object.

The Jones polynomial and its daughters (quantum-type invariants) can be developed by using knot diagrams to encode matrix multiplication. The Witten-Reshetikhen-Turaev invariants [**65, 78**] demonstrate that the diagrammatic axioms of the Kirby calculus are mirrored by representations of quantum groups [**44**]. These ideas are touched upon in Chapter 9.

In the process of moving from matrix to diagram, observations have been made that large portions of quantum mechanics can be subsumed into a diagrammatic context. See for example [**1, 5, 18**]. Coecke, in many collaborations, is extending the diagrammatic metaphor into the study of linguistics and the ZX-calculus, e.g. [**17**].

Many of the advances within the categorical, physical, and diagrammatic realms have been made in the context of computer science [**25, 26**]. This perspective is interesting from a historical context. Dodgson developed diagrammatic descriptions of logic at Oxford, and their computer science department is continuing the traditions of diagrammatic axiomatizations.

The success of quantum topology is predicated upon the ability to use diagrams to encapsulate algebra. Moreover, it gave further methods to study relations in arbitrary categories, and better yet, in higher dimensional categories. As we mentioned above, the notion and the notation that is used to describe higher dimensional knots has grown in parallel with the notation that is used to describe higher dimensional categories.

One aspect of diagrammatic algebra that we only make tangential reference to is that of categorification. This word has changed meanings since its introduction in [**20**]. Those authors proposed that a categorical analogue of a Hopf algebra could be used to construct a 4-dimensional topological quantum field theory. Khovanov's [**48**] description of his theory of knot homology is dependent upon the ideas in which representations are lifted into a more broad categorical context. Khovanov's homology is described as a "categorification of the Jones polynomial." Moreover, Khovanov is quite talented in his ability to move algebraic axioms into a variety of diagrammatic realms. Therefore, diagrammatic algebra is used in many different contexts. In Chapter 9, Section 4, the definition of the Jones polynomial is outlined. The Jones-Wentzl idempotents are related to the quantum group representations of $sl_2(q)$. Chapter 9, Section 6 also contains a brief outline of Khovanov homology, and in Section 6.3 a diagrammatic version of a Hopf algebra is presented.

In our own experience, the judicious use of diagrams and diagrammatic manipulations allows us both to predict algebraic identities and to verify them. In many cases, algebraic strings of symbols can be interpreted as a diagram in another province, and the diagram suggests methods for simplification. Some computer programs, such as GLOBULAR.SCIENTIFIC or her smarter younger sister HOMOTOPY.IO allow for diagrammatic manipulations, and even are equipped to show the path of simplification so that an author can remove unnecessary steps.

Chapter-by-chapter overview. The main purpose in Chapter 2 is to establish notation for Penrose's [**62**] abstract tensors. Here a morphology from font to glyph is demonstrated. The definitions of an algebra, in general, and a Frobenius algebra, specifically, are presented. Diagrammatic methods show that the set of $(n \times n)$-matrices form a Frobenius algebra. Section 5 defines some general principles that should be taken as philosophical axioms for the remainder of the text.

Chapter 3 abstracts the multiplication/co-multiplication, pairing/co-pairing, and unit/co-unit of a Frobenius algebra into a category in which arrows are diagrams. Most importantly, it defines the notion of an exchanger and discusses the naturality of exchanges.

In Chapter 4, the algebraic axioms of a Frobenius algebra are transformed first into surface-like diagrams and next into 2-arrows in a multi-category. So here the diagrammatic process is continued. Exchangers are discussed in more detail. Multiplication and co-multiplication, unit and co-unit, and pairing and co-pairing are all interpreted as a pair of arrows in which one is the reverse of the other. The composition of following an arrow and its reverse is compared, by means of a higher order arrow, to doing nothing. On the other hand, one could first follow the reverse of an arrow, and then follow the arrow. This too can be compared to doing nothing. Those comparisons move the discussion from the algebraic towards the topological. Glyphographic notation for the algebraic axioms is introduced.

Chapter 5 continues moving Frobenius algebras to a more topological and diagrammatic arena. There are two main ideas in this chapter. The first has to do with the idea of composing an arrow and its reverse in either order and comparing these compositions to identities. The second has to do with describing diagrammatic proofs as higher order arrows. Handles are introduced.

Chapter 6 leaves the notion of a Frobenius algebra and studies a multi-category that has two objects and a pair of generating arrows between them. This situation is the prototypical example of an arrow and its reverse. A careful analysis of that situation results in an interesting topological context. In this chapter, there are objects, arrows between objects, double arrows between compositions of arrows, triple arrows between compositions of these, and quadruple arrows among compositions of these. So the level of abstraction has grown. However, the context is mathematically familiar: the categorical/diagrammatical aspects describe embedded surfaces in 3-dimensional space.

Chapter 7 abstracts the multi-category that was defined in Chapter 6 even further. It first exemplifies the axioms of the previous chapter in the contexts that were associated to Frobenius algebras, and it establishes an inductive construction.

The relations among births, deaths, saddles, forks and cusps that were described in Chapter 6 hold among higher dimensional arrows. Herein the idea that an arrow and its reverse need not be inverses is continued. Compositions of higher order derived arrows can be compared and some, but not all, of these compositions are thought of as equality. We believe that this approach is novel and non-standard. It abstracts concepts from singularity theory to the category theory that has been presented. At the end of Chapter 7 several different topological realms are presented by means of restricting the collection of double arrows. All of these can be thought of as sub-realms of the multi-category that is constructed from the abstraction of Frobenius algebras that was initiated in Chapter 3.

Chapter 8 begins with a brief discussion of Kauffman's diagrammatic description of the Temperley-Lieb algebra. It is a tour of relationships among associated products and triangulations of manifolds that is informed by several items that are counted by means of Catalan numbers. A representation of the Temperley-Lieb algebra that gives rise to the Jones polynomial in Chapter 9 is presented. This representation circles back upon ideas that were presented in Chapter 2.

Since much of diagrammatic algebra originated with knot theory, the tour that began in Chapter 8 continues in Chapter 9. Oriented framed tangles are described categorically. The Reidemeister moves, the fundamental group, and the fundamental quandle are defined. Examples are computed. The Jones polynomial is defined and the representation of the braid group into the Temperley-Lieb algebra that is given by the bracket relation is demonstrated. Hopf algebras are discussed briefly. An example of a calculation in Khovanov homology is presented. Knotted trivalent graphs and handlebody knots are contrasted. The Reidemeister moves and related isotopies are abstracted to being quadruple arrows in another context, and a method to construct glyphs to represent these isotopies is given. Thus Chapter 9 is a sampler of many of the delightful, diagrammatic topics that both classical and modern knot theory has to offer.

Chapter 10 uses the categorical and diagrammatic constructions to describe isotopies of surfaces that are embedded in 4-dimensional space. In addition, glyphographic descriptions of some of the isotopies of embedded foams in 4-space are given. There are eight relations among embedded foams that deserve special attention. Most are related by means of the naturality of exchangers. Quandles are abstracted to shalgebras. The axioms for that structure were discovered by means of diagrammatic methods. A homology theory for shalgebras is defined. The glyphs that describe foams that are embedded in 4-space are used to represent abstract tensor relations and the homology theory of shalgebras is used to solve these tensorial relations. Again the ideas circle back to those of Chapter 2.

Chapter 11 contains an outline and summary of the work of the authors upon surface braids and their isotopies, but rephrased as a methodology in diagrammatic algebra. Immersed surface braids are included within the discussion. Here the diagrammatic algebraic aspect allows us to further describe braided 3-manifolds in 5-space and 4-manifolds in 6-space. The end of the chapter describes intricate examples of these.

Chapter 12 is a coda. It recapitulates the description of n-arrows for small values of n in the context of globular categories in which horizontal compositions are mitigated by exchanger isomorphisms. The dihedral group is described by this method.

Remarks. The text develops the story gradually. So, forgive us, sometimes it takes several passes on idea before we get it to a point of reasonable cohesion. Two items come to mind immediately. First, those things that are double arrows in Chapters 3, 4, and 5 become triple arrows later since a dot, $\multimap\!\bullet\!-$, is promoted to lying in a plane rather than along a line. Second, adjoint relationship as described explicitly in Chapter 10, Section 1.4 are described as fish hooks in Chapter 7, Section 8, but this type of relation appears much earlier in Chapter 4, Fig. 4. The expository fashion in which the authors were reared would require us to define the general relationship first, and then point out instances of it when it occurs. However, the text grew as we developed our understanding. So we rebel against that aspect of our training because the gradual process of abstraction helps the readers' understanding to develop in parallel with the authors'.

Here is a small note on the index. The expressions "double arrow" and "2-arrow" are synonyms. Similar remarks apply to higher arrows. We attempted to reference both expressions when either appeared. In general the word "arrow" holds at least two meanings. The predominant meaning is a morphism in a category. But an arrow is also a vector indicator in a diagram. Sometimes, but not always, when a word is used in more than one context, the less important context is not included as an index entry. Specifically, "fold" serves as a noun: the set of points at which the derivative drops rank; it serves as a verb: a sheet of surface can fold in front of another sheet; or as an adjective: the 3-fold branched cover. The adjectival use is not indicated within the index. A casual reader who has made it this far into the introduction should stop here to look at the index. Word frequency helps indicate what this book is about.

Authors' expectations of the reader. This book is aimed towards graduate students and researchers who are working or studying in any area of mathematics. Even so, our primary focus is topological applications. We expect the reader to have knowledge of some standard notions in algebra and topology; for example, linear algebra, groups and their presentations using generators and relations, rings and modules, categories, homeomorphisms, quotient spaces, manifolds, and homology groups.

Non-standard notations such as the use of glyphs, rather than alpha-numeric characters, to represent functions or more generally arrows are employed throughout. The text develops these notations slowly. So we assume that the readers' experience with these ideas is new. Suggestive alpha-numeric characters foreshadow their glyphographic counterparts. As the diagrammatic methods develop, pointers are given towards past and future sections in which concepts or notation are elucidated. We don't mind recalling notation even when it is standard. When lists that comprise definitions are compiled, the location of their introduction is referenced. On the other hand, we assume that readers are more comfortable with the ideas, notation, notions, and abstractions after they have been announced.

We have attempted to keep the text self-contained. Some very basic ideas of homology are introduced (Chapter 9 and Chapter 10). A group presentation for the braid group is introduced in Chapter 9. Other groups are also introduced in the same chapter and in Chapter 12.

Continuous functions, homeomorphisms, and quotient spaces are approached from the usual intuitive manner. All of the spaces that are exemplified in the text are regarded as non-pathological. They are either (smooth) low-dimensional manifolds or quotient spaces of manifolds. Terms such as "isotopy" and "homotopy" are used informally.

The language of categories that includes notions of objects and collections of arrows between objects is recurring. More precisely, the text develops the notions of higher categories in which the collection of arrows (morphisms) between a pair of objects also is a category. The text examines a very few examples of multi-categories and builds them from one or two objects and a small number of generating arrows between them.

Throughout the text, we have interspersed exercises. Some of these fore-shadow up-coming sections. Others allow the reader to construct variations on constructions that were not explicitly presented. At least one is a small research project.

The subject of the book, how to use diagrams in an algebraic fashion, is quite new. Metaphorically, the book is not a guide to all latin based languages, but it is a treatise written in some latin language. By studying here, you will gain facility with many other diagrammatically inspired approaches to categories, language, or quantum mechanics.

Our goal is to demonstrate that embeddings and immersions of n-dimensional objects into both $(n+1)$- and $(n+2)$-dimensional spaces can be described by means of a finite set of generating "glyphs" that are subject to a finite set of relations among them. The glyphs correspond to labeled graphs with specific edge and vertex types. They can be composed according to rules that mimic the forms which they represent. Moreover, topological equivalences, such as isotopy and cobordism, are represented by means of relations among glyphs, as well as, their own descriptive glyphs. These glyphographic representations are models for certain, quite specific, higher dimensional categories.

As is now understood, the theory of classical knots and quantum-type invariants can be used to exploit the glyphographic so that a functor between a category of tangles and, say, a category of representations of a quantum group can be constructed. Most of the work developed here, describes other knotted objects (for example surfaces and foams) in a higher categorical language. The hope is that other, more algebraic, concepts obey analogous rules. In some instances, both in the classical case, and the cases of knotted surfaces and foams, outlines of those algebraic concepts are presented by means of the homology theory of shalgebras.

Furthermore, the glyphography of surface braids, can be extrapolated to describe braidings of 3-dimensional solids in 5-dimensional space, and 4-dimensional hyper-solids in 6-dimensional space. In this way, the diagrammatic knot theory that was so successful in the classical purview can be extended into the unseen worlds of 5- and 6-dimensional space.

Acknowledgements. A long list of acknowledgements is in order. For continued and unending emotional and morale support we thank our wives, respectively, Huong and Naoko.

Much of this work was developed during two periods. From December 2012 through August 2013, JSC spent time at Kyungpook National University, Daegu,

South Korea. At that time he gave a series of lectures on material that this manuscript has evolved into. This visit was supported by Brain Pool Trust. Several times during this period, he and SK was able to visit each other. During the period, August 1, 2018 through May 31, 2019, JSC was an International Research Fellow supported by The Japanese Society for the Promotion of Science grant number L-18511. SK was supported by JSPS KAKENHI Grant numbers JP26287013 and JP19H01788. During Fall 2019, JSC was supported by ICERM.

We also would like to thank our colleagues and others who inspire: Yongju Bae, Seonmi Choi, Roger Fenn, Robert Fripp, Amar Hadzihasanovic, Kengo Kawamura, Byorhi Kim, Jieon Kim, Seongjeung Kim, Sera Kim, Louis Kauffman, Akio Kawauchi, Victoria Lebed, Insook Lee, Sang Youl Lee, Tony Levin, Sam Lomanaco, Joaõ Martins, Shin Satoh, Masahico Saito, Dan Silver, Seung Yeop Yang, and Susan Williams.

Thanks are over-due to Eriko Hironaka who entertained the idea of this book for many years. She and her colleagues at the AMS pushed it along. It is better because of their editorial acumen.

CHAPTER 2

Elements

1. Sets, relations, and functions

From the point of view of the authors, math books start with sets. That might make us seem old fashion, or it might make the discussion seem arcane. But we have to start somewhere. We will define relations and functions, domains, ranges, and codomains. The notion of category is introduced only to state that the collection of sets forms a category. Subsequent chapters discuss categories more thoroughly. We won't discuss Russell's paradox.

Let X, Y, Z and W denote sets. A subset

$$R \subset X \times Y = \{(x, y) : x \in X \;\&\; y \in Y\}$$

is called a *relation from* X *to* Y. A relation R is a *function* if for each $x \in X$ there is a unique $y \in Y$ such that $(x, y) \in R$. We write $y = f(x)$. The *domain* of the function is the entire set X, the *codomain* is the set Y while the *range* is the set

$$\mathrm{Range}(f) = \{y \in Y : y = f(x) \text{ for some } x \in X\}.$$

The expression $Y \xleftarrow{f} X$ denotes such a function. The same notation will be used in an arbitrary category when f is a morphism (or arrow) from X to Y. The set of all such functions is denoted $Y^X = \{Y \xleftarrow{f} X\}$.

When encountering an expression that appears as an exponent, it is often worth the trouble to puzzle through its identification with a class of functions. For example, in writing $\mathbb{R}^2$ to denote the set of ordered pairs of real numbers, we are thinking of such a pair (x^1, x^2) as the image of a function $\mathbb{R} \xleftarrow{x} \{1, 2\}$ and writing the values $x(1)$ and $x(2)$ as superscripts. Take care; in this context the superscript is *not* an exponent. How should one consider an exponent from the current perspective?

For any set X there is the identity function $\mathsf{I}(x) = x$ which is written as $X \xrightarrow[\quad]{\mathsf{I}} X$. The arrow is not included since the identity is tautologically invertible and bidirectional.

Given functions $Z \xleftarrow{f} Y$ and $Y \xleftarrow{g} X$, the composite $f \circ g$ is the function $Z \xleftarrow{f \circ g} X$ that is defined by $(f \circ g)(x) = f(g(x))$. Its domain is X and its range is $\{z \in Z : z = f(y) \text{ where } y = g(x) \text{ for some } x \in X\}$. The identity functions on Y and X serve as left and right identities for composition, respectively as indicated.

$$\left[Y \overset{\mathsf{I}_Y}{\text{———}} Y \xleftarrow{f} X\right] = \left[Y \xleftarrow{f} X\right] = \left[Y \xleftarrow{f} X \overset{\mathsf{I}_X}{\text{———}} X\right].$$

Furthermore, composition is associative. If $W \xleftarrow{f} Z$, $Z \xleftarrow{g} Y$, and $Y \xleftarrow{h} X$, then the equation $(f \circ g) \circ h = f \circ (g \circ h)$ holds. Thus the center expression in the following string of equalities is unambiguous.

$$\left[W \xleftarrow{(f\circ g)} Y \xleftarrow{h} X\right] = \left[W \xleftarrow{(f\circ g\circ h)} X\right] = \left[W \xleftarrow{f} Z \xleftarrow{(g\circ h)} X\right].$$

These axioms are schematized. The composition of two functions is written as 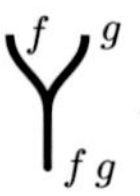 . The identity function from X to X is written as ┃. The equation that expresses this as the identity is written as

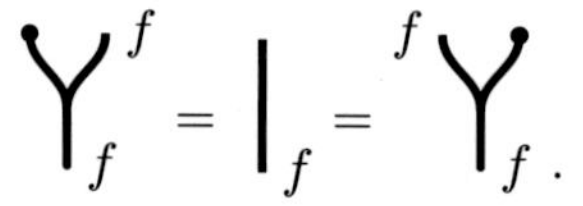

The associative rule for composition is written in diagrammatic form as

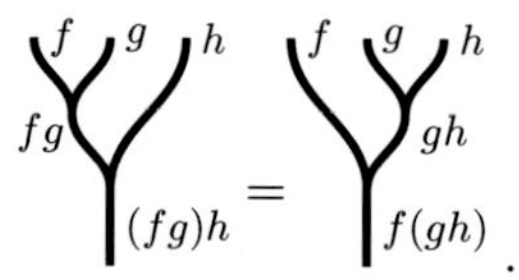

These two diagrammatic equations are, in fact, foundational to this book.

The collection of sets forms a *category* — there is a collection of objects, there is a collection of arrows between any two objects, each object has an identity arrow, and arrows can be composed. The composition is associative, and the identity arrow serves as a unit under composition.

The idea of a category will be re-introduced subsequently and repeatedly. Furthermore, diagrammatic descriptions will also be developed in more detail and in other contexts soon.

2. Diagrammatics of linear algebra

The gist of the material here was originally introduced by Penrose as the notion of abstract tensors [**62**]. Over the last half-century, the ideas of abstract tensors has flourished, not only into the opus that you are reading, but also in general into the ideas that evolved into string diagrams that represent (higher order) arrows in a category.

The material presented here works over an arbitrary field and often over an arbitrary integral domain, but we will state things explicitly over the real numbers $\mathbb{R}$. When the Kauffman bracket is introduced (Chapter 9, Section 4), it is better to work over the complex numbers $\mathbb{C}$. We assume that the reader has a working knowledge of these fields.

A bit of caution is encouraged. Two primary purposes of linear algebra are to find solutions for linear systems of equations and to transform spaces via linear maps. These purposes are interrelated, but the diagrammatics in this section do not help us achieve such ends. Instead, they serve as an example of diagrammatic thinking that will be expanded upon throughout the book.

Let n denote a positive integer, and let $[n] = \{1, 2, \ldots, n\}$ denote the (ordered since $1 < 2 < \cdots < n$) set of positive integers from 1 to n. The vector space $\mathbb{R}^n$ consists of the set of functions $\{\mathbb{R} \xleftarrow{\vec{x}} [n]\}$. The value of $\vec{x}$ is written as a superscript: $\vec{x}(j) = x^j$, and called the jth coordinate. Vector addition and scalar multiplication are pointwise: $(\vec{x} + \vec{y})(j) = x^j + y^j$ and $(\alpha\vec{x})(i) = \alpha x^i$, for $\alpha \in \mathbb{R}$. In a vector space, addition is commutative and associative, there is a zero vector $\vec{0}$ that serves as an additive identity, and the negative of a vector, $-\vec{x}$, unambiguously[1] describes the additive inverse. Vector addition is distributive over scalar multiplication: $\alpha(\vec{x} + \vec{y}) = \alpha\vec{x} + \alpha\vec{y}$, and vice versa $(\alpha + \beta)\vec{x} = \alpha\vec{x} + \beta\vec{x}$. Scalar multiplication satisfies the associative rule $(\alpha\beta)\vec{x} = \alpha(\beta\vec{x})$.

An ordered *vector space basis* is an ordered set $(\vec{v}_1, \vec{v}_2, \ldots, \vec{v}_n)$ such that any vector $\vec{x}$ can be written uniquely as $\vec{x} = \sum_{j=1}^n \alpha_j \vec{v}_j$. The characteristic functions

$$e_j(i) = \begin{cases} 1 & \text{if } i = j \\ 0 & \text{otherwise} \end{cases}$$

(for $j = 1, 2 \ldots, n$) form the *standard basis* for $\mathbb{R}^n$. We don't bother to put arrow-like hats upon these elements, and they are indexed by subscripts. Any vector $\vec{x}$, then, is written uniquely as $\vec{x} = \sum_{j=1}^n x^j e_j$. The jth *coordinate* x^j is a specific real number for each $j = 1, 2, \ldots, n$.

The standard basis $\{e_j\}_{j \in [n]}$ is tacitly ordered. That is to say when choosing matrix representations of linear maps and so forth, the ordering is used, but we don't usually make a big deal about the ordering. The adjective *ordered* will be dropped unless additional emphasis is needed.

The basis vector e_j is written as a column

$$e_j = \begin{bmatrix} 0 \\ \vdots \\ 0 \\ 1 \\ 0 \\ \vdots \\ 0 \end{bmatrix} \leftarrow j\text{th row.}$$

With the exception of the basis vectors $\{e_j\}_{j \in [n]}$ and their duals $\{e^j = e_j^*\}_{j \in [n]}$, which will be defined shortly, superscripts indicate row numbers and subscripts indicate column numbers for the coefficients of the associated characteristic functions or a matrix presentation $A = [a_j^i]_{i \in [m], j \in [n]}$ of a linear map $\mathbb{R}^m \xleftarrow{A} \mathbb{R}^n$. But be aware, multiply indexed objects will soon appear in the narrative[2]. In this case, it is amusing and instructive to consider these as hyper-rectangular arrays of real numbers. Also, the "contraction of indices," as it relates to linear and multilinear transformations will be discussed.

The vector $\vec{x}$ will also be written as $\vec{x} =$ 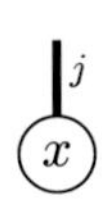

or $\vec{x} =$ 

.

[1]Unambiguous, in the sense that $-\vec{x} = (-1) \cdot \vec{x}$

[2]A less experienced reader might take comfort in the fact that the authors had to check and recheck the range of row and column indices in these expressions!

A *linear map* $\mathbb{R}^m \overset{A}{\leftarrow} \mathbb{R}^n$ is one for which $A(\alpha\vec{v} + \beta\vec{w}) = \alpha A(\vec{v}) + \beta A(\vec{w})$. It is determined by its action on the basis $\{e_j\}_{j=1,2,\ldots,n}$. If $\{e'_i\}_{i=1,2,\ldots,m}$ is the corresponding basis for $\mathbb{R}^m$, then $A(e_j) = \sum_{i=1}^m a^i_j e'_i$. So we write, $A =$ 

or simply $A =$ 

. Computing

$$A(\vec{x}) = A(\sum_{j=1}^{n} x^j e_j) = \sum_{j=1}^{n} x^j A(e_j)$$

$$= \sum_{j=1}^{n} x^j \left(\sum_{i=1}^{m} a^i_j e'_i \right) = \sum_{i=1}^{m} \left(\sum_{j=1}^{n} a^i_j x^j \right) e'_i,$$

notice that the coefficient of e'_i is $\left(\sum_{j=1}^n a^i_j x^j\right)$, and the sum in this case occurs over a repeated upper/lower indexed pair. So the situation is schematized as this 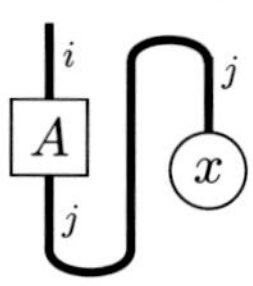

. In a few paragraphs zig-zags will be removed in an algebraic fashion, but for now 

denotes the action of the linear operator A upon the vector $\vec{x}$. That is to say that the box above the circle indicates the quantity $A(\vec{x})$; moreover the internal line segment in this figure, tacitly supposes that an inner index ($j \in [n]$) is being summed over. In general such an index is called a "dummy variable," and as such it does not require a name.

The vector $\vec{x}$ also represents a linear map $\mathbb{R}^n \leftarrow \mathbb{R}$ that is given by the diagram below. Namely, $1 \in \mathbb{R}$ is mapped to the end point of the arrow that points from 0 to the point in $\mathbb{R}^n$ whose coordinates are x^j, for $j = 1, 2, \ldots, n$.

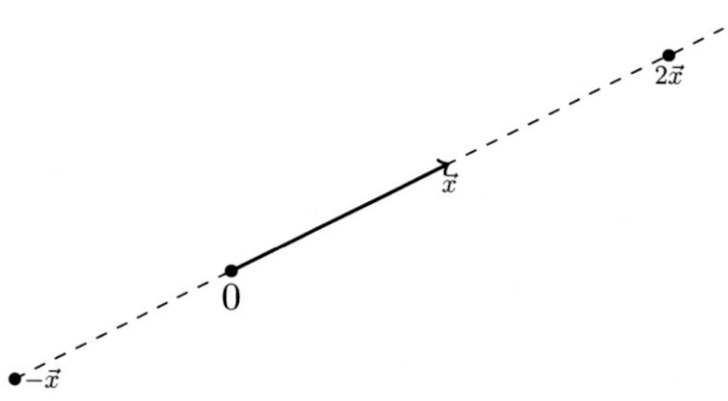

In this way we write 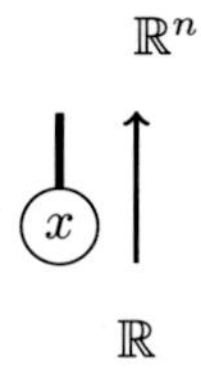

. Now notice that even though $\mathbb{R} = \mathbb{R}^1$, there is no need to specify the action of this linear map upon a basis. The basis consists of the unique function that sends the single element set [1] to the point $1 \in \mathbb{R}$; no subscripting is necessary.

Suppose that $\mathbb{R}^\ell \xleftarrow{A} \mathbb{R}^m$, $\mathbb{R}^m \xleftarrow{B} \mathbb{R}^n$, and $\mathbb{R}^n \xleftarrow{C} \mathbb{R}^p$ are linear maps. Let $\{e_f\}_{f\in[p]}$, $\{e'_j\}_{j\in[n]}$, $\{e''_i\}_{i\in[m]}$, and $\{e'''_k\}_{k\in[\ell]}$ denote the standard basis of $\mathbb{R}^p$, $\mathbb{R}^n$, $\mathbb{R}^m$, and $\mathbb{R}^\ell$, respectively. Then for $i \in [m]$, $j \in [n]$, $k \in [\ell]$, and $f \in [p]$, there are real numbers a_i^k, b_j^i, and c_f^j such that

- $$A(e''_i) = \sum_{k=1}^{\ell} a_i^k e'''_k,$$
- $$B(e'_j) = \sum_{i=1}^{m} b_j^i e''_i,$$
  and
- $$C(e_f) = \sum_{j=1}^{n} c_f^j e'_j.$$

As these functions are composed, we first get

$$(A \circ B)(e'_j) = A(B(e'_j)) = A\left(\sum_{i=1}^{m} b_j^i e''_i\right) = \sum_{i=1}^{m} b_j^i \left(\sum_{k=1}^{\ell} a_i^k e'''_k\right)$$
$$= \left(\sum_{i=1}^{m}\sum_{k=1}^{\ell} a_i^k b_j^i e'''_k\right) = \left(\sum_{k=1}^{\ell}\left[\sum_{i=1}^{m} a_i^k b_j^i\right] e'''_k\right).$$

So for fixed j, the coefficient of the kth characteristic function e'''_k in this composition is $\left[\sum_{i=1}^{m} a_i^k b_j^i\right]$. The situation is schematized as such: 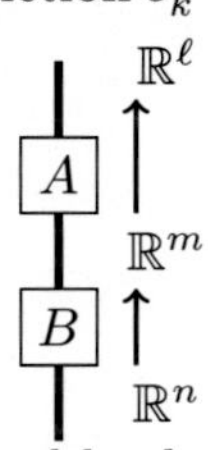

. The lower string can be indexed by j and the upper string can be indexed by k. The interior string is indexed by i and sums occur over all the admissible values of i. Indices, however, will be dropped, and we turn to composing with 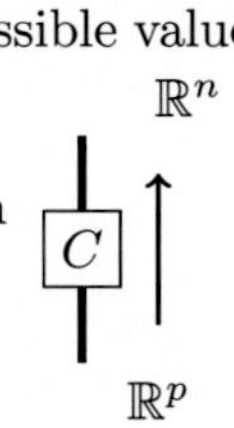

.

The three-fold composition $A \circ (B \circ C)$ is computed on the basis element e_f.

$$\begin{aligned}
A \circ (B \circ C)(e_f) &= A((B \circ C)e_f) \\
&= A\,(B\,(C(e_f))) \\
&= A\left(B\left(\sum_{j=1}^{n} c_f^j e_j'\right)\right) \\
&= A\left(\sum_{j=1}^{n} c_f^j B\left(e_j'\right)\right) \\
&= A\left(\sum_{j=1}^{n} c_f^j \sum_{i=1}^{m} b_j^i e_i''\right) \\
&= \sum_{j=1}^{n} c_f^j \sum_{i=1}^{m} b_j^i A\left(e_i''\right) \\
&= \sum_{i=1}^{m}\sum_{j=1}^{n} \left(b_j^i c_f^j\right) A\left(e_i''\right) \\
&= \sum_{i=1}^{m}\sum_{j=1}^{n} \left(b_j^i c_f^j\right) \sum_{k=1}^{\ell} a_i^k e_k''' \\
&= \sum_{k=1}^{\ell} \left[\sum_{i=1}^{m}\sum_{j=1}^{n} a_i^k \left(b_j^i c_f^j\right)\right] e_k'''.
\end{aligned}$$

Thus the coefficient of e_k''' under the image of e_f is the double sum $\left[\sum_{i=1}^{m}\sum_{j=1}^{n} a_i^k \left(b_j^i c_f^j\right)\right]$. Moreover, this is equal to the composition $(A \circ B) \circ C$ since associativity (and commutativity) holds at the level of the coefficients. At a more fundamental level each of A, B, and C represents a function, as we saw in the previous section, the composition of functions is associative.

The associativity law for compositions of linear functions is indicated here

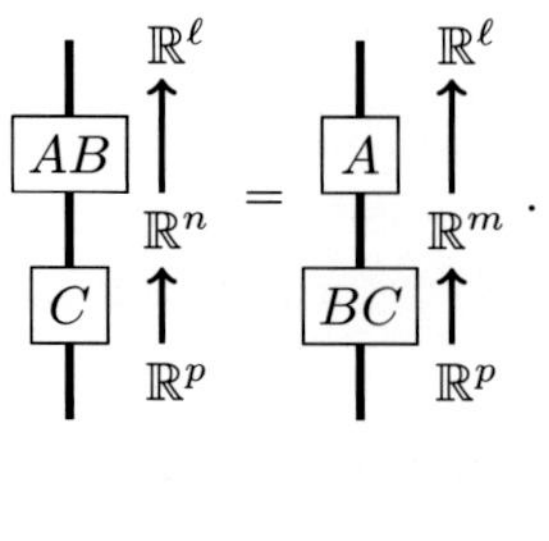

The diagram above is combined with the diagram for associativity for functions to produce the detailed equation:

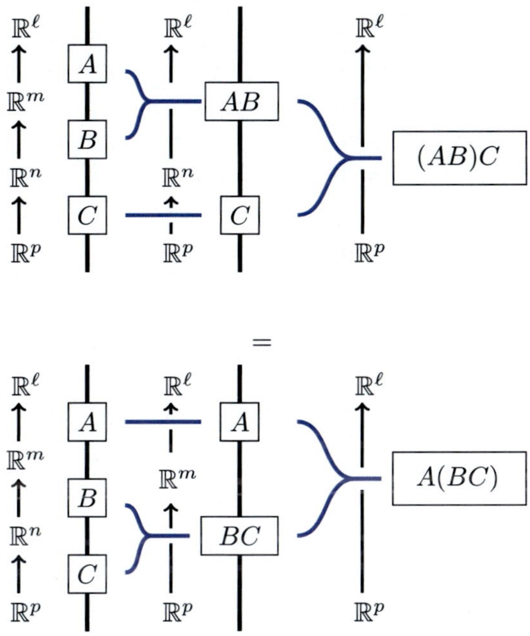

which also helps illustrate the order of composition. Domains for functions appear on the bottom of diagrams. This convention is keeping with morphisms written as leftward pointing arrows. There is an initial conceptual cost in our conventions that we believe is offset by the ease in left-to-right or up-to-down transcriptions that results. We regret that bottom/top and right/left arrow conventions will not be maintained throughout. At times, other conventions are more suitable.

A *bilinear form* on $\mathbb{R}^n$ is a function $\mathbb{R} \xleftarrow{\Lambda} \mathbb{R}^n \times \mathbb{R}^n$ that is defined with domain the Cartesian product of $\mathbb{R}^n$ with itself, and that satisfies the condition

$$\Lambda(\alpha_1\vec{v_1} + \alpha_2\vec{v_2}, \beta_1\vec{w_1} + \beta_2\vec{w_2})$$
$$= \alpha_1\beta_1\Lambda(\vec{v_1}, \vec{w_1}) + \alpha_1\beta_2\Lambda(\vec{v_1}, \vec{w_2}) + \alpha_2\beta_1\Lambda(\vec{v_2}, \vec{w_1}) + \alpha_2\beta_2\Lambda(\vec{v_2}, \vec{w_2}).$$

Such a form is determined by its actions upon bases. So there is a doubly subscripted matrix $[\Lambda_{i,j}]_{i,j\in[n]}$ defined by $\Lambda_{i,j} = \Lambda(e_i, e_j)$.

Since Λ is bilinear, it can be thought of as being defined upon the tensor product of $\mathbb{R}^n$ with itself. The *tensor product* of the vector spaces $\mathbb{R}^m$ and $\mathbb{R}^n$ is the set of functions $R \xleftarrow{\tau} [m] \times [n]$. Thus a vector τ in the tensor product is a doubly superscripted quantity $\tau(i,j) = \tau^{ij}$ for $i = 1, \dots, m$ and $j = 1, \dots, n$. The rectangularly arranged[3] basis is the set of characteristic functions $e_{i,j}$ where the value $e_{i,j}(k,\ell) = 1$ if and only if $(i,j) = (k,\ell)$ and is 0 in all other cases. When e_j is thought to be a column vector as above, then $e_{i,j}$ is the rectangular array determined as the matrix product $e_i \cdot e^j$ where $e^j = (e_j)^T$ is the row vector

$$e^j = [0, \dots, 0, \underbrace{1}_{j\text{th}}, 0, \dots, 0]$$

[3]It can be said that $\{\tau^{i,j}\}$ is lexically ordered: $(\tau^{1,1}, \tau^{1,2}, \dots, \tau^{1,n}, \tau^{2,1}, \dots, \tau^{2,n}, \tau^{m,1}, \dots, \tau^{m,n})$.

that is the transpose, T, of e_j. So

$$e_i \cdot e^j = i\text{th} \rightarrow \begin{bmatrix} 0 \\ \vdots \\ 0 \\ 1 \\ 0 \\ \vdots \\ 0 \end{bmatrix} \cdot \begin{bmatrix} 0 & \dots & 0 & \underbrace{1}_{j\text{th}} & 0 & \dots & 0 \end{bmatrix}$$

$$= \begin{bmatrix} 0 & \dots & 0 & 0 & 0 & \dots & 0 \\ \vdots & & \vdots & \vdots & \vdots & & \vdots \\ 0 & \dots & 0 & 0 & 0 & \dots & 0 \\ 0 & \dots & 0 & 1 & 0 & \dots & 0 \\ 0 & \dots & 0 & 0 & 0 & \dots & 0 \\ \vdots & & \vdots & \vdots & \vdots & & \vdots \\ 0 & \dots & 0 & \underbrace{0}_{j\text{th}} & 0 & \dots & 0 \end{bmatrix} \begin{matrix} \\ \\ \\ \leftarrow i\text{th} \\ \\ \\ \\ \end{matrix} .$$

This matrix is written as $e_i \otimes e_j$ or $e_{i,j}$. A vector $\vec{\tau} \in \mathbb{R}^m \otimes \mathbb{R}^n$ is expressed, then, as $\vec{\tau} = \sum_{i\in[m]} \sum_{j\in[n]} \tau^{i,j} e_{i,j} = \sum_{(i,j)\in[m]\times[n]} \tau^{i,j} e_{i,j}$. As a doubly superscripted matrix, $\vec{\tau}$ is also represented as a box with a pair of strings that emanate from its top. For example, we write 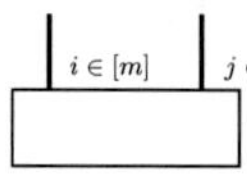

, or when $\mathbb{R}^m$ and $\mathbb{R}^n$ are understood, we write 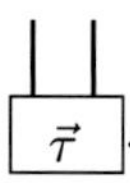

.

In general, a multi-indexed box such as

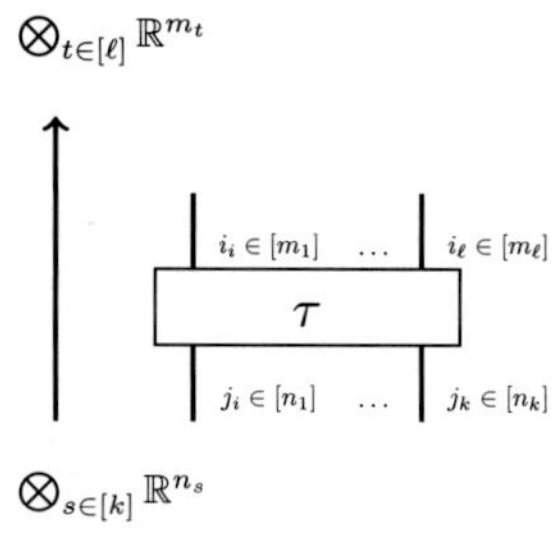

represents a multi-linear map, τ, from $\bigotimes_{s\in[k]} \mathbb{R}^{n_s} = \mathbb{R}^{n_1} \otimes \cdots \otimes \mathbb{R}^{n_k}$ to $\bigotimes_{t\in[\ell]} \mathbb{R}^{m_t} = \mathbb{R}^{m_1} \otimes \cdots \otimes \mathbb{R}^{m_\ell}$, or what is traditionally called a *tensor operator*. In practice, the number of tensor factors (*tensorands*) is rather small (usually less than three or four on each side of the box).

It is well-known among mathematicians and physicists that the variables with many superscripts and subscripts in these expressions present tedium in the extreme for our work. To represent a tensor operator as such, follows the same procedure that was used for matrices: choose bases in the domain and range, and express the coefficient of the range basis element in terms of the action of the operator upon the domain basis. This process is conceptually easy only after having succeeded in implementing the strategy it in several small examples. Within the text below, we will run through the process a few times. The so-called *abstract tensor notation* — in which (1) variables are replaced by boxes, (2) Einstein summation notation is employed, and (3) diagrams are manipulated according to particular rules — simplifies calculations in these arenas. These simplifications are being developed as the discussion moves forward.

Since $[1] \times [n] = [n] = [n] \times [1]$, the tensor products $\mathbb{R} \otimes \mathbb{R}^n = \mathbb{R}^n = \mathbb{R}^n \otimes \mathbb{R}$ are identical. This remark further justifies considering 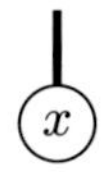 as a linear map $\mathbb{R}^n \leftarrow \mathbb{R}$.

In general a bilinear form 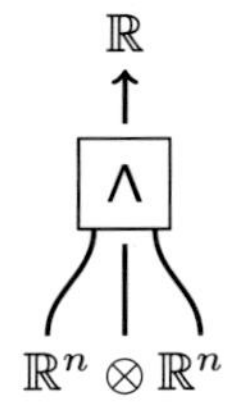

is thought of as a map $\mathbb{R} \xleftarrow{\wedge} \mathbb{R}^n \otimes \mathbb{R}^n$, and may be written as .

The *standard inner product* $\cap(\vec{x}, \vec{y}) = \sum_{i \in [n]} x^i y^i$ can be expressed in subscripted notation by declaring that

$$\cap_{i,j} = \begin{cases} 1 & \text{if } i = j; \\ 0 & \text{otherwise} \end{cases}$$

is a *Kronecker delta function.* The dual tensor operator $\mathbb{R}^n \otimes \mathbb{R}^n \xleftarrow{\cup} \mathbb{R}$ is expressed as $1 \mapsto \sum_{j=1}^{n} e_j \otimes e_j = \sum_j e_{j,j}$. In matrix notation, this is the $(n \times n)$ identity matrix.

Here and elsewhere, the same font may denote an operator's name and its *structure constants*, which are the coefficients of basis elements under the action of the operator upon basis elements. Usually the geometric shape of that font is meant to reflect the tensorial nature of the operator — the number of inputs and the number of outputs of the tensor operator. Thus the font chosen for the operator and the font chosen to express the structure constants often coincide (or differ as upper and lower case). The notational ambiguity is intentional.

For example, the operator $\cup$ serves to raise subscripts and $\cap$ can serve to lower superscripts. At a primitive level, when vectors $\vec{x}$ are column vectors, their dual functionals $\mathbb{R} \xleftarrow{\vec{x}^*} \mathbb{R}^n$ are row vectors. The expression $\vec{x}^{\,T} \cdot \vec{y}$ that is used to compute the inner product of $\vec{x}$ and $\vec{y}$ is conveniently drawn as 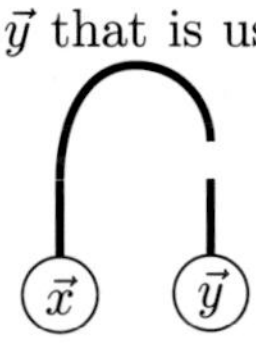. For each $\vec{x}$, the expression $\cap(\vec{x}, -)$ is a real-valued linear function, so that the dual vector $\vec{x}^*$ is represented as 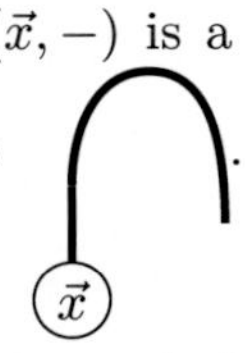. This is to say that $\cap$ has the effect of turning a superscript into a subscript and thereby transposing a column vector into a row vector. Meanwhile $\cup$ does the opposite; it transposes a row vector into a column vector. Of course, ∪ and ∩ have the opposite effect upon basis vectors e_j and their dual basis vectors $e^j = e_j^*$.

Imagine the tensor product of linear maps 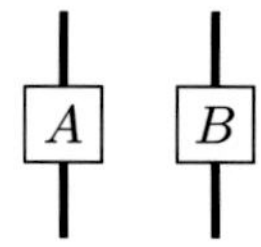 (or more generally the tensor product of linear operators). First, observe that within the diagrammatic, we drop the ⊗. So that the tensor product, $A \otimes B$, of linear maps $\mathbb{R}^m \xleftarrow{A} \mathbb{R}^n$ and $\mathbb{R}^p \xleftarrow{B} \mathbb{R}^q$ is represented via the left to right juxtaposition of the corresponding boxes. As variables with subscripts and superscripts $(A \otimes B)^{i,k}_{j,\ell}$ can be thought of as a 4-dimensional right-rectangular parallelepiped (hypercube), where the entry at the (i, j, k, ℓ) position is the product entry $a^i_j b^k_\ell$. Or upon lexically ordering the bases, the matrix

$$\begin{bmatrix} a^1_1 B & a^1_2 B & \dots & a^1_n B \\ a^2_1 B & a^2_2 B & \dots & a^2_n B \\ \vdots & \vdots & & \vdots \\ a^m_1 B & a^m_2 B & \dots & a^m_n B \end{bmatrix}$$

$$= \begin{bmatrix} a^1_1 \begin{bmatrix} b^1_1 & \dots & b^1_q \\ \vdots & & \vdots \\ b^p_1 & \dots & b^p_q \end{bmatrix} & a^1_2 \begin{bmatrix} b^1_1 & \dots & b^1_q \\ \vdots & & \vdots \\ b^p_1 & \dots & b^p_q \end{bmatrix} & \dots & a^1_n \begin{bmatrix} b^1_1 & \dots & b^1_q \\ \vdots & & \vdots \\ b^p_1 & \dots & b^p_q \end{bmatrix} \\ \vdots & \vdots & & \vdots \\ \vdots & \vdots & & \vdots \\ a^m_1 \begin{bmatrix} b^1_1 & \dots & b^1_q \\ \vdots & & \vdots \\ b^p_1 & \dots & b^p_q \end{bmatrix} & a^m_2 \begin{bmatrix} b^1_1 & \dots & b^1_q \\ \vdots & & \vdots \\ b^p_1 & \dots & b^p_q \end{bmatrix} & \dots & a^m_n \begin{bmatrix} b^1_1 & \dots & b^1_q \\ \vdots & & \vdots \\ b^p_1 & \dots & b^p_q \end{bmatrix} \end{bmatrix}$$

is formed.

That approach is used to examine in detail the two compositions that are represented in the diagram below.

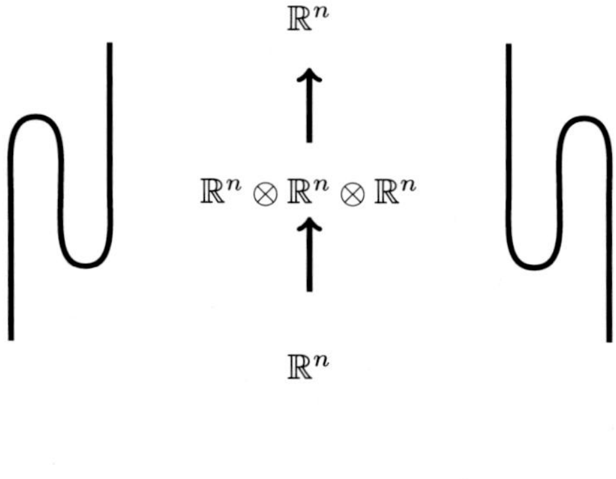

First consider the case $n = 2$. The map $\mathbb{R}^2 \otimes \mathbb{R}^2 \xleftarrow{\mathsf{U}} \mathbb{R}$ is represented as a column vector

$$\begin{bmatrix} \begin{bmatrix} 1 \\ 0 \end{bmatrix} \\ \\ \begin{bmatrix} 0 \\ 1 \end{bmatrix} \end{bmatrix}.$$

Then $\mathbb{R}^2 \otimes \mathbb{R}^2 \otimes \mathbb{R}^2 \xleftarrow{\mathsf{I} \otimes \mathsf{U}} \mathbb{R}^2$ is represented as an (8×2)-matrix

$$\begin{bmatrix} \begin{bmatrix} \begin{bmatrix} 1 \\ 0 \end{bmatrix} \\ \\ \begin{bmatrix} 0 \\ 1 \end{bmatrix} \end{bmatrix} & \begin{bmatrix} \begin{bmatrix} 0 \\ 0 \end{bmatrix} \\ \\ \begin{bmatrix} 0 \\ 0 \end{bmatrix} \end{bmatrix} \\ \\ \begin{bmatrix} \begin{bmatrix} 0 \\ 0 \end{bmatrix} \\ \\ \begin{bmatrix} 0 \\ 0 \end{bmatrix} \end{bmatrix} & \begin{bmatrix} \begin{bmatrix} 1 \\ 0 \end{bmatrix} \\ \\ \begin{bmatrix} 0 \\ 1 \end{bmatrix} \end{bmatrix} \end{bmatrix}.$$

Meanwhile, the tensor operator $(\mathsf{\cap} \otimes \mathsf{I})$ is represented by the matrix

$$\begin{aligned} &\left[1 \begin{bmatrix} 1 & 0 \\ 0 & 1 \end{bmatrix}, \ 0 \begin{bmatrix} 1 & 0 \\ 0 & 1 \end{bmatrix}, \ 0 \begin{bmatrix} 1 & 0 \\ 0 & 1 \end{bmatrix}, \ 1 \begin{bmatrix} 1 & 0 \\ 0 & 1 \end{bmatrix} \right] \\ &= \begin{bmatrix} 1 & 0 & 0 & 0 & 0 & 0 & 1 & 0 \\ 0 & 1 & 0 & 0 & 0 & 0 & 0 & 1 \end{bmatrix}. \end{aligned}$$

So

$$\begin{aligned} (\mathsf{\cap} \otimes \mathsf{I}) \circ (\mathsf{I} \otimes \mathsf{U}) &= \begin{bmatrix} 1 & 0 & 0 & 0 & 0 & 0 & 1 & 0 \\ 0 & 1 & 0 & 0 & 0 & 0 & 0 & 1 \end{bmatrix} \cdot \begin{bmatrix} 1 & 0 \\ 0 & 0 \\ 0 & 0 \\ 1 & 0 \\ 0 & 1 \\ 0 & 0 \\ 0 & 0 \\ 0 & 1 \end{bmatrix} \\ &= \begin{bmatrix} 1 & 0 \\ 0 & 1 \end{bmatrix}. \end{aligned}$$

In general, the composition $(\mathsf{\cap} \otimes \mathsf{I}) \circ (\mathsf{I} \otimes \mathsf{U})$ is considered to be a linear map $\mathbb{R}^n \leftarrow \mathbb{R}^n$ as follows. We have $(\mathsf{\cap} \otimes \mathsf{I}) \circ (\mathsf{I} \otimes \mathsf{U})$, where I indicates the identity operator on $\mathbb{R}^n$, and the tensor product of operators is indicated by $\otimes$. Since $\mathsf{U}(1) = \sum_{j \in [n]} e_j \otimes e_j$, we compute

$$\begin{aligned} \mathsf{\cap} \otimes \mathsf{I} \left(\sum_{i \in [n]} x^i e_i \otimes \left(\sum_{j \in [n]} e_j \otimes e_j \right) \right) &= \mathsf{\cap} \otimes \mathsf{I} \left(\sum_{i \in [n]} \sum_{j \in [n]} x^i \left(e_i \otimes e_j \right) \otimes e_j \right) \\ = \left(\sum_{i \in [n]} x^i \sum_{j \in [n]} (\mathsf{\cap} \otimes \mathsf{I}) \left((e_i \otimes e_j) \otimes e_j \right) \right) &= \left(\sum_{i \in [n]} x^i \left((1) \otimes e_i \right) \right) \end{aligned}$$

$$= \sum_{i \in [n]} x^i e_i.$$

Thus the composition on the left side of the diagram is the identity map. A similar computation is achieved on the right. Evaluate $\sum_{i,j \in [n]} x_i \, (\mathsf{I} \otimes \mathsf{\cap}) \, (e_j \otimes (e_j \otimes e_i))$. Since $\mathsf{\cap}$ is the standard inner product, one sum collapses. Thus $(\mathsf{\cap} \otimes \mathsf{I}) \circ (\mathsf{I} \otimes \mathsf{U}) = \mathsf{I} = (\mathsf{I} \otimes \mathsf{\cap}) \circ (\mathsf{U} \otimes \mathsf{I})$. This is the first instance, of many, of the *zig-zag identity* that we will encounter.

Each of $\mathsf{U}^{i,j}$, I^i_j, and $\mathsf{\cap}_{i,j}$ can be thought of as a matrix or a nesting of matrices whose (i,j)-th entry is the value of the Kronecker $\delta(i,j)$ — the value is 1 if $i = j$ and 0 otherwise.

More generally, suppose that $\mathbb{R} \xleftarrow{\Lambda} \mathbb{R}^n \otimes \mathbb{R}^n$ is an inner-product and $\mathbb{R}^n \otimes \mathbb{R}^n \xleftarrow{\mathsf{V}} \mathbb{R}$ is also linear. We examine $(\Lambda \otimes \mathsf{I}) \circ (\mathsf{I} \otimes \mathsf{V})$. The entries in the matrix representation of the map V are labeled as $\mathsf{v}^{i,j}$ and those for Λ as $\wedge_{i,j}$ where $i, j \in [n] = \{1, 2, \ldots, n\}$. Then the (p,q)th entry of the matrix that represents $(\Lambda \otimes \mathsf{I}) \circ (\mathsf{I} \otimes \mathsf{V})$— that is the entry in the pth row and qth column is the element $\sum_{i,j,k \in [n]} \wedge_{i,j} \cdot \mathsf{I}^p_k \cdot \mathsf{I}^i_q \cdot \mathsf{v}^{j,k}$. Since I^p_k and I^i_q are Kronecker deltas, this element is $\sum_{j \in [n]} \wedge_{q,j} \cdot \mathsf{v}^{j,p}$ — or the (p,q)-entry of the matrix product of Λ and V.

Here is an alternate derivation. Let $\vec{\mathsf{v}}$ denote the $(n^2 \times 1)$-column vector

$$\left[(\mathsf{v}^{1,1}, \mathsf{v}^{1,2}, \ldots, \mathsf{v}^{1,n}), (\mathsf{v}^{2,1}, \mathsf{v}^{2,2}, \ldots, \mathsf{v}^{2,n}), \ldots, (\mathsf{v}^{n,1}, \mathsf{v}^{n,2}, \ldots, \mathsf{v}^{n,n})\right]^T,$$

and let $\vec{0}$ denote the $(n^2 \times 1)$-column vector that consists entirely of 0s. Then the $(n^3 \times n)$-matrix

$$\begin{bmatrix} \vec{\mathsf{v}} & \vec{0} & \ldots & \vec{0} \\ \vec{0} & \vec{\mathsf{v}} & \ldots & \vec{0} \\ \vdots & \vdots & & \vdots \\ \vec{0} & \vec{0} & \ldots & \vec{\mathsf{v}} \end{bmatrix}$$

rerpresents $\mathsf{I} \otimes \mathsf{V}$ while the $(n \times n^3)$-matrix

$$[(\wedge_{1,1} I_n, \ldots, \wedge_{1,n} I_n), (\wedge_{2,1} I_n, \ldots, \wedge_{2,n} I_n), \ldots, (\wedge_{n,1} I_n, \ldots, \wedge_{n,n} I_n)]$$

represents $\Lambda \otimes \mathsf{I}$. Here parentheses have been added to help parse the matrix. Compute the product $(\Lambda \otimes \mathsf{I}) \circ (\mathsf{I} \otimes \mathsf{V})$ of these matrices. The (p,q) entry is $\sum_{j \in [n]} \wedge_{q,j} \cdot \mathsf{v}^{j,p}$ as we described in the previous paragraph.

Let us say that $\mathbb{R} \xleftarrow{\Lambda} \mathbb{R}^n \otimes \mathbb{R}^n$ is a *non-degenerate* bilinear form if there is a dual operator $\mathbb{R}^n \otimes \mathbb{R}^n \xleftarrow{\mathsf{V}} \mathbb{R}$ such that $(\Lambda \mathsf{I}) \circ (\mathsf{I} \mathsf{V}) = \mathsf{I} = (\mathsf{I} \Lambda) \circ (\mathsf{V} \mathsf{I})$. Thus a bilinear form is non-degenerate if and only if its representing matrix is invertible.

3. Algebras

Consider the vector space of $(n \times n)$-matrices, $M(n,n)$. The set $\{e_{i,j}\}_{i,j \in [n]}$ is a basis. The product, $A \cdot B$, of $(n \times n)$ matrices, A and B, is also an $(n \times n)$ matrix. The identity matrix $I = \mathsf{U}(1) = \sum_{i \in [n]} e_{i,i}$ is an identity for the product: $I \cdot A = A = A \cdot I$ for any $(n \times n)$ matrix $A \in M(n,n)$. And the product is associative $A(BC) = (AB)C$.

A vector space, V, that possesses an associative, unital, binary, and bilinear[4] product is called *an algebra*. The set $M(n,n)$ of $(n \times n)$-matrices is an example of

---

[4]All these properties will be reviewed below.

an algebra that has a non-degenerate associative bilinear pairing $\wedge$. Thus $M(n,n)$ is a *Frobenius algebra.* The pairing $\wedge$ is *associative* if $\wedge(AB,C) = \wedge(A,BC)$. In the case of $M(n,n)$, the pairing $\wedge$ is given by $\wedge(A,B) = \text{tr}(AB)$ where the trace of a matrix A is the sum of the elements along the diagonal.

The existing diagrammatics and a few more are developed to succinctly describe the structure in purely diagrammatic terms. In addition, diagrammatics will be used to find additional structures and to prove that $M(n,n)$ is a Frobenius algebra. First, the principal ideas are described using bases.

The algebra structure on a vector space W can be described in terms of its structure constants. In general, suppose that $\{w_j\}_{j\in[m]}$ is a basis for the vector space W. A product $W \xleftarrow{\Lambda} W \times W$, is *bilinear* in the sense that

$$\Lambda\left(\sum_{i\in[m]} x^i w_i, \sum_{j\in[m]} y^j w_j\right) = \sum_{i,j\in[m]} x^i y^j \Lambda(w_i, w_j).$$

Therefore, its value is determined by the *structure constants* $\{\lambda_{i,j}^k\}_{i,j,k\in[m]}$ where $\Lambda(w_i,w_j) = \sum_{k\in[m]} \lambda_{i,j}^k w_k$. A *unit* for the product is an element $I = \sum_{i\in[m]} u^i w_i$ such that $\Lambda(I,\vec{w}) = \vec{w} = \Lambda(\vec{w}, I)$. In terms of the structure constants,

$$\Lambda\left(\sum_{i\in[m]} u^i w_i, w_j\right) = \sum_{i,k\in[m]} u^i \lambda_{i,j}^k w_k = w_j,$$

and

$$\Lambda\left(w_i, \sum_{j\in[m]} u^j w_j\right) = \sum_{j,k\in[m]} u^j \lambda_{i,j}^k w_k = w_i.$$

Therefore, $\sum_{i\in[m]} u^i \lambda_{i,j}^k = \delta_j^k$ and $\sum_{j\in[m]} \lambda_{i,j}^k u^j = \delta_i^k$. Where $\delta_j^k = |_j^k$ and $\delta_i^k = |_i^k$ are Kronecker delta functions. The associative law for Λ also determines a system of equations for the structure constants when Λ is evaluated upon the basis elements. The associative law reads

$$\Lambda(\Lambda(w_i,w_j),w_k) = \Lambda(w_i, \Lambda(w_j,w_k)).$$

We compute

$$\begin{aligned}
\Lambda(\Lambda(w_i,w_j),w_k) &= \Lambda\left(\sum_{p\in[m]} \lambda_{i,j}^p w_p, w_k\right) \\
&= \sum_{p\in[m]} \lambda_{i,j}^p \Lambda(w_p, w_k) \\
&= \sum_{p\in[m]} \lambda_{i,j}^p \left(\sum_{\ell\in[m]} \lambda_{p,k}^\ell w_\ell\right) \\
&= \sum_{p\in[m]} \sum_{\ell\in[m]} \lambda_{i,j}^p \lambda_{p,k}^\ell w_\ell.
\end{aligned}$$

And the right-hand-side of associativity gives

$$\begin{aligned}
\Lambda(w_i, \Lambda(w_j, w_k)) &= \Lambda\left(w_i, \sum_{p\in[m]} \lambda^p_{j,k} w_p\right) \\
&= \sum_{p\in[m]} \lambda^p_{j,k} \Lambda\left(w_i, w_p\right) \\
&= \sum_{p\in[m]} \lambda^p_{j,k} \sum_{\ell\in[m]} \lambda^\ell_{i,p} w_\ell \\
&= \sum_{p\in[m]} \sum_{\ell\in[m]} \lambda^p_{j,k} \lambda^\ell_{i,p} w_\ell \\
&= \sum_{p\in[m]} \sum_{\ell\in[m]} \lambda^\ell_{i,p} \lambda^p_{j,k} w_\ell .
\end{aligned}$$

So upon equating the coefficients of w_ℓ, the following relationship holds among the structure constants in an associative algebra:

$$\sum_{p\in[m]} \lambda^p_{i,j} \lambda^\ell_{p,k} = \sum_{p\in[m]} \lambda^\ell_{i,p} \lambda^p_{j,k} .$$

By adopting *Einstein summation convention*, in which an unarticulated sum occurs over all repeated indices that appear once as a subscript and once as a superscript, this reads as

$$\lambda^p_{i,j} \lambda^\ell_{p,k} = \lambda^\ell_{i,p} \lambda^p_{j,k} .$$

Thus on either side of the above equation, all values of p are summed over. As usual, the structure constant version of this relationship can be expressed in terms of functions.

Assume that the vector space W is, in fact, $\mathbb{R}^m$. Then the associative law for a bilinear multiplication $\mathbb{R}^m \xleftarrow{\curlywedge} \mathbb{R}^m \otimes \mathbb{R}^m$ reads in diagrammatic form as:

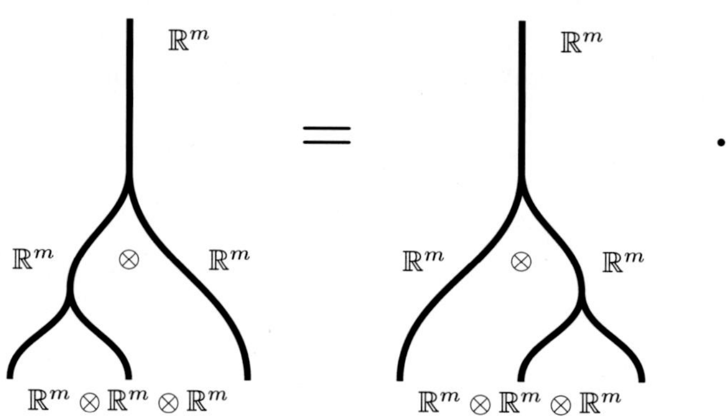

As an alternative formulation we could write $\curlywedge \circ [\curlywedge \otimes \mathsf{I}] = \curlywedge \circ [\mathsf{I} \otimes \curlywedge]$. The unit axiom reads

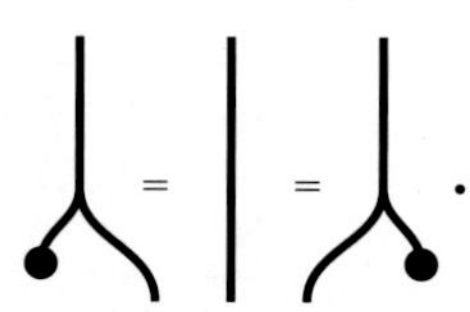

The point in this notation is that the unit in the algebra $\mathbb{R}^m$ is a choice of a vector, $\vec{1}$, therein where $\vec{1}\vec{v} = \vec{v} = \vec{v}\vec{1}$. Furthermore, once such a vector has been chosen, then a copy of the real line $\mathbb{R}$ has been distinguished within $\mathbb{R}^m$. So the notation suggests such a function.

The choice of vector appears as $\overset{|}{\bullet}$. The axiom can also be understood as an equality of the compositions

$$\mathbb{R}^m \xleftarrow{\curlywedge} \mathbb{R}^m \otimes \mathbb{R}^m \xleftarrow{\overset{|}{\bullet}\,|} \mathbb{R}^m,$$

$$\mathbb{R}^m \xleftarrow{\curlywedge} \mathbb{R}^m \otimes \mathbb{R}^m \xleftarrow{|\,\overset{|}{\bullet}} \mathbb{R}^m,$$

and

$$\mathbb{R}^m \xleftarrow{|} \mathbb{R}^m.$$

The up/down convention of composition is reversed from the conventions discussed in Section 1 of this chapter. At differing times, differing up/down or left/right conventions will be used. The inconsistency probably reflects an unconscious and unarticulated prejudice of thought with respect to the algebraic structure in question.

A pairing $\mathbb{R} \xleftarrow{\wedge} \mathbb{R}^m \otimes \mathbb{R}^m$ is said to be *an associative pairing* with respect to a multiplication $\curlywedge$ if $\wedge(\curlywedge(\vec{x},\vec{y}),\vec{z}) = \wedge(\vec{x},\curlywedge(\vec{y},\vec{z}))$. In terms of the structure constants, $\wedge_{\ell,k}\lambda^{\ell}_{i,j} = \wedge_{i,\ell}\lambda^{\ell}_{j,k}$ when summation conventions are used.

Recall from the introductory remarks to this section, that a Frobenius algebra is a vector space W which admits an associative, unital, multiplication, and a non-degenerate associative bilinear pairing. The axioms of the algebra are depicted in diagrammatic form within Fig. 1 (this chapter). The diagrammatic definitions of the Frobenius algebra structure on $M(n,n)$ that is defined in terms of the standard inner product $\cap$ and its diagrammatic dual $\cup$ are outlined in Fig. 2 of this chapter. Within the figure, the maps $\cup$, $\cap$, and $|$ are also identified with the structure constants that define them. For example, to the right of the inner product $\mathbb{R} \xleftarrow{\cap} \mathbb{R}^n \otimes \mathbb{R}^n$ the Kronecker delta function $\delta_{i,j}$ is indicated. That is to say that the value of the inner product upon basis elements is $\cap(e_i,e_j) = \delta_{i,j}$. Similarly, the composition $(\cap) \circ (|\cap|)$ is computed via the structure constant $\delta_{p,q}\delta^p_i\delta_{j,k}\delta^q_\ell$.

At the bottom of Fig. 2, the Frobenius form, $\wedge$, is computed upon a pair of basis elements $e_{i,j} = e_i \otimes e_j$ and $e_{k,\ell} = e_k \otimes e_\ell$ of the vector space $M(n,n)$. Let us discuss this further. The matrix product $e_i \cdot e^j \cdot e_k \cdot e^\ell$ can be viewed as a composition of linear maps

$$\mathbb{R}^n \xleftarrow{e_i} \mathbb{R} \xleftarrow{e^j} \mathbb{R}^n \xleftarrow{e_k} \mathbb{R} \xleftarrow{e^\ell} \mathbb{R}^n,$$

and therefore is a linear map $\mathbb{R}^n \longleftarrow \mathbb{R}^n$ that is represented by the $(n \times n)$-matrix product. The product is parsed as $(e_i \cdot e^j)\cdot(e_k \cdot e^\ell)$ in order to compute the structure constant $\lambda^{(p,q)}_{(i,j),(k,\ell)}$ of the algebra. On the other hand, matrix multiplication has been shown to be associative — both because it is a special case of the composition of functions and associativity of matrix multiplication holds at the level of the entries of the matrix product: $\left[\sum_{i=1}^m \sum_{j=1}^n a^k_i \left(b^i_j c^j_f\right)\right] = \left[\sum_{i=1}^m \sum_{j=1}^n \left(a^k_i b^i_j\right) c^j_f\right]$. So upon reassociation, the matrix product is $e_i \cdot ((e^j \cdot e_k) \cdot e^\ell)$. Then the expression

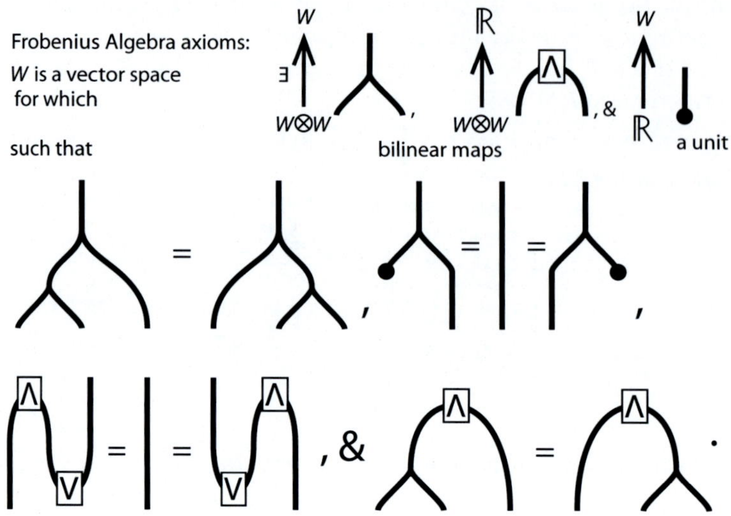

FIGURE 1. The definition of a Frobenius algebra

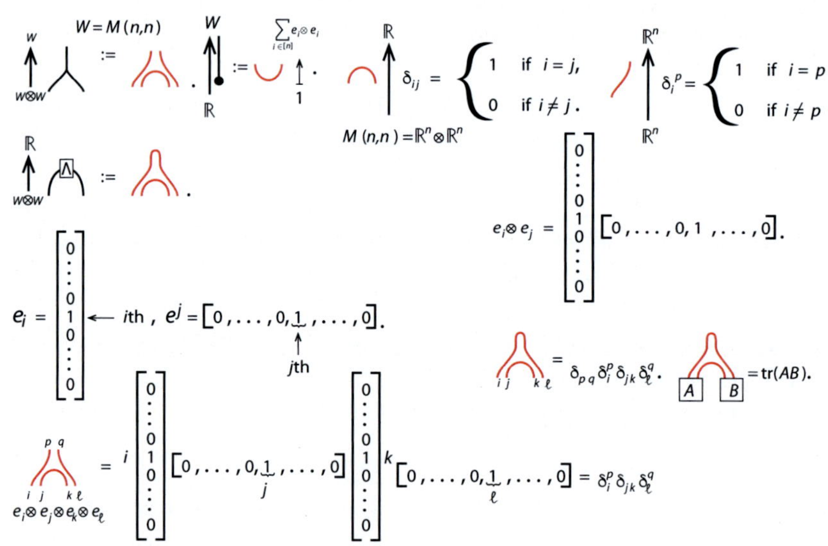

FIGURE 2. The Frobenius algebra structure on $M(n,n)$

$(e^j \cdot e_k)$ is the inner product $\cap(e_j, e^k) = \delta_{j,k}$. This value of 0 or 1 is multiplied by e^ℓ. Then the final result is the "outer product" $e_i(\delta_{j,k}e^\ell) = \delta_i^p \delta_{j,k} \delta_\ell^q$. Thus this gives the computation of $\curlywedge$.

In order to present diagrammatic proofs that a Frobenius algebras also has a co-associative and co-unital co-multiplication and some additional properties, one more observation about the tensor product of tensor operators is needed. Namely, the equality below holds.

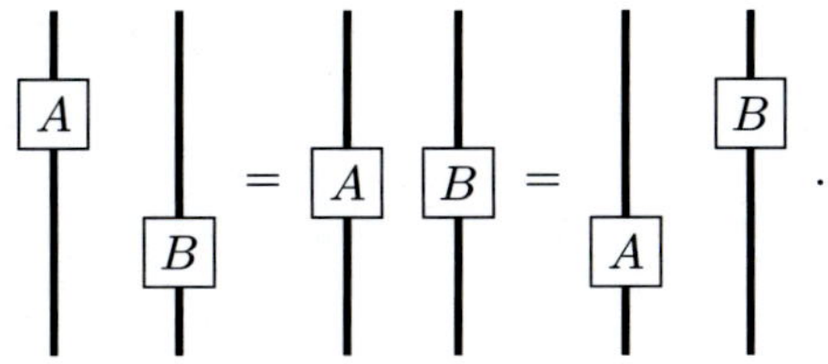

Now here, this has been expressed in terms of linear maps A and B, but by bundling strings together, and reindexing the tensor bases to a single source basis for each of A and B, and a single target basis for each, we can see that the commutation relations $(A \otimes \mathsf{I}) \circ (\mathsf{I} \otimes B) = (A \otimes B) = (\mathsf{I} \otimes B) \circ (A \otimes \mathsf{I})$ hold for any tensor operators A and B.

THEOREM 2.1. *The set of* $(n \times n)$*-matrices* $M(n, n)$ *is a Frobenius algebra.*

PROOF. Multiplication $\curlywedge$ is defined by $\curlywedge(A, B) = A \cdot B$. This is an associative operation, and the unit is the $(n \times n)$-identity matrix whose (i, j)-entry is $\delta^{i,j}$. The inner-product $\wedge$ is defined by $\wedge(A, B) = \mathrm{tr}(A \cdot B)$. These are depicted diagrammatically in Fig. 2. In Fig. 3, diagrammatic proofs that $\curlywedge$ is associative and unital, and that $\wedge$ is non-degenerate and associative are illustrated. □

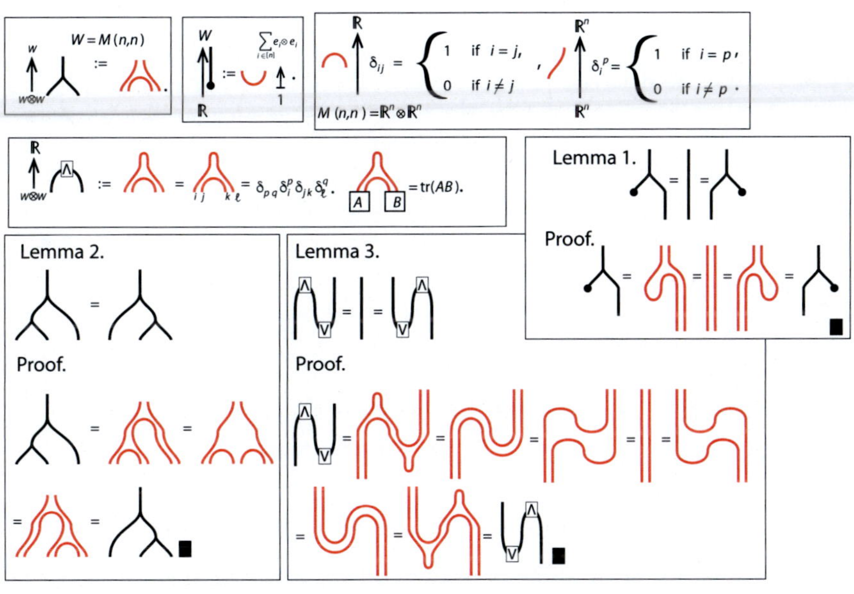

FIGURE 3. $M(n, n)$ is a Frobenius algebra

THEOREM 2.2. *A Frobenius algebra* W *(vector space with an associative, unital, bilinear multiplication* $\curlywedge$ *and an associative non-degenerate inner product* $\wedge$*) has a co-associative, co-unital, co-product and a co-associative co-pairing* Y.

PROOF. We will establish the proof diagrammatically by defining $\mathsf{Y} = (\mathsf{I} \otimes \curlywedge) \circ (\mathsf{V} \otimes \mathsf{I})$ where $(\mathsf{I} \otimes \wedge) \circ (\mathsf{V} \otimes \mathsf{I}) = \mathsf{I} = (\wedge \otimes \mathsf{I}) \circ (\mathsf{I} \otimes \mathsf{V})$ (so V is the matrix inverse of Λ) and showing this is equal to $(\curlywedge \otimes \mathsf{I}) \circ (\mathsf{I} \otimes \mathsf{V})$. The co-unit Y is defined as $\cap \circ (\,\downarrow \otimes \mathsf{I})$.

See Fig. 4 for the details. □

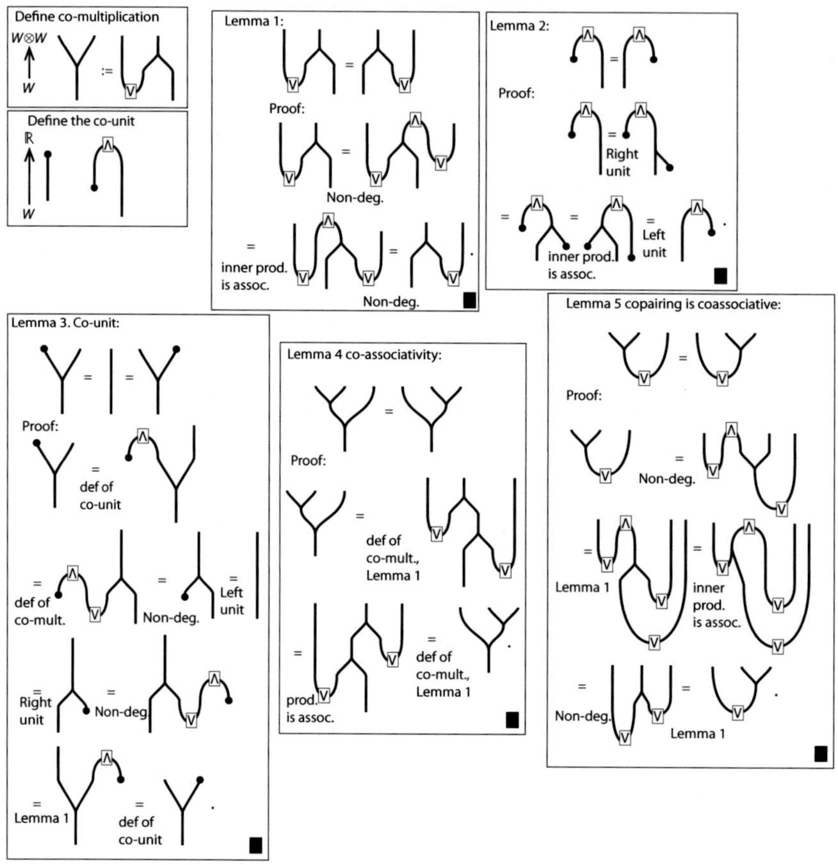

FIGURE 4. A Frobenius algebra is also a coalgebra.

4. Simplfying, clarifying, and abstracting

In the text development above, we started from a vector space over the field of real numbers $\mathbb{R}$, and we pointed out that everything works over an arbitrary field. But in fact, none of the diagrammatics so far have depended upon being able to divide by constants. In the future sections, we will have to divide, and we also will have to be careful when we divide by 2 if the ground field has characteristic 2.

The diagrammatic codification of matrix multiplications, abstract tensors, and tensor operators did not depend upon the field axioms per se. So often, we can work over an integral domain without worrying about smaller details. As always, we encourage readers to worry about smaller details when generalizing, and we will take caution as needed. It is fairly safe to discuss Frobenius algebras as modules over an *integral domain* — a commutative ring for which $ab = 0$ implies that $a = 0$ or $b = 0$.

The painstaking work that is needed to keep track of indices at the subscript and superscript levels soon will allow us to shift the notation from vectors, matrices, and tensors to more general categorical and diagrammatic contexts. We will do so gradually, but before we enter more deeply into the diagrammatic kingdom, we state some general principles about categorification.

5. General categorical principles

Within roughly the decade around 1994, the ideas of "categorification" were becoming popular as an approach to resolving the main problems of modern physics [**20**]. Categorification as a mathematical technique has changed meanings since its inception, and here we only use the word in passing. However, as we approached the work expressed within this volume, we came back to a list of five principles that guide the narrative.

Five categorical principles.

(1) Different things are not the same. They may be naturally isomorphic.
(2) Arrows in a category are used to compare things.
(3) When an action (or arrow) is applied and the process is reversed, the result may, or may not, be the same as inaction. So an arrow followed by its reverse can be compared (by an arrow of larger degree) to staying still. Colloquially, "doing" followed by "undoing" may, or may not, be the same as not doing.
(4) Simultaneity is illusory. This leads to the idea of exchange.
(5) Change followed by exchange is comparable to exchange followed by change via an arrow of higher degree.

At a very basic level, to say that different things are not the same is to deny the veracity of statements such as $1+3=4$, or $1+3=3+1$. If we are counting candy between the two of us, then you and I will have a different sense of value when one has three pieces and the other has one. At a mathematical level, the commutativity of addition involves the process of adding, and it involves a geometric aspect in which the numbers one and three are exchanged.

More relevant to the work herein, is the application of the idea, "different things are not equal," to the processes of doing and undoing. As we continue, certain processes are named birth and death. So a birth followed by a death, even when brief, seems substantial. Or imagine two coins on a table. Their position is interchanged by a clockwise rotation, and then they are returned into place by another clockwise rotation. Those who are trained in knot theory will recognize this as a full twist. Thus the process of doing followed by undoing is not the same as not doing. There is, indeed, a subtle aspect to that example since the rotation direction remained constant.

We write that arrows in a category are a method of comparison. That is a grand assertion. But let us begin with a pair of finite sets, for example. The morphisms in the category of sets are functions, and in understanding a function between two finite sets, we ask, "Is it injective? Is it surjective? Do two points have the same image?" The function is, indeed, a method of comparison. In an algebraic category, a homomorphism describes that the algebraic structures are analogous. The kernel and cokernel of the homomorphism quantify the strength of the analogy.

Returning to the comparison of $1 + 3$ and 4, the connection between them is a graph λ, and this graph is an arrow from the two objects 1 and 3 to the object 4. So if we have things that appear to be equal, we can compare them with an arrow in a category.

As authors, we have chosen the less highfalutin word "arrow" instead of "morphism." The English words that better convey our meaning in a number of categories are "comparison," "metaphor," "simile," or "analogy." The word "arrow," also suggests a trajectory or a path. As we continue, we will have arrows of different multiplicities. In this case, arrow as a path with trajectory is what is being suggested. Then a 2-arrow or double arrow is a disk and a 3-arrow or triple arrow is a ball-like region of space. In both of these cases, there are source and target hemispheres along their boundaries.

The assertion that "simultaneity is illusory" is fanciful, at best. It is clearly influenced by a twentieth century relativistic philosophy. The context here has to do with the nature of arrows as processes. Two independent processes that appear to occur at the same time can be re-examined using a different coordinate system so that they occur at distinct times. Having distinguished their occurrences, we can compare two such distinctions. The comparisons between distinguishing the events is an arrow of sorts that we call an *exchanger*.

The final principle is that time exchanges can be considered natural with respect to other changes. We will not dwell on this idea here, but we remark that it is the bedrock of a variety of categorical constructions of knot theoretic invariants. In our works, we often compare change followed by exchange to exchange followed by change. Such comparisons are of a higher order, and they lead us to categorical and quantifiable expressions that are associated to knots in higher dimensions. The text will take some time to get to these ideas.

On a final note, we will use the term "multi-category" to express a category for which the $(k-1)$-arrows form a category, but for which a horizontal composition (Chapter 4, Section 1.2) is absent. The notions and notations here are not standard among category theorists, and we make no claims to having a deep understanding of category theory. Instead, we hope to understand the material presented here, and some sort of categorical language is useful to express things.

CHAPTER 3

Planar trivalent diagrams

The natural numbers $\mathbb{N} = \{0, 1, 2, \ldots\}$ are discussed briefly as a monoid under addition. Then, also briefly, small categories are defined. The category of most interest has the set $\mathbb{N}$ as its collection of objects. Natural numbers will be expressed in unary notation; for example, the number 5 will consist five points along a line segment. For graphical convenience, a notation for identity morphisms will be developed. Then a category of trivalent planar graphs will be defined. In this category, a method is developed to exchange the heights of trivalent vertices and critical points when these are sufficiently far apart. The method involves introducing a notion of double arrows. Succinctly, double arrows are a method to relate different compositions of 1-arrows. The chapter finishes with the idea of the exchanger double arrows, and that will be a segue into other possible double arrows.

1. Small categories

The natural numbers $\mathbb{N} = \{0, 1, 2, \ldots\}$ is the first example of a monoid that many of us encountered. A *monoid* is a set that has an associative binary operation and a unit. In the case of $\mathbb{N}$, the binary operation is $+$, and the unit is 0. All of mathematics is abstracted from this structure on $\mathbb{N}$. A monoidal category is an abstraction of a monoid — there is an associative method of combining pairs of objects and pairs of morphisms or arrows. The binary operation on arrows is metaphysically orthogonal to composition. A monoidal category is a special type of 2-category, but here an axiomatically different structure is considered. It has objects, arrows, double arrows, and so forth and methods for composing these. One structural difference is articulated by the fourth categorical principle of Chapter 2, Section 5 wherein 2-arrows (or double arrows) are not composed directly via horizontal composition (Chapter 4, Section 1.2). In this case, the structure is called a sesqui-category.

The prefix "sesqui" indicates "one and a half" or not quite two. Each collection of arrows is a category, but the lack of simultaneity (fourth categorical principle Chapter 2, Section 5) avoids compositions between categories. And even though this is awkward, it is diagrammatically desirable. All the multi-categories that will be considered or constructed will, by their nature, be incomplete in the sense that horizontal compositions at each level are mitigated by exchanges. The authors do not know of standard terminology for such structures. So we have adapted an ambiguous terminology: multi-category.

A *small category* consists of a set of *objects* such that between any two objects a and b, there is a set of *morphisms* or *arrows* $\{b \stackrel{f}{\leftarrow} a\}$ so that the *source* of an arrow f is the object a, so $s(f) = a$, and the target of an arrow f is the object b, so $t(f) = b$. For any object a, there is an arrow $(a \overset{\mathsf{I}}{-\!\!-} a)$ called the identity on a. When necessary, we will write I_a.

If $c \xleftarrow{g} b$ and $b \xleftarrow{f} a$ are arrows so that $s(g) = t(f)$, then their composition is an arrow $c \xleftarrow{g \circ f} a$ with source $s(g \circ f) = a$ and target $t(g \circ f) = c$. Compositions of arrows is associative:

$$\left[d \xleftarrow{h} c \xleftarrow{g \circ f} a\right] = \left[d \xleftarrow{h \circ g} b \xleftarrow{f} a\right].$$

For any object a the arrow $a \overset{\mathsf{I}}{-\!\!-} a$ behaves as an identity under compositions:

$$\left(b \overset{\mathsf{I}}{-\!\!-} b \xleftarrow{f} a\right) = \left(b \xleftarrow{f} a\right) = \left(b \xleftarrow{f} a \overset{\mathsf{I}}{-\!\!-} a\right).$$

2. Trivalent graphs

In many of the small diagrammatic categories to be studied, the collection of objects is the set of natural numbers. In this case, we write 1 as -•-, and in general a natural number, α, will be given as a sequence of α evenly placed points along a line segment. Throughout this section, α, β, γ, etc. will denote natural numbers.

2.1. Objects in FA. The sesqui-category FA that is modeled upon the Frobenius algebra axioms will be built from objects, arrows, double arrows, and so forth. The objects in the category FA correspond to the elements of $\mathbb{N} = \{0, 1, 2, \ldots\}$.

In any category, there is an identity arrow for any object. The identity on $\alpha \in \mathbb{N}$ will be denoted by α vertical line segments. Sometimes such a segment will not be entirely vertical, but involve a shift to the left or right. In any case, a predominantly vertical arc that has no critical points will serve as an identity on -•-. The integer 0 will be denoted by an empty set of points, and its identity is an empty set of vertical arcs. We will also use the graphic 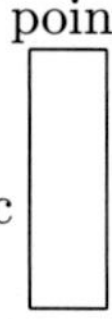

to mean the identity on the integer object α. Namely, 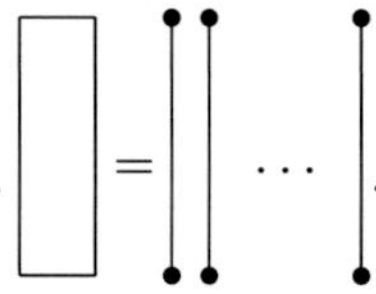

. Such a vertical band is analogous to a wire strip of jumper cables that were commonly found inside a computer towards the end of the 20th century. Thus

, 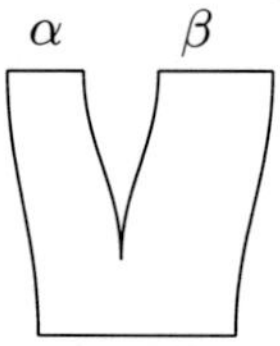

, and 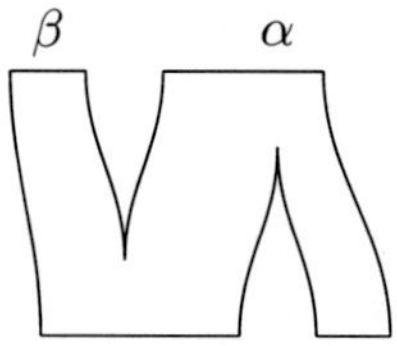

are schematics that denote the identities upon $\alpha + \beta$. It is amusing that the schematic on the right indicates that the commutative law for addition is a consequence of the associative law. The reader should imagine vertical arcs running through the ribbons, and the ribbons as being non-existent.

2.2. Arrows in the category FA. The rules for describing arrows are:

(1) the diagrams here

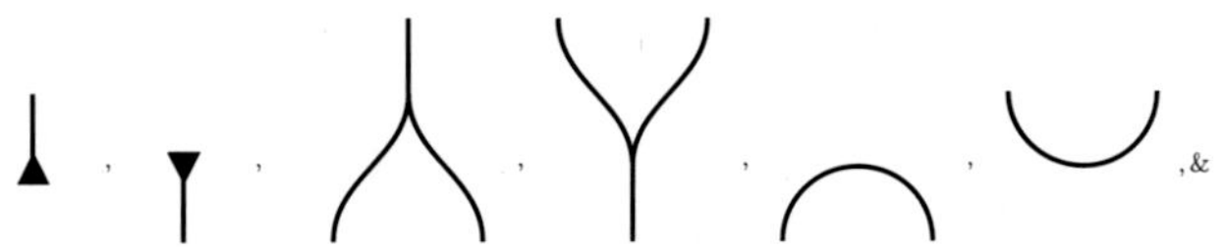

are arrows,[1]

(2) if A and B are arrows with suitable sources and targets, then each of the diagrams here

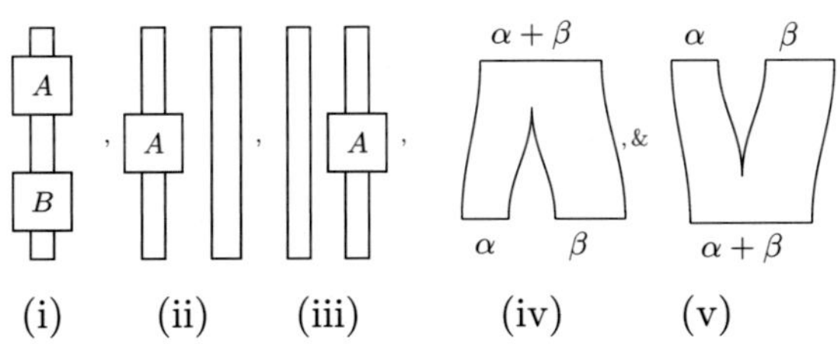

is an arrow, and

(3) the composition of arrows can always be rescaled.

Precisely speaking items (iv) and (v) are understood as identities that may be nested between other identities as indicated:

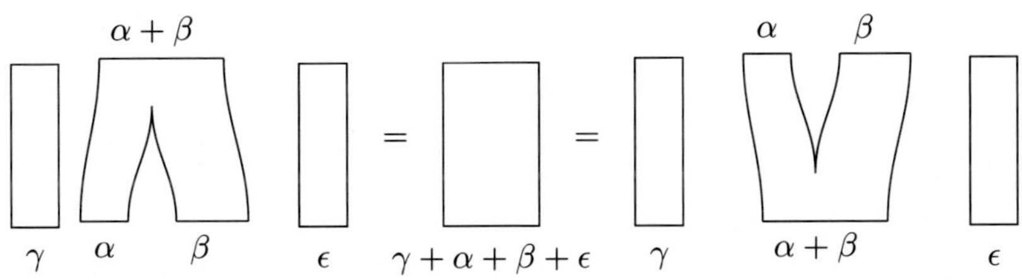

In these illustrations, it is possible that either γ or ϵ is 0. Thus, in using different parameters, there are identity arrows of the form given below.

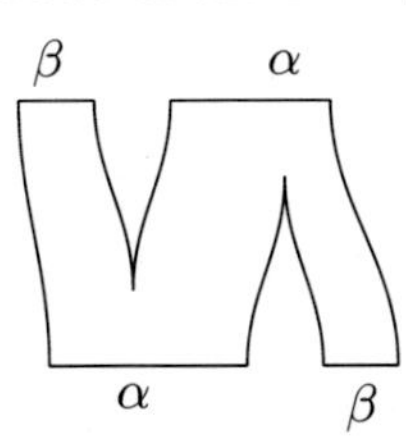

The source of the arrow is at the bottom of the diagram, and the target is at the top. We will call the arrows in item 1 the *generating* arrows. In item 2, arrows cannot be composed willy-nilly, but rather the strings that go out from, or go into,

[1]We have replaced  with  to avoid confusion with the objects of the category.

A and those that go out from, or go into, B are all accounted for: either they are ultimate sources or targets or they interact between A and B. Thus when the boxes that represent A and B overlap, the source of the top box intersects the target of the bottom arrow. Item 2, implies that we can form arrows as indicated here.

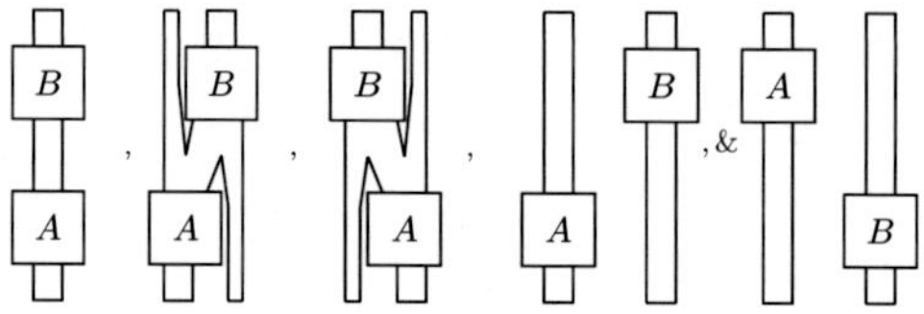

Rescaling is used so that associativity holds. For example, suppose that each of the generating arrows occurs within a square of unit (1×1) size. Then $A \circ B$ will be of size (1×2). So we have to rescale $A \circ B$ back to unit size to compose with a morphism C. And even further the scaling factors for $(A \circ B) \circ C$ and $A \circ (B \circ C)$ differ. The technicalities involved in multiple rescalings present difficulties that divert attention from the narrative, and so rescaling will occur without fanfare.

No relations are assumed to hold among the arrows. So a composition of arrows is an embedding of a graph that has mono-valent and trivalent vertices such that at each horizontal level there is at most one vertex or non-degenerate critical point of an edge. However, the purpose in defining this category is to provide a categorical model for Frobenius algebras. In many categorical contexts, relationships among arrows are encoded as commutative diagrams in which one compositional sequence of arrows is deemed to be equal to another compositional sequence. On the other hand, from the first categorical principle (Chapter 2, Section 5) different things are not equal.

This principle will be encoded by formulating the posited equalities between two (compositions of) arrows as a *double arrow*, 2-arrow, or 2-morphism. The situation is analogous to the construction of a group via a presentation given by generators and relations. In fact, the group construction is a special case of this higher categorical principle. To complete the analogy, consider the collection of diagrammatic morphisms as the analogue of a free group. Then the relations to be imposed between morphisms are analogous to the group relators (Chapter 12).

For the moment, the idea of a double arrow will only be used to describe a categorical method for exchanging the heights of distant critical points or crossings. Chapter 4 discusses double arrows in detail.

2.3. The exchanger between 1-arrows. Recall that if F and G are arrows, there are arrows

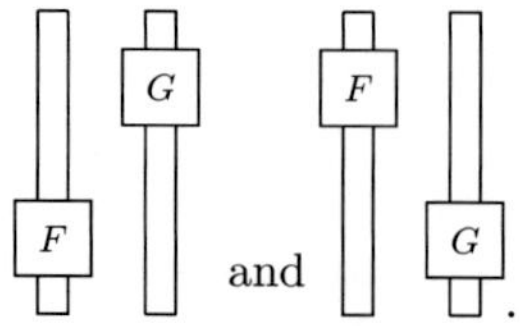

By our conventions, these are considered to be different unless, of course, F or G is an identity arrow. Since simultaneity is illusory, (Fourth principle Chapter 2, Section 5), different events occur at different times. That is, the vertical direction in our figures is time-like. In this sense, there is no simultaneity. When a choice of relative heights for F and G is chosen and when F and G act upon sufficiently distant objects, an axiomatization of interchanging the heights occurs as follows.

Exchanger axiom. *Suppose that* $\gamma \xleftarrow{F} \alpha$ *and* $\zeta \xleftarrow{G} \beta$ *are arrows. There is a 2-arrow*

$$\mathsf{X} : (F \otimes \mathsf{I}_\zeta) \circ (\mathsf{I}_\alpha \otimes G) \Rightarrow (\mathsf{I}_\gamma \otimes G) \circ (F \otimes \mathsf{I}_\beta)$$

which is a 2-isomorphism. The arrow X *depends naturally upon the arrows* F *and* G. Here $(F \otimes \mathsf{I}_\zeta) \circ (\mathsf{I}_\alpha \otimes G)$ and $(\mathsf{I}_\gamma \otimes G) \circ (F \otimes \mathsf{I}_\beta)$ are respective algebraic expressions of the graphics:

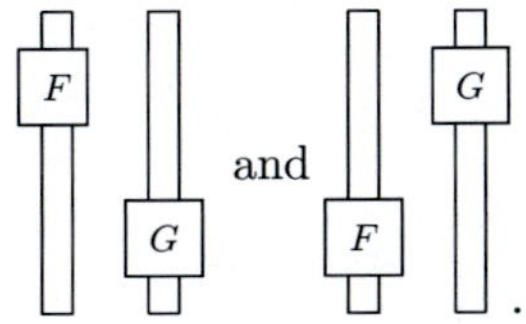

In Figs. 1 and 2, the exchange axiom is illustrated in several graphical forms. These help define the natural dependence upon F and G. In Fig. 1, the exchange

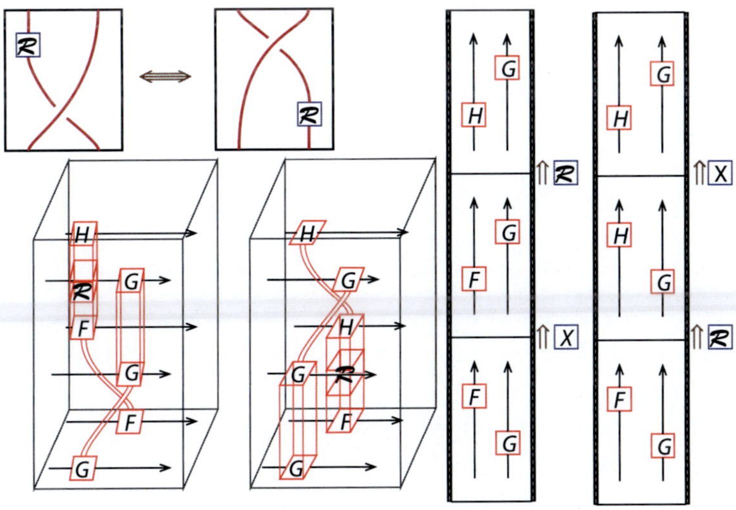

FIGURE 1. The exchanger family is natural

between arrows F and G occurs, and F is transformed into an arrow H via a 2-arrow $\mathcal{R}$. Naturality indicates that these processes commute. The commutation is presented from three different points of view. At the bottom left, the differing processes are enclosed in boxes following the convention of Ch. 4, Sec. 1.3, Fig. 1. On the right-hand-side of the illustration, a movie move version is presented. There are two movies. On the left film-strip from the bottom to the top, the heights of F and G are exchanged, and then F is transformed into H via the 2-arrow $\mathcal{R}$. On the right, F is transformed to H via $\mathcal{R}$, and then G and H have their relative heights exchanged. At the top of the illustration, the illustrations within the boxes are projected into the plane of the page, and schematized. The catch phrase for the naturality condition of the exchanger axiom is, "Change followed by exchange is equivalent to exchange followed by change."

The illustration in Fig. 2 serves several purposes. First, the direction of the exchanger is established. Second, to state that this is an isomorphism is to state

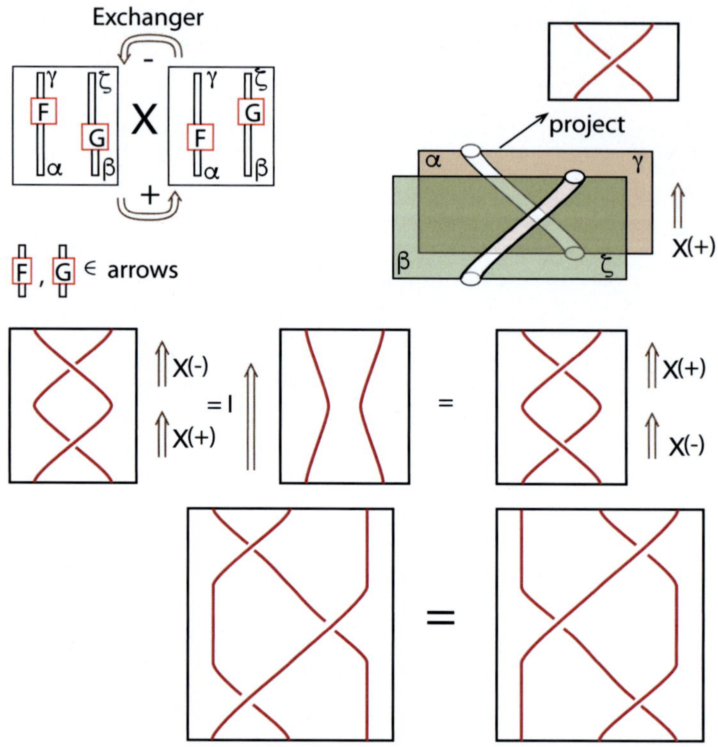

FIGURE 2. The exchanger family of natural isomorphisms

that there is an inverse double arrow. We denote the exchanger as $\mathsf{X}(+)$ and its inverse as $\mathsf{X}(-)$, or more simply as X and $\overline{\mathsf{X}}$. That these are inverses is to assert that there are equivalences among $\mathsf{X}(+) \circ \mathsf{X}(-)$, $\mathsf{X}(-) \circ \mathsf{X}(+)$, and an identity upon $(F \otimes \mathsf{I}_\zeta) \circ (\mathsf{I}_\alpha \otimes G)$. These equivalences are illustrated in the middle of the illustration. But remain vigilant. As they are equivalences, they will be connected with a triple arrow that will also be an isomorphism. Finally at the bottom of the illustration, a braid-like relation is indicated among the exchangers. This relation is a consequence of the naturality condition. The notion of composition ($\circ$) of double arrows will be described in general in Chapter 4, Section 1 below. For now, it is safe to consider the double arrow as living in a box, and the composition to be a vertical stacking of the boxes.

2.4. Further remarks on the exchanger. The exchanger is considered in more detail. First, given any 1-morphisms, F and G, there is an exchanger

$$\mathsf{X}^{F,G}_{G,F}(+) : (F \otimes \mathsf{I}_\zeta) \circ (\mathsf{I}_\alpha \otimes G) \Rightarrow (\mathsf{I}_\gamma \otimes G) \circ (F \otimes \mathsf{I}_\beta).$$

Thus the exchange laws require a *family*

$$\{\mathsf{X}^{F,G}_{G,F} : F \in \{\gamma \leftarrow \alpha\} \ \& \ G \in \{\zeta \leftarrow \beta\}\}.$$

It is, however, cumbersome to indicate F and G within the notation. As before, we are asserting that X is a 2-isomorphism. Explicitly, we write $\mathsf{X}(+)$ to indicate the forward arrow, and $\mathsf{X}(-)$ to indicate its double arrow inverse. Within the literature it is common to assert that $\{\mathsf{X}_{G,F}^{F,G}\}$ is a *natural family of isomorphisms.* The isomorphisms are natural in the sense of Fig. 1. The discussion of this family in terms of naturality of functors is beyond the scope of text.

Within the discussion, hardly any mention of the nature of F and G is used. Indeed, examining the axioms for arrows (1), (2), and (3) in Section 2.2 (this chapter), we can drop axiom (1), and assume that the generating set of arrows is whatever set that we like. Furthermore, in Chapter 4, Section 1.2 exchangers for double arrows will be indicated. Indeed in the future, the existence of a natural family of $(n+1)$-arrows that are exchangers between n-arrows will be asserted. Also, as we proceed, we will establish an $(n+2)$-relation among exchangers.

CHAPTER 4

The multi-category FA

In this chapter, the axioms for a Frobenius algebra and the definitions of co-multiplication and co-units will be interpreted in two ways. In one instance, surface-like diagrams interpolate between the graph-like diagrams that represent the compositions of multiplication, pairings, units and their reverse arrow counterparts. In the other instance, formal symbols, or glyphs, are introduced to label the relationships between the pairs of compositions that express the Frobenius algebra axioms. Both the surfaces and the glyphs are double arrows in their respective multi-categories. A diagrammatically obvious functor[1] between these two descriptions will be constructed.

The discussion will start in a formal categorical setting by abstracting relationships among arrows as double arrows. When the collection of arrows in a category is itself a category, then the composition of double arrows needs to be defined. There are two types of compositions: one that is called globular, and another that is called skew. These ideas are first presented in abstract, and then in the context of the Frobenius algebra.

1. Composition of double arrows

In a category, it is common to assert that identities between compositions of arrows occur. In this case, we say that the diagram

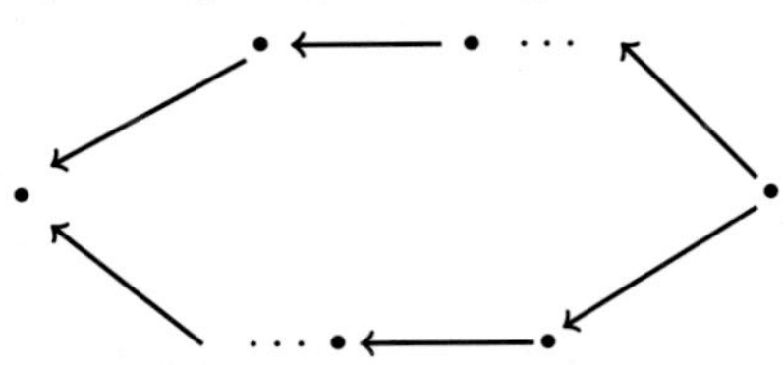

commutes. However, when different things are not equal, this diagram is replaced by a double arrow

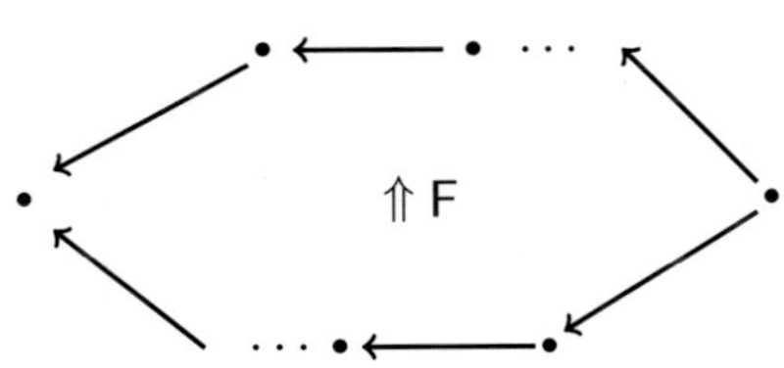

Often, since the double arrow was once thought of as an equality, the double arrow, F, pretends to be an isomorphism. In that case, its source and target are interchanged to form a new double arrow Ⅎ. The double arrow Ⅎ is the *reverse* of F. These are composed by gluing the polygonal diagrams together along a consistently

[1]The notion of functor is given in Chapter 6, Section 5.1.

oriented sequence of arrows in one of two possible ways. In brief, write $\begin{bmatrix}\mathsf{F}\\ \text{Ⅎ}\end{bmatrix}$ or $\begin{bmatrix}\text{Ⅎ}\\ \mathsf{F}\end{bmatrix}$ to denote these compositions. Either is asserted to be related to an identity double arrow: $\begin{bmatrix}\mathsf{F}\\ \text{Ⅎ}\end{bmatrix}$ is comparable to the identity upon the composition of arrows that forms the target of F while $\begin{bmatrix}\text{Ⅎ}\\ \mathsf{F}\end{bmatrix}$ is comparable to the identity upon the composition of arrows that forms its source. When both of these comparisons are thought of as equality and other conditions hold, then we will say that the double arrow F is invertible.

There is a related subtle notion of a weakly invertible 2-arrow. It will be explored in detail in Sections 1.6 and 2 of this chapter, and fully defined in Chapter 7, Section 2. In general, doing followed by undoing is comparable to not doing. However, we often make a distinction between comparing the process of executing an action followed by its reverse to doing nothing. In a mathematical realm, absolute value provides a metaphor. For example, when I swim to and fro, I wind up in the same place. Did I do nothing?

The collection of double arrows will form a category. Therefore a double arrow has a 1-arrow as a source and a 1-arrow as a target. In a schematic form, the structure of the double arrow is depicted here:

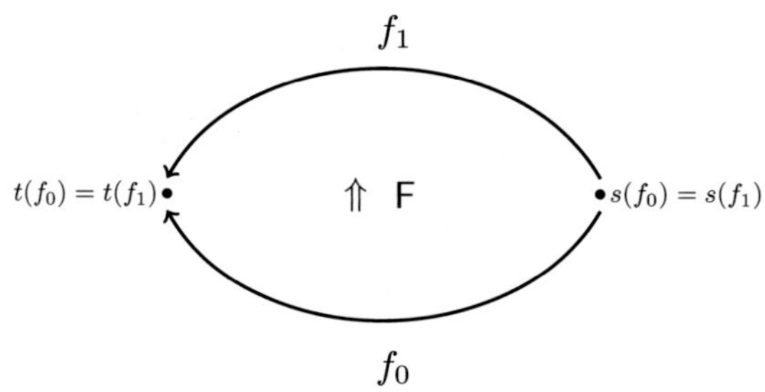

In the figure, $S(F) = f_0$, and $T(F) = f_1$. Often, we will not need to specify the source and target of single arrows.

An alternative representation of a double arrow is a *rectangular diagram* or a *rectangular description of a double arrow* such as this:

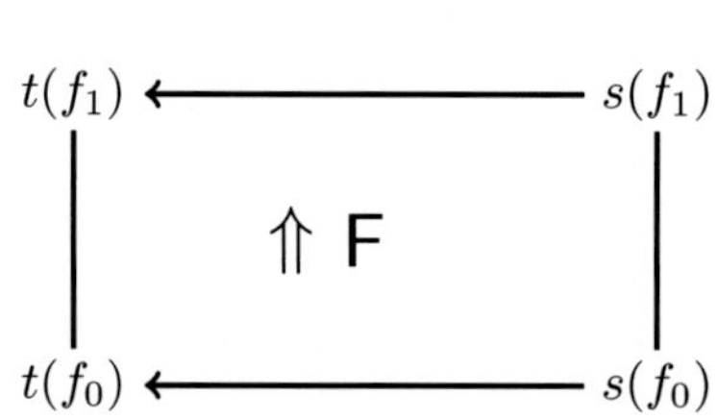

The vertical segments are identity arrows upon $s(f_0) = s(f_1)$ and $t(f_0) = t(f_1)$.

1.1. Globular and skew compositions of double arrows. The *vertical composition* or *globular composition* of double arrows is indicated.

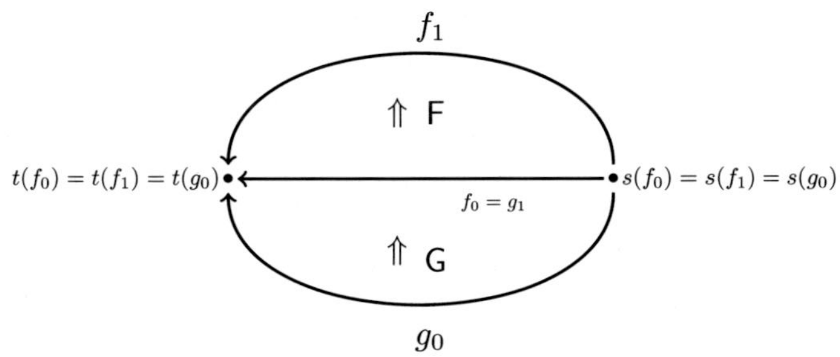

Often, the source or target of a double arrow can be written as the composition of more than one single arrow. By composing all the arrows in the source, and leaving two arrows in the target we obtain the configuration here.

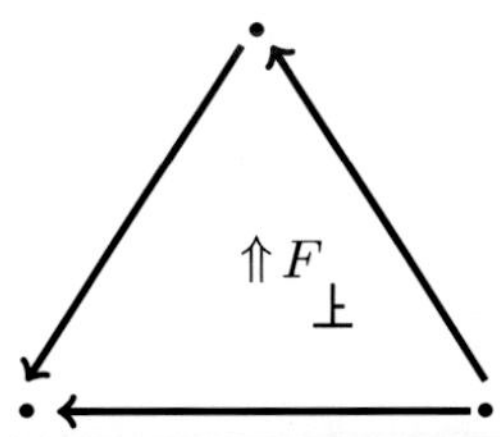

We can also compose all the arrows in the target and leave two in the source to obtain the configuration here.

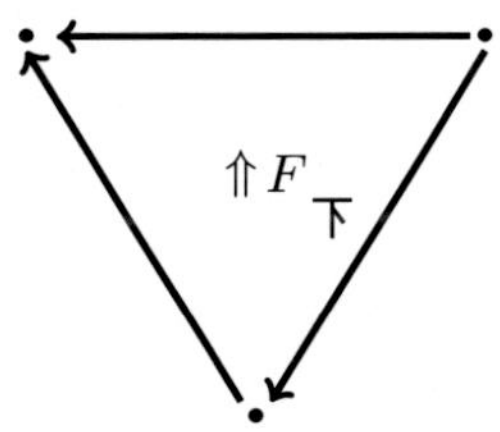

The triangles are labeled with up/down [上, 下] to make the following indications. In the down [下] case, there is one target arrow, and in the up [上] case, there is one source arrow. The orientation of these Chinese characters or *kanji* has the horizon of the character along the single arrow source or target. In many subsequent diagrams these kanji provide methods to identify triangles in skew compositions with the ones above. Instead of the kanji, we often denote double arrows by F or Ⅎ.

In these cases, the triangular pictures are redrawn as rectangular diagrams by augmenting 1-arrows by identity arrows so that other forms of composition can be described. Below objects, a,b,c, and d are named, and the double arrows are redrawn using identities upon some of these. The source arrows appear at the bottom of the diagram.

In these configurations, the double arrows have a common arrow, one as a portion of the source, and another as a portion of the target. In cases such as these, the double arrows can also be augmented by identity double arrows to form *skew compositions* such as these indicated below.

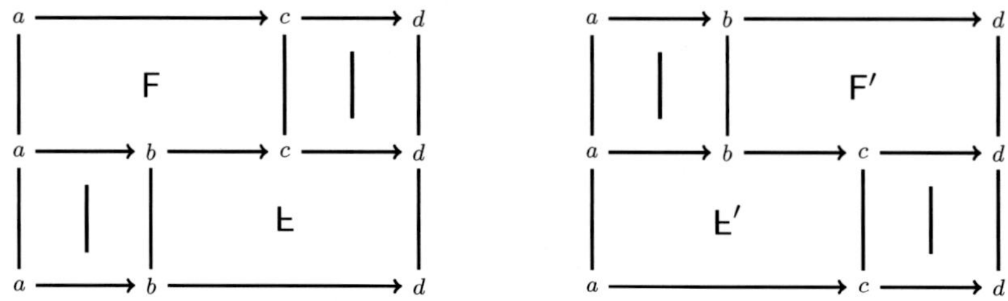

Here the vertical arcs indicate the identity double arrows on the 1-arrows $a \to b$ and $c \to d$. The skew diagrams above can also be expressed in a simplified manner by means of the triangular diagrams indicated here

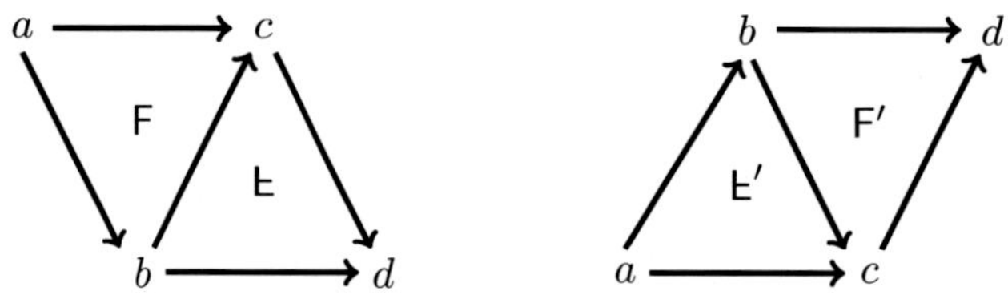

Exercise. Consider a pair of up 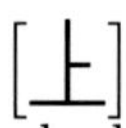 and down  triangles. Join them in all possible manners along an oriented edge so that a parallelogram results. In these gluings, one or both triangles might be reflected. Which of these parallelograms have oriented boundaries for which there is a single source object, a single target object, a source composition of 1-arrows, and a target composition of 1-arrows? Conclude that up to reflection, there are only two possible skew-compositions.

1.2. Horizontal composition. In the multi-categories that we consider, there are objects, arrows, and double arrows. The collection of arrows between two objects $\{b \xleftarrow{f} a\}$ is also a category. The arrows between arrows are called *double arrows*, 2-*morphisms*, or 2-*arrow*. For every 1-arrow f, there is an identity double arrow, and globular and skew compositions occur between pairs of double arrows.

Those who are familiar with 2-categories, might expect a horizontal composition as indicated.

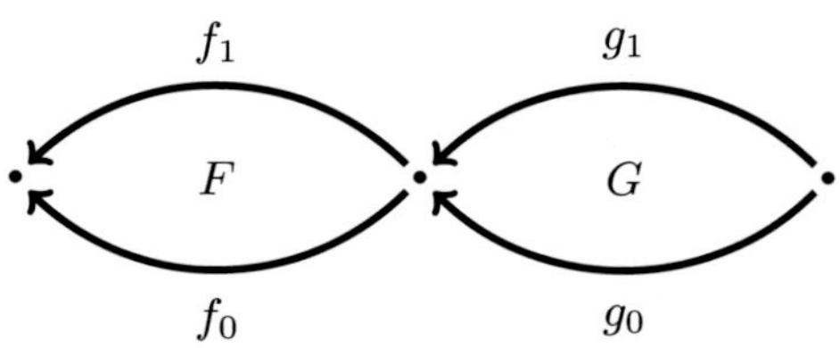

In our situation this is disallowed, since double arrows are required to have a portion of their 1-dimensional boundary coincident in order to be composed. So we replace

this with the composition indicated here. As above, double arrows are represented in rectangular form.

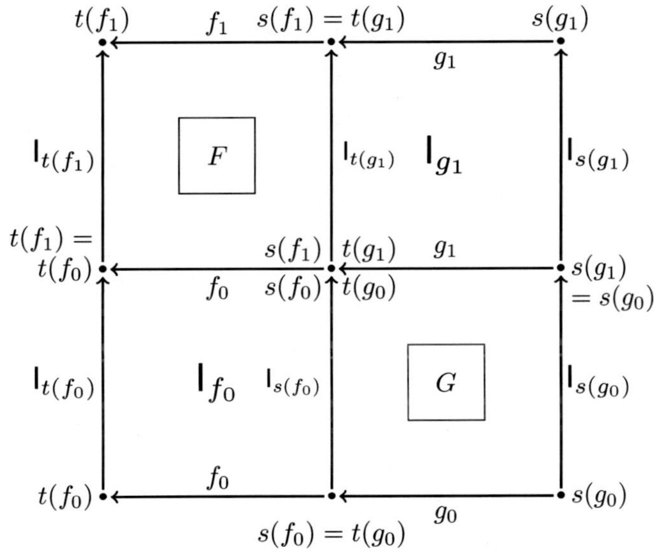

This choice of resolution is, of course arbitrary, for we could have chosen to replace with the figure below.

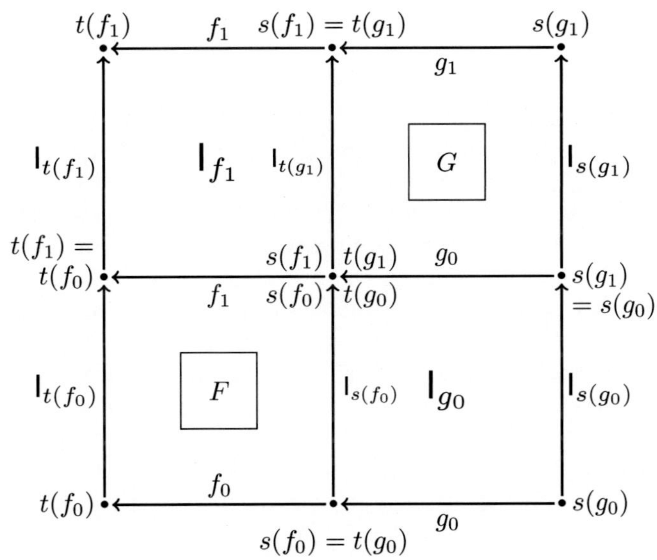

We make the initial choice and then suppose that there is a higher order relation that connects them.

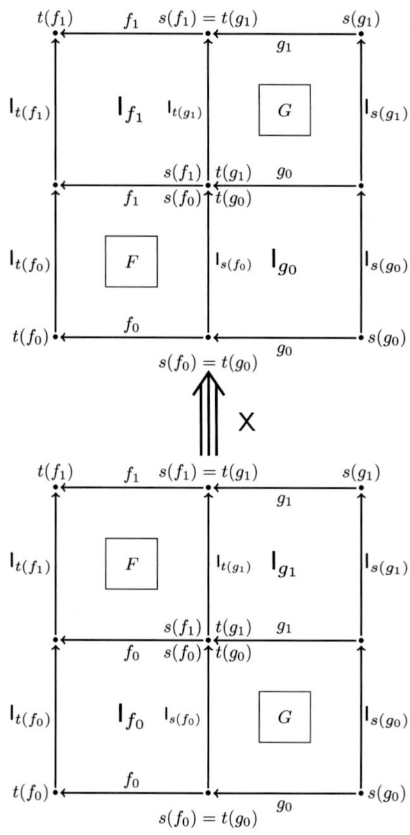

This triple arrow X is called an *exchanger*. As before, it depends naturally upon F and G, and so the family $\{\mathsf{X}_{F,G}\}$ is a natural family of isomorphisms in the sense that change followed by exchange is exchange followed by change.

A *sesqui-category* is a (small) category such that the collection of arrows between any pair of objects is a category. Arrows between 1-arrows are called *double arrows* or *2-arrows*. These can be composed in a globular or skew manner. A horizontal composition is replaced by the first diagram above, and for any pair of double arrows F and G there is an exchanger triple arrow that is an isomorphism as described above and which is natural with respect to any other triple arrow.

The exchanger axiom is a key aspect of all the higher dimensional categories that are constructed. In particular, in Chapter 3, Section 2.3 there is also an exchanger that is a double arrow which is defined to exchange the height of a pair of 1-arrows in the current sesqui-category. The diagrammatic similarity between the exchanger as a double arrow and an exchanger as a triple arrow occurs since the monoidal structure on the object set ($\mathbb{N}$) actually means that the object set is a category, and therefore the 1-arrows are, in fact, double arrows. We will return to this point later.

1.3. Double arrows. To construct the categorical analogue of a Frobenius algebra, we think of it as an example of a sesqui-category. Let us call this sesqui-category FA.[2] The object set is $\mathbb{N} = \{0, 1, 2, \ldots\}$. The arrows are constructed as above (Chapter 3, Section 2.2) in items 1, 2, and 3. Here we will abstract the Frobenius algebra axioms Ch. 2, Fig. 1 and their consequences Ch. 2, Fig. 4, but we will drop the notation $\wedge$ and $\vee$ that indicate the pairing and its dual. Also for more notational clarity, the unit is denoted by an upward pointing triangle in the figures.

We establish a correspondence between pairs of planar diagrams that are connected by an equality and surface-like diagrams that interpolate between the diagrams of the arrows and that are placed inside cubes via Fig. 1. Copies of this illustration adorn subsequent figures as reminders. Herein the top, bottom, left, and right as well as the left and right of the equal signs are established. In a sesqui-

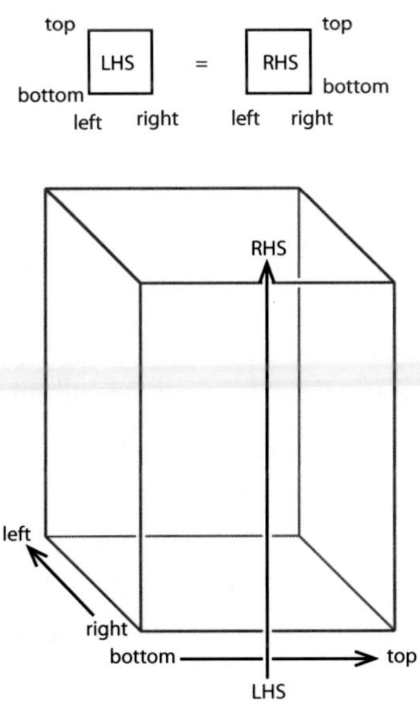

FIGURE 1. Pairs of planar diagrams related by an interpolating surface in a box

category, the collection of arrows between any two objects is also a category. Thus we will need to describe the double arrows in an analogous fashion to that of the 1-arrows, and we need to define composition.

(1) The surface-like diagrams that are depicted in Fig. 2 are double arrows.
(2) The surface-like diagrams that are depicted in Figs. 3 through 8 are double arrows.
(3) If double arrows F and G appear in boxes, then the juxtaposition of the boxes that are depicted in Fig. 9 are double arrows.

[2]As we add structure, this category FA will have higher order arrows. In that case it will be called a multi-category.

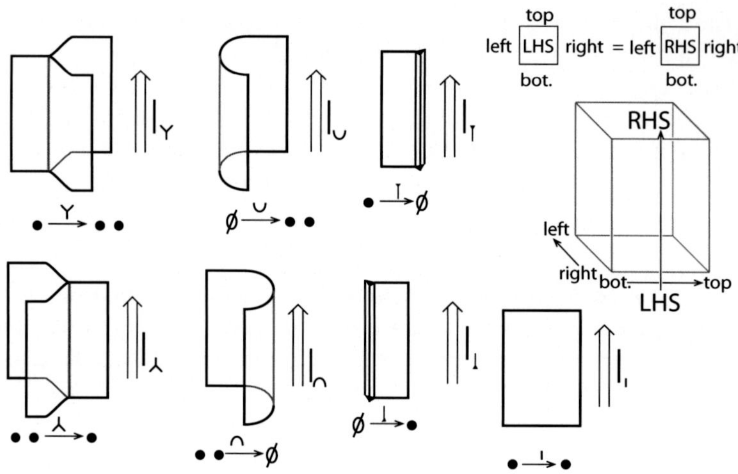

FIGURE 2. The identity 2-arrows upon the generating arrows

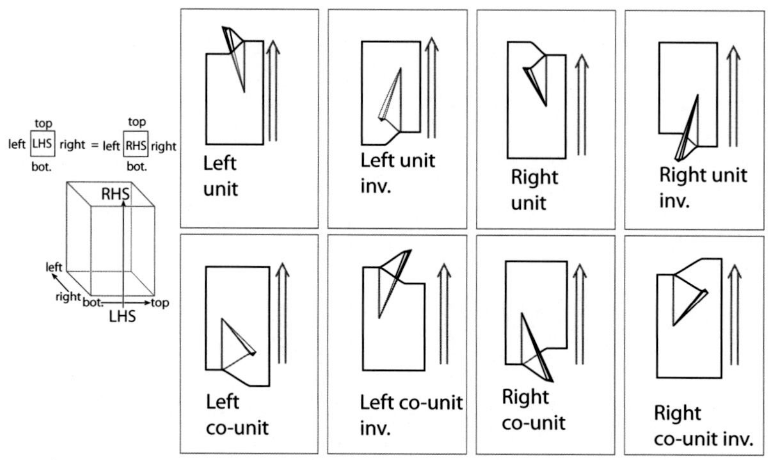

FIGURE 3. The 2-arrows that give the properties of unit and co-unit

The composition of double arrows is achieved by stacking boxes on top of each other as in Fig. 9. Here as in item 2 of Chapter 3, Section 2.2, boxes can be stacked when the source of the top box has a region that coincides with a region on the target of the bottom box. After boxes are stacked, the remaining unmatched sources and targets are extended by identities. The front and back faces of boxes are empty, and the left and right faces of any box is an identity on a collection of dots $\alpha = (\underbrace{\bullet\!-\!\bullet \cdots \bullet}_{\alpha})$. A horizontal juxtaposition of two boxes is forbidden (as indicated on the bottom left of the illustration) unless at least one is an identity upon the 1-arrow on the bottom of that box. Stacking boxes in the fashion indicated can also be interpreted from the point of view of Section 1.4 of this chapter.

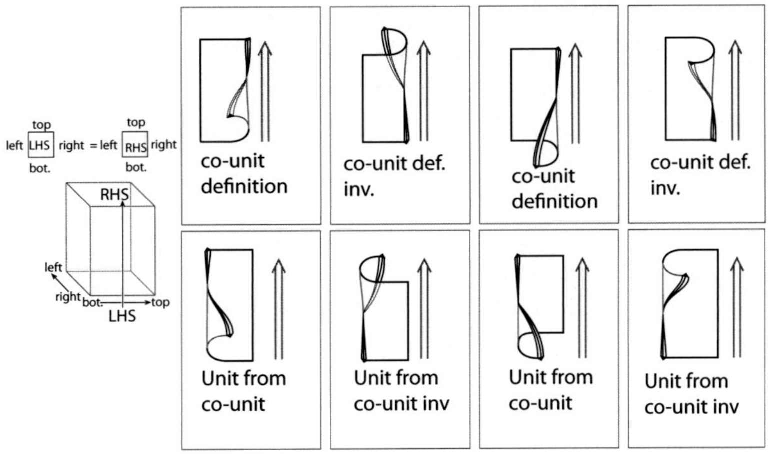

FIGURE 4. The double arrows that relate units and co-units

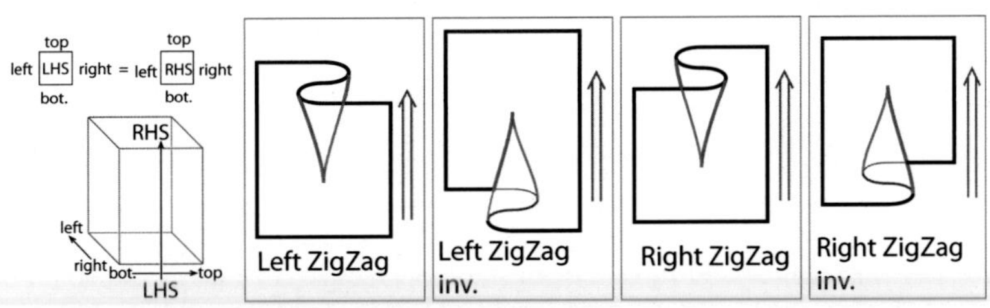

FIGURE 5. The double arrows that correspond to the non-degeneracy of the pairing

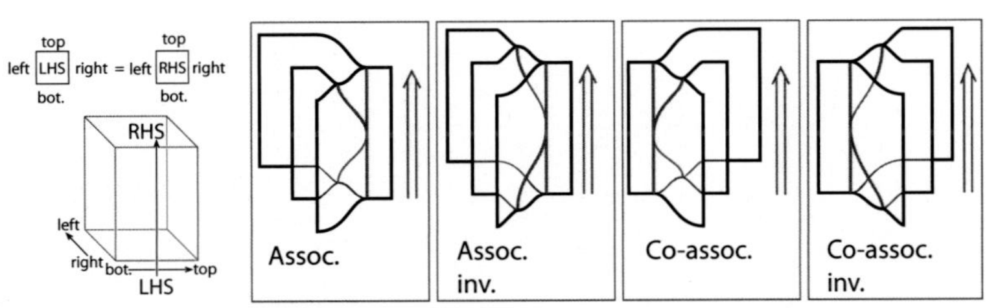

FIGURE 6. The double arrows that correspond to associative and co-associative axioms

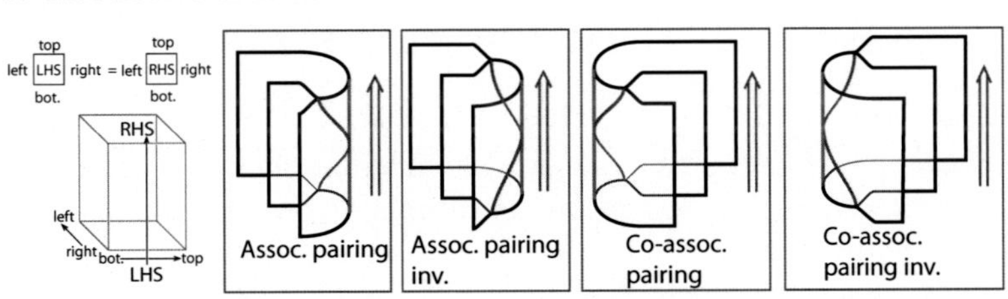

FIGURE 7. The double arrows that correspond to associativity of the pairing and co-pairing

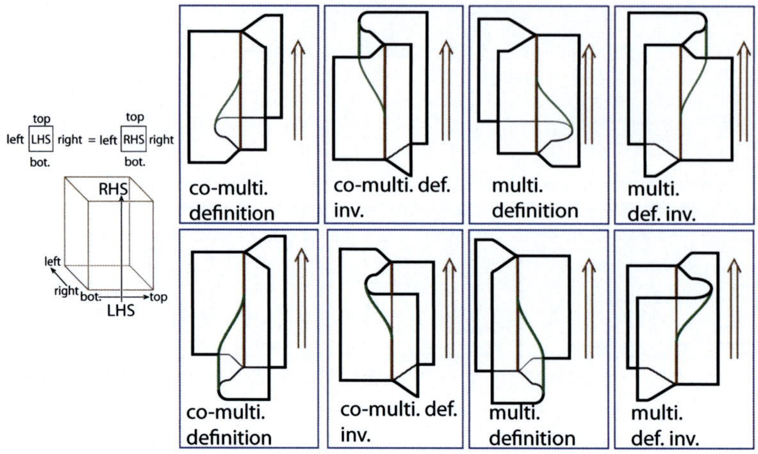

FIGURE 8. The double arrows that relate multiplication and co-multiplication

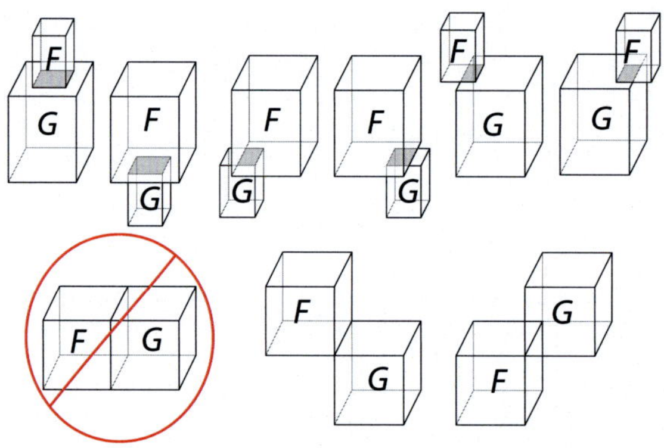

FIGURE 9. The compositions of double arrows

(4) For any two 1-arrows F and G there is a double arrow $\mathsf{X} = \mathsf{X}_{G,F}$ called an *exchanger* such that the Exchanger axiom holds. In particular, $\mathsf{X}_{G,F}$ depends naturally upon F and G. See also Chapter 3, Sections 2.3 and 2.4.
(5) Boxes are rescaled as needed for composition.

In brief, we are asserting that the surface-like diagrams that appear in Fig. 10 are double arrows, and that reflections of double arrows are also double arrows. Specifically, suppose that a double arrow lies in the box $\{(x, y, z) \in [0, 1]^3]\}$.[3] We can think of the surface-like diagram given by a set of parametrizations $\{(x(s_i, t_i), y(s_i, t_i), z(s_i, t_i)) : i \in S\}$ where S is a finite set and $x, y, z \in [0, 1]$. Then

[3]Unless specifed otherwise the x-axis points out of the page, the y-axis points right, and the z-axis points up. The unit interval is the point set $[0, 1] = \{t \in \mathbb{R} : 0 \leq t \leq 1\}$

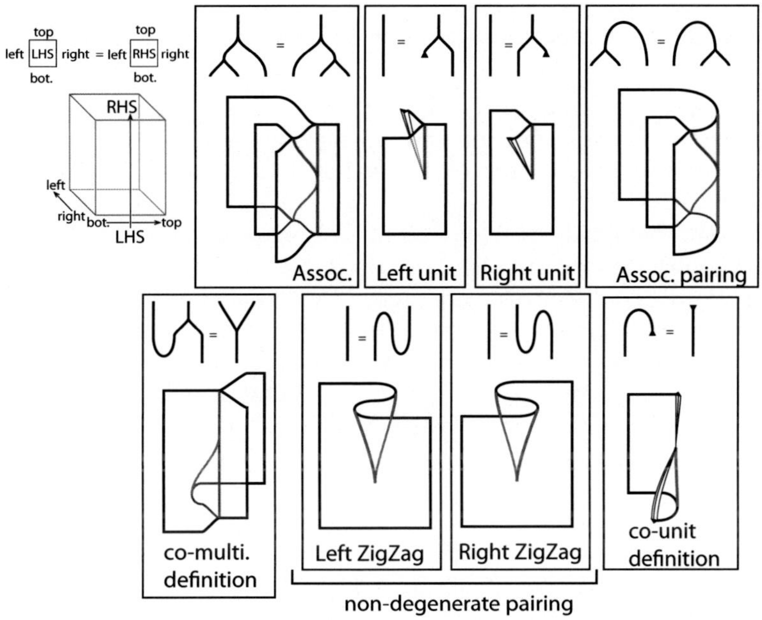

FIGURE 10. The axioms of a Frobenius algebra as surface-like diagrams

each of $\{(1-x(s_i,t_i)),y(s_i,t_i),z(s_i,t_i)) : i \in S\}$, $\{(x(s_i,t_i)),1-y(s_i,t_i),z(s_i,t_i)) : i \in S\}$, and $\{(x(s_i,t_i)),y(s_i,t_i),1-z(s_i,t_i)) : i \in S\}$ also represents a double arrow. From this point of view, we expect that if a double arrow corresponds to an axiom of a Frobenius algebra, then there should be a total of eight double arrows associated. However, the associativity axiom, the non-degeneracy axioms, and the associativity of the pairing have additional symmetries. So in these cases, only four double arrows arise.

In Fig. 2, 1-arrows now point to the right. Right-pointing single arrows allow the identity double arrows that are depicted in that illustration to be represented by folds, edges, and seams in the associated surface-like diagrams. They also accommodate an easy reading of the compositions of double arrows that will occur.

A horizontal reflection of a Frobenius algebra axiom interchanges multiplication and co-multiplication or pairings and co-pairings. Similarly, unital axioms reflect to co-unital axioms. At the level of the axioms themselves, the double arrows are those that are depicted in Fig. 10, but within the illustrations Fig. 3 through 8, we have indicated some reflections of these figures with the same moniker. For example in Fig. 8 two distinct surfaces are named as the definition of co-multiplication. We know from the Lemma 1 portion of Fig. 4 (Ch. 2) that the two source arrows in those double arrows are equal as a result of the axioms. So rather than renaming the double arrow, we give the two double arrows that are reflections in the (y,z)-plane $(x \mapsto (1-x))$, the same name. In the same spirit, multiplication can be defined in terms of co-multiplication, and vice versa or via caps or cups on the left or right of the trivalent vertex. So we name double arrows "multiplication definition" to indicate that given a co-multiplication a multiplication can be defined.

1.4. Notion and notation. At an abstract level, a double arrow is only a method used to relate a pair of compositions of 1-arrows that share the same ultimate source and target. For example, the pairing in a Frobenius algebra is associative in the sense that $\cap \circ (\curlywedge \otimes |) = \cap \circ (| \otimes \curlywedge)$. When the difference between the left and right hand side of an equation needs emphasis, the equality is replaced by a pair of double arrows: $\cap \circ (\curlywedge \otimes |) \Leftarrow \cap \circ (| \otimes \curlywedge)$, and $\cap \circ (\curlywedge \otimes |) \Rightarrow \cap \circ (| \otimes \curlywedge)$. Traditionally, such double arrows are named by alphabetic characters. One might call the relationship AP for associative pairing, and furthermore label as AP : $\cap \circ (\curlywedge \otimes |) \Rightarrow \cap \circ (| \otimes \curlywedge)$ and $\mathrm{AP}^{-1} : \cap \circ (\curlywedge \otimes |) \Leftarrow \cap \circ (| \otimes \curlywedge)$. Instead of alphabetic characters, we will construct *glyphs* to describe label the double arrows. The glyphs will be suggestive of the occammatic[4] surfaces that are bounded by the trivalent diagrams that represent the compositions of 1-arrows on either side of the algebraic axiom.

Much later in Chapter 9, Section 7.2, the general recipe for constructing glyphs will be presented. They will be given in three parts, (1) vertices at which edges converge, (2) edges that mimic seams, folds, and later double curves, and (3) serifs that restate the type of edges that are being considered. Some glyphs have additional information encoded upon them.

The glyphs are the names of the double arrows. The surface-like diagrams that lie within the 3-dimensional boxes and that interpolate between two diagrammatic representations of compositions of 1-arrows abide beyond the explicit realm of the algebraic theory. In fact, there is a metaphor[5] between the algebraic realm and the diagrammatic realm. This metaphor will be presented by means of the compilation of 2-arrows that will be given.

Such glyphs are, unfortunately, unpronounceable. Moreover, similar looking glyphs represent different singularities. Therefore, it is our duty now to compile the glyphs. We will describe them as accurately as possible, but phrases such as "the inverse relation to the co-associativity of the co-pairing" are even more cumbersome than the glyph is. In the tabulation, the glyph is given, an English phrase is presented as a descriptor, the double arrow with source 1-arrow and target 1-arrow is presented, and the associated surface-like diagram that the glyph represents is illustrated and described. In the case of the *shark's fin* (Table 5), the surface description is more fanciful than other cases, but the pictures suggest the metaphor. Several tables will appear. For example, a unit can be expressed in terms of a co-unit either by composing with a cup on the left or right and vice-versa. So eight variations of that glyph appear. Another table contains the variations on the relations between multiplication and co-multiplication.

The authors will try to be careful in using a glyph or a singularity with an accompanying description. We try to write, "the unit expressed in terms of co-unit on the left 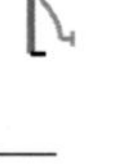." In the diagrammatic realm, "the bottom left corner is folded

[4]The most simple interpolating surface-like object in 3-space.

[5]Technically the metaphor is a functor between two sesqui-categories that preserves arrows of higher multiplicities.

back:  (−) . In this case, the sign (−) is meant to emphasize left versus right. Because of the figure placement, items on the left appear in the back of the box.

Within this section, the names of certain double arrows will be of the form "such-and-such inverse," in some cases these are what are called "weak-inverses." Sections 1.6 and 2 of this chapter and Chapter 7, Section 2 discuss weak invertibility in detail. At this point, the difference between inverses and weak inverses will not concern us.

A given glyph has arcs with color and thickness, and at the top and bottoms of a glyph are serifs. Color choices are as follows. Brown lines correspond to seams in the associated surface. A *seam* is the junction of three sheets in the surface-like diagram that interpolates between two planar trivalent graphs. A brown line often is duplicated by a thinner black core. This is meant to emphasize the junction of three sheets, and the black line will appear more strongly in a paper print of the text. Black and gray lines correspond to the edges of the diagram that correspond to the identities upon the units and co-units. Graying and thinning is meant to suggest the edge is veiled by one or more sheets between the edge and the viewer. Green lines correspond to folds. Locally, folds correspond to the identity upon $\cup$ or $\cap$. The serif on a green line is a short horizontal arc. An arc that extends to the right of the line indicates that the green line is the identity upon $\cup$. One that points to the left indicates the identity upon $\cap$. The serifs upon brown lines are of the form Y or $\curlywedge$, but these lie sideways and indicate the identity upon the multiplication $\curlywedge$ or the co-multiplication Y. Graying of the legs of the serif helps indicate the corresponding face is veiled by another sheet. The serif  on the black lines is a stand-in for the black triangle in . It indicates a unit (resp. co-unit) if it points right (left).

The shark's fins as double arrows are not direct consequences of the structure of a Frobenius algebra. Instead, they follow as theorems. The proofs may be obtained by examining the "units and pairing" diagrams in Chapter 7, Section 3.2. At the algebraic level $\langle A, B\rangle = \langle 1 \cdot A, B\rangle = \langle 1, A \cdot B\rangle = \eta(A \cdot B)$ where η denotes the co-unit.

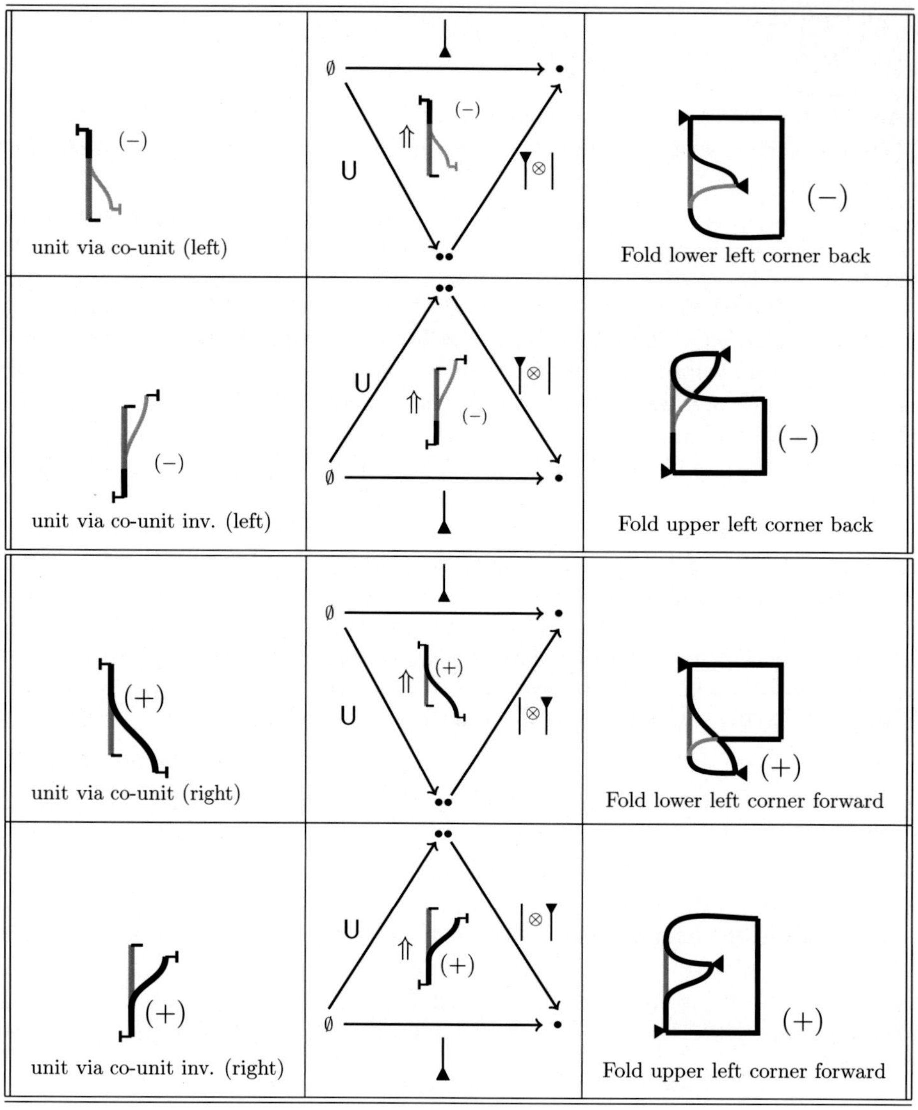

TABLE 1. Writing units in terms of co-units

1.5. Structural recap. The structure in the sesqui-category FA that is the categorical version of a Frobenius algebra has not yet been completely described, but it deserves review. The natural numbers $\mathbb{N} = \{0, 1, 2, \ldots\}$, written in unary notation, forms the set of objects. The object 0 is an empty collection of dots, and the positive integer α consists of dots arranged on a line $\underbrace{\bullet\!-\!\bullet\!-\cdots-\!\bullet}_{\alpha}$. There are generating arrows I, Y, ⅄, U, ∩, ⊥ and ⊤, and arrows are constructed from these by attaching vertical strings among them. The double arrows correspond to the algebra and coalgebra axioms, the coincidences between multiplication and

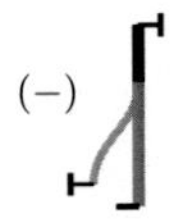 co-unit via unit (left)	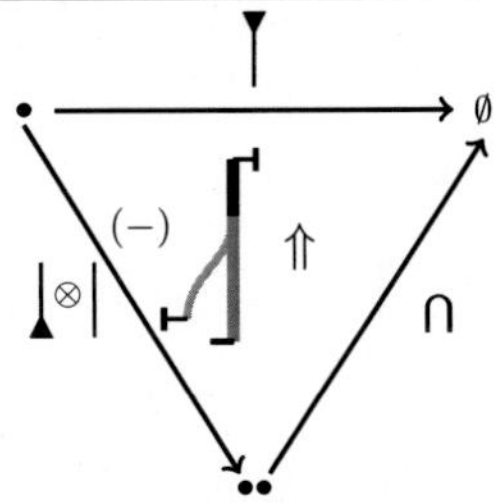	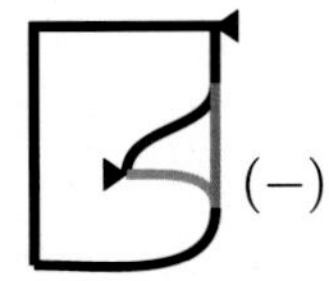 Fold lower right corner back
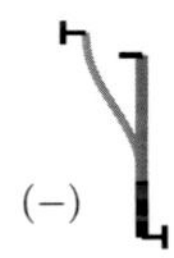 co-unit via unit inv. (left)	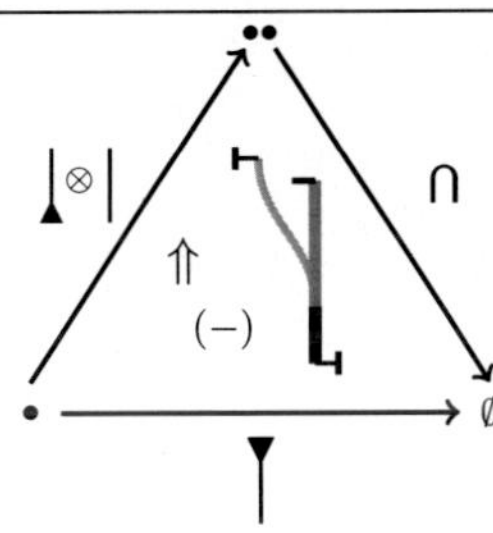	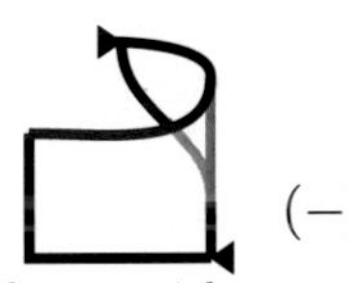 Fold upper right corner back
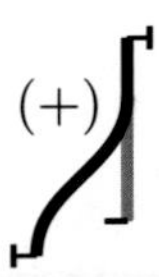 co-unit via unit (right)	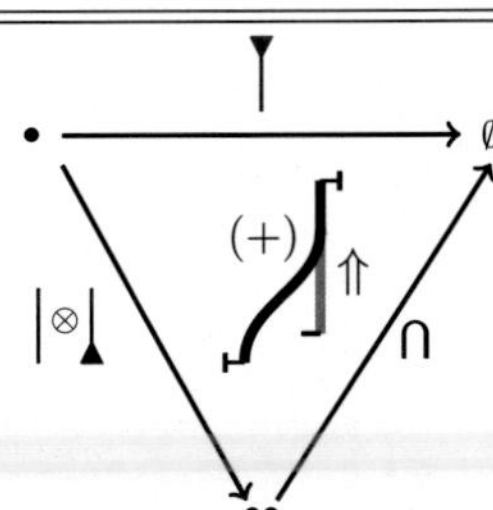	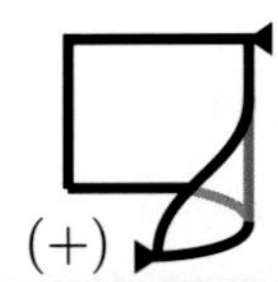 Fold lower right corner forward
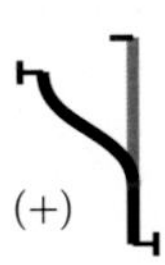 co-unit via unit inv. (right)	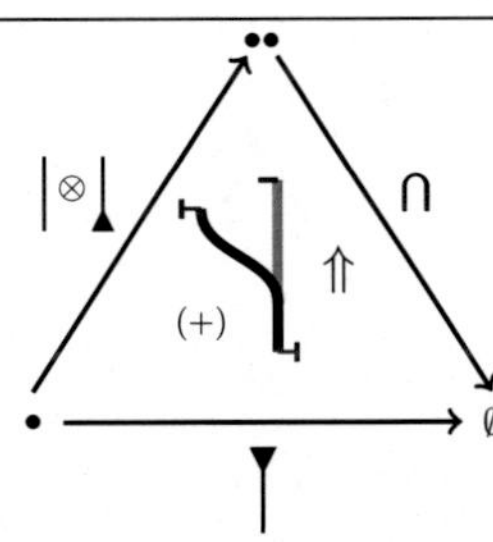	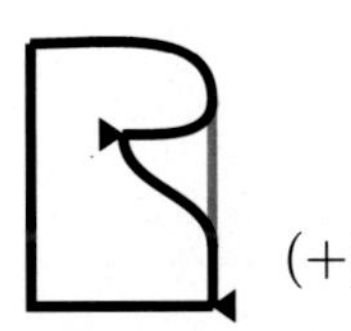 Fold upper right corner forward

TABLE 2. Writing co-units in terms of units

co-multiplication, the ability to define units or co-units in terms of the other, and the exchanger isomorphism between distant arrows. With the exception of the exchanger axiom, these generating double arrows are depicted in the left-hand or center columns of the Tables 1 through 10.

These tables suggest a metaphor, or more precisely a functor, between the algebraically defined sesqui-category that appears in the center, and the topologically defined sesqui-category whose arrows are the planar diagrams of trivalent graphs,

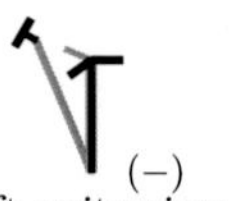 Left unit axiom	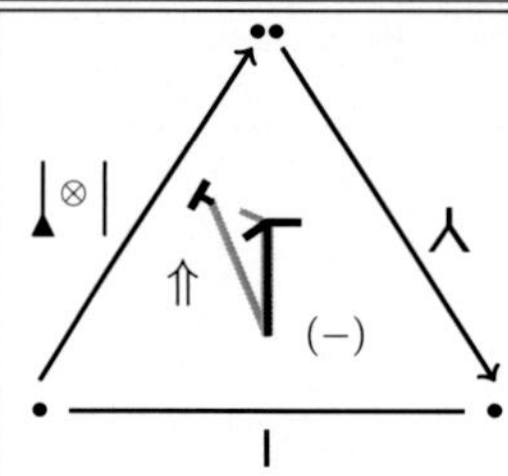	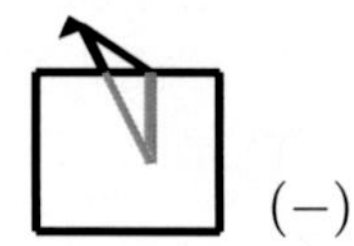 upper left triangular flap back
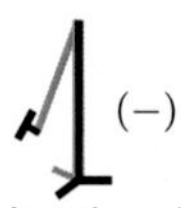 Left unit axiom inverse	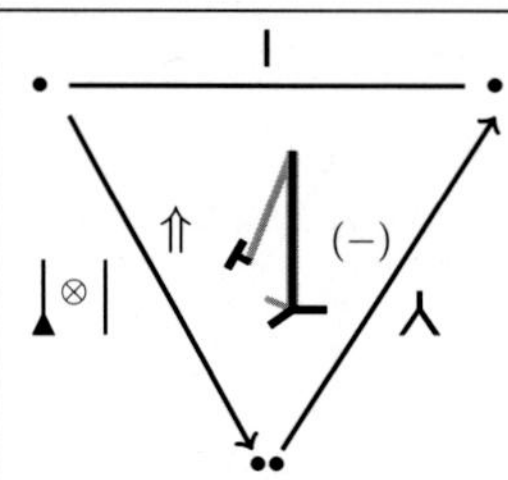	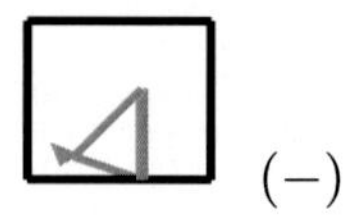 lower left triangular flap back
 Right unit axiom	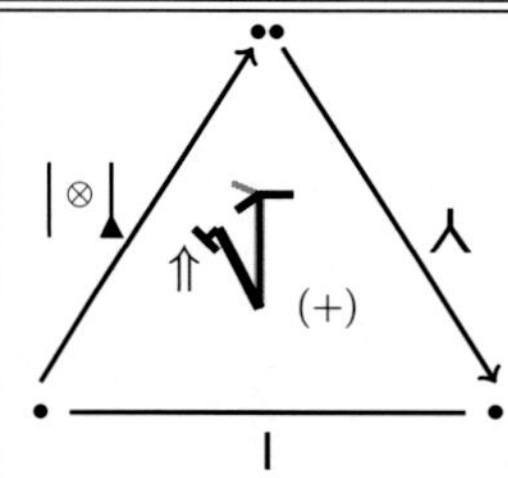	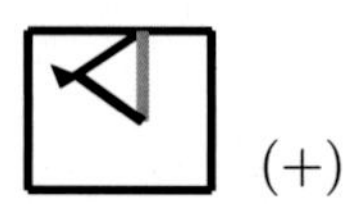 upper left triangular flap forward
 Right unit axiom inverse	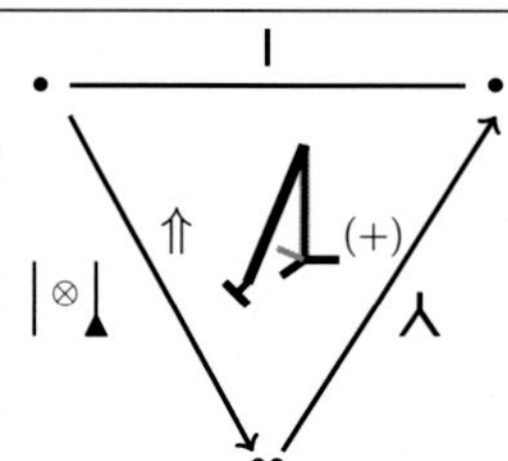	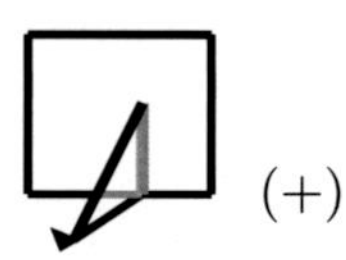 lower left triangular flap forward

TABLE 3. Units

and whose double arrows are the surface-like foams[6] that interpolate between these graphs. The foams are illustrated in the right-hand column of the tables. To go from the foam to the algebraic setting, the pieces of the foam that appear inside rectangular boxes are projected to the plane, and the projections of the seams, folds, and edges (along the units and co-units) provide pen strokes that one would draw when illustrating the glyphs. To go the other direction, the surfaces that interpolate between the two sides the Frobenius algebra identity are constructed as in Section 1.3.

While they are not, strictly speaking, part of the algebraic structure of the Frobenius algebra, twelve additional double arrows will be introduced for both the

[6]The definition of a foam appears at the end of Section 1.6 in this chapter. It is a space in which neighborhoods of points are modeled by neighborhoods of points on the right-hand side of Table 8.

 Left co-unit axiom	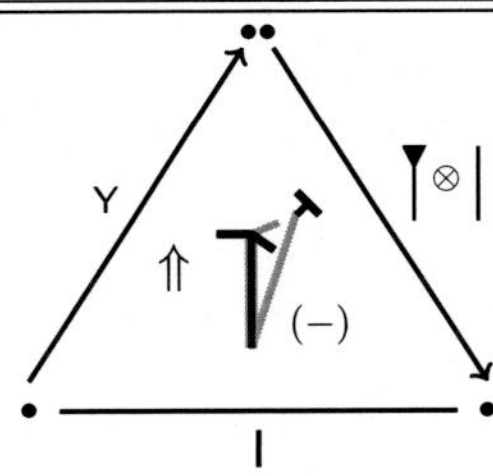	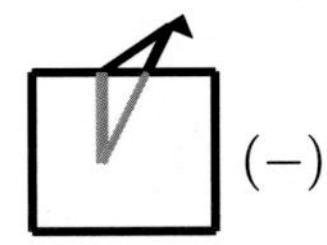 upper right triangular flap back
 Left co-unit axiom inverse	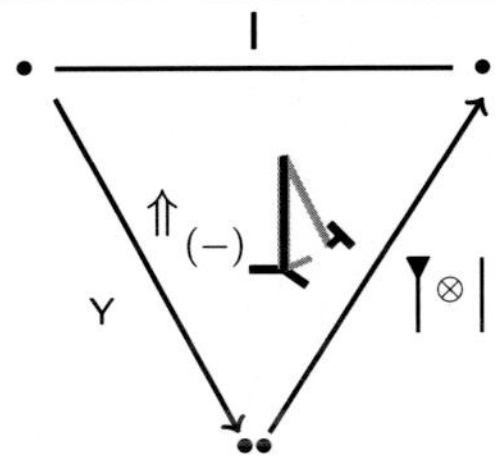	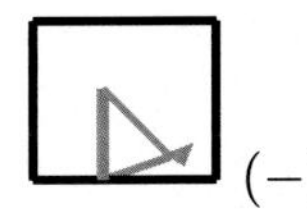 lower right triangular flap back
 Right co-unit axiom	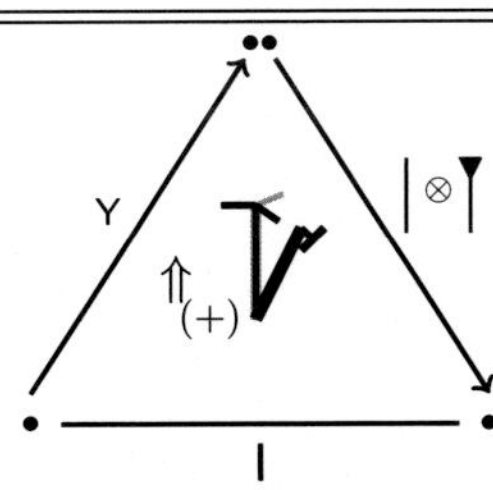	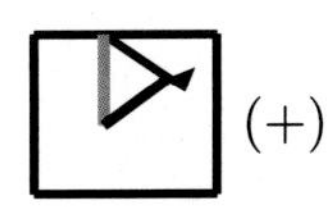 upper right triangular flap forward
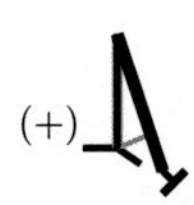 Right co-unit axiom inverse	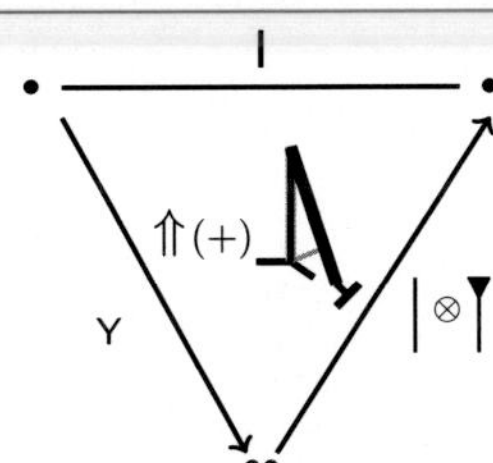	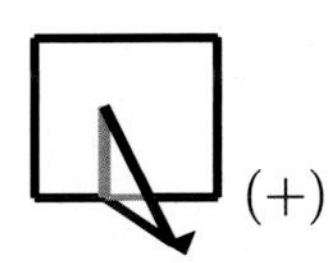 lower right triangular flap forward

TABLE 4. Co-units

algebraic and the diagrammatic settings in Section 2 of this chapter. The tabulation will continue in that section, and the functor suggested by the tables extends.

In the sequel, some triple arrows will be introduced. Any functorial relation between higher categories must also extend over the higher arrows. In the case at hand, they do.

1.6. Isomorphism and invertibility. Relationships among compositions of double arrows are triple arrows. For example, there is a natural family of invertible exchanger triple arrows. In this section, we establish triple arrows that encode the weak and strong inveritibility of the double arrows that were articulated in Tables 1 through 10. These arise when considering the globular composition between a generating double arrow and its reflection in a horizontal plane —$(x, y, z) \mapsto (x, y, 1-z)$.

 pairing compared to multiplication and co-unit	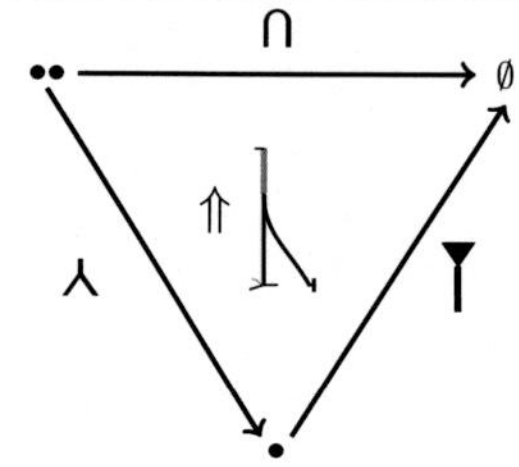	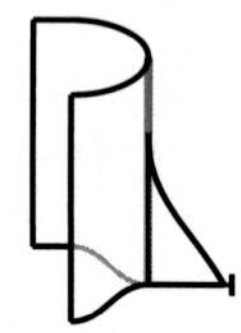 right bottom shark's fin
 multiplication and co-unit compared to pairing	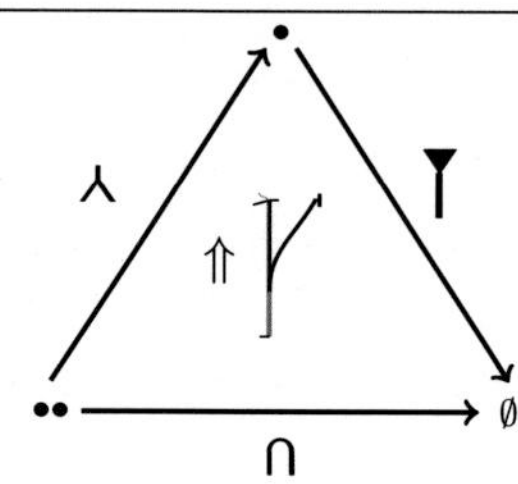	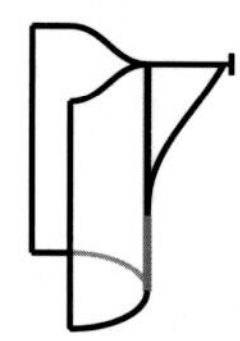 right top shark's fin
 co-pairing compared to unit and co-multiplication	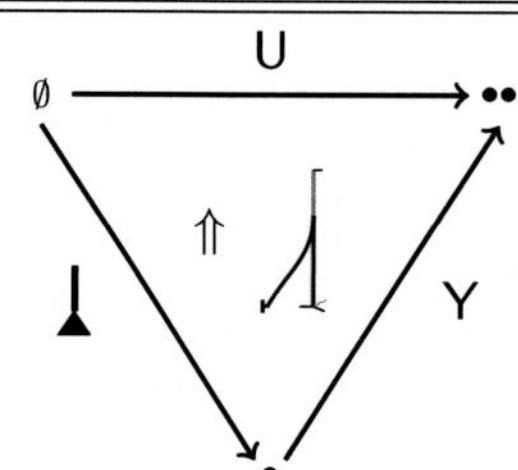	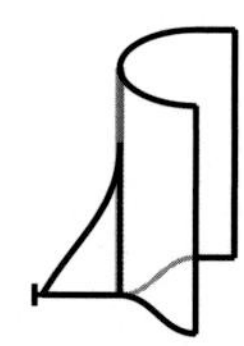 left bottom shark's fin
 unit and co-multiplication compared to co-pairing	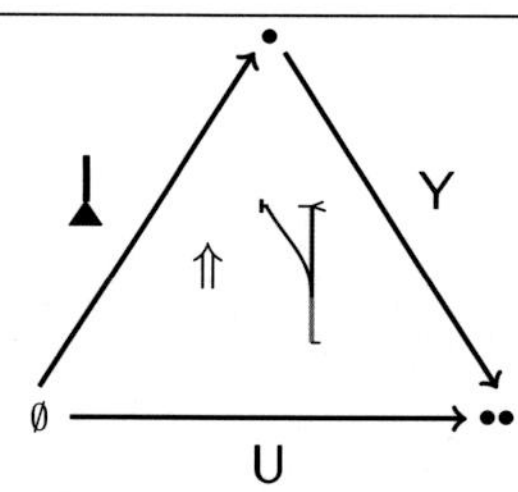	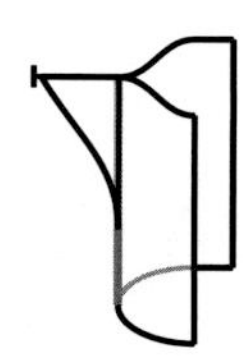 left top shark's fin

TABLE 5. Shark's fins

Specifically, when the surface-like representatives of a double arrow F is reflected and globularly composed with its reverse Ⅎ, the compositions $\left[\begin{matrix}\text{Ⅎ}\\ \mathsf{F}\end{matrix}\right]$ and $\left[\begin{matrix}\mathsf{F}\\ \text{Ⅎ}\end{matrix}\right]$ are formed, and these are compared with $\mathsf{I}_{s(F)}$ and $\mathsf{I}_{t(F)}$, respectively. In both the weakly invertible and the stongly invertible case, a triple arrow between the stacked Fs and the identity on the source or target is given. In the weakly invertible case, that triple arrow and its reverse compose, again in two ways; each composition is to be compared with another identity arrow (Chapter 7, Section 2).

Unfortunately, the distinctions that we wish to make between weak invertibility and strong invertibility are subtle. When, say, $\left[\begin{matrix}\text{Ⅎ}\\ \mathsf{F}\end{matrix}\right]$ is compared to $\mathsf{I}_{s(F)}$, there

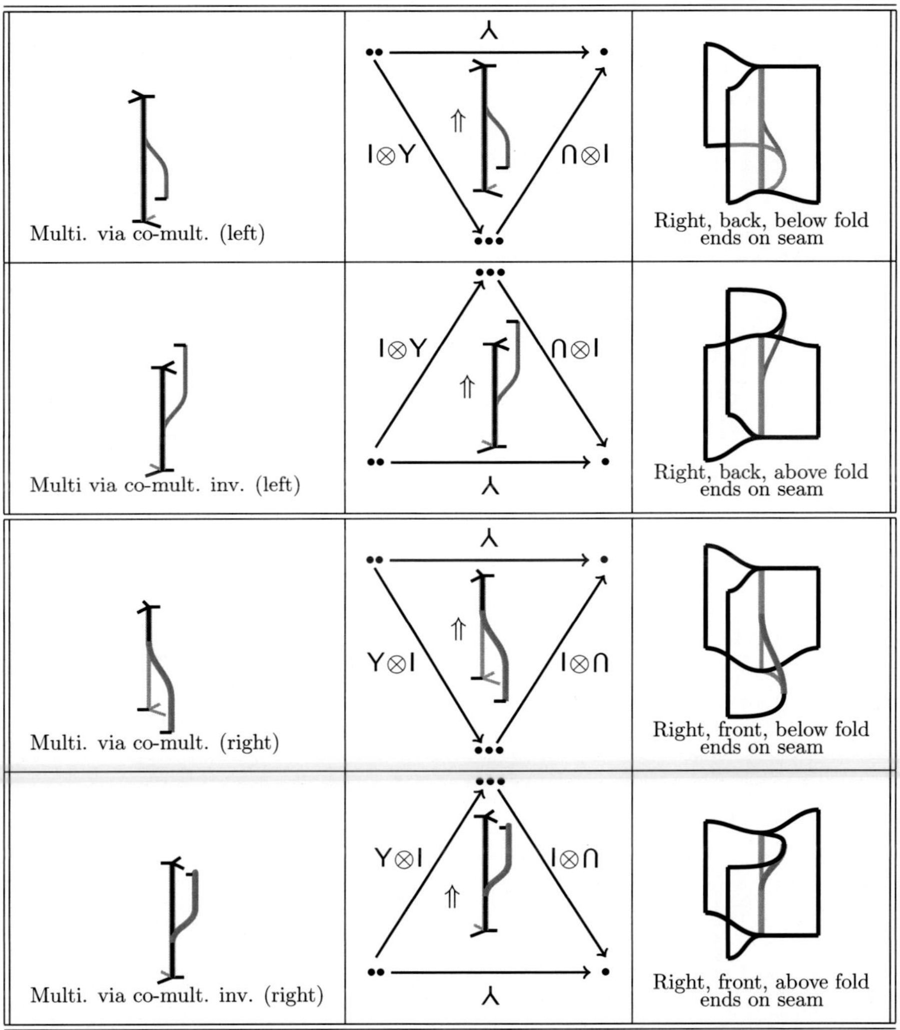

TABLE 6. Writing multiplication in terms of co-multiplication

are triple arrows $\left[\begin{array}{c} \text{Ⅎ} \\ \mathsf{F} \end{array}\right] \to \mathsf{I}_{s(F)}$ and $\mathsf{I}_{s(F)} \to \left[\begin{array}{c} \text{Ⅎ} \\ \mathsf{F} \end{array}\right]$. These triple arrows, which are comparisons, also have reverses. A 3-arrow and its reverse can be composed, again to-to-from or from-to-to, and those compositions are compared to appropriate identities. The process continues. From the authors' point of view, a choice is made whether the comparisons being made are thought to be equality.

We will say that double arrows F and Ⅎ are *strongly invertible*, with each being the strong inverse of the other, if the surface-like compositions of the form $\left[\begin{array}{c} \text{Ⅎ} \\ \mathsf{F} \end{array}\right]$ or $\left[\begin{array}{c} \mathsf{F} \\ \text{Ⅎ} \end{array}\right]$ are isotopic to $\mathsf{I}_{s(F)}$ or $\mathsf{I}_{t(F)}$, respectively. However, there are other relations

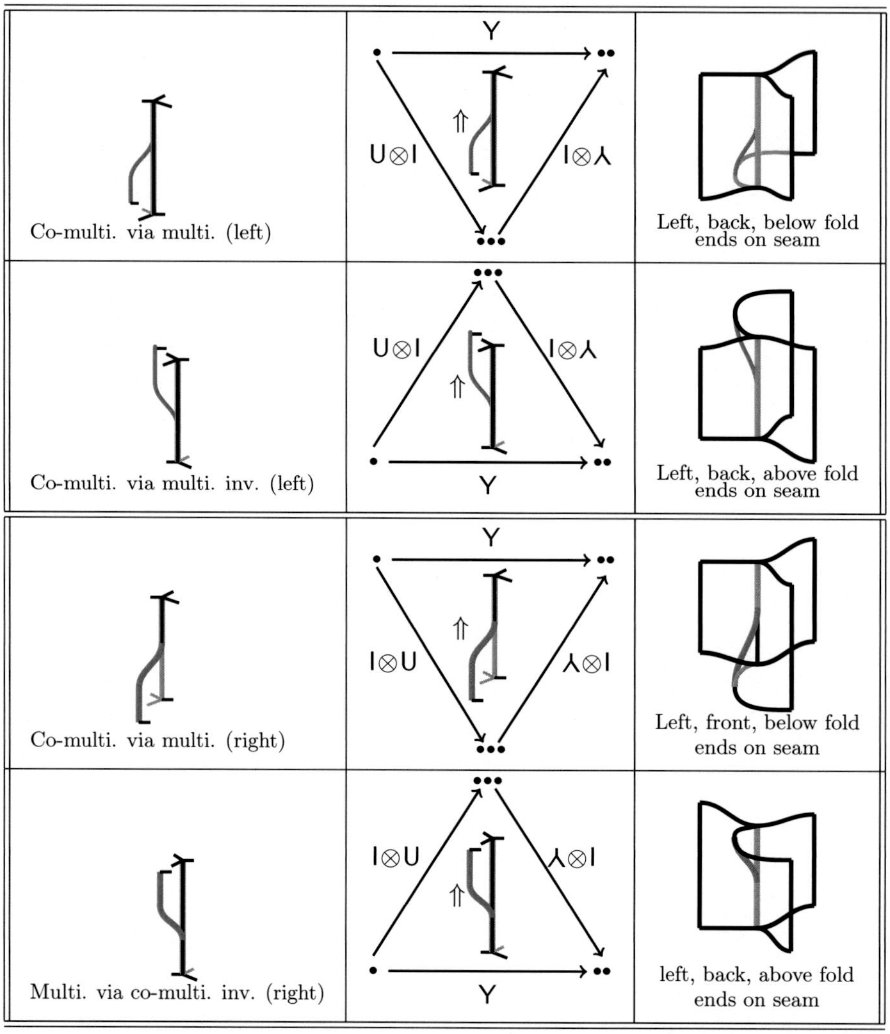

TABLE 7. Writing co-multiplication in terms of multiplication

that necessarily hold in both the strong and the weak inverse case (Chapter 7, Section 2).

First, consider the double arrows from Table 10 that represent the non-degeneracy axioms. In Figs. 11 and 12, the non-degeneracy of the pairing is represented by the fact that the pairing/co-pairing composition that is given via *zig-zags* is equal to the identity. In this way *cusps* occur in the surface description of the axiom. Such a cusp is called left (right) if the *fold* that is closer to the viewer of the page is on the viewer's left (right). The triple arrows on the right of the illustrations represent topological equivalences. In singularity theory they are called the *lips* and *beak-to-beak* singularities between folds.

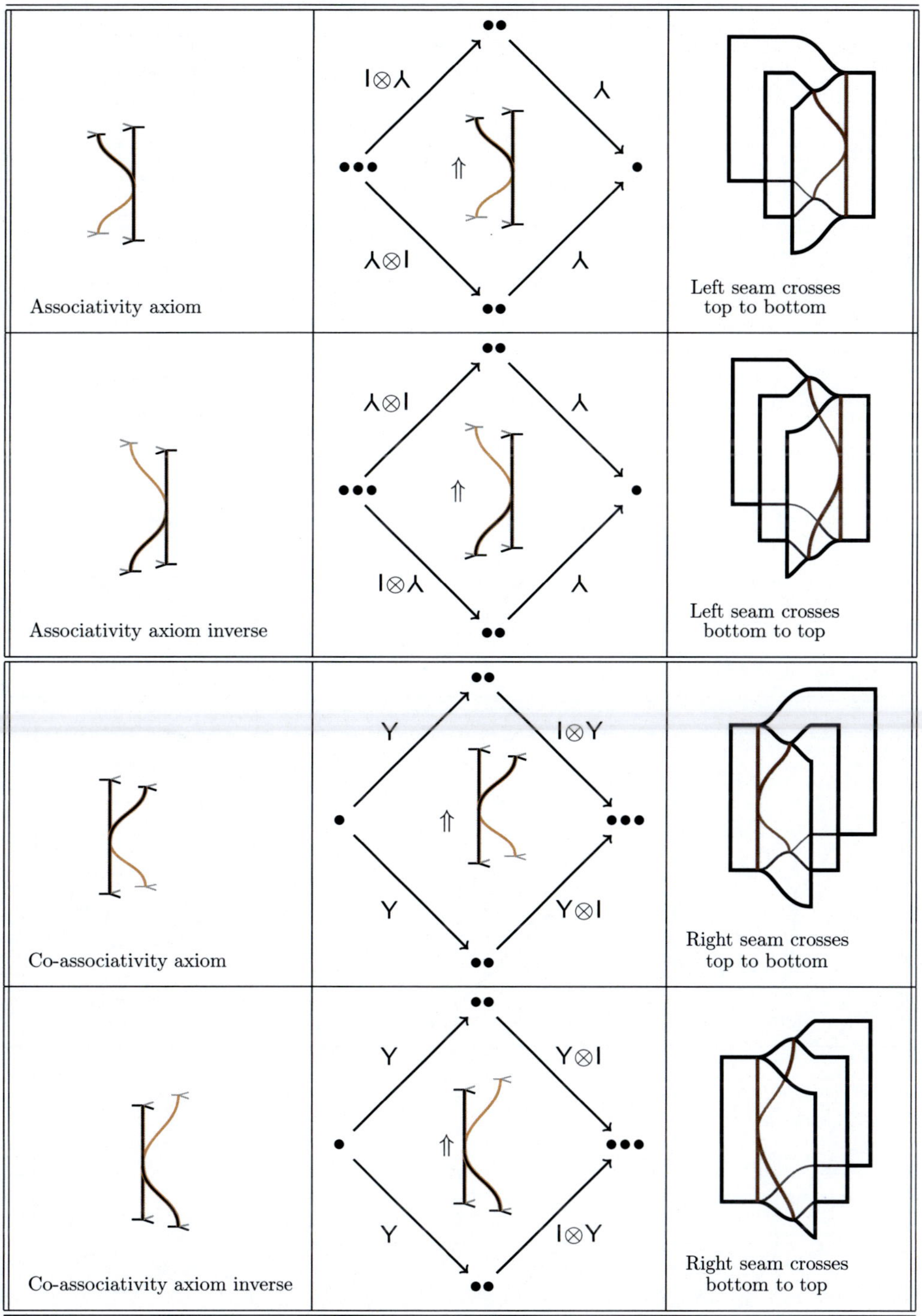

TABLE 8. Assocativity and co-associativity

 Associative pairing	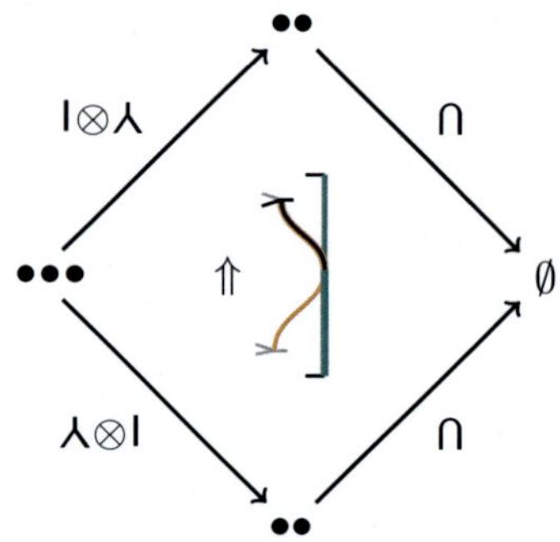	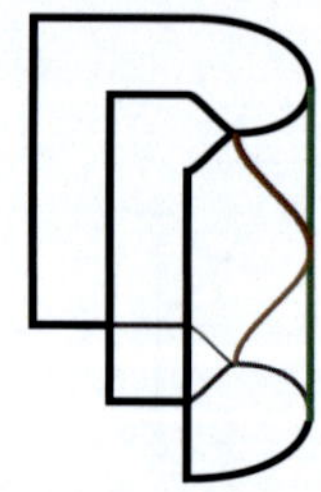 Left seam crosses a fold on the right top to bottom
 Associativity pairing inv.	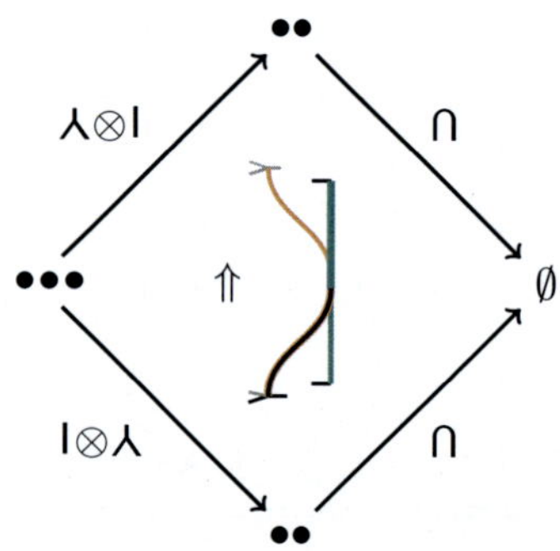	 Left seam crosses a fold on the right bottom to top
 Co-associative co-pairing	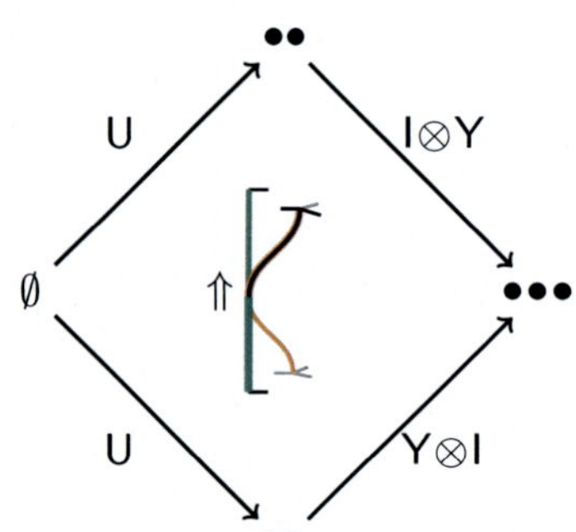	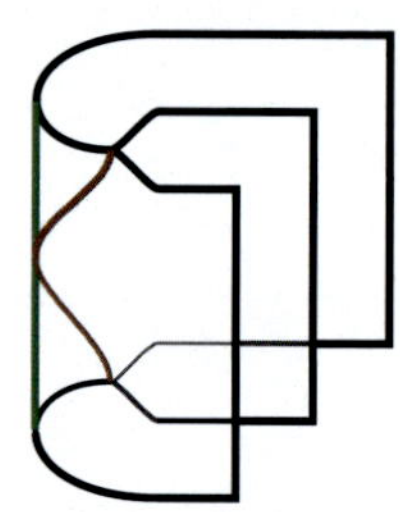 Right seam crosses a fold on the left top to bottom
 Co-associative co-pairing inv.	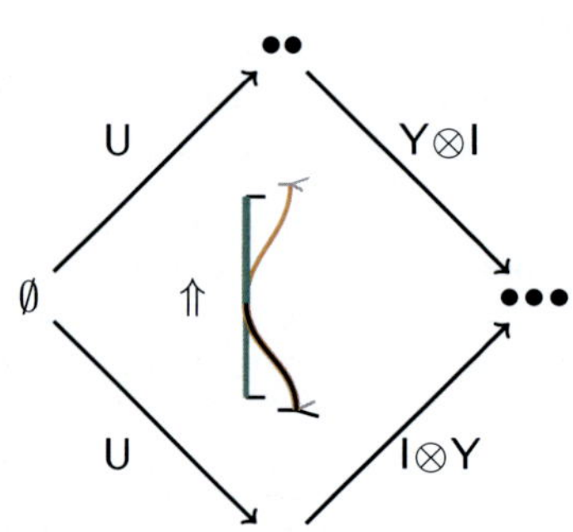	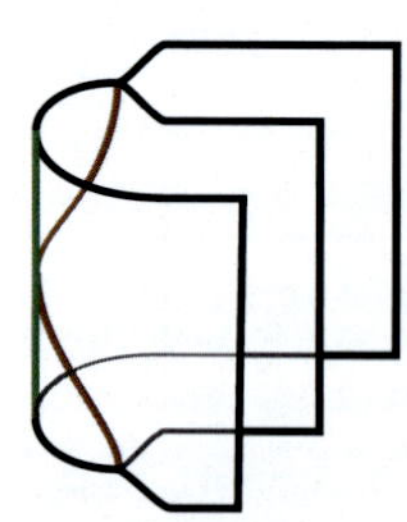 Right seam crosses a fold on the right bottom to top

TABLE 9. Assocativity and co-associativity of the pairing

 Left down cusp	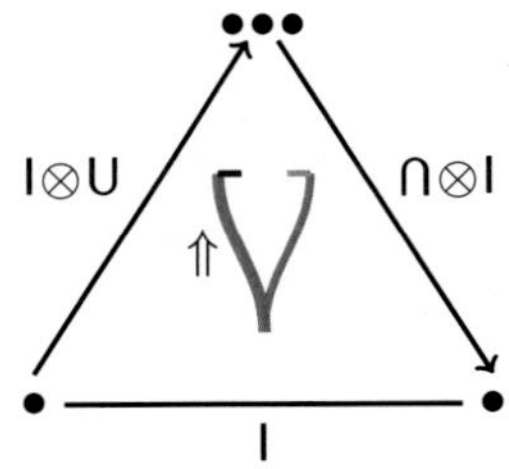	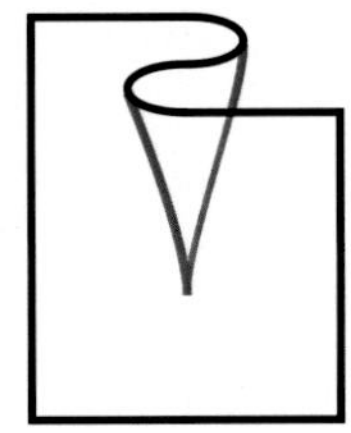 left zig zag (top)
 Left up cusp	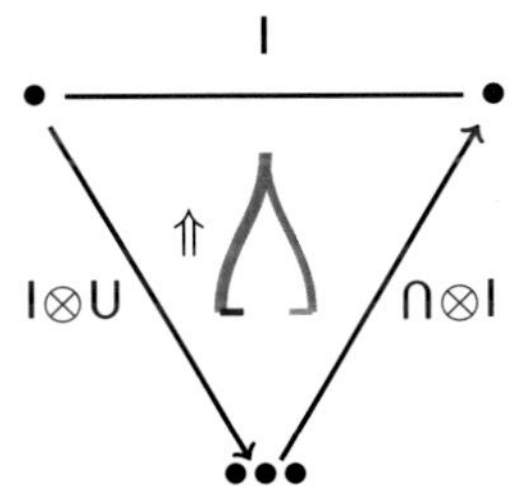	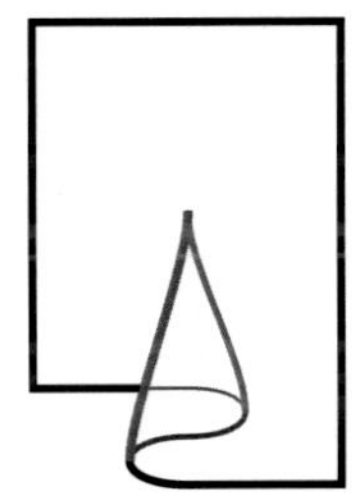 left zig zag (bottom)
 Right down cusp	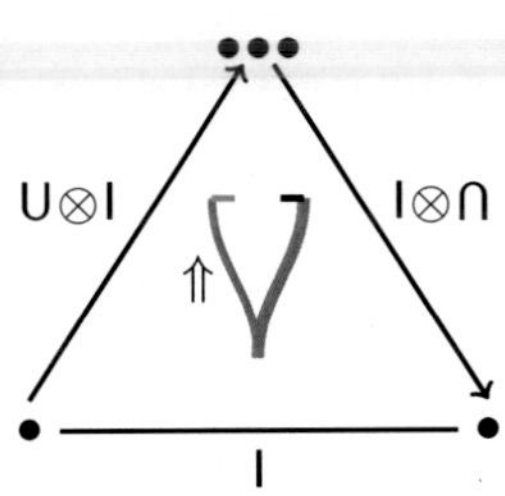	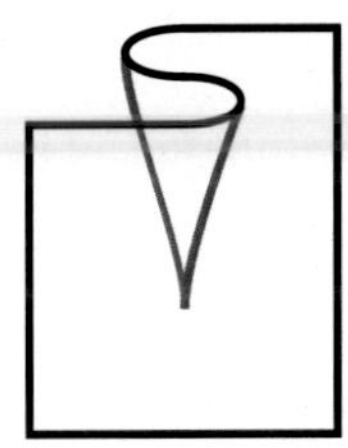 Right zig zag (top)
 Right cusp up	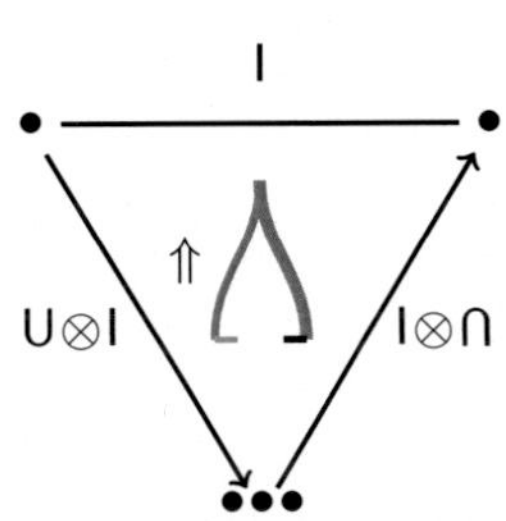	 Right zig zag (bottom)

TABLE 10. Non-degeneracy of the pairing and co-pairing

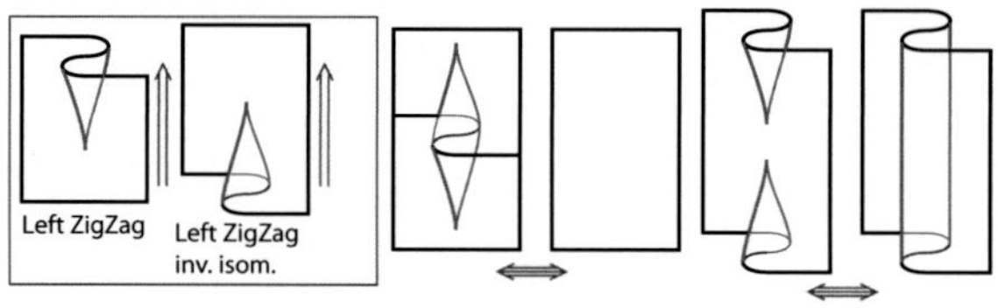

FIGURE 11. The non-degenerate isomorphism, part 1

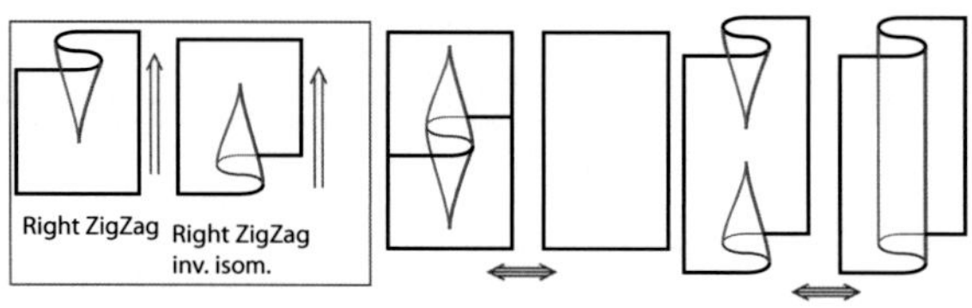

FIGURE 12. The non-degenerate isomorphism, part 2

In a laundry metaphor, the singularites are resolved by ironing the fabric more carefully. Mathematically, the surface that appears is projected to the plane of the page. Generic singularities (points of the surface along which the rank of the differential drops) for maps from the plane to itself are cusps and folds. Along folds, the rank of the differential is exactly 1; the rank drops to 0 at the cusps. In the case of foams, folds also end along *seams*, which are the junctions among three separate sheets, or at the edges which are defined by idenitities upon units or co-units.

The lip and beak-to-beak triple arrows represent isotopies between embedded surfaces in 3-dimensional space. Thus the zig-zags double arrows are strongly invertible.

Next we consider the double arrows in which co-mulitplication is written in terms of multiplication (Table 7) and vice versa (Table 6), and we consider the co-unit written in terms of the unit (Table 2) and vice versa (Table 1). In these cases, the folds end at seams or edges. So they share some properties with the cusps that were considered above. In Figs. 13 and 14 triple arrows indicate isotopies between foams embedded in 3-dimensional space. The isotopies are achieved when the foams are ironed. As above, these triple arrows are strongly invertible — higher dimensional compositions of the form $\begin{bmatrix} \text{Ⅎ} \\ \mathsf{F} \end{bmatrix}$ or $\begin{bmatrix} \mathsf{F} \\ \text{Ⅎ} \end{bmatrix}$ are isotopic to $\mathsf{I}_{s(F)}$ or $\mathsf{I}_{t(F)}$. The figures illustrate the cases in which the fold is veiled by the foam in the projection. Similar triple arrows exist in case the fold lies towards the observer.

The same assertion (strong invertibility) can be made about the associativity of the pairing and the co-associativity of the co-pairing (Table 9). Fig 15 illustrates the triple arrows when the fold is on the right — associativity of the pairing. The co-associativity pictures are reflections of these. In these cases, the *seam*, which is the junction of three surface sheets, passes over the fold.

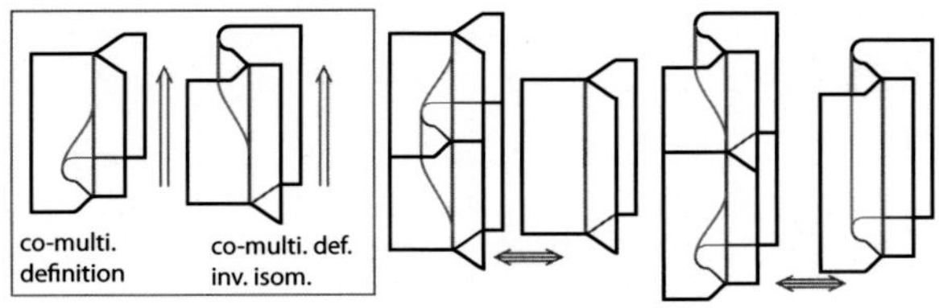

FIGURE 13. The co-multiplication isomorphism

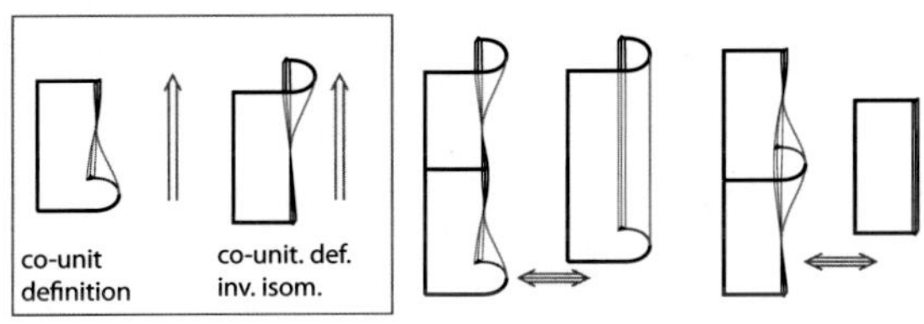

FIGURE 14. The co-unit isomophism

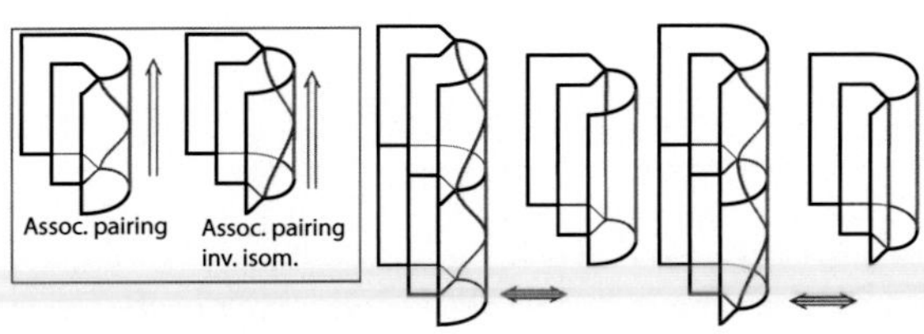

FIGURE 15. The associative pairing isomorphisms

Because the unit (Table 3) and co-unit double arrows (Table 4) involve grafting a triangular flap to a surface along a seam, the triple arrows that relate $\begin{bmatrix} \text{Ⅎ} \\ \mathsf{F} \end{bmatrix}$ or $\begin{bmatrix} \mathsf{F} \\ \text{Ⅎ} \end{bmatrix}$ to $\mathsf{I}_{s(F)}$ or $\mathsf{I}_{t(F)}$, are not topological equivalences. See Figs. 16 and 17. The double arrows that represent the unit and co-unit axioms are said to be *weakly invertible at level* 3.[7] There are several additional criteria that need to be met for a 2-morphism F and its *reverse* Ⅎ to be weakly invertible. Other examples of (level 3) weakly invertible double arrows will occur shortly.

The associativity axiom is *weakly invertible at level* 3 in the sense that the triple arrows depicted in Fig. 18 are not topological equivalences as foams (see below) since a pair of vertices is cancelled or introduced in the process.

The associativity axiom, when drawn as a surface-like diagram is the model for a 2-*dimensional foam.* It consists of a *vertex* at which four seams and six faces are incident. In a neighborhood of a point on a *seam*, three faces are incident. Here

[7]In Chapter 5, Section 3, weak invertibility at level $(k+1)$ is defined for all k. Level 3 indicates that the compositions of F and its reverse are compared to identities via triple arrows. See also Chapter 7, Section 2.

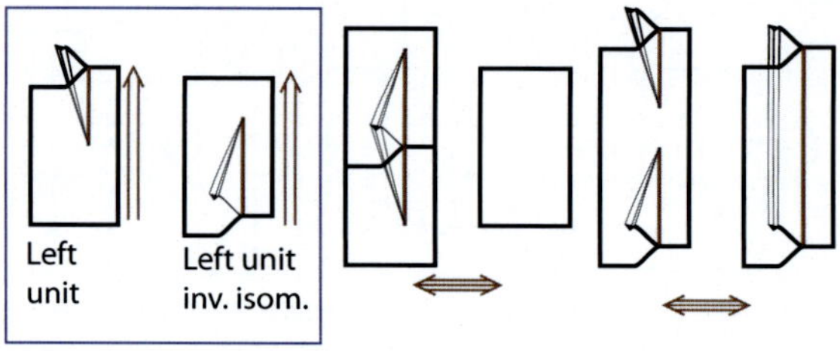

FIGURE 16. The left unit isomorphism

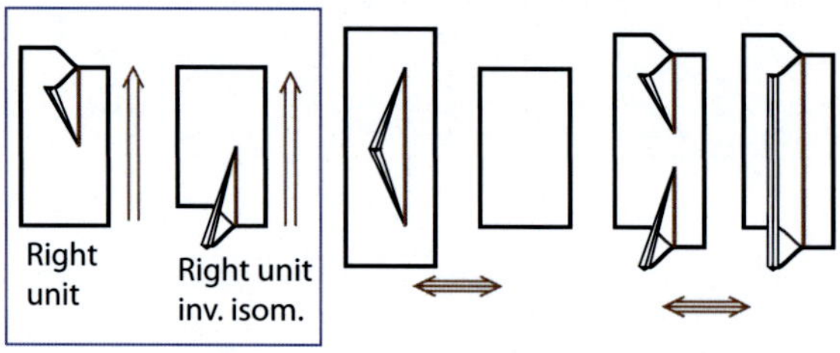

FIGURE 17. The right unit isomorphism

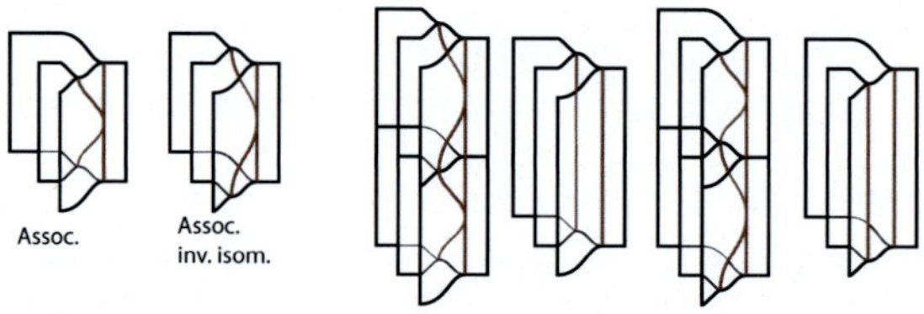

FIGURE 18. The associative isomorphism

the foam appears to be of the form $\mathsf{Y} \times \mathsf{I}$. An *edge*, which corresponds to the double arrows that serve as the identities upon the units or co-units, has one face incident.

The boundary of a properly embedded foam is a graph with monovalent and trivalent vertices. The monovalent vertices correspond to units or co-unit arrows and they are the end points of edges.[8] The trivalent vertices correspond to multiplication and co-multiplication arrows. In a 2-dimensional foam every point has a neighborhood that is homeomorphic[9] to a neighborhood of a point in the right-hand side of either Table 3, 4, or 8. The illustrations in Tables 3 and 4 indicate a type of vertex at which a (unit or co-unit) edge emerges from the foam. An edge and a seam are incident at such a vertex. The surfaces in Table 8 or in Fig. 18 include interior vertices at which four seams and six faces are incident.

[8]Strictly speaking, the edges are also part of the boundary, but these edges join the monovalent vertices to each other or to the seams.

[9]There is a continuous bijective function between the neighborhoods that has a continous inverse.

2. Reversible arrows and additional double arrows

Starting from the diagrammatic description of the Frobenius algebra operators and axioms, diagrammatically defined arrows for multiplication, units, pairing, co-multiplication, co-units, and co-pairings have been constructed. Double arrows correspond to the axioms in the algebra. They are labeled by glyphs that mimic surface-like foams embedded in 3-dimensional space. Compositions of double arrows corresponds to gluing foams together along portions of their boundaries.

In Chapter 5, theorems about Frobenius algebras will be represented as triple arrows that can be interpreted as isotopies between foams. The relations between 1-arrows form 2-dimensional cells. Relations among these cells form globes. The glyphs that label the 2-cells are composed until well-defined source and target cells are articulated. The globes will then be labeled by additional glyphs.

Before introducing triple arrows in a more detailed setting, we examine weak invertibility from the point of view of 1-arrows.

2.1. An arrow and its reverse. In the current section, some double arrows will be presented that come from both topological and categorical, rather than algebraic, considerations. On the categorical side, we examine the weak invertibility of the 1-arrows ⅄, Y, $\cap$, $\cup$, $\perp$, and $\top$. From the point of view of a Frobenius algebra, the compositions ⅄ $\circ$ Y, Y $\circ$ ⅄, $\cup \circ \cap$, $\cap \circ \cup$, $\perp \circ \top$, and $\top \circ \perp$ are not identity maps, but we can compare the diagrammatic compositions to the identities upon $\emptyset$, -•-, and -•—•-. We say that the elements in the pairs $\{\perp, \top\}$, {⅄, Y}, and $\{\cup, \cap\}$ are *reverses* of each other. In this section, compositions of pairs of reverses are formed, and double arrows are created to compare these compositions to identities.

1. Birth and Death. Compare the composition

$$\emptyset \xrightarrow{\cup} \bullet\bullet \xrightarrow{\cap} \emptyset$$

to the identity arrow upon the empty set

$$\emptyset \underline{\quad |_{\emptyset} \quad} \emptyset.$$

The 1-arrows that are being compared are different. In the former case, a pair of identical particles is born and immediately annihilated. In the latter, nothing occurs. The existence of the particles, albeit brief, is notable. The differences are described by asserting that there are double arrows called *birth* $\backslash\!\cdot\!/$ and *death* $/\!\cdot\!\backslash$. They are compiled in Table 11.

On occasion, birth will be denoted as $\underset{0}{\overset{0}{\backslash_2/}}$ and death by $\underset{0}{\overset{0}{/_2\backslash}}$ when an emphasis on the source, target, and intermediate objects is needed.

2. Saddle and fork.[10] Compare the composition

$$\bullet\bullet \xrightarrow{\cap} \emptyset \xrightarrow{\cup} \bullet\bullet$$

to the identity arrow upon two dots

$$\bullet\bullet \underline{\quad || \quad} \bullet\bullet.$$

Two identical particles are annihilated, and subsequently two are born, or two particles continue to exist in the same unit of time. The double arrows that relate them are called *saddle* denoted as $\underset{\bullet}{\cup}$ and *fork* $\dot{\cap}$. As above, alternative glyphs

[10] We are told that British tailors refer to the junction of trouser legs as the "fork."

emphasize the source, target, and intermediate objects: $\underset{2}{\overset{0}{\cup}}$ and $\overset{2}{\underset{0}{\cap}}$. These also are compiled in Table 11.

These four comparisons suggest that ∪ and ∩ are not, strictly speaking, inverses. At the object level a pair of points is annihilated and a subsequent pair is created, or creation is followed by annihilation. So the net result is an identity, but the intermediate situation draws our attention. Thus these arrows are *weak inverses*, and each is *weakly invertible*. There are additional relationships among double arrows that involve ∪ and ∩. These additional relations help clarify weak invertibility (Chapter 7, Section 2).

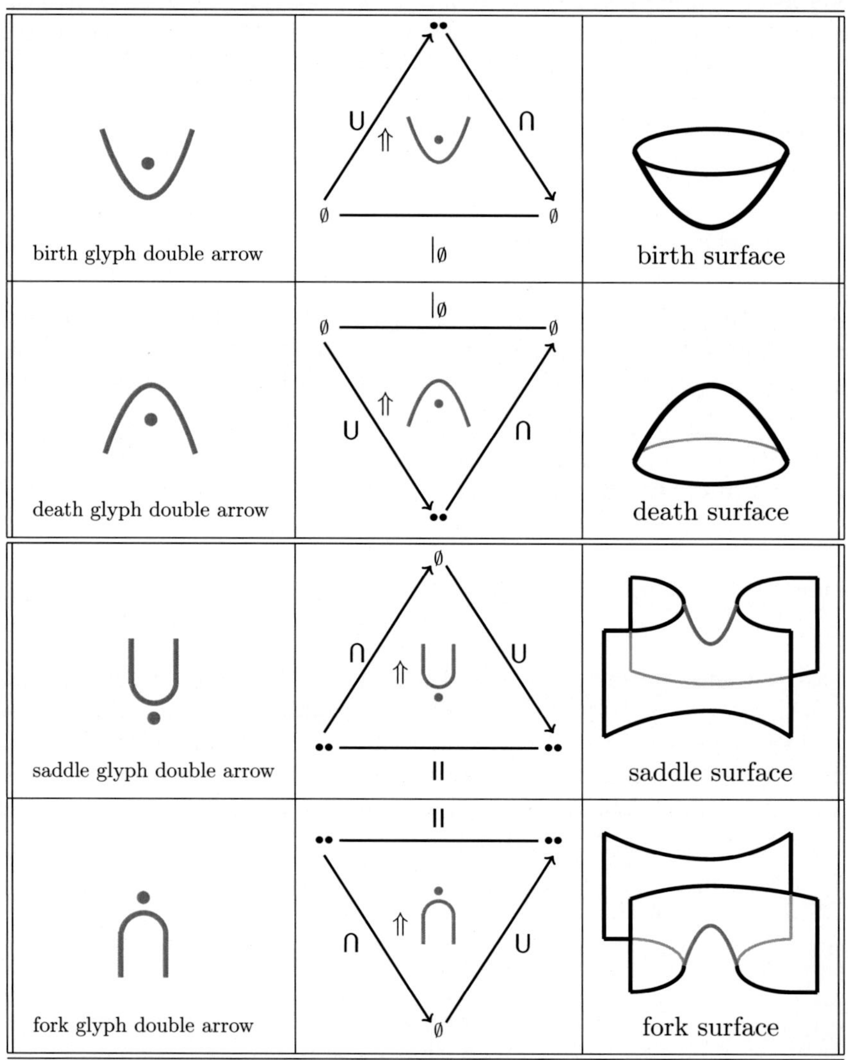

TABLE 11. Birth, death, saddle, and fork double arrows

3. Zipping purses (Elliptic case) Compare the composition

$$\bullet \xrightarrow{\mathsf{Y}} \bullet\bullet \xrightarrow{\curlywedge} \bullet$$

to the identity arrow upon the object $-\!\!\bullet\!\!-$:

$$\bullet \underline{\quad \mathsf{I} \quad} \bullet.$$

As above, the 1-arrows being compared differ. In the former case, a particle splits into two and these merge. The differences are compiled as the top two rows of the Table 12 found in this chapter. We refer to these colloquially as zipping purses. At the level of the foams, minimal and maximal points of the seams occur. In the top row the purse appears to be full, and the second it is empty.

4. Zipping purses (hyperbolic case). Compare the composition

$$\bullet\bullet \xrightarrow{\curlywedge} \bullet \xrightarrow{\mathsf{Y}} \bullet\bullet$$

to the identity arrow upon the object that corresponds to the natural number 2:

$$\bullet\bullet \underline{\quad \mathsf{II} \quad} \bullet\bullet.$$

The 1-arrows being compared differ. In the former case, a pair of identical particles merge and then split; in the latter they don't interact. The double arrows that relate them are illustrated in the bottom two rows of Table 12 in this chapter.

In the case of these four zippers, the represented pieces of foams are all homeomorphic to the topological space $\mathsf{Y} \times \mathsf{I}$. This should be no surprise: births, deaths, and saddles are all homeomorphic to disks.

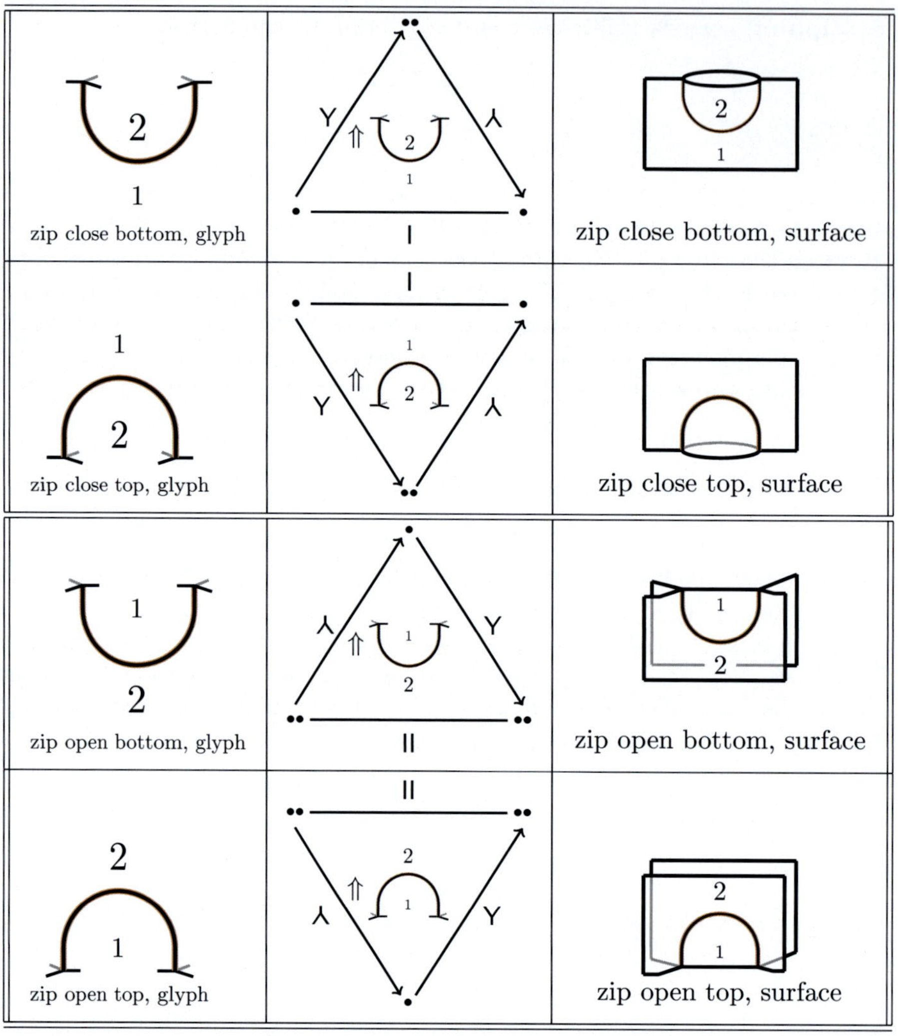

TABLE 12. Optimal points on seams

Canceling and creating unit/co-unit pairs. Compare the composition

$$\emptyset \xrightarrow{\perp} \bullet \xrightarrow{\top} \emptyset$$

to the identity arrow upon the empty set. And also compare the identity

$$\bullet \xrightarrow{\mathsf{I}} \bullet$$

to the composition

$$\bullet \xrightarrow{\top} \emptyset \xrightarrow{\perp} \bullet.$$

As in the previous cases, there are differences between the items being compared, so we propose that there are double arrows that relate them. They come in a

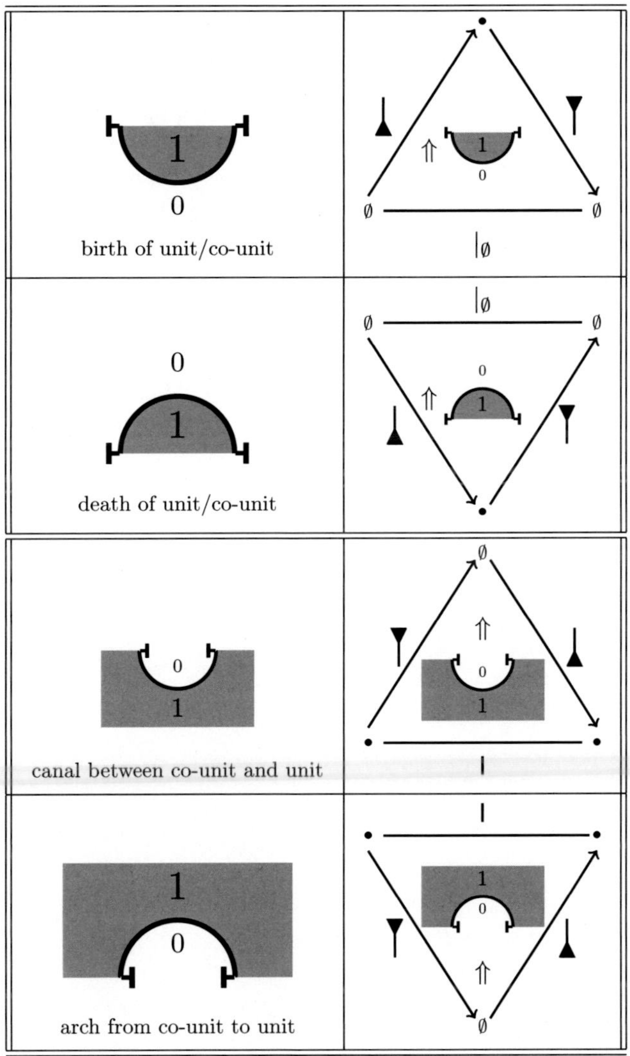

TABLE 13. Birth, death, arch, and canal for unit/co-unit pairs

birth/death pair, and in a canal/arch pair. Thus also in this case, the single arrow  has the arrow  as a weak inverse. Weak invertibility means that the double arrows in Table 13 exist and they satisfy some higher order relations that are not yet articulated (Chapter 7, Section 2). The glyph and the foam that is defined by these double arrows coincide. For example, the foam at a unit/co-unit birth is the southern hemisphere of a disk. So Table 13 does not need a third column.

In all these table, the top two rows are separated by a pair of double lines from the bottom two rows. Between the first and second row there are two possible globular compositions of the triangular double arrows. The same is true for the bottom two rows. In either case, the resulting compositions can be compared with the identity double arrow on the ultimate source and target 1-arrows. Higher order births and saddles occur. These are among the triple arrows to be included.

There are other skew compositions that can also occur as analogues to the compositions comprizing the non-degeneracy axiom (zig-zag rules) for the pairing and co-pairing. These also result in additional triple arrows. In the case of weakly invertible 1-arrows, we will assert many additional relations among the higher order arrows. These will be of several types: lips, beak-to-beak, swallow-tail, and horizontal cusps. The births, deaths, and higher order saddles comprise additional cuspidal relations. Sorting these out will be a major theme throughout the text.

CHAPTER 5

Triple arrows for FA

Before the chapter embarks upon the general notion of triple arrows, a property of non-degenerate pairings that has an interesting diagrammatic interpretation is worth contemplating. After this example is better understood, it will be abstracted to describe triple arrows in general.

1. Algebraic identities as double arrows

We take a short respite from the diagrammatics. But as this section continues, it will be convenient to move back into thinking diagrammatically.

Suppose that a non-degenerate inner product $\mathbb{R} \longleftarrow \mathbb{R}^m \otimes \mathbb{R}^m$ in an m-dimensional vector space is represented by an $(m \times m)$-matrix A. The inner product of column vectors $\vec{x}$ and $\vec{y}$ is $\vec{x}^{\,T} A \vec{y}$ where $\vec{x}^{\,T}$ denotes the transpose of $\vec{x}$. We are operating at the algebraic, rather than the categorical level, so "=" will be used, but will soon be replaced by $\Rightarrow$. Since A is non-degenerate, there is a matrix B such that $BA = AB = I$ — the identity matrix, The following string of equalities holds:

$$\vec{x}^{\,T} A \vec{y} = \vec{x}^{\,T} A I \vec{y} = \vec{x}^{\,T} A(BA) \vec{y} = \vec{x}^{\,T} (AB) A \vec{y} = \vec{x}^{\,T} I A \vec{y} = \vec{x}^{\,T} A \vec{y}$$

Let's consider the steps more slowly. We won't distinguish A from AI or IA. By replacing the identity with the product BA, an operation such as this $\curlyvee : I \Rightarrow (BA)$ occurs. After the associative law is applied, the next equality is thought of as an operation like this: $\curlywedge : (AB) \Rightarrow I$.

The associative law can be thought of a prioritization. It tells the reader which operation occurs first. It corresponds to an operation $X : A(BA) \Rightarrow (AB)A$. For added clarity, write $X : A_1(BA_2) \Rightarrow (A_1B)A_2$. In these operations, B is fixed. In the source, $A_1(BA_2)$, the matrix B is acted upon by A_2 on the right; in the target, A_1 acts on the left of B.

So the string of equalities can be replaced by double arrows

$$A \Rightarrow AI \overset{\curlyvee}{\Rightarrow} A(BA) \overset{X}{\Rightarrow} (AB)A \overset{\curlywedge}{\Rightarrow} IA \Rightarrow A.$$

Unfortunately, the notation here is still not rich enough to capture all of the meaning that we wish to extract. So let us enhance it.

First, represent the matrices by means of their indexed entries. So $A = [\wedge_{i,j}]_{i,j \in [m]}$ and $B = [\vee_{i,j}]_{i,j \in [m]}$ while $\vec{x} = [x^i]_{i \in [m]}$ and $\vec{y} = [y^j]_{j \in [m]}$ where, recall, $[m] = \{1, 2, \ldots, m\}$. Let I^{ℓ}_{q} denote a Kronecker delta function. We adopt summation conventions:

$$x^i \wedge_{i,j} y^j = x^i \wedge_{i,j} \mathsf{I}^j_\ell y^\ell \Rightarrow x^i \wedge_{i,j} (\vee^{j,k} \wedge_{k,\ell}) y^\ell$$

$$\Rightarrow x^i (\wedge_{i,j} \vee^{j,k}) \wedge_{k,\ell} y^\ell \Rightarrow x^i \mathsf{I}^k_i \wedge_{k,\ell} y^\ell = x^k \wedge_{k,\ell} y^\ell$$

Replace the structure constants by their diagrammatic counter parts $\cup$ and $\cap$ and write the double arrows as follows

$$\cap \Rightarrow [\cap\circ(\mathsf{II}\cap)\circ(\mathsf{I}\cup\mathsf{I})] \Rightarrow [\cap\circ(\cap\mathsf{II})\circ(\mathsf{I}\cup\mathsf{I})] \Rightarrow \cap$$

Each double arrow can be expressed as a glyph in a disk whose boundary is decomposed as the compositions of source and target 1-arrows. To that end we write

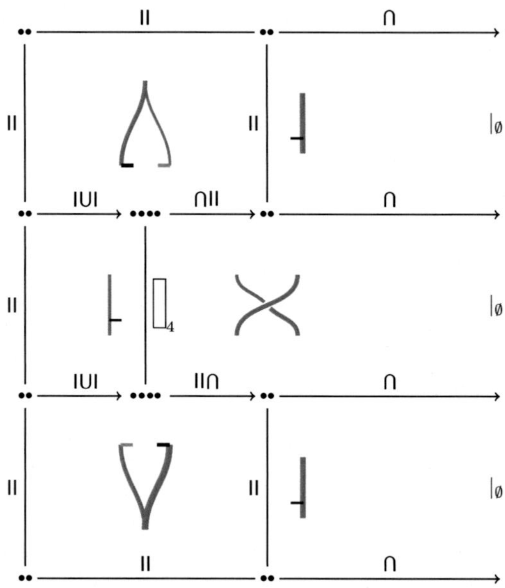

Undirected vertical and horizontal edges are identity 1-arrows. Vertical lines with short indicators in the interior of the diagram are the identity double arrow upon the source = target. These horizontal 1-arrows are of the form $\mathsf{I}_\alpha\cup\mathsf{I}_\beta$ or $\mathsf{I}_\alpha\cap\mathsf{I}_\beta$ where $\alpha, \beta \in \mathbb{N}$. The X shaped double arrow in the middle right of the diagram is an exchanger.

The vertical identities on the left and right can be contracted, and the resulting composite double arrow has $\bullet\bullet \xrightarrow{\cap} \emptyset$ as both its source and its target. Therefore, this double arrow is comparable to the identity double arrow upon $\bullet\bullet \xrightarrow{\cap} \emptyset$. At the algebraic level, where the discussion began, these are equal since each double arrow is an equality. But at the diagrammatic level, we don't yet know.

In fact, these double arrows are equal at the level of the surface or foam that is used to represent them. That equality is called the *swallow-tail singularity.* You can see a simulation of it as you straighten or bend your knee. See Fig. 1.

Two double arrows that share the same source and target 1-arrows are compared by means of triple arrows. Section 2 gives an informal discussion about the categorical structure upon double arrows. An arrow that runs between a pair of double arrows is a triple arrow. An associative and unital composition is defined on the set of triple arrows whose sources and targets align. Some triple arrows are introduced in Sections 3 and 4. Then in Chapter 6, we change focus to studying a related multi-category[1] so that generalities about higher arrows derived from pairs of reverse arrows can be made. Chapter 7 returns to a catalogue of triple arrows (and higher) in FA and some other multi-categories that are both restrictions and extensions of FA. Even though the multi-category FA that we are building is not the main focus of this book, many analogues will be modeled upon it.

2. The category of double arrows

In the previous example, we suggested that the identity double arrow upon $\bullet\bullet \xrightarrow{\cap} \emptyset$ be compared with the swallow-tail arrow. To that end, suppose that the set of double arrows that share the same source and target 1-arrows is a category. In this way, triple arrows can be thought of as relations among double arrows. A given pair of double arrows might be comparable in more than one way. That would be a good thing since then the category of triple arrows between these double arrows has more structure.

Since there is a category, a pair of double arrows with suitable source and target have a triple arrow between them. For any double arrow $F \Uparrow$, there is a triple arrow with source $\mathcal{S} = F$ and target $\mathcal{T} = F$ that acts as an identity under the associative globular composition of triple arrows.

As above, double arrows are composed in a skew or globular manner, and a composition of double arrows corresponds to a disk in the plane as we now explain. Let a composition of double arrows, F, be given. Two oppositely oriented arcs —$S(F)$, the source, and $T(F)$, the target — constitute the boundary of the disk that represents F. The common initial point of these arcs is the source object, $s = s(S(F)) = s(T(F))$. The common end point is the target object, $t = t(S(F)) = t(T(F))$. In the example of Section 1, the source and target were $\bullet\bullet \xrightarrow{\cap} \emptyset$ — the 1-arrow that corresponds to a non-degenerate pairing.

Suppose that $F \Uparrow$ and $G \Uparrow$ are double arrows whose source and target *objects* agree. Suppose that the 1-arrows $S(F)$ and $S(G)$ are equal as are $T(F)$ and $T(G)$. A triple arrow between F and G is a globe whose boundary is decomposed into hemispherical disks. Say S and T lie close to the equator with their source s at the location of Equatorial Guinea and (for geographic convenience) their target t at the Solomon Islands. Say that F is the double arrow in the south and G is the double arrow in the north. Then a triple arrow between them lies in the interior of the globe.

Composition of triple arrows is defined in a globular manner that mimics the globular composition of double arrows. A Θ-*sphere* is comprised of three disks: $\{(x,y,z) : z = \sqrt{1-x^2-y^2}\}$ — the northern disk, $\{(x,y,z) : z = -\sqrt{1-x^2-y^2}\}$ — the southern disk, and $\{(x,y,z) : x^2+y^2 \leq 1 \ \&\ z = 0\}$ — the equatorial disk. The equatorial disk is the target of one triple arrow and the source of the other. The *globular composition* of these triple arrows is the result of merging the globes that represent them along the equatorial disk. It is as if two dumplings in a pot are stuck together along a portion of the wrapper.

[1]A *multi-category* is a category for which the set of k-arrows is also a category. So there are 1-arrows, double arrows, etc.

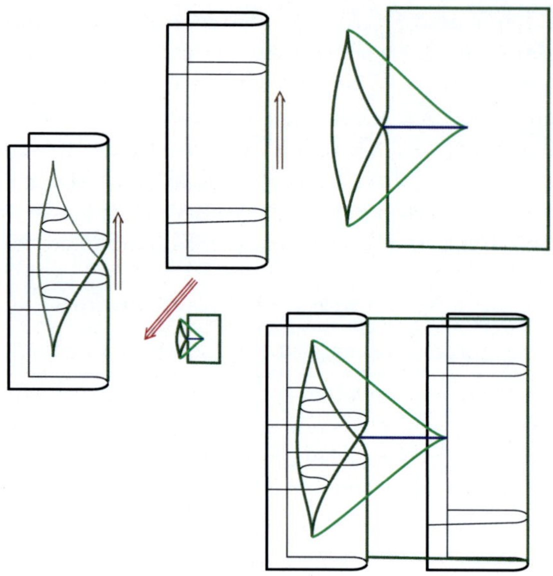

FIGURE 1. The swallow-tail

There is also an analogue to skew composition, but that discussion will be postponed for a while. We leave it to the reader as an exercise to develop a dumpling, mandu, or pierogi metaphor for the associative composition of triple arrows.

Exercise. There are four variations on the swallow-tail triple arrow. Describe the other three in terms of the matrix representation of the pairing or co-pairing, the structure constants, and as triple arrows between disk diagrams of the composition of glyphs. As the text continues (Chapter 6, Section 4.5), these variations will be presented.

3. Critical aspects of weakly invertible 1-arrows

$\emptyset \xrightarrow{\blacktriangle} \bullet$	$\bullet \xrightarrow{\blacktriangledown} \emptyset$
$\bullet\bullet \xrightarrow{\curlywedge} \bullet$	$\bullet \xrightarrow{\mathsf{Y}} \bullet\bullet$
$\bullet\bullet \xrightarrow{\cap} \emptyset$	$\emptyset \xrightarrow{\cup} \bullet\bullet$

TABLE 1. Generating 1-arrows and their reverses

We begin by calling attention to the generating 1-morphisms that are depicted in Table 1. The pair of arrows in any row are called *reverses* of each other. To reverse an arbitrary 1-arrow reflect its diagrammatic description in a horizontal line. The reverses of the arrows in Table 1 are unique and are horizontal reflections of each other. So for an arbitrary 1-arrow, the horizontal reflection presupposes that each generating 1-arrow is reflected to its unique reverse.

In the Tables 11, 12, and 13 of Chapter 4, necessary criteria that these arrows were weakly invertible were established. The weak inverse of any entry in Table 1 is the other entry in the same row. Specifically,

- The reverse of the unit arrow is the co-unit arrow. A necessary criterion for this to be a weak inverse is the existence of the double arrows that are listed as reverse pairs.
  - unit/co birth 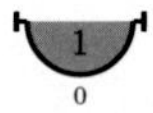, unit/co death 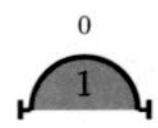;
  - unit/co canal 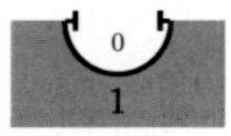, and unit/co arch 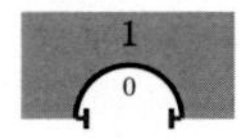.
- The reverse of the multiplication arrow ⋏ is the co-multiplication arrow Y. A necessary criterion for this to be a weak inverse is the existence of the zipper double arrows that are listed as reverse pairs.
  - 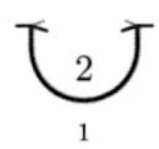 and 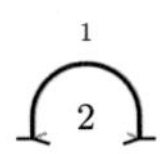;
  -  and 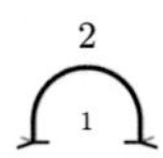.
- The reverse of the pairing arrow ∩ is the co-pairing arrow U. A necessary criterion for this to be a weak inverse is the existence of the double arrows that are listed as reverse pairs.
  - birth and death;
  - saddle and fork.

We suppose, inductively, that the set of $(k-1)$-arrows is a category with an associative, globular, unital composition defined. Suppose that $s \xrightarrow{\mathsf{p}}_{k-1} t$ is a $(k-1)$-arrow and $t \xrightarrow{\mathsf{b}}_{k-1} s$ is its inductively defined reverse: an arrow that interchanges source and target. Then we have the four k-arrows F_s, F_t, E_s, and E_t that are indicated. Here and below, the symbol $\|_*$ indicates an identity arrow on a multiple arrow $*$ where the multiplicity of $*$ is greater than 0.

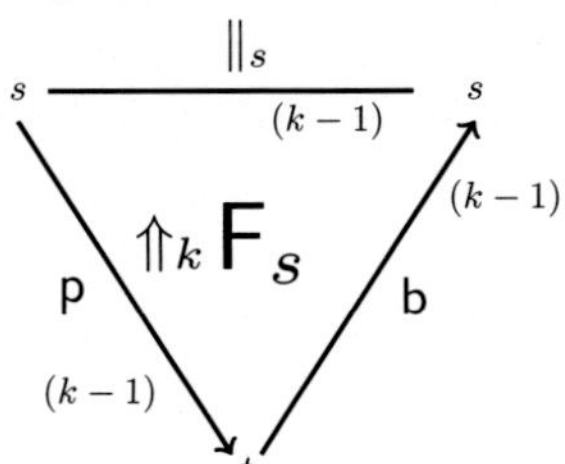

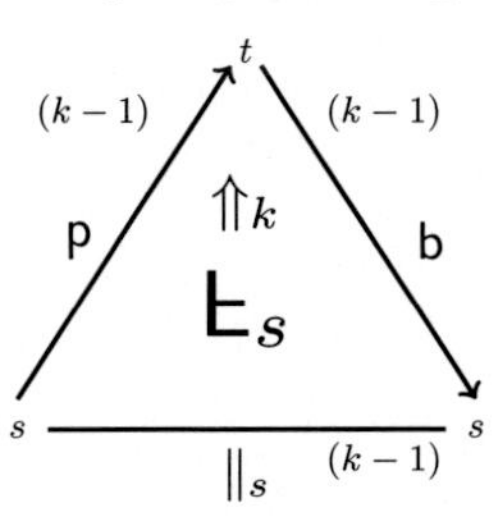

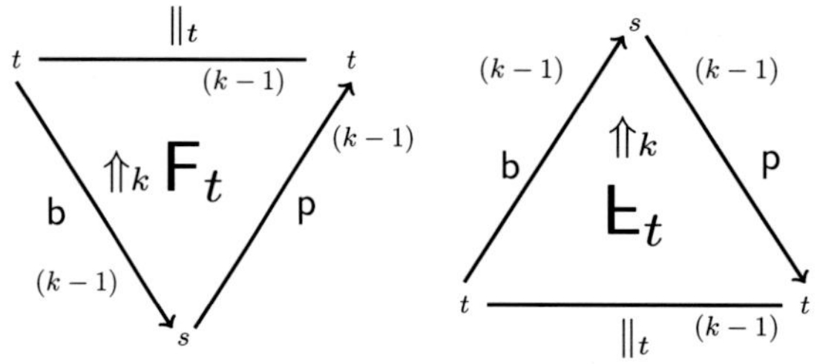

We'll say that F_s and F_t are of type ⊤. Their reverses Ⱶ$_s$ and Ⱶ$_t$ are of type ⊥. In this case F_s and its reverse Ⱶ$_s$ are *weak inverses at level* $(k+1)$ when there is a $(k+1)$-arrow from the globular composition $\left[\begin{array}{c}Ⱶ_s\\ \mathsf{F}_s\end{array}\right]$ along $\|_s$ to the identity on $(\mathsf{p} \circ_{k-1} \mathsf{b})$ and a $(k+1)$-arrow from the globular composition $\left[\begin{array}{c}\mathsf{F}_s\\ Ⱶ_s\end{array}\right]$ along $(\mathsf{b} \circ_{k-1} \mathsf{p})$ to the identity along $\|_s$. Similarly, F_t and its reverse Ⱶ$_t$ are weak inverses.

Necessary criteria that the pairs (⅃, ⊤) (⅄, Y), *and* (∩, ∪) *are weak inverse pairs is that for all* $k = 2, \ldots$ *the derived k-arrows* (F_s, Ⱶ$_s$) *and* (F_t, Ⱶ$_t$) *are weak inverses at level* $(k+1)$.

In order to develop some further understanding of the arrows of higher multiplicity, we will present a case-by-case analysis of the induced triple arrows. The reader is encouraged to construct the quadruple arrows from the given triple arrows. These analyses will help indicate why these aspects are "critical." That is, in their geometric realizations, they correspond to points at which derivatives vanish.

3.1. Triple arrows from unit and co-unit. Birth and death. Consider the composition 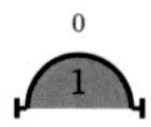 $\circ$ . This can be achieved via a vertical juxtaposition of the glyphs.

The double arrow 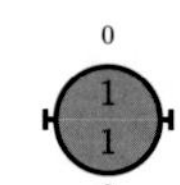 can be compared with the double arrow $\|_{|_\emptyset}$ which is the identity double arrow from the identity arrow on the empty object $\emptyset$ back to itself. The empty object corresponds to the natural number 0. An arrow from it to itself is vacuous as is the identity double arrow upon it.[2]

In the comparisons, triple arrows that are reverses of each other are defined. Recall from Section 2 of this chapter that a triple arrow lies within a globe. It points away from the double arrow that engulfs the southern hemisphere and towards the double arrow that engulfs the northern hemisphere. The source and target 1-arrows agree and lie along the equator. In movie descriptions of triple arrows, the source double arrow appears in the bottom still of the movie and the target double arrow appears as the top still.

[2]Much ado about nothing.

The triple arrow on the left is of birth-type; that on the right is of death-type.

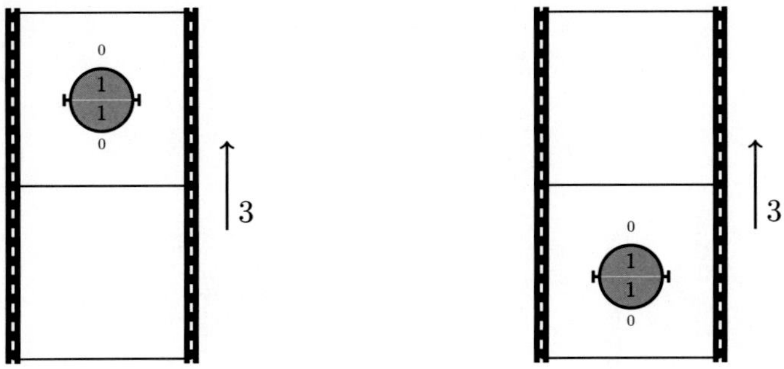

Saddle and fork. The composition 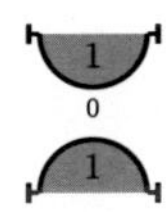 is to be compared to the identity on the composition of 1-arrows $\emptyset \longrightarrow \bullet \longrightarrow \emptyset$. To that end, the triple arrows

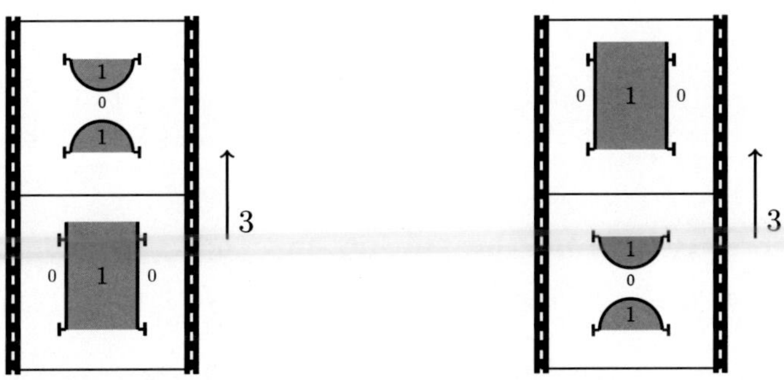

are represented. These are reverses of each other, the triple arrow on the left is of saddle-type, and the one on the right is of fork-type.

Hole filling. The first composition of arch and canal that will be considered is the one in which the composition of glyphs results in a hole. To wit, triple arrows between 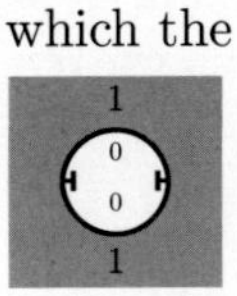 and the identity $||_{|}$ on the identity on $-\bullet-$ are represented as below.

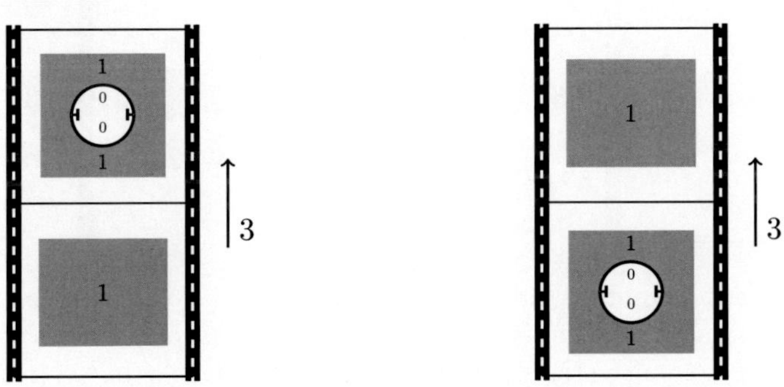

These are complementary to the birth-type and death-type triple arrows at the beginning of this section.

Excavating. The other composition between canal and arch 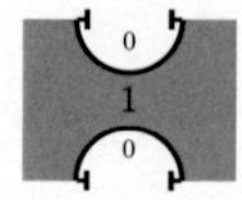 is to be compared to the identity upon the composition of 1-arrows $\bullet \xrightarrow{} \emptyset \xrightarrow{} \bullet$. The triple arrows that result appear.

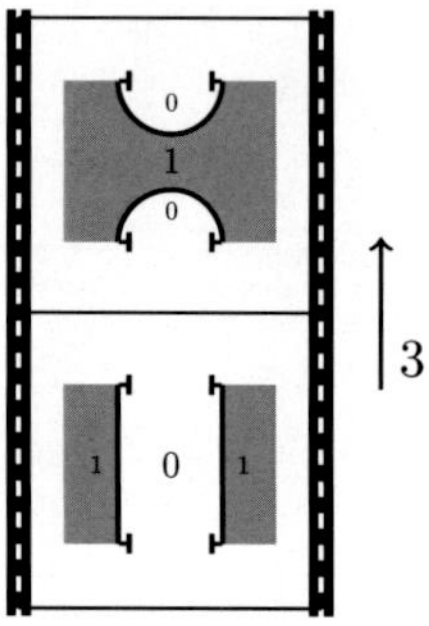

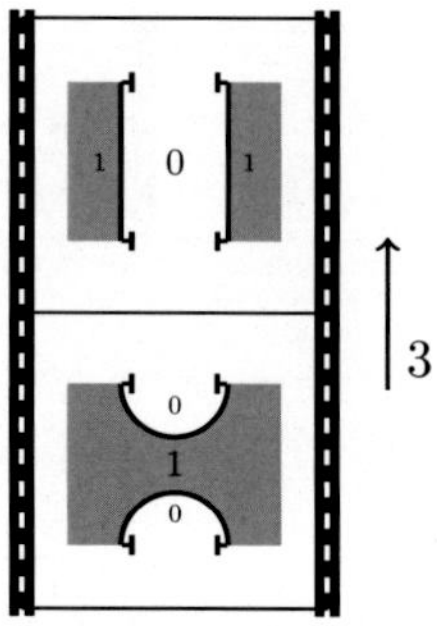

These are complementary to the saddle-type and fork-type triple arrows given above.

Exercise. Perform the next step in the induction and construct the eight quadruple arrows from the composition of pairs of these triple arrows and their reverse. See Chapter 7.

3.2. Triple arrows from the pairing and the co-pairing. Birth and death. In the cases in which the double arrows as glyphs 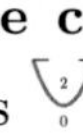 and 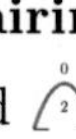 are mapped to their foam representatives  and , the composition 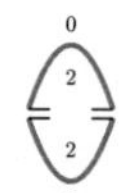 corresponds to a sphere that is schematized as . The triple arrows that are depicted below compare it to the identity double arrow on the identity 1-arrow on the empty object.

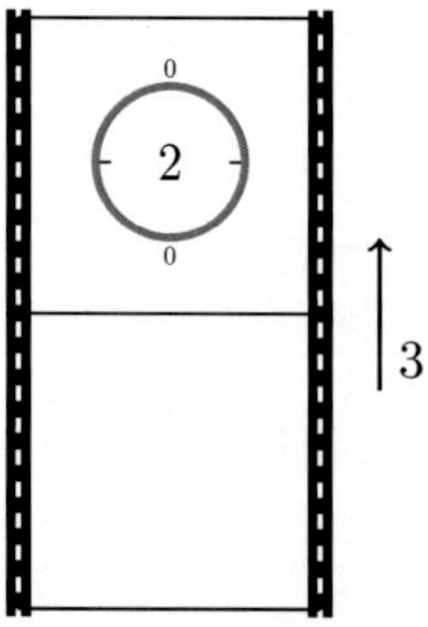

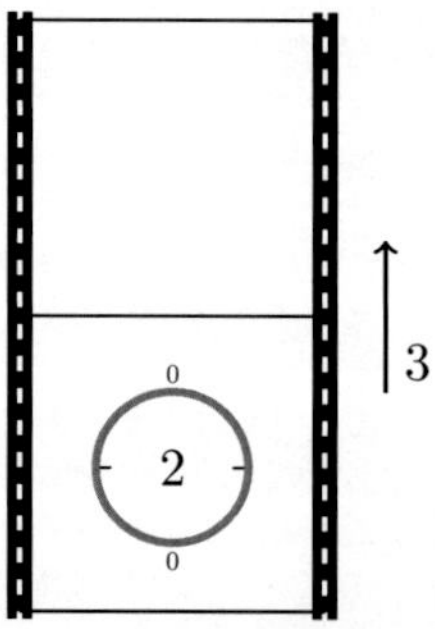

A spherical surface bounds a ball that is expressed in a time-like metaphor as a balloon filling with air. The triple arrow on the left, then, corresponds to the balloon inflating, and the one on the right correspond to it deflating[3]. In the former case, we

[3]This metaphor was told to us by Thomas Banchoff.

imagine the $w = 0$ and the $w = 2$ level surfaces of $w = x^2 + y^2 + z^2 + 1$. In the latter, the movie stills correspond to the $w = 0$ and $w = 2$ levels of $w = 1 - x^2 - y^2 - z^2$.

Handles. In the cases in which the double arrows as glyphs  and 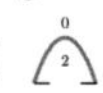 are mapped to their foam representatives  and  the composition 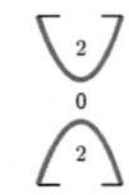 corresponds to a hyperboloid of two sheets. The triple arrows that are depicted below compare it to the identity double arrow on the composition  $= \cap \circ \cup$ of 1-arrows.

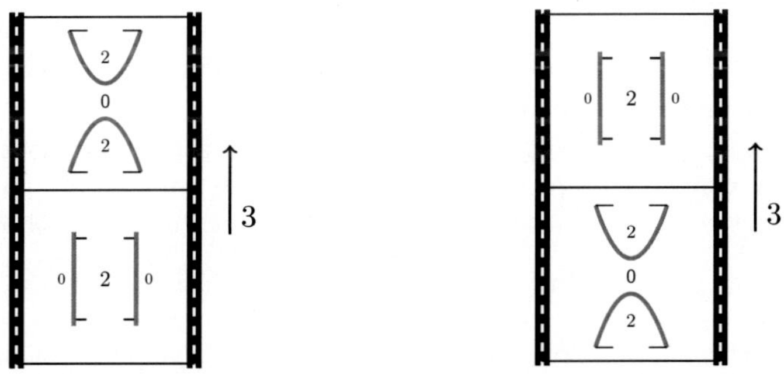

When passing to triple arrows between foams, these triple arrows correspond to 3-dimensional solids that can be given in parametric form as

$$\{(x, y, z, w) : w = \pm(z^2 - x^2 - y^2) \ \& \ -1 \leq w \leq 1\}.$$

In the topological context that these categorical ideas are meant to model, the movie on the left corresponds to attaching a 2-handle and the one on the right corresponds to attaching a 1-handle. These ideas will recur subsequently in Chapter 7.

Exercise. In the cases above the triple arrows depicted in companion movies are reverses of each other. Compose them and determine the corresponding quadruple arrows that relate the compositions to the identities on source and target. Again, see Chapter 7.

Handles part 2. In the cases in which the double arrows as glyphs  and  are mapped to their foam representatives  and 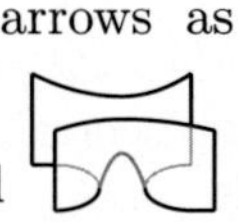 the composition  corresponds to a hyperboloid of one sheet that is turned sideways from the viewer. The triple arrows that are depicted below compare it to the identity double arrow on the composition  $= \cup \circ \cap$ of 1-arrows.

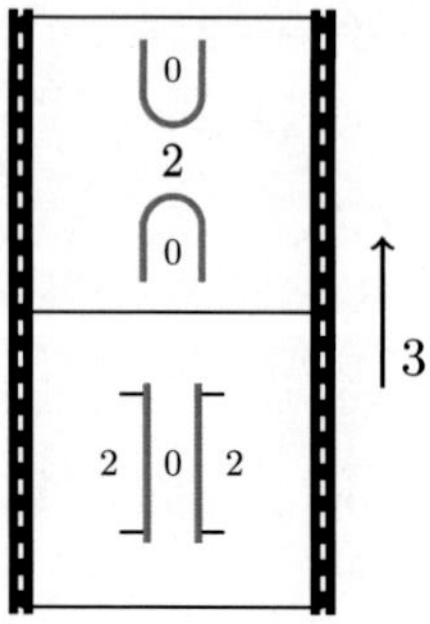

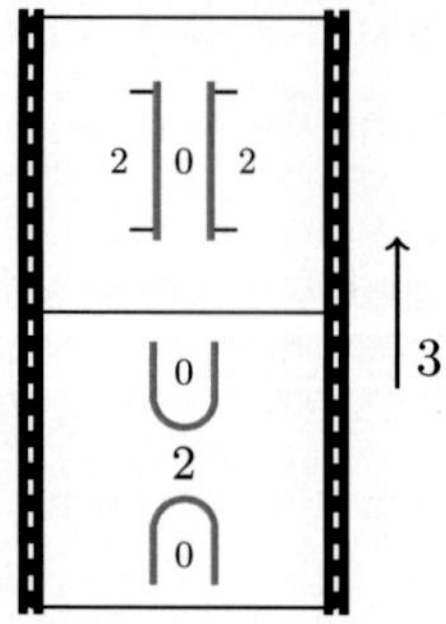

When passing to triple arrows between foams, these triple arrows also correspond to 3-dimensional solids.

In the topological context that these categorical ideas are meant to model, the movie on the left corresponds to attaching a 1-handle and the one on the right corresponds to attaching a 2-handle.

Exercise. In the cases above the triple arrows depicted in companion movies are reverses of each other. Compose them in both orders and determine the corresponding quadruple arrows that relate the compositions to the identities on source and target.

Meanwhile, the composition  also corresponds to a hyperboloid of one sheet, but for which the axis points towards the observer.

The triple arrows that are depicted below compare it to the identity double arrow on II which, in turn, is the identity on $-\bullet\!-\!\bullet-$. This double arrow is denoted by 2.

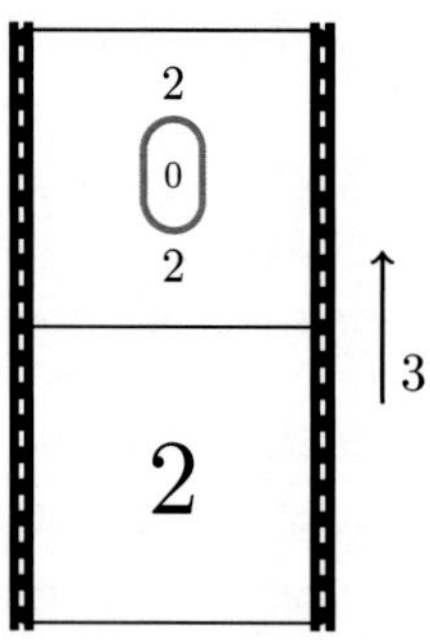

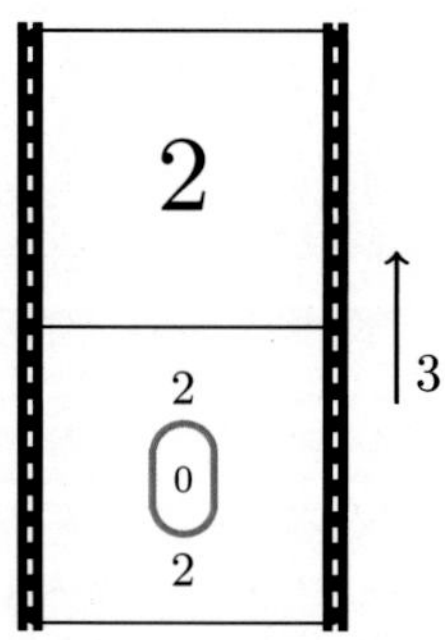

When passing to triple arrows between foams, these triple arrows also correspond to 3-dimensional solids. In the topological context that these categorical ideas are meant to model, the movie on the left corresponds to attaching a 1-handle and the one on the right corresponds to attaching a 2-handle.

Exercise. In the cases above the triple arrows depicted in companion movies are reverses of each other. Compose them in both orders and determine the corresponding quadruple arrows that relate the compositions to the identities on source and target.

3.3. Zipping and re-zipping. The double arrows that are associated to the reversible pairs Y and ⅄ are labeled as zippers. There are four pairs of triple arrows that result.

Zippers closing. There are two compositions of the closing double arrows 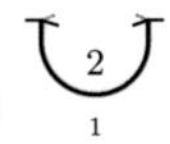 and 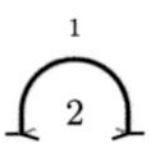. First, triple arrows are constructed by comparing 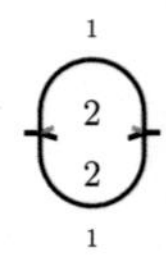 to the identity upon the identity $\bullet \xrightarrow{|} \bullet$ which will be denoted by 1.

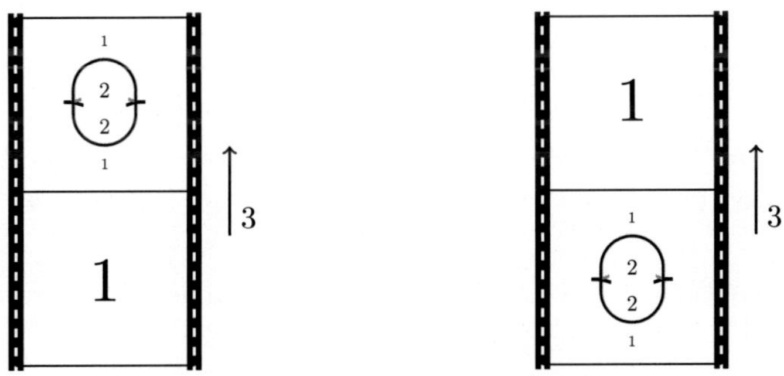

These correspond to adding or subtracting a bubble to a foam.

Next compare 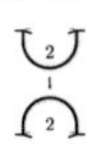 to 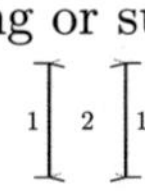. The triple arrows that result follow.

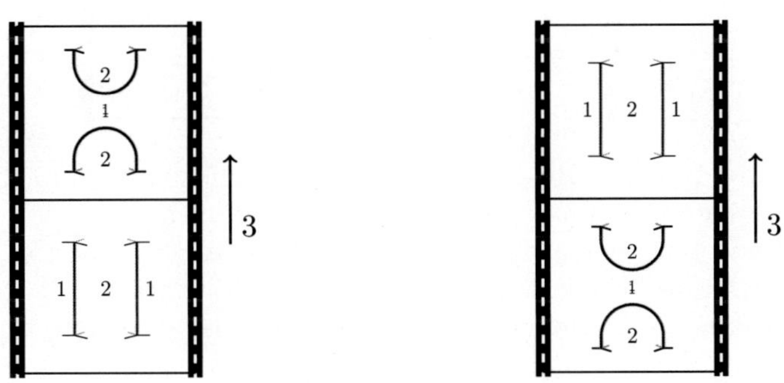

Zippers opening. When the opening zippers  and 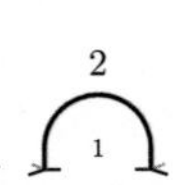 are composed as 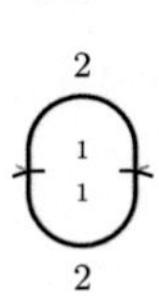, the result should be compared with the identity upon the 1-arrow $\bullet\bullet \xrightarrow{||} \bullet\bullet$

which will be denoted by 2.

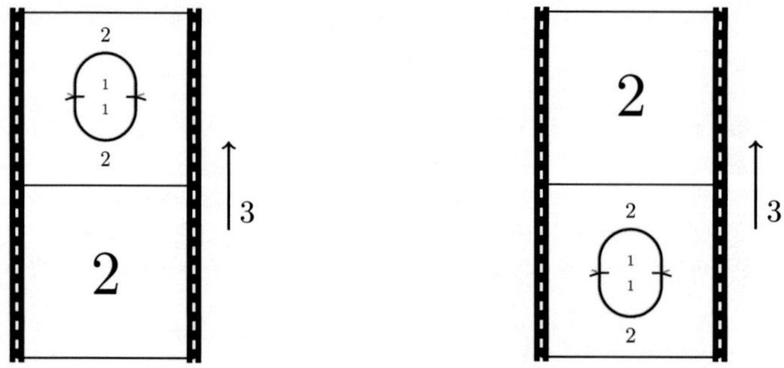

In the foam picture, a disk bounded by a seam is removed.

When these same two opening zippers are composed as 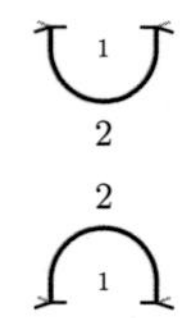 , the result should be compared with the identity upon the composition 1-arrow 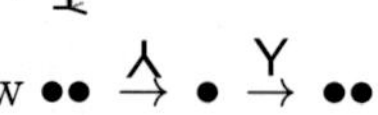 which will be denoted as 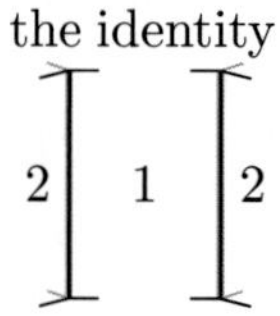 . The following triple arrows result.

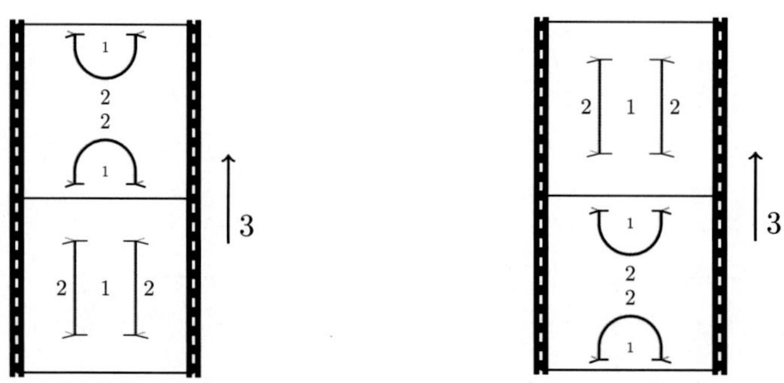

3.4. Introducing handles. At the level of a Frobenius algebra, that we used to motivate the higher category FA, the composition of a unit with a co-unit is hardly ever an identity. For example, in the case of $(n \times n)$ matrices, the unit is the identity matrix and the co-unit applied to this is its trace. The value of the composition in $\mathbb{R}$ is n. In the composition in the other direction, any matrix A will be mapped to the diagonal matrix $\mathrm{tr}(A) \cdot I$ which has very little to do with A itself. Similar computations apply to compositions of multiplication and co-multiplication and to compositions of pairings with co-pairings. Consequently at the algebraic level, there is no reason to believe that a 1-morphism composed with its reverse should be an identity. Nevertheless, comparisons can, and should be, made at the categorical level.

The necessary criterion established for reversible arrows to be weakly invertible is: the next higher arrows — comparing the compositions of the arrow and its reverse to an identity — give rise to higher arrows still. In some cases, for example with zig-zags, these next higher arrows can be thought of as equality. The criterion for equality is ad hoc since it relies upon topological considerations of a functorially equivalent category — the sesqui-category of foams (see Chapter 6, Section 5.1 for the definition of a functor). In fact, there are other categorical reasons for strong invertibility, but more machinery is needed before these are stated.

The triple arrows above are not topological equivalences (isotopies or homeomorphisms). They are, however, atomic pieces used to construct the next higher dimensional analogue. That is the triple arrows represent handles and critical changes. In the next few paragraphs, a few more pieces will be assembled from the kinematic (or cinematic) point of view in the topological context.

In the discussion above, handles were mentioned. For now, it is safe to assume that the positive integer $n \leq 4$ so that intuition can be gained. Suppose that an integer k is given with $0 \leq k \leq n$. The k-dimensional *disk* is

$$D^k = \{(x_1, x_2, \dots, x_k) \in \mathbb{R}^k : \sum_{i=1}^{k} x_i^2 \leq 1\}.$$

The $(k-1)$-dimensional sphere

$$S^{k-1} = \{(x_1, x_2, \dots, x_k) \in \mathbb{R}^k : \sum_{i=1}^{k} x_i^2 = 1\}$$

is the boundary, ∂D^k, of the disk[4]. Let $i + j = n$. An i-handle is a space homeomorphic to $D^i \times D^j$. The boundary of an i-handle is decomposed (via the Leibniz rule) as the union of an *A-region*, $S^{i-1} \times D^j$, and a *B-region*, $D^i \times S^{j-1}$. The A-region is called *the attaching region*, and the B-region is called the *belt region*. More details will be provided in Chapter 7, Section 1.3.

Suppose that an $(n-1)$-dimensional topological space M is given. Usually, M is a *manifold* so that every point has a neighborhood that is homeomorphic to an open subset of $\mathbb{R}^{n-1}$, and some other natural conditions hold[5]. Suppose that there is a subset of M that is homeomorphic to the A-region of an i-handle. Then the space M' that is obtained from M by cutting out the A-region, and replacing it with the B-region of a handle is said to be obtained from M by attaching an i-handle.

Here the language choices are imprecise. The interior of the A region is removed, and is replaced by the interior of the B-region. There is a toroidal region $S^{i-1} \times S^{j-1}$ that is common to both M and M'. A more precise description is achieved by speaking of the quotient topology, and discussing things in the smooth or PL-category (See for example [**69**]).

The space W that is the quotient of $(M \times [0,1]) \cup (D^i \times D^j)$ is an n-dimensional space that has $M \times \{0\}$ as part of its frontier, and $M' = (M \times \{1\} \setminus (S^{i-1} \times D^j)) \cup (D^i \times S^{j-1})$ as the other. Here the handle is attached at $M \times \{1\}$, and M was thickened so that W is n-dimensional. It is not just a pair of plates with a tube attached between them.

Now pass from the categorical context in which double arrows are merely labeled by suggestive glyphs to the topological context in which double arrows are represented as foams.

---

[4]Subscripts are used to denote coordinates since exponents (squares) are used in the description.

[5]The conditions can be abstracted from the definition of a surface with boundary that is given in Chapter 6, Section 5.2.

Observation. *In the translation from glyph to foam, each of the movies presented above represents a transition between foams that corresponds to attaching a handle. So do the higher categorical* $(k+1)$*-arrows that are constructed inductively to relate a* k*-arrow to its reverse.*

Moreover, these movies and their inductively defined generalizations are the building blocks that will both constitute higher dimensional arrows and their geometric realizations.

Exercise. Describe each of the triple arrows that are presented in movie format above as a handle attachment keeping track of the dimension of the ambient space and the handle type.

4. Coalgebra axioms as triple arrows

In Ch. 2, Fig. 4 diagrammatic proofs were presented that established that a Frobenius algebra is also a coalgebra. These proofs depended upon defining the co-unit in terms of the unit and defining the co-multiplication in terms of multiplication. In both of these cases, the co-pairing was used either on the left or on the right. Let us analyze the foams that appear as statements and proofs in Fig. 2.

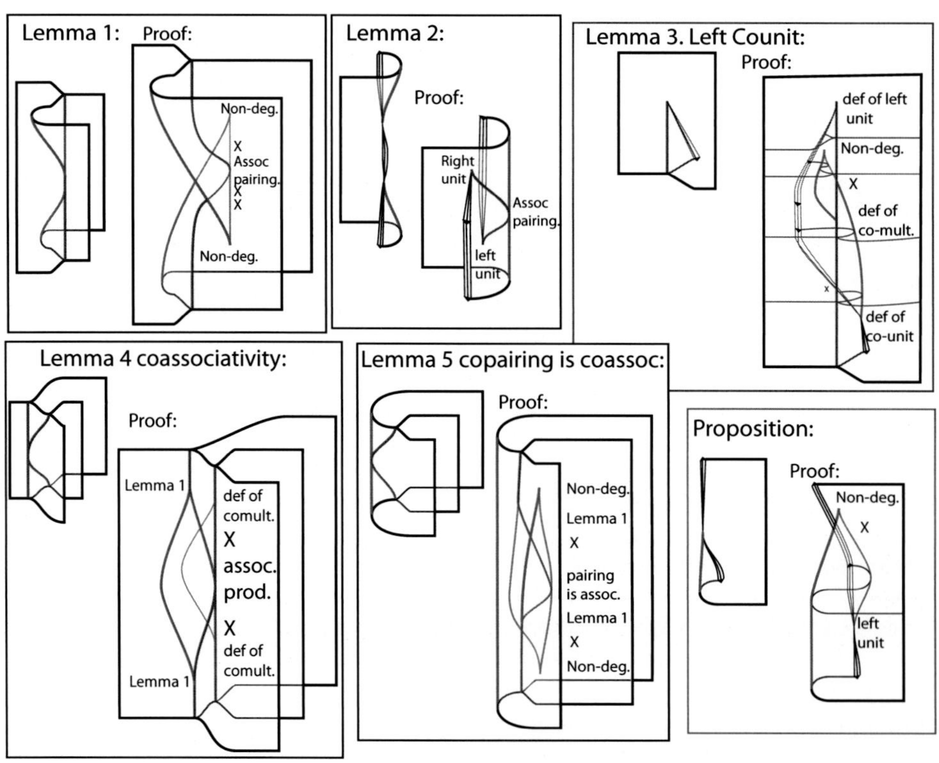

FIGURE 2. Surface-like proofs of coalgebra axioms

The statement of Lemma 1 in the top left-hand corner of the illustration consists of a foam that represents a composition of the double arrows 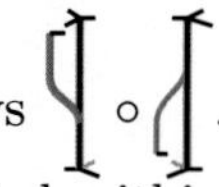. The proof consists of the composition of double arrows that are labeled within the figure. That proof coincides with an algebraic proof that does not use diagrams. Now we imagine that the connection between the statement and its proof is a triple arrow. At the level of foams, the authors imagine a geometric process that is analogous to a swallow-tail in which the two cusps that appear in the proof are brought close together and are coincident at an instant with the point at which the seam passed over the fold. At any rate, we assert that there is an invertible triple arrow whose source is the statement and whose target is the proof.

In the lower right corner of Fig. 2 a proposition appears. It corresponds to the statement that the co-unit can be defined via the unit and the co-pairing by placing the unit on the left leg of the co-pairing. The proof appeals to the reflection of the Lemma 2 that appears in the figure, and the use of a cusp so that a fold can change direction. An ironing type isotopy can be used to connect these. Hence we assert that the statement and its proof are connected by a triple arrow.

The proof of Lemma 2 involves the introduction of a pair of vestigial flaps caused by the introduction and subsequent deletion of unit. A swallow-tail like relation also suggests an isotopy, so the existence of an invertible triple arrow that connects the statement to its proof is asserted.

Lemma 3 asserts that the co-unit is a co-unit for co-multiplication. The proof, as it is presented, is a generic perturbation of a door opening or closing. Again we assert that there is an invertible triple arrow that connects statement to proof.

In Lemma 4 and Lemma 5, swallow-tail like perturbations relate multiplication to co-multiplication or pairing co-pairing. As above invertible triple arrows connect statements to proofs.

4.1. Review and preview. There are several more types of triple arrows that should be part of FA— the categorical construction that encompasses Frobenius algebras. Furthermore, necessary criteria for weak invertibilty have been given, and there are a few families of criteria that need to be added so that this idea is fully characterized.

In the next chapter, rather than studying foams, we study a categorical characterization of isotopy classes of oriented surfaces that are embedded in 3-dimensional space. Moreover, another level of abstraction is necessary. Namely, it is better to think of U and ∩ as double arrows. A surface in 3-space separates space. So the complement of a surface in space is checkerboard colorable. The objects in the corresponding higher category correspond to the black and white regions of space. The singularities of surface projections and of surface isotopies will provide models for relationships among higher arrows in FA. They are source and target. Let us proceed.

CHAPTER 6

Surfaces in 3-space

In this chapter, a multi-categorical description of isotopy classes of properly embedded surfaces in the 3-dimensional slab $\mathbb{R}^2 \times [0,1]$ will be developed. Unfortunately, the initial method will appear to be arcane. A two object category, for which there is a reversing pair of arrows between them, is the starting point. Then double arrows, triple arrows, and quadruple arrows will be constructed from these. The construction is a camouflaged version of surfaces that are embedded in 3-dimensional space and for which the complement is colored black and white. Such a shading induces a normal orientation to the surface: an arc transverse to the surface that passes from a white region to a black region is called "positive," and one that passes from black to white is called "bad." By the usual orientation conventions, the normal orientation induces an orientation of the surface.

On the other hand, this camouflaged situation applies to other contexts. So it will help define higher arrows that are to be included in FA and related multi-categories.

Section 1 provides a guide to the ideas. In Section 2, a category is introduced that has two objects: t and ɟ and a pair of weakly invertible arrows: p : ɟ → t and b : t → ɟ between them. These are reverses of each other and they generate the set of arrows. That the fonts are reflections of each other is intentional and meant to suggest an analogy with the generating object of FA. The multi-category generated by the arrows b and p will be developed in abstraction, but vocabulary is used that is derived from the functorially equivalent multi-category embedded surfaces in 3-space and their isotopies. Subsequent sections define double, triple, and quadruple arrows.

1. Guide to terminology

There are two objects t and ɟ in the category to be constructed. The letter ɟ is the vertical reflection of the letter t; it is a poorly drawn f. Loosely, we call ɟ "from," and we call t "to." The generating arrow p : ɟ → t is called "positive" since it goes from "from" to "to" while b : t → ɟ is "bad" since it goes from "to" to"from".

Abstract properties of smooth closed surfaces that are embedded in 3-space motivate the discussion. In that case, the complement of the surface can be black and white colored in a way that the color changes upon passing through the surface. An arc that passes once through the surface is said to be *positive* (p) if it passes from a white region to a black region, and it is *bad* (b) if it passes in the other direction. This induces a normal orientation of the surface, and consequently a tangent orientation.

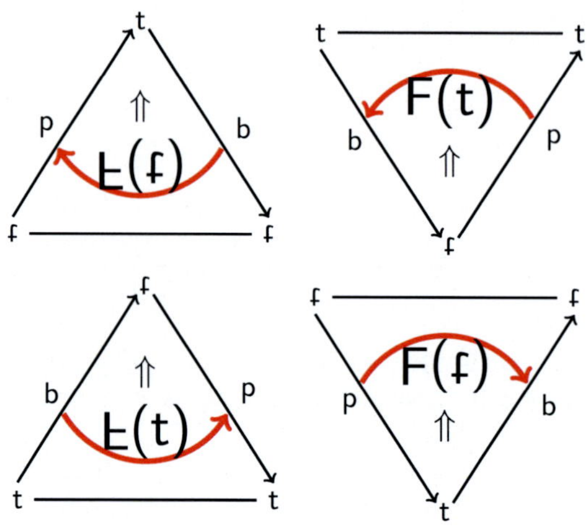

FIGURE 1. The generating double arrows for $\mathcal{S}$

2. Objects, 1-arrows, and double arrows

Let $\mathcal{S}$ denote the multi-category that has two objects ɟ, t, and generating arrows p : ɟ → t and b : t → ɟ. There are also identity arrows[1] ɟ —— ɟ and t —— t. We will say that p : ɟ → t and b : t → ɟ are *reverses* of each other. In general, a non-identity arrow is a finite sequence pbpb ⋯ b, pbpb ⋯ p, bpbp ⋯ b, or bpbp ⋯ p . In these expressions, instances of identity arrows have been contracted. Also, juxtaposition is used to indicate composition of 1-arrows.

The *generating double arrows* are depicted in Fig. 1. In each diagram there is a red curve that points upwards from the arrow labeled p and downwards at the one labeled b when these 1-arrows are considered to be pointing towards the right. Such red curves are a short hand notation for the double arrows, and they also indicate the functor to the category of surfaces: they map the double arrows to embedded disks that intersect folds of the surface. Alternatively, they may be thought of as oriented versions of ∪ and ∩.

For convenience, the identities upon the identity arrows ɟ —— ɟ and t —— t as well as the identities upon p and b are indicated.

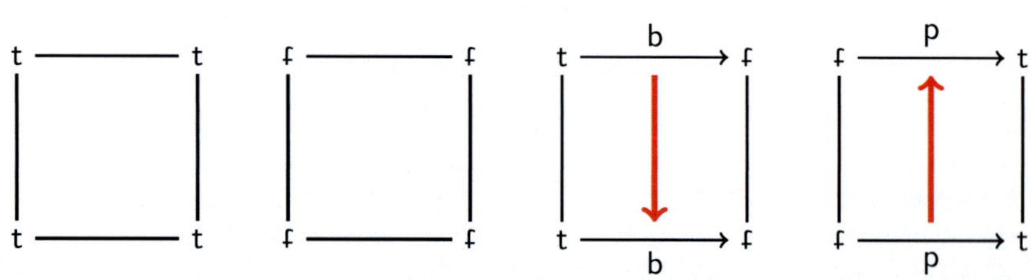

2.1. Alternative descriptions. It will be convenient to express (composites of) the double arrows F(t), F(ɟ), Ⅎ(t), and Ⅎ(ɟ) in terms of rectangular diagrams.

[1] All identity arrows, 2-arrows, etc. are thought of as strict in the sense that they may be inserted or contracted as needed and without fanfare.

Below several globular and skew compositions are redrawn in this fashion. A number of the remaining double arrows can be obtained by reflecting these illustrations in a horizontal line on the page.

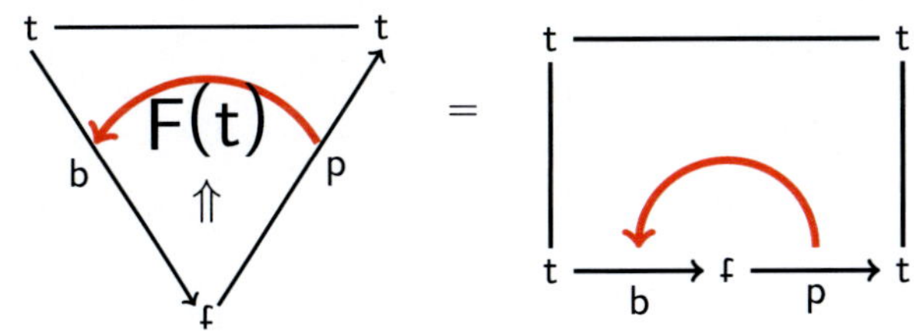

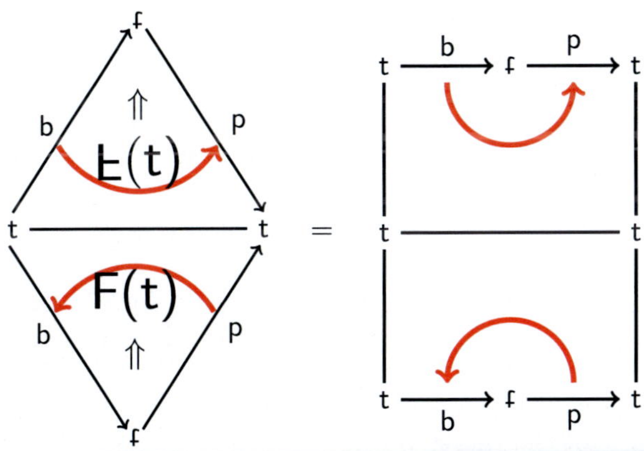

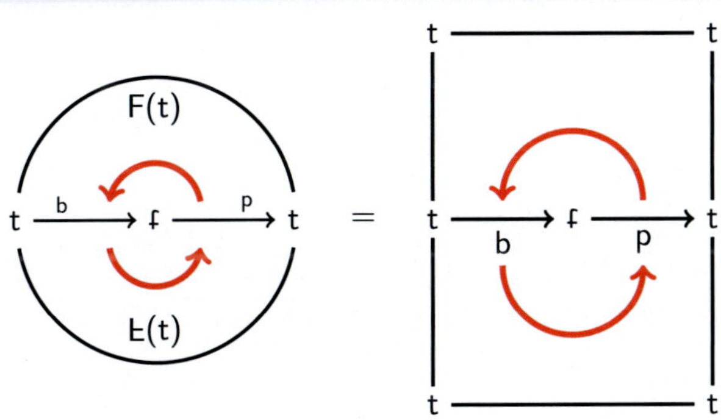

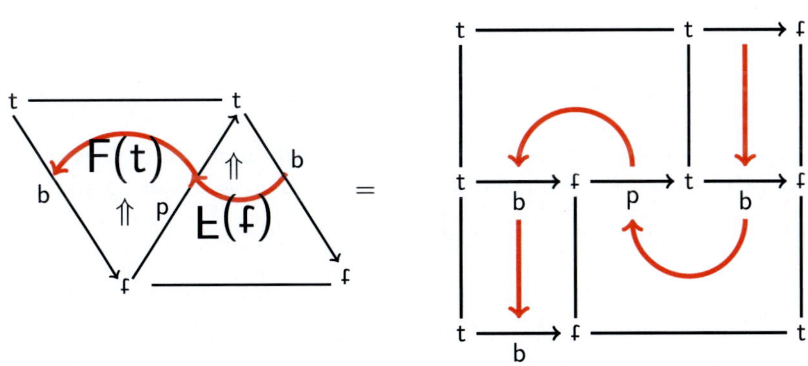

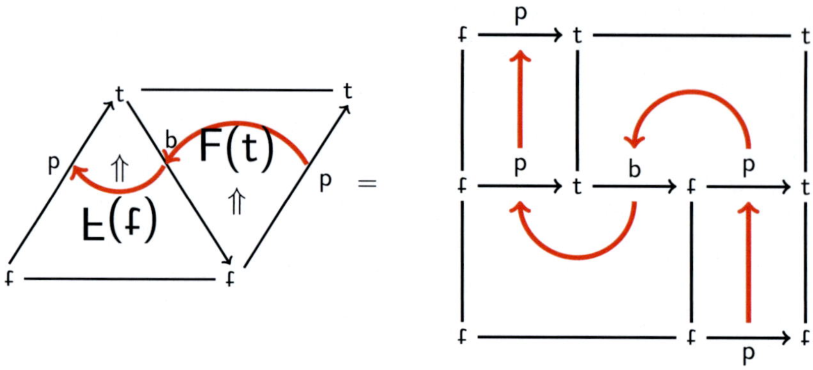

2.2. Decategorification, part 1. *The collection of* 2*-arrows in* $\mathcal{S}$ *are the arrows in a category.* It is a close analogue of the category FA that was described in Chapter 3, Section 2.

The objects of the corresponding category are finite length words that are of the form pbpb ⋯ b, pbpb ⋯ p, bpbp ⋯ b, and bpbp ⋯ p .

The collection of arrows between any two objects is empty unless the words both start and end with the same letter. Otherwise, their word length is arbitrary. In this way, there are many arrows between b and itself, between b and bpb, or between b and bpbpb, for example. The source and target objects of the source word must coincide with those of the target word.

To construct identities in the category use the "jumper cable" model modified so that every other identity is either an up "↑" (the identity upon p) or down "↓" (the identity upon b). Imagine current running through the parallel identity wires. In any adjacent pair of wires the directions of the current are opposed.

Mimic the axioms from Chapter 3, Section 2.2. The first axiom is replaced with "Ⅎ(ɟ), Ⅎ(t), F(ɟ), and F(t) are arrows." The schematics of items (2) and (3) are identical except for the nature of the identity cablings. Thus in the diagrammatics, neither Ⅎ nor F lies upon the same horizontal level. Because different things are not equal, a natural family of exchanger 3-arrows is supposed.

When a multi-category is *decategorified*, k-arrows in the original multi-category are considered to be $(k-j)$-arrows, for some $j \geq 1$, in the decategorified version. So initially in FA, a non-negative integer was thought of as an object. In the *categorified* version of FA, -•- is a generating arrow from an object back to itself. Those things that were arrows become 2-arrows, and so forth.

A little more is going on in this example. Instead of one object, we have two: t and ɟ. And the arrow -•- is replaced with a reverse pair of arrows: p and b. Finally, in the current setting, none of Y, ⅄, ⊥, or ⊤ are arrows.

3. Triple arrows in $\mathcal{S}$

In addition to exchanger triple arrows, several families of generating triple arrows will be defined.

3.1. Abstract births and deaths. The triple arrows that are analogues of births and deaths are illustrated via the four film strips below. Recall from Chapter 5, Section 3.1 that a two-frame movie description of a triple arrow contains the source double arrow in the bottom still of the movie, and the target double arrow in the top frame. The triple arrows are presented side-by-side in reverse pairs. Thus, the reverse of a death is a birth.

In all the movie expressions, the glyph that will be used to indicate the triple arrows is drawn immediately to the right of the film strip. An indicator p or b that appears on the glyph suggests that the identity double arrow I_p or I_b appears in front upon the surface realization. Here an observer is located at the right of a still and is looking left; Cf. Fig. 1 (Ch. 4). Additional redundant orientation information is supplied. Rightward or leftward pointing arrows occur as serifs when the indicator is p or b, respectively. An encircled dot ⊙ coincides with the indicator p; a crossed circle ⊗ coincides with the indicator b. These are suggestive of the usual way that is used to depict normal orientations to surfaces in space.

The source of the *positive death* — depicted in the left film strip below and target of the *positive birth*, in the right film strip — is the globular composition $\begin{bmatrix} \mathsf{F(t)} \\ \mathsf{Ⅎ(t)} \end{bmatrix}$ of generating double arrow $\mathsf{F(t)}$ and its reverse $\mathsf{Ⅎ(t)}$. These double arrow are composed along the 1-arrow bp. The target of a positive death and the source of a positive birth is the identity upon the identity on t which is indicated by $||_{\mathsf{I}_\mathsf{t}}$. Geometrically this double arrow will correspond to the empty portion of 3-space that is colored by t. A similar remark holds for the identity upon the identity on ɟ which is denoted by $||_{\mathsf{I}_ɟ}$.

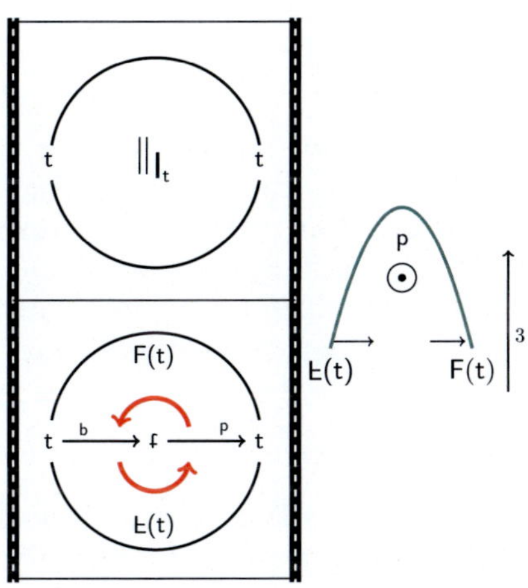

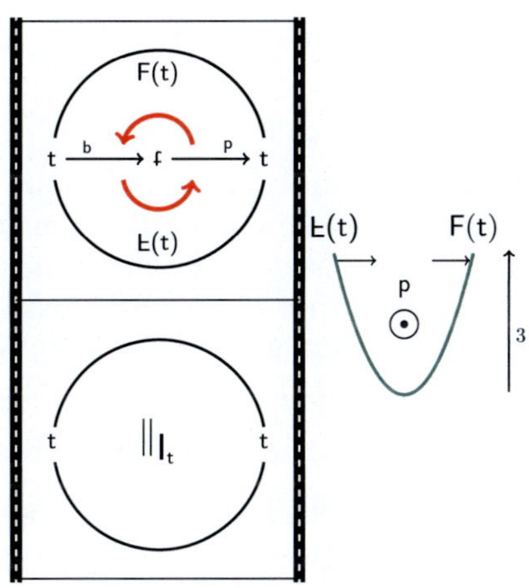

The glyphographic notation 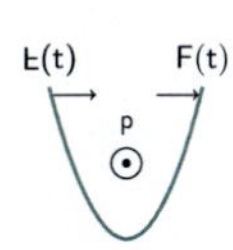 for a positive birth, for example, is meant to suggest the fold lines of an embedding of a surface as it projects to a plane. However at first reading, it is better to think of this as a glyph or an icon that serves as a name of the transition.

A *negative death* is indicated below by the film strip on the left; its reflective companion a *negative birth* is illustrated on the right. The source of the negative death and target of the negative birth is the globular composition $\left[\begin{matrix}\mathsf{F}(\mathsf{f})\\ \mathsf{Ⅎ}(\mathsf{f})\end{matrix}\right]$ of the generating double arrow $\mathsf{F}(\mathsf{f})$ and its reverse $\mathsf{Ⅎ}(\mathsf{f})$. These double arrow are composed along the 1-arrow pb.

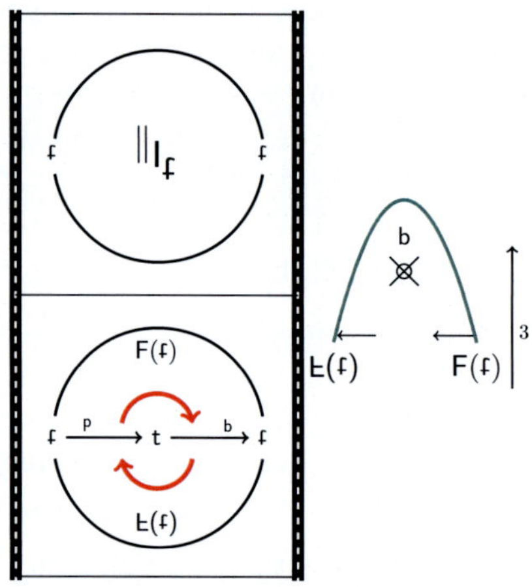

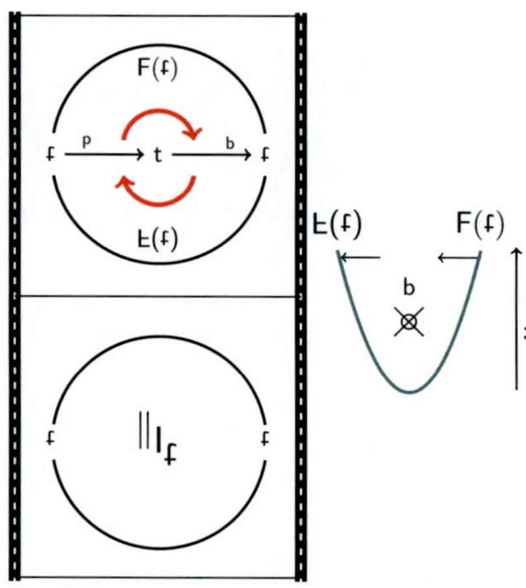

Since b and p are not only reverses, but they are intended to be weak inverses, the double arrows F and $\mathsf{Ⅎ}$ are also weak inverses, as are the birth/death triple arrows that are constructed as above. More is true. The saddle and fork type triple arrows constructed in the next section are also listed as pairs of reverses, and these too are weak inverses. On the other hand, the higher order multiple arrows that express that these are weak inverses fall within the realm of higher dimensional topological properties. As such, they don't concern us now.

3.2. Abstract forks and saddles. The triple arrow analogues of forks and saddles are illustrated within the four film strips below. The source double arrow is the bottom still in the strip, the target is at the top. The triple arrows are presented side-by-side in reverse pairs. The reverse of a fork is a saddle.

The triple arrow that is depicted below in the film strip on the left is called a *positive fork*; that on the right is its reflective companion, a *positive saddle*. The source of a positive fork and target of a positive saddle is the globular composition $\left[\begin{matrix}\mathsf{Ⅎ}(\mathsf{t})\\ \mathsf{F}(\mathsf{t})\end{matrix}\right]$ of the generating double arrow $\mathsf{Ⅎ}(\mathsf{t})$ and its reverse $\mathsf{F}(\mathsf{t})$. These are composed in a $\overset{\Delta}{\underset{\nabla}{=}}$ fashion along l_{t}. The source and target of this composition is the 1-arrow bp. The target double arrow of the fork triple arrow and the source double arrow of the saddle triple arrow is the identity $\|_{\mathsf{bp}}$.

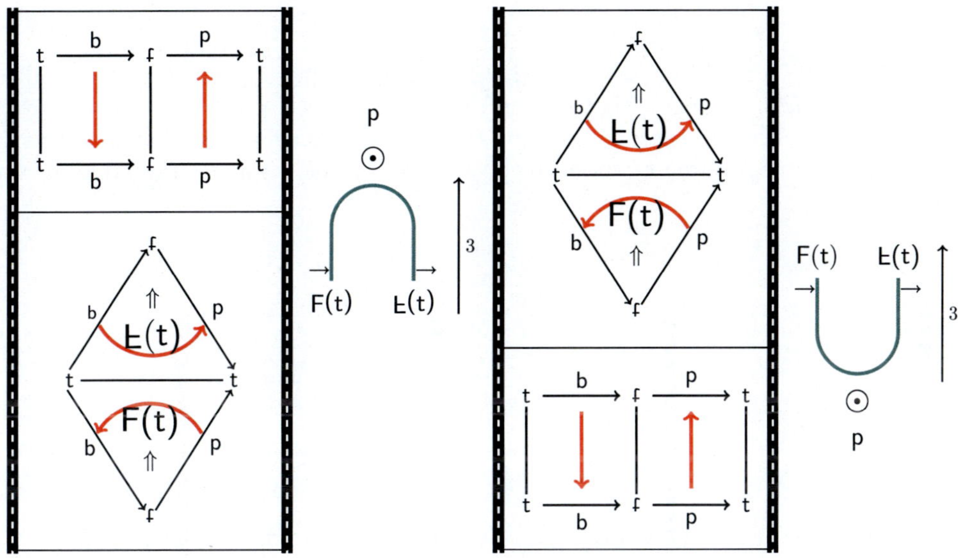

The triple arrow that appears below in the left film strip is called a *negative fork*; that on the right is its reflective companion, a *negative saddle.* The source of a negative fork and target of a negative saddle is the globular composition $\left[\begin{array}{c} \text{Ⅎ(ɟ)} \\ \text{F(ɟ)} \end{array}\right]$ of the generating double arrow Ⅎ(ɟ) and its reverse F(ɟ). These are composed in a $\frac{\triangle}{\triangledown}$ fashion along $\mathsf{I}_{\text{ɟ}}$. The source and target of this composition is the 1-arrow pb. The target double arrow of the fork triple arrow and the source double arrow of the saddle triple arrow is the identity $\|_{\mathsf{pb}}$.

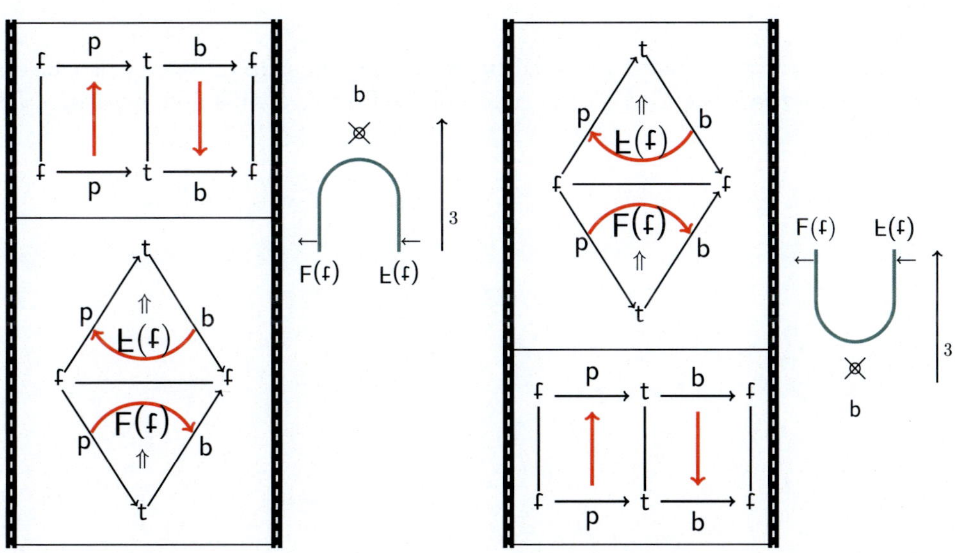

3.3. Abstract cusps. The triple arrow analogues of cusps or zig-zags are illustrated within the eight film strips below. The source double arrow is the bottom still in the strip, the target is at the top. The triple arrows are presented in reverse pairs. The reverse of an upward pointing cusp is a downward pointing cusp.

The triple arrow that is depicted below in the film strip on the left is called a *positive right down cusp*; that on the right is its reflective companion, a *positive right up cusp.* The indicator p that appears on the glyph suggests that the identity I_{p} appears as the source of the strip on the left or the target of the strip on the right.

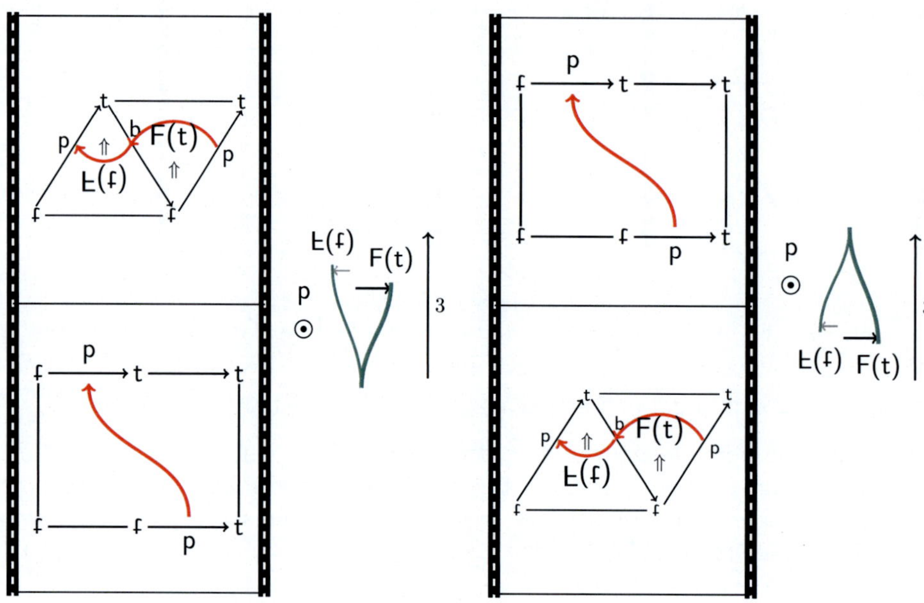

The triple arrow that is depicted below in the film strip on the left is called a *positive left down cusp*; that on the right is its reflective companion, a *positive left up cusp.*

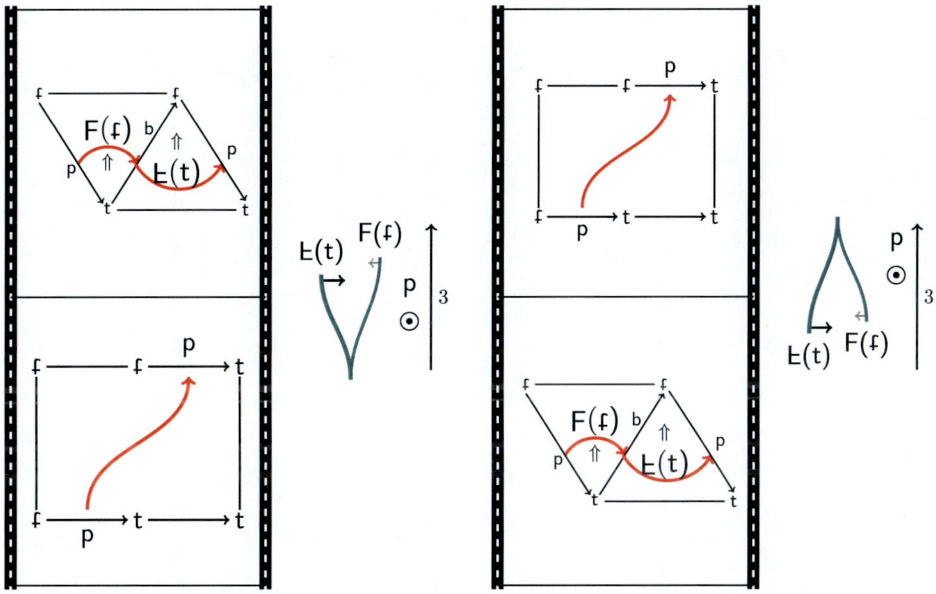

The triple arrow that is depicted below in the film strip on the left is called a *negative right down cusp*; that on the right is its reflective companion, a *negative right up cusp*. The orientation is “bad” since the source double arrow of the down cusp is the identity upon b.

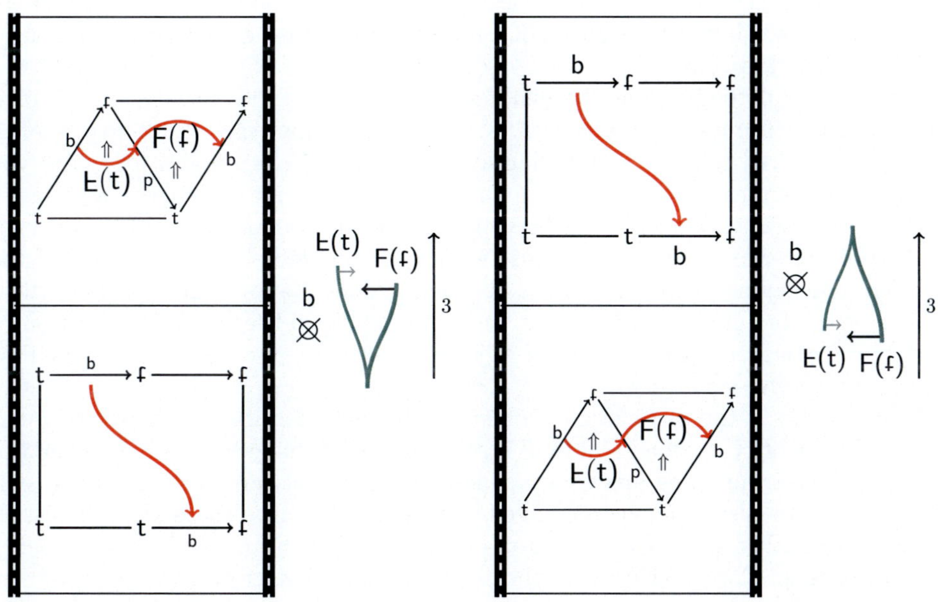

The last two generating triple arrows are shown below. That in the film strip on the left is called a *negative left down cusp*; that on the right is its reflective companion, a *negative left up cusp*.

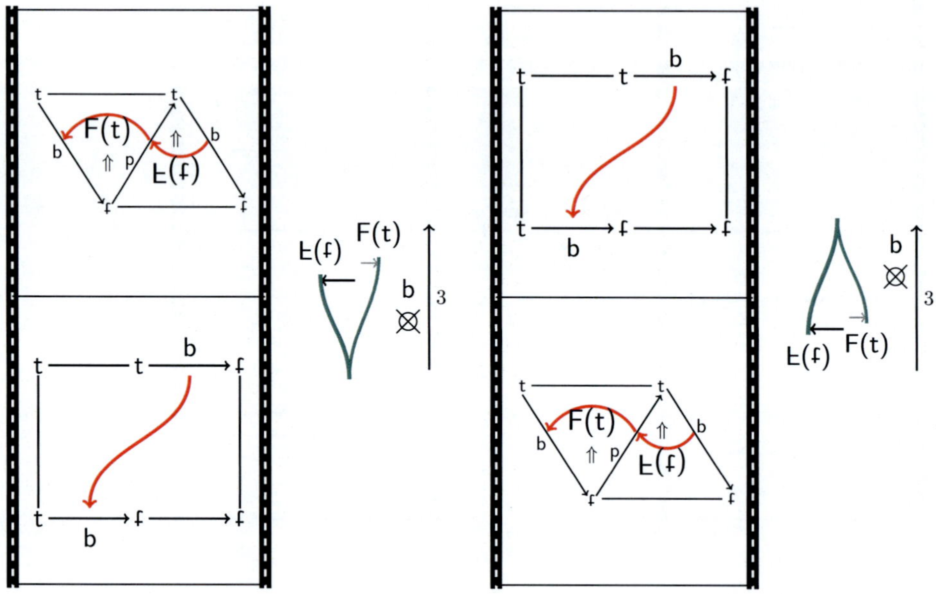

3.4. Decategorification, part 2. *The collection of* 3*-arrows in* $\mathcal{S}$ *are the arrows in a category.* This section is a close analogue to Section 2.2. A category is described. The object set is the collection of 2-arrows described in Section 2.2. Care is needed here because we are decategorifying. So the things that once were called 2-arrows will become objects of a new category. In the next paragraph, they will be called "2-arrows."

It is helpful to describe the generating 2-arrows of the form Ⅎ and F in a glyphographic form as oriented red ∩s and ∪s. In this way, the 2-arrows of $\mathcal{S}$ correspond to red-colored circuit diagrams embedded in rectangles for which the current flows alternately up or down along the bottom and top edges. An edge is adorned by a sequence such as pbpbp which is schematized as $(+)\ (-)\ (+)\ (-)\ (+)$. If the bottom edge begins (resp. ends) with a p, then so does the top edge. If the bottom edge begins (resp. ends) with a b, then so does the top edge. No two critical points of the oriented red arcs lie at the same horizontal level. The collection of such circuit diagrams form the set of objects in the category mentioned in the first paragraph above.

The arrows in the new category, of course, include identities upon any such diagram. The identity upon an Ⅎ or an F is schematized glyphographically as a vertical green arc. The remaining arrows are generated by exchangers, births, deaths, saddles, forks, and cusps. The green-edged glyphs that describe these may be nestled or veiled on either side by identities upon identities upon an alternating sequence such as bpbpbp. The description of the arrows follows that given in Chapter 3, Section 2 with the generating arrows being represented by the glyphs for exchangers, births, deaths, saddles, forks, and cusps with the identities upon ⊢ and F serving the roles of the jumper cable wires.

Compositions of these new arrows are given by vertically stacking when the source of the top arrow coincides with the target of the bottom arrow. These source and target objects are the oriented red circuit diagrams.

Since different things are not equal, the next section will describe the relationships among these arrows. First, we begin with a recap.

4. Quadruple arrows in the multi-category $\mathcal{S}$

The multi-category $\mathcal{S}$ that is being described here has the following properties:

(1) It has two objects ɟ and t.
(2) There are generating arrows p : ɟ → t — "positive" and b : t → ɟ — "bad." The arrows p and b are reverses and weak inverses of each other.
(3) The objects ɟ and t, identity arrows upon them, and the arrows p and b generate a category.
(4) Since the arrows p and b are weak inverses, there are two reverse pairs of generating double arrows ⊢(ɟ), F(ɟ) and ⊢(t),F(t). These are pairwise weak inverses.
(5) The 1-arrows and double arrows form a sesqui-category. Double arrows can be composed globularly and in a skew fashion, but horizontal compositions are mitigated by a natural family of triple arrow exchangers.
(6) Since the generating double arrows appear as weak inverse pairs ⊢(ɟ), F(ɟ) and ⊢(t), F(t), there are triple arrows (or 3-arrows) of the form (±)-birth, (±)-death, (±)-saddle, and (±)-fork triple arrows.
(7) The eight cusp like triple arrows that are listed in the previous section are present.
(8) Triple arrows are composable in globular and skew fashions.
(9) The quadruple arrows that are described in this section generate the last level of the multi-category. These are deemed to be strongly invertible. So they are used to define an equivalence relation upon the set of triple arrows. We *should* write the quadruple arrows in the form LHS $\rightarrow_4$ RHS with its reverse and strong inverse of the form LHS $_4\!\leftarrow$ RHS, but instead, we write LHS = RHS.

The generating set of quadruple arrows will be completely catalogued by using glyphs and their graphical compositions to depict them. There are six families of generating quadruple arrows: (1) critical cancelations, (2) lips, (3) beak-to-beak, (4) horizontal cusps, (5) swallow-tails, and (6) naturality isomorphisms for exchangers. The names of the first five are derived from their functorial image when considered as the projection of isotopies of surfaces that are embedded in 3-space. These singularities can be found, for example, in [**70**]. Each of the first five types of quadruple arrow will be exemplified by a movie move. Let us explain this concept.

A movie depicts a triple arrow by indicating the source double arrow in the bottom still and the target double arrow in the top. Within each still of a movie, source and target 1-arrows are read around the boundary of the disk that depicts the double arrow, and these boundaries will coincide from frame to frame. To facilitate compositions, additional identity 1-arrows may be inserted or contracted. Similarly, identity double arrows may be inserted or contracted without explicit mention from still to still. In a multiple frame movie the globular composition of triple arrows is suggested by gluing the dumpling shaped triple arrows along the double arrows that are represented by the intermediate stills in a given movie.

A *movie move* provides the one dimensional higher analogue of triple arrow. Two triple arrows are represented by a pair of movies that have coincident sources at their bottom stills and coincident targets at their tops. The source and target 1-arrows in all stills agree up to insertions or deletions of identity 1-arrows. The two movies are compared by means of a quadruple arrow that has its source triple arrow depicted by the left movie and its target triple arrow depicted by the right movie. The movie move represents a quadruple arrow that is deemed to be an equivalence and that will be denoted by "=" within the illustrations.

The triple arrow exchangers satisfy naturality conditions. There is a short list of these that generate naturality with respect to any triple arrow. These will be listed in a formal manner.

4.1. Critical cancelations. Critical cancelations occur in two types. Birth and fork triple arrows cancel as do death and saddle triple arrows. All of the triple arrows birth, death, saddle and fork have signs that are associated. So there are positive and negative birth/fork cancelations and positive and negative death/saddle cancelations. Each signed cancelation can occur in one of two ways. For example, in a positive death/saddle cancelation the death and saddle triple arrows overlap on either the double arrow Ⱡ(t) or on F(t). A careful count indicates that there are eight possible critical cancelations. In pairs, they are related to the identity on one of the four generating double arrows F(t), F(ʇ), Ⱡ(t), or Ⱡ(ʇ). They are depicted below.

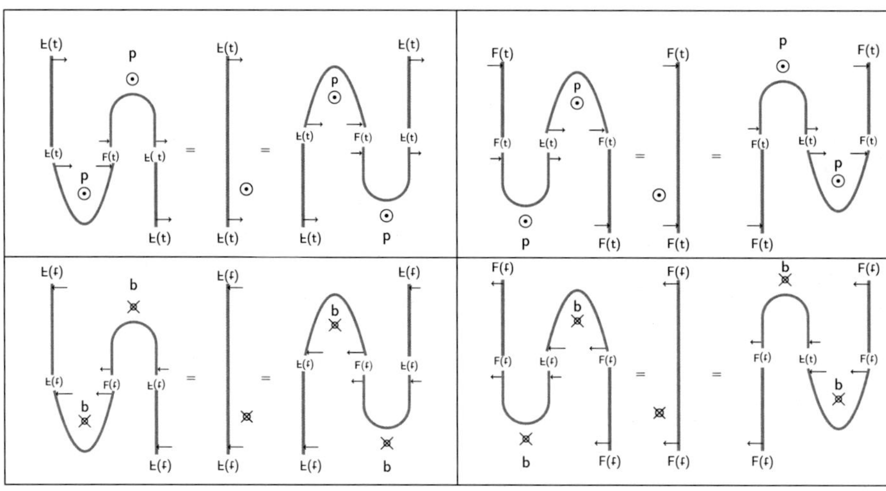

The movie move depiction of the quadruple arrow that appears as the first entry on the lower left of the table above is presented below this paragraph. On the right-hand movie, the identity upon the double arrow Ⱡ(ʇ) is illustrated via three stills each of which contains the diagram of that double arrow. On the left-hand movie, the bottom and top still are also the same double arrow, but on the top, it is augmented by some identity double arrows. These are drawn so that it is easier to decipher the relationship with the middle still. The reader may elect to replace these double arrows by means of their rectangular representatives; see, for example

Section 5.4 of this chapter. The change from middle to top is a negative fork. The two triangles of the form Δ over ∇ share the identity upon ɟ along their horizontal edges. Their union is replaced by a pair of rectangles $\mathsf{I}_{\mathsf{p}} \otimes \mathsf{I}_{\mathsf{b}}$. To move from the bottom still to the middle still a negative birth occurs. But here the identity upon the identity on ɟ is not drawn.

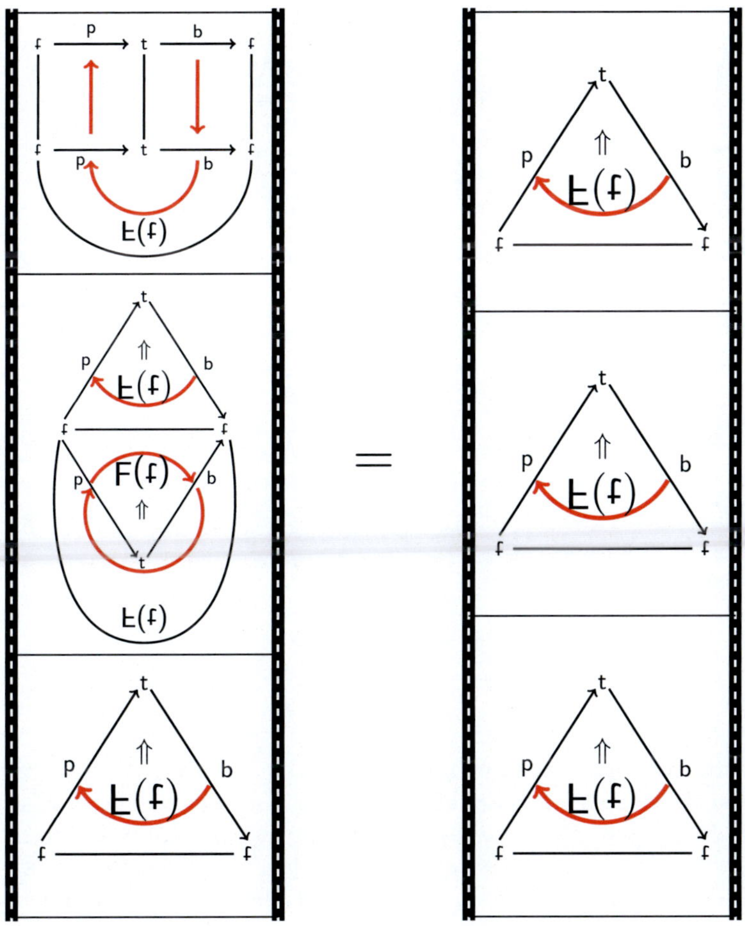

Exercise. Develop the remaining seven movie moves for the remaining 4-isomorphisms that represent critical cancelations.

4.2. The lips quadruple arrows. In the next table, the *lips* quadruple arrows are depicted. They occur because cusp triple arrow are, by definition, strongly invertible. So a negative left down cusp that lies below a negative left up cusp cancel. In this case the right-hand triple arrow is the identity upon the identity of the arrow b. There are four such possible cancelations: $(\pm$,left/right$)$. Compare these illustrations to those given in Ch. 4, Figs. 11 and 12. In the prior discussion, the representative surfaces were not oriented.

The terminology "lips" comes from singularity theory. The composition of the glyphs when rotated ninety degrees reminded an author of lips. It is our belief that the terminology originated with R. Thom.

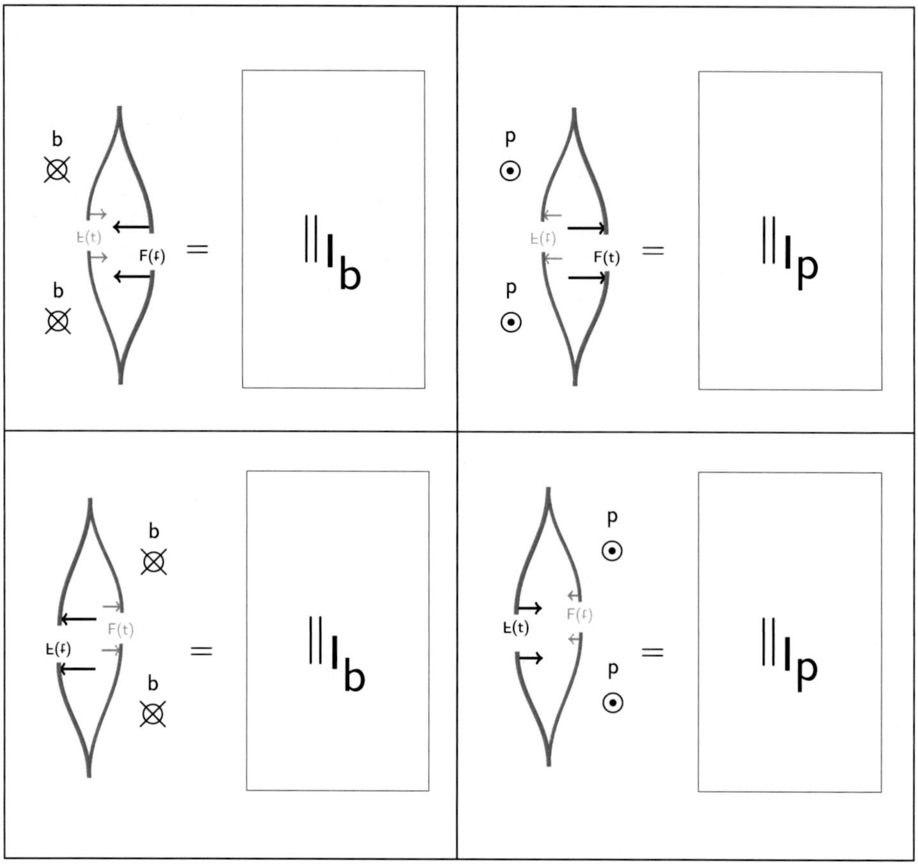

The movie move version of the quadruple arrow that appears in the lower left corner of the table above appears below. On the right-hand side movie, each still is an identity on the 1-arrow b. On the left-hand side the bottom still differs from the middle still by means of a negative left down cusp triple arrow. The difference from the middle to the top is the reverse cusp — a negative left up. The lip quadruple arrows and the beak-to-beak quadruple arrows express the strong invertibility of cusps with the reverse and inverse of an up cusp being the corresponding down cusp.

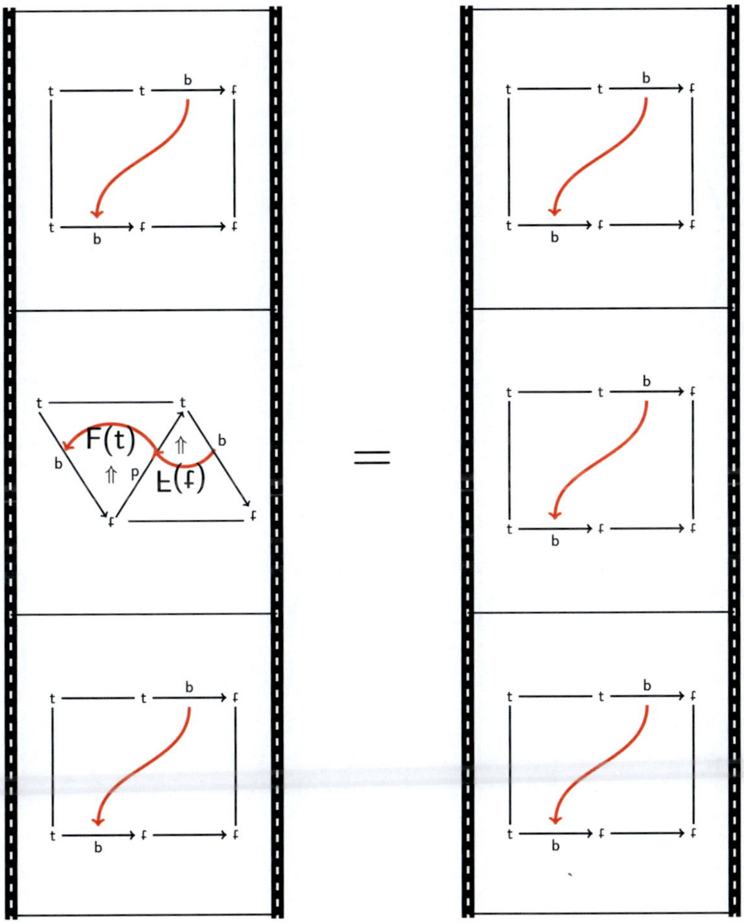

Exercise. Construct movie move versions of the remaining lips 4-isomorphisms.

4.3. Beak-to-beak quadruple arrows. The beak-to-beak quadruple arrows correspond to the remaining identities imposed by the assertions that the cusps are strongly invertible. Here the order of composition of the triple arrows is reversed from those in the lip relations. There are four types of beak-to-beak quadruple arrows: ($\pm$,left/right). Compare these illustrations to those given in Ch. 4, Figs. 11 and 12 on the right end of those figures. In those figures, orientations are absent. Alternatively, to relate the current situation to the prior one, the objects t and ɟ are identified, and the generating object -●- in FA now is considered to be a 1-arrow from the unique object back to itself. Objects in the prior situation are 1-arrows here. Arrows of the prior situation are double arrows here, etc. So the arrows' degrees from the prior situation are increased by one in the current situation.

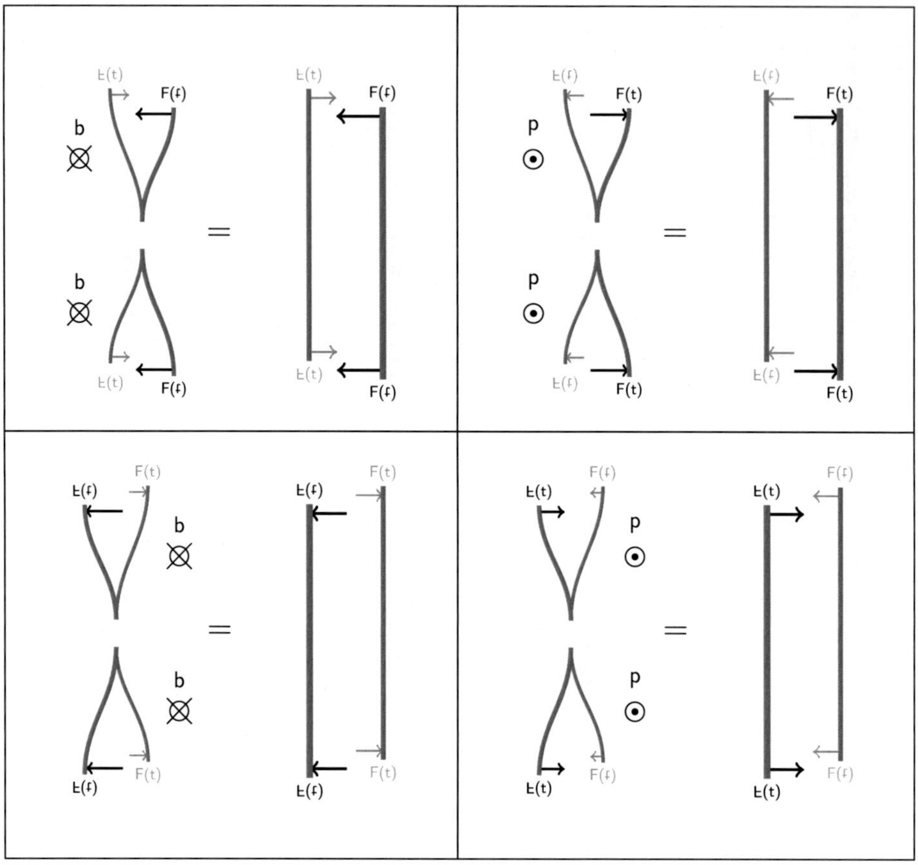

The movie move that appears below is a representation of the quadruple arrow that appears in the lower left-hand box of the table above. On the right-hand side of the movie move the identity upon a composition of F(t) and Ⅎ(ʇ) appears by indicating this composition in each of the three stills of the movie. On the left of the movie, the triple arrow that connects the bottom still to the middle still is the negative left up cusp. The triple arrow between the middle and the top is a negative left down cusp. As was mentioned above, the order of composition for the triple arrows is reversed from that given in the movie move expression for the lip quadruple arrow.

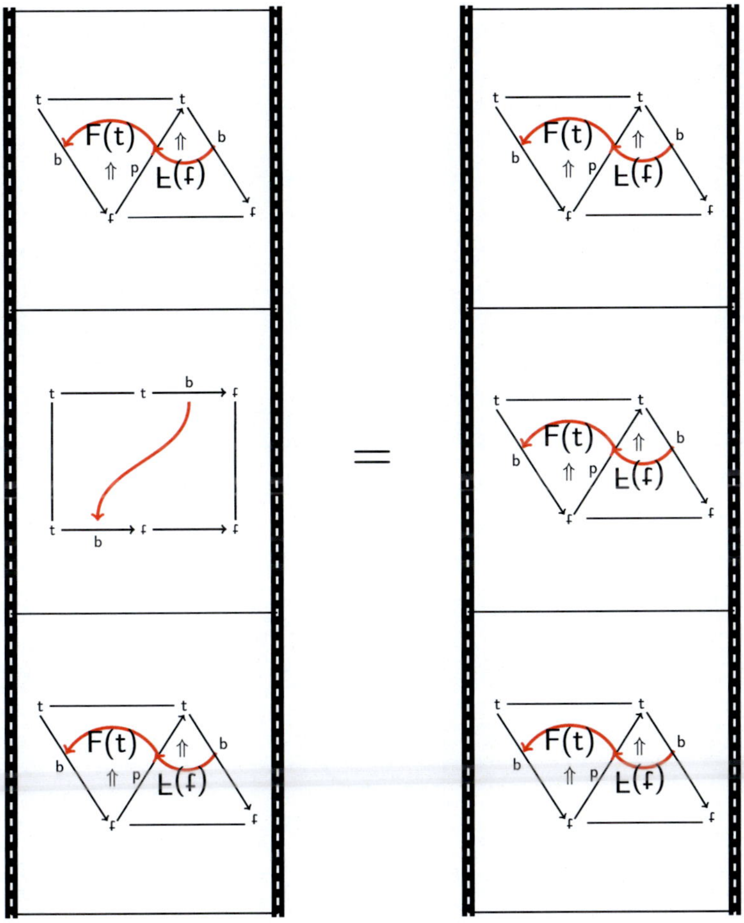

Exercise. Construct movie move versions of the remaining beak-to-beak 4-isomorphisms.

4.4. Horizontal cusps. There are eight different horizontal cusp quadruple arrows that are indicated in the two tables that appear below. Perhaps the easiest way to characterize them is to consider four pairs of 1-arrows, (pbp,p), (bpb,b), (p,pbp), and (b,bpb). Consider the first 1-arrow in any of these pairs as a source arrow, and the second as a target. In case of (pbp,p), the 1-arrow pbp can be grouped as (pb)p or as p(bp). In the first case, F(ɟ) cancels the pair (pb). So write F(ɟ) : (pb)p $\Rightarrow$ p. In the second case, apply F(t) to the pair (bp). For each source/target pair, there is an initial grouping that determines a simple double arrow to be applied to it. So there are two such groupings for each *aba*-type single arrow. Thus the eight horizontal cusps arise.

The double arrows F(ɟ) : (pb)p $\Rightarrow$ p and F(t) : p(bp) $\Rightarrow$ p can be related by means of different triple arrows. The two illustrations on the right-hand side of the first table provide the triple arrows that relate them. The triple arrows are either of the form down cusp followed by a fork or a saddle followed by an up cusp. The quadruple arrows that relate the differing triple arrows appear below.

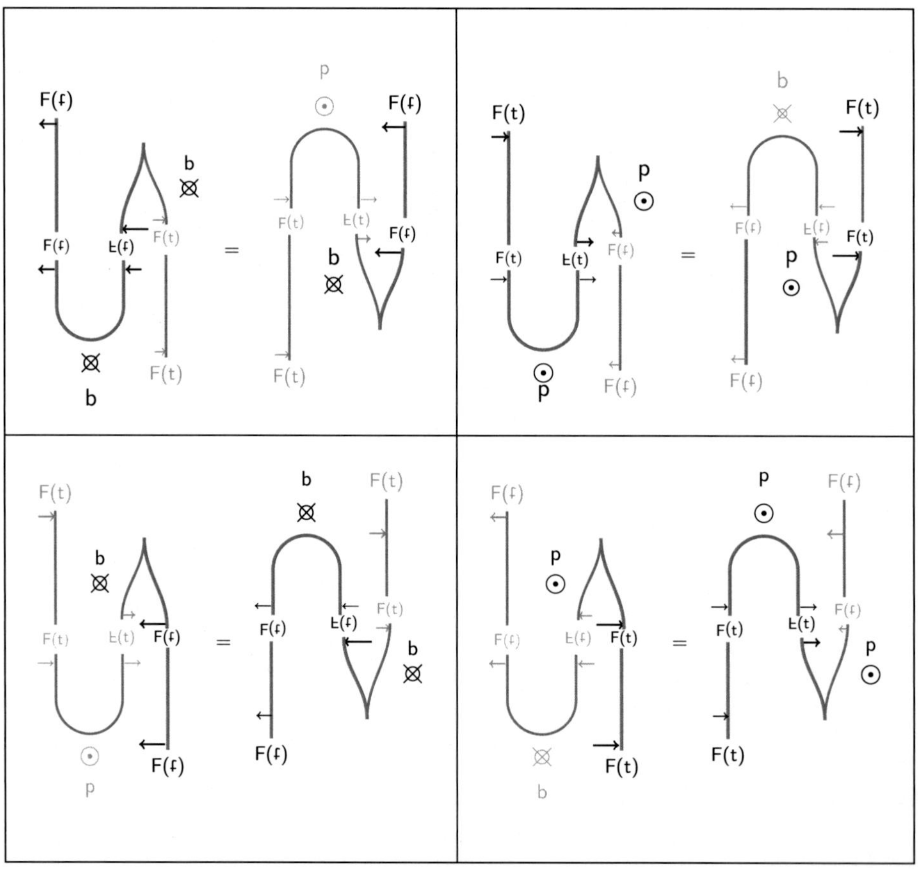

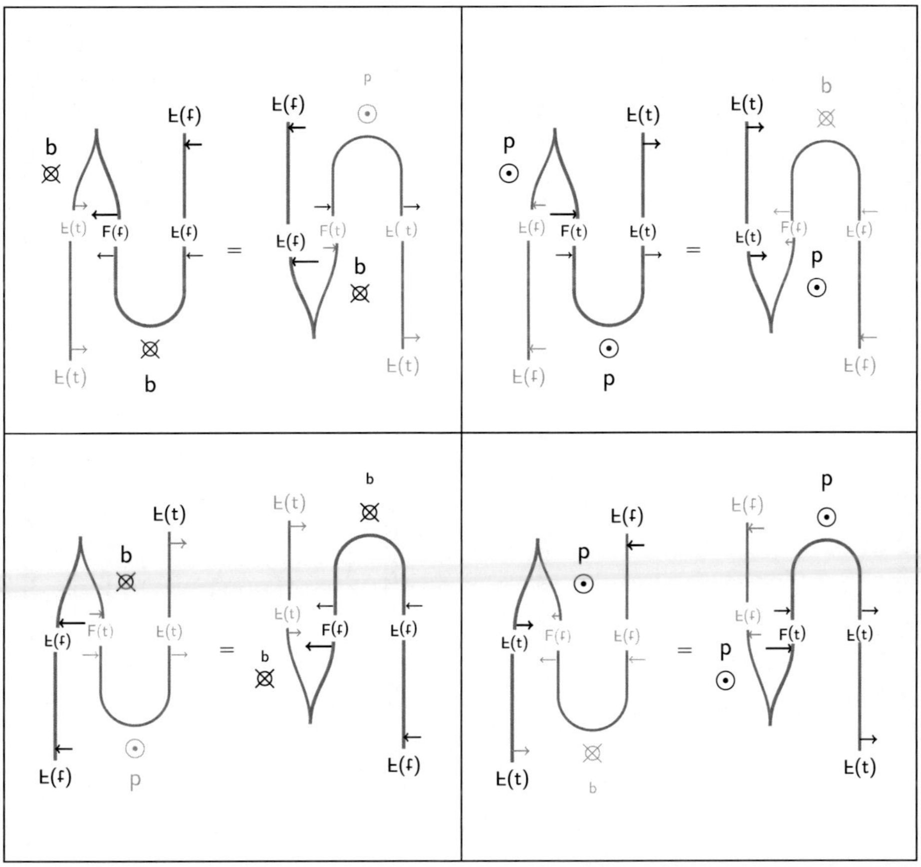

The movie move that fully articulates the horizontal cusp quadruple arrow in the upper right entry of the first table appears below the tables. We analyze the two movies. The source double arrow in either movie corresponds to the grouping ((pb)p,p). That is, the double arrow F(ɟ) : (pb)p $\Rightarrow$ p appears as the source of either movie while F(t) : p(bp) $\Rightarrow$ p appears as the target.

The middle still of the movie on the left is related to the bottom still by means of a positive saddle. It is related to the top by means of positive left up cusp. On the right-hand side, the middle still is related by a positive right down cusp, and the top is related by a negative fork.

Exercise. Develop movie moves for the remaining horizontal cusp 4-isomorphisms.

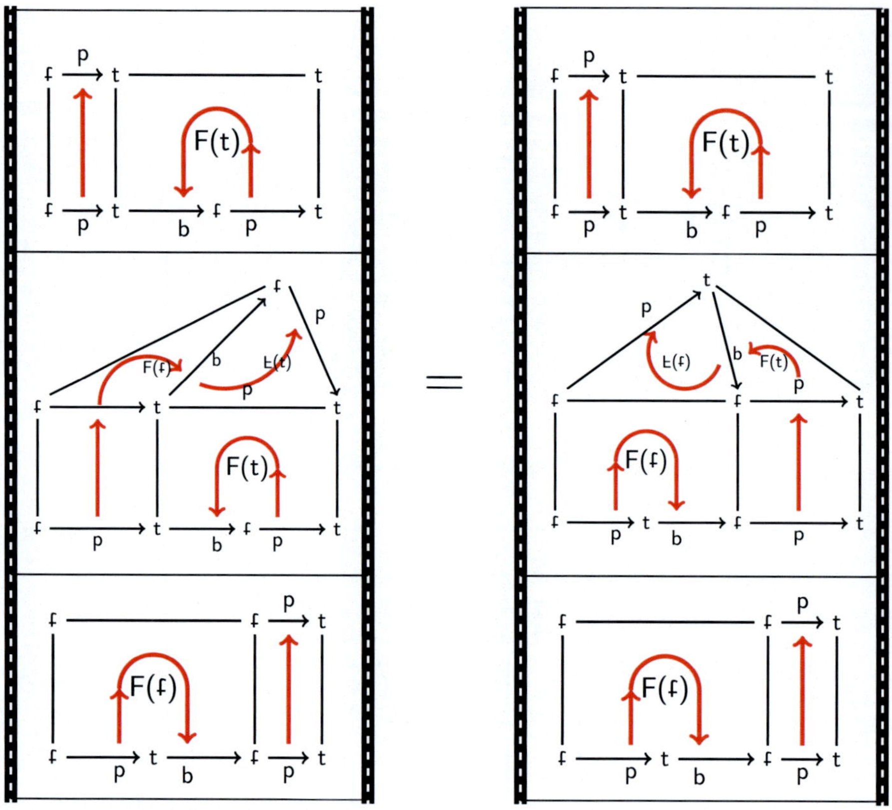

4.5. Swallow tails. Compare the eight swallow-tail moves depicted in the table below to the swallow-tail that appears in Ch. 5, Fig. 1. That swallow-tail is an unoriented version of the two swallow-tails that appear on the left hand side of the top row in the table below. The reader is also referred to Chapter 5, Section 1 which motivates the swallow-tail by means of algebraic considerations. But double arrows of FA correspond to triple arrows here. The description that was given of the multi-category FA was a decategorified version of the multi-category given here. See Section 2.2 for more discussion of decategorification.

Within *the current* decategorification, a quotient category is also involved. The eight swallow-tails here depend on the four variations of F and the choice of prioritization that was described in Chapter 5, Section 1.

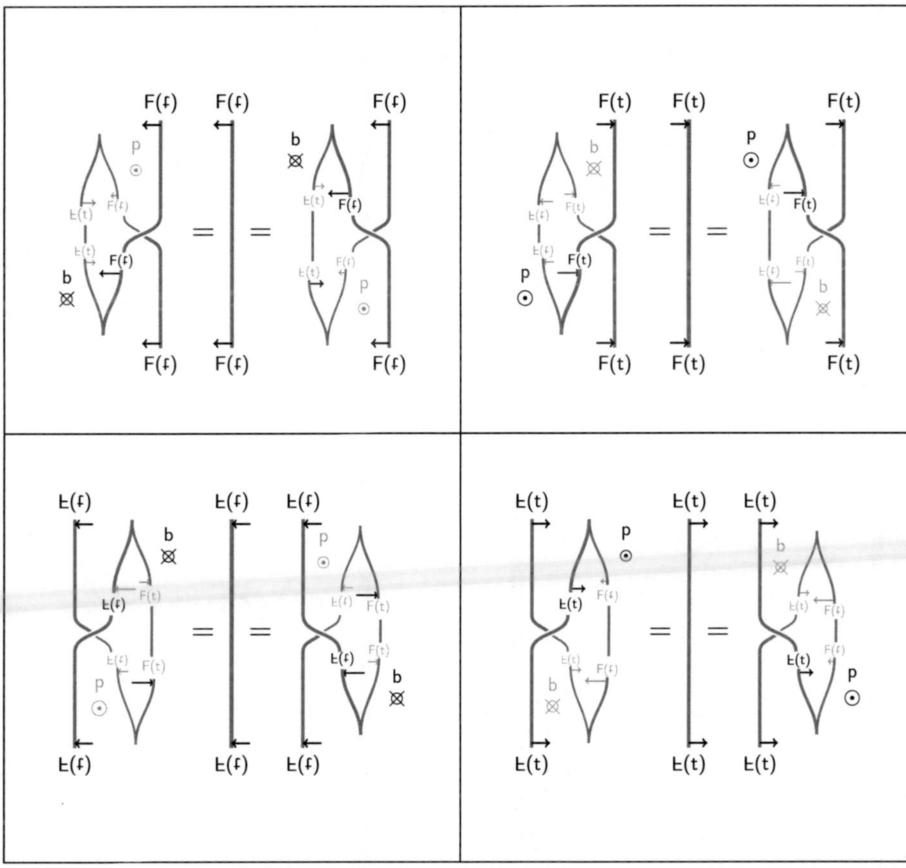

The movie move description below corresponds to the swallow-tail on the left of the top row, right column of the table above. The description given in Chapter 5, Section 1 about the order of operations applies here, and the change in tilt of the trapezoids in the two middle stills is meant to suggest that priority. Alternatively, the trapezoids are tilted to avoid the double arrows being horizontally aligned. And tilting back and forth is a short-hand notation for the application of the exchanger triple arrow.

Exercise. Develop movie move versions of the remaining swallow-tail 4-isomorphisms.

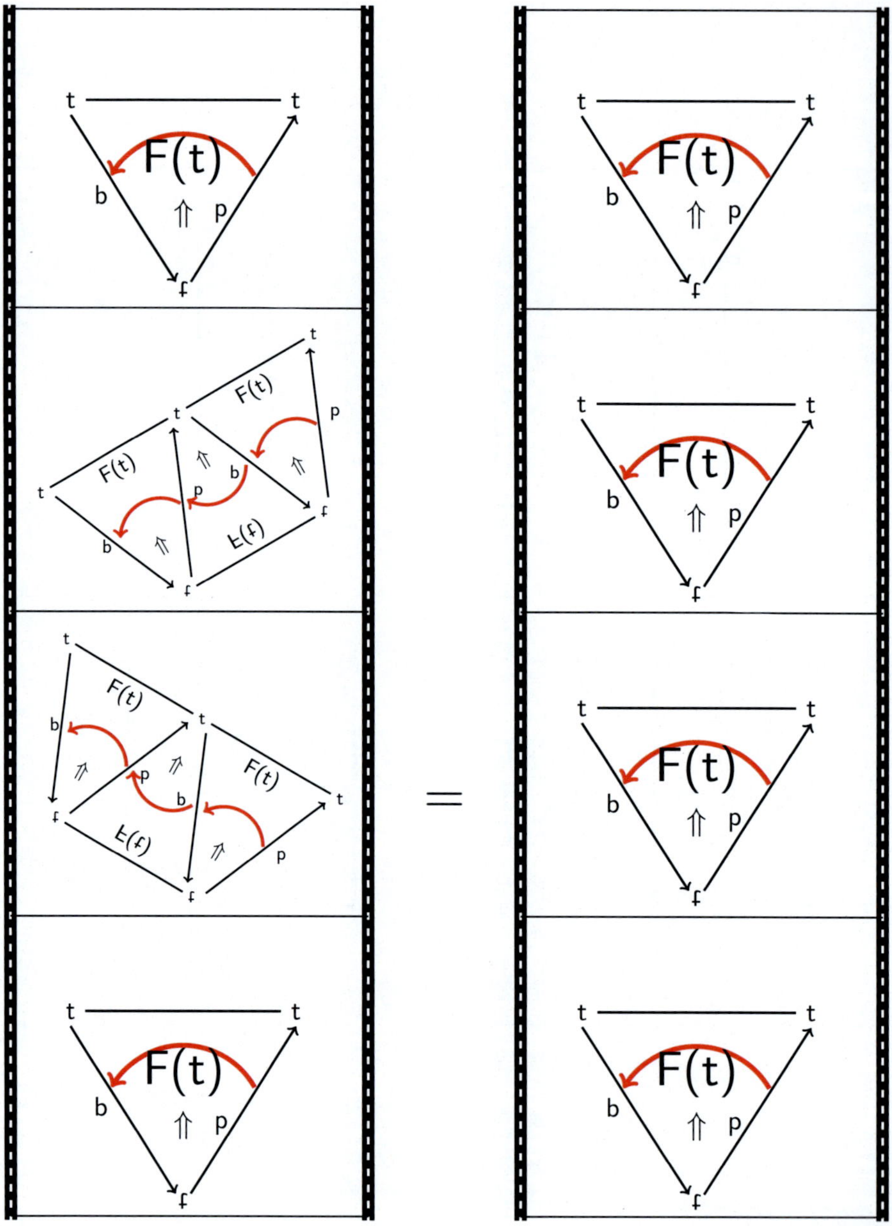

4.6. Naturality of the exchanger. The triple arrow exchanger X is a natural family of triple arrows that is used to interchange the priority of double arrows. As in the previous section, it can be schematized by rotating the disk containing a pair of interchangeble double arrows. The naturality condition of the exchanger — described in Chapter 3, Section 2.3 as "change followed by exchange is the same as exchange followed by change" — is easy to express using the glyphographic representation of the birth, death, saddle, fork, and cusp triple arrows.

In the illustration below, the box labeled Q is meant to indicate any triple arrow in the form of a birth, death, saddle, fork, or cusp. The green lines represent the identity triple arrows upon any of F(t), F(ք), Ⱡ(t), or Ⱡ(ք). Normal orientations in the form of arrow serifs are omitted since the naturality of the exchangers X apply regardless of the orientations. The illustrations below as well as that given at the bottom of Fig. 1 (Ch. 3) demonstrate naturality of X(±) in all cases.

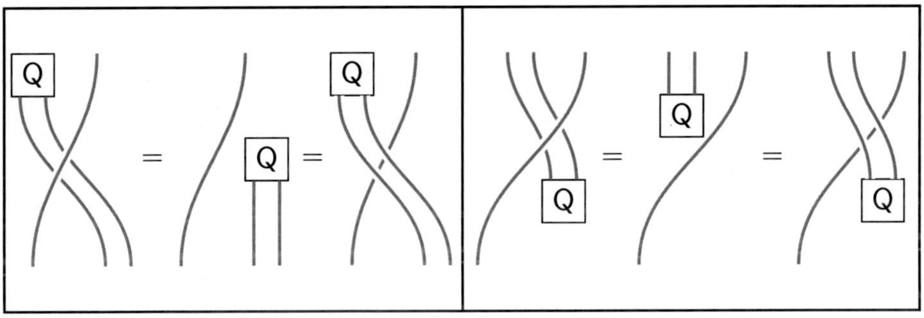

4.7. The quadruple arrow exchanger. As a final axiom to the multi-category $\mathcal{S}$ that is being constructed in this chapter, we suppose that there is a natural family of quadruple arrows that allow the order of evaluation for triple arrows to be exchanged. Graphically, the expressions for the exchangers will coincide with those given for 1-arrows and double arrows as presented in Chapter 3, Section 2.3. See also Figs. 1 and 2 of Chapter 3.

Here is an additional metaphor for the quadruple arrow exchanger. A movie representation for the composition of a pair of triple arrows consists of three stills. The bottom still differs from the middle still on a 2-disk region. Similarly, the middle still differs from the top. Suppose these two disks in the middle region do not overlap. Then imagine the two triple arrows as if they are pistons in a two cylinder engine. The exchange of the triple arrows coincides with a half-cycle of the engine. The pistons exchange height in the same way the quadruple exchange occurs.

The authors are accustomed to using the phrase, "distant critical points commute." Upon the functorial image of $\mathcal{S}$ into the multi-categorical description of smooth surfaces that are embedded in 3-space, the quadruple exchanger is the interchange of distant critical points. Let us turn now to describing that functor.

5. Functorial equivalent multi-categories

The multi-category $\mathcal{S}$ has two objects t and ք, 1-arrows generated by the weakly invertible arrows b and p, double arrows generated by F(t), F(ք), Ⱡ(t), and Ⱡ(ք), triple arrows generated by births, deaths, saddles, forks, cusps, and natural exchangers. The triple arrows are related by the quadruple arrows that are called critical cancelations, lips, beak-to-beak, horizontal cusps, swallow-tail and exchangers. Section 5.4 of this chapter will indicate how to construct the quadruple arrows from a categorical viewpoint. Now a functorial equivalence is established between $\mathcal{S}$ and a multi-category that is associated to isotopy classes of properly embedded surfaces in 3-dimensional slabs of space of the form $\mathbb{R}^2 \times [0, 1]$.

Section 5.1 outlines the definition of a multi-functor. The notion of a multi-category that we have been working with in an informal manner is presented. Section 5.2 describes the multi-category that is associated to properly embedded surfaces. Section 5.3 constructs the categorical equivalence. Section 5.4 recapitulates the quadruple arrows from a categorical point of view.

5.1. Multi-categories and multi-functors. Recall from Chapter 3, Section 1, that a category has objects and 1-arrows. In a sesqui-category, the set of arrows between any two objects is also a category. In this case, the morphisms between a pair of 1-arrows are called double arrows. A double arrow is a generalization of a commutative diagram. However, a double arrow is usually not an equality instead it is a method to compare two 1-arrows that have the same source and target.

Double arrows are composed globularly. Horizontal composition[2] is strictly disallowed, and instead it is mitigated by a natural family of triple arrows that are called exchangers. The collection of double arrows between a pair of 1-arrows is also a category, and the morphisms that connect a pair of double arrows are called triple arrows. These are also composed globularly, and induce quadruple arrows that are exchangers. When triple arrows form a category, the arrows between globes are quadruple arrows.

It is convenient to express a composition of double arrows as a disk with two oppositely oriented segments along its boundary as source and target 1-arrows. In this way a triple arrow is represented by the interior of a ball that is bounded by a pair of hemispherical disks. Alternatively, in the multi-categories that are being discussed, a triple arrow can be represented as a movie in which each still of the movie contains a disk that represents a double arrow. The triple arrow, then, represents the difference between successive stills. There can be sub-disks in each still that differ while their complements remain unchanged.

So far, quadruple arrows are experienced as equivalences between a pair of movies and are dubbed movie moves. But be aware, arrows are not always equivalences.

A *multi-functor* or simply a *functor* between multi-categories is method of comparing them. A functor F from the multi-category A to the multi-category B assigns an n-arrow $F(a) \stackrel{F(f)}{\to_n} F(b)$ in B to any n-arrow $a \stackrel{f}{\to_n} b$ in A. Here $n = 0, 1, 2, \ldots$, and a 0-arrow indicates an object. As part of the assertion here, the source, a, and target, b, of an n-arrow, f, are sent to the source, $F(a)$, and target, $F(b)$, of the n-arrow $F(f)$. Relations among n-arrows in A map to relations among n-arrows that hold in B.

5.2. Properly embedded surfaces. A *surface, M, with boundary, ∂M*, is a compact Hausdorff space for which every point has a neighborhood homeomorphic to the neighborhood of a point in the closed upper half-plane $\mathbb{R}^2_{0\le y} = \{(x, y) \in \mathbb{R}^2 : 0 \le y\}$. The boundary, ∂M, consists of those points that map to $\{(x, 0) \in \mathbb{R}^2\}$. Two such neighborhoods in M that overlap have separate homeomorphisms. Starting in the image in $\mathbb{R}^2_{0\le y}$ of one such function, moving up to the surface via the

[2]Chapter 4, Section 1.2

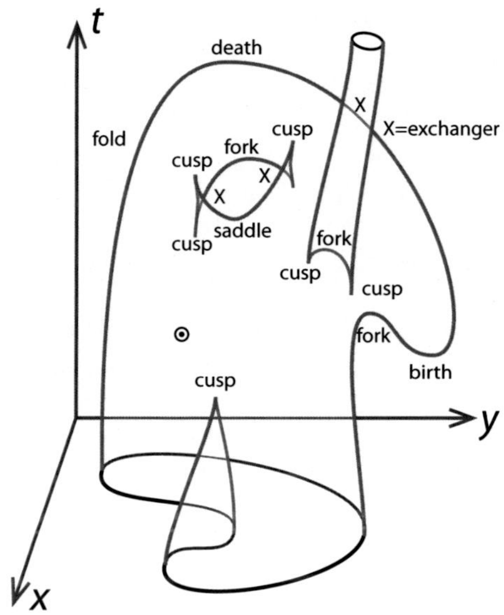

FIGURE 2. A properly embedded surface and its chart description

inverse homeomorphism, and going back to the plane provides a homeomorphism between open subsets of $\mathbb{R}^2_{0\le y}$. The surface M is *smooth* if these overlap functions are infinitely differentiable. See [**29**], for a thorough discussion of n-dimensional manifolds.

A detailed description of properly and smoothly embedded surfaces in $\mathbb{R}^2 \times [0,1]$ follows. Recall that the unit interval is the set $[0,1] = \{t \in \mathbb{R} : 0 \le t \le 1\}$. For M to be *properly embedded*, the boundary ∂M is decomposed into the union two subsets $(\partial M)_0 \subset \mathbb{R}^2 \times \{0\}$ and $(\partial M)_1 \subset \mathbb{R}^2 \times \{1\}$. Be aware, one or both of these subsets may be empty. Moreover, the boundary, as a subset of M has a normal direction that points inward at $(\partial M)_0$ and outward at $(\partial M)_1$. These normal directions map non-trivially in the positive t direction of $((x,y),t)$. Furthermore, the interior, $M \setminus \partial M$ is contained in the interior, $\mathbb{R}^2 \times (0,1)$, of $\mathbb{R}^2 \times [0,1]$. Briefly, the surface is mapped into space as nicely as you might imagine. An illustration is furnished in Fig. 2 that will be used to guide the discussion.

We may assume that $M \subset \mathbb{R}^2 \times [0,1]$ lies within the region of space for which $0 \le x$, $0 \le y \le N$, and $0 \le t \le 1$ for some sufficiently large integer N. Furthermore, we may assume that $M \subset \mathbb{R}^2 \times [0,1]$ is in a good position, as described in the next few paragraph, with respect to projections onto the (y,t)-plane and onto the vertical t-axis. These assumptions are not restrictive since M may be adjusted by some small perturbations so that they are true. Many facts about general position maps and surface singularities that are found, for example, in [**11, 24, 33, 66, 70, 77**] underlie these assumptions.

Consider an embedded surface M in $\mathbb{R}^2 \times [0,1]$, and let $g : M \to \mathbb{R}^2 \times [0,1]$ denote the inclusion map. For each point of M consider a local coordinate system (u,v) about the point. The inclusion map g is written locally as $g : (u,v) \mapsto (x(u,v), y(u,v), t(u,v))$. First assume that for almost all points of M, the differential of the projection $(u,v) \mapsto (y(u,v), t(u,v))$ onto the (y,t)-plane is of full rank. This assumption is justified by means of perturbing the map appropriately (See [**24**]). Thus many coordinate patches of M map diffeomorphically onto open subsets of the (y,t)-plane. Such coordinate neighborhoods are the portion of the surface that a viewer might see, and they are also the portions of the surface that are occluded by the visually apparent pieces.

The points of M for which this projection is singular form the *fold set* of the embedding. There is an ambiguity in the language that describes folds. We can consider the image of the folds in the (y,t)-plane, or the pre-image in M. For precision, the *image folds* and *pre-image folds* can be distinguished.

The pre-image folds form a properly embedded collection of arcs and loops in M. The projection of M onto the vertical is given locally by $(u,v) \mapsto t(u,v)$. Parameterize the pre-image fold set. Then the projection restricted to the pre-image fold set provides a real-valued function of a real variable whose critical points may be assumed to be non-degenerate. So the first derivative vanishes, but the sign of the second derivative remains unchanged at the critical point. These critical points correspond to the surface singularities of $(y(u,v), t(u,v))$ that are called cusps, births, deaths, saddles, and forks. At a *cusp*, a tangent vector along a preimage fold maps to a vector that is parallel to the x-axis. Meanwhile at a saddle, fork, birth, or death, the tangent to the pre-image fold is parallel to the y-axis. Fig. 2 contains examples of all of these. In the latter case, the critical points of the fold set are also critical points of the surface map $(u,v) \mapsto t(u,v)$.

At non-critical points of the pre-image fold set, the tangent to the fold has a component in the t-direction. Two distinct segments on the fold set can cross in the (y,t)-plane. The image folds are assumed to have isolated transverse crossings. Otherwise they do not intersect. The crossings of folds are called *fold exchanges* and are labeled X in the figure.

Fold exchanges, births, deaths, saddles, forks, and cusps may be assumed to all occur at different times. Here the variable t is considered time-like. That is to say that distinct critical points have distinct critical values.

To summarize, let p denote the projection of $\mathbb{R}^2 \times [0,1]$ to the (y,t)-plane, and let $g : M \to \mathbb{R}^2 \times [0,1]$ denote the inclusion map. The composition $p \circ g$ is chosen to be a smooth proper map from the surface M onto $\mathbb{R} \times [0,1]$ that has generic isolated singularities which occur at distinct times.

According to the Alexander duality theorem [**27**], the surface separates the slab of space $\mathbb{R}^2 \times [0,1]$ into disjoint pieces. These can be colored, say black and white, so that an arc that starts in the black region, and that passes once through M, ends in a white region, and vice versa. To standardize, we will label any region that is white to be a "from"-region while the black colored regions are the "to"-regions. An oriented arc that intersects the surface once and that goes from white to black is said to be "positive." One that goes from black to white is "bad." So the black and white coloring induces a normal orientation to the surface. The oriented normals point into the black regions of space.

A small $\odot$ is drawn upon a visually apparent piece of M in Fig. 2. This is meant to indicate an outward pointing normal direction. It induces normal pointing serifs (not depicted) to the outer most folds. Furthermore by convention, the origin $(0,0,0)$ in this example is colored black. That coloring is omitted for obvious reasons. A small crossed circle ⊗ indicates an inward pointing normal direction. No such label appears in the Figure.

5.3. The equivalence. Here are some last fragments of terminology. Two embeddings g_a and g_b of the surface M are said to be *properly isotopic* if there is an isotopy between them that is constant on a neighborhood of the boundary ∂M. A *homogeneous triple arrow in* $\mathcal{S}$ is one for which its source and target double arrows have source and target 1-arrows that are the identities upon ʇ or t. If the triple arrow is a composition, then the source and target 1-arrows of any intermediate double arrow are also identity 1-arrows. Triple arrows that are related up to an application of quadruple arrows will be said to be *equivalent.*

THEOREM 6.1. *A generic properly and smoothly embedded surface M in $\mathbb{R}^2 \times [0,1]$, that has boundary $\partial M = (\partial M)_0 \sqcup (\partial M)_1$ with $(\partial M)_i \subset \mathbb{R}^2 \times \{i\}$, for $i = 0, 1$, corresponds to a globular composition of homogenous triple arrows in $\mathcal{S}$.*

Properly isotopic surfaces correspond to equivalent homogenous triple arrows in $\mathcal{S}$. Moreover, given a globular composition of homogeneous triple arrows in $\mathcal{S}$, a representative proper embedding can be constructed. Equivalent triple arrows give rise to properly isotopic surfaces.

PROOF. The Fig. 2 that is given in Section 5.2 outlines the proof of the first statement. Perturb M slightly so that the generic surface singularities of the projection of M into the (y,t)-plane are isolated and they occur at distinct times. Associate to each surface singularity, the triple arrow of the same name, where the incoming and outgoing fold lines are associated to the identities upon F(t), F(ʇ), Ⅎ(t), or Ⅎ(ʇ). Furthermore use the normal orientation of the surface to induce serif arrows along these folds.

This process is demonstrated in some finer detail by examining the double arrows that are represented in a few intermediate non-critical slices in the illustration given below. These are arranged as a movie on the left, and then reinterpreted as a movie in $\mathcal{S}$ on the right, with the glyphs associated to the differences between stills included. The bottom-most still is taken just below the lowest fork on the right of Fig. 2.

The arrangement of the glyphs on the right is given to suggest that they can be inter-connected via identities upon F(ʇ) and Ⅎ(ʇ). Segments in the glyphs are thinned within the illustration because they are occluded by the identity upon the identity on p. Other identities on double arrows F(t) and Ⅎ(t) are not drawn. But they can be filled in by comparing the surface view to the triple arrow view in $\mathcal{S}$. We recall that a still in a movie is rotated clockwise by ninety degrees when converting to a surface or a glyphographic representation.

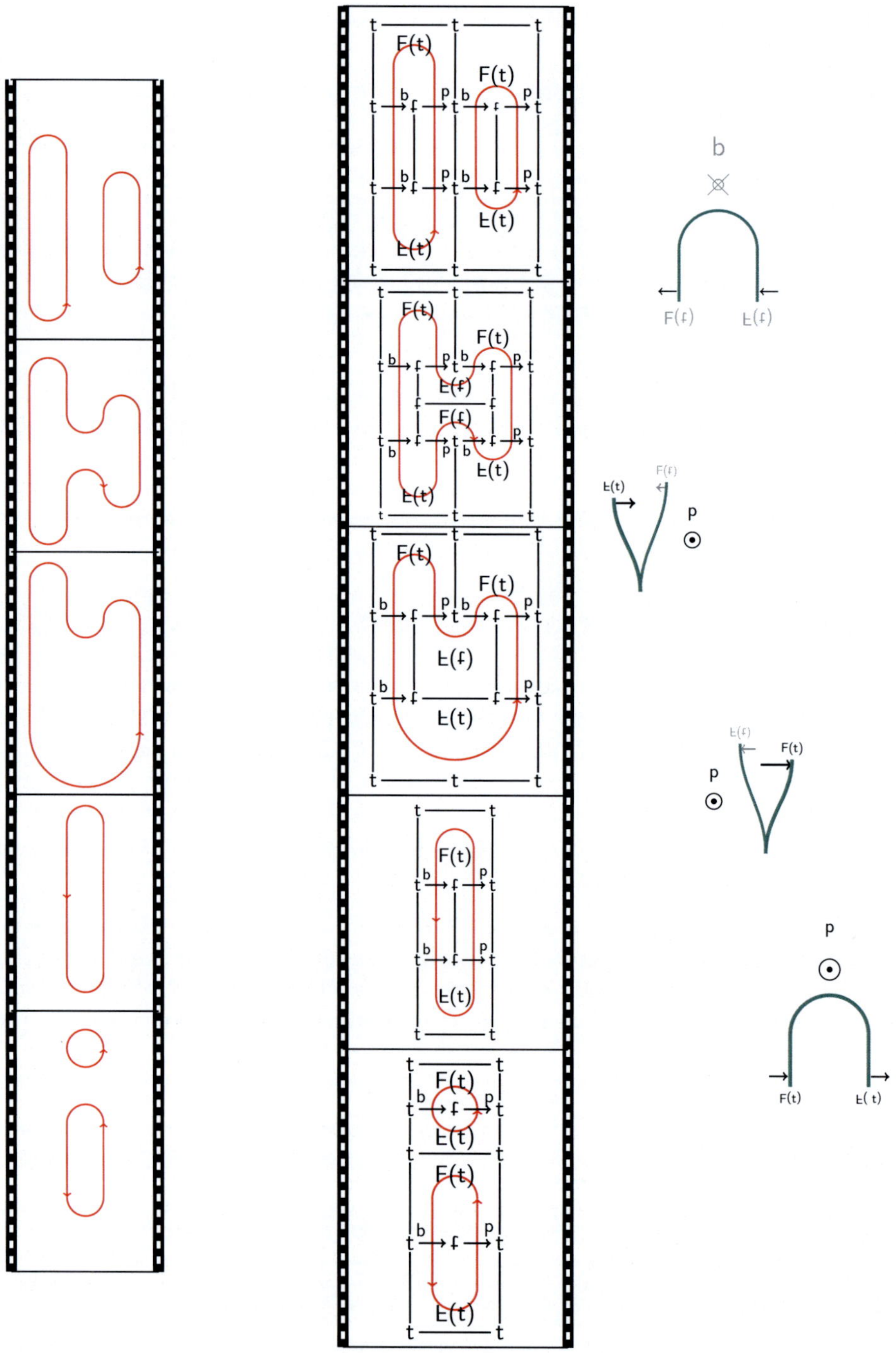

Here the method for constructing the triple arrows is described. First cut M by a sequence of (x, y)-planes that pairwise lie on either side of a critical point, fold exchange, or cusp. Any such plane intersect M in a collection of closed curves. The y-axis serves as a generic height function on these curves in such a plane. That is, all critical points of the curves projected towards y are non-degenerate, and they each have distinct critical y-values.

In each cross-sectional (x, y)-plane we can associate a composition of double arrows. The closed curves in a horizontal cross-section separate the (x, y)-plane into a collection of regions. Label the outside region with t, and alternate labels between t and ʇ upon crossing the boarder between two regions. Now consider a pair of points in the outside region on the same horizontal (y = constant) level below the lowest local minimum in the y direction. These points are both labeled t. A horizontal arc between them represents an identity arrow t —— t. Immediately above the local minimum draw two left-to-right arrows $\mathsf{t} \xrightarrow{\mathsf{b}} \mathsf{ʇ} \xrightarrow{\mathsf{p}} \mathsf{t}$. Then orient the local minimum in an anti-clockwise direction, and connect the horizontal arrows by vertical segments that correspond to the identity upon t. Then this critical point corresponds to the double arrow Ⅎ(t).

A local minimal point for which the outside region is labeled ʇ will correspond to a double arrow Ⅎ(ʇ) by a similar process. Local maxima determine double arrows of the form F(t) or F(ʇ) depending on the label outside the maximum. The fourth still from the bottom in the illustration from the previous page contains all four critical point types. In general, each optimal point can be oriented by considering the labels on either side of it.

Meanwhile, arcs with no critical points upon them correspond to identity maps on p or b. In this way, each non-critical cross-sectional plane is decomposed into a collection of rectangles, each of which corresponds to a double arrow in $\mathcal{S}$.

A critical level or cusp in the time-like direction, then, induces a triple arrow whose name or identifying glyph corresponds the type of oriented critical point or cusp that occurs at the critical level. The movie above contains several examples of critical changes.

Proper isotopies of embedded surfaces in $\mathbb{R}^2 \times [0, 1]$ can be chosen to be generic. So these can be decomposed into a finite sequence of changes of the form critical cancelation, lip, beak-to-beak, horizontal cusp, swallow-tail, and critical exchanges [**33, 66**]. The movie move that describes such a change at the surface level, translates mutatis mutandis into the quadruple arrows in $\mathcal{S}$.

In the other direction, both the glyphographic descriptions of the triple arrows given in Section 3.1 of this chapter, and the inclusion of the thin (red) arcs in the disk representations of the double arrows given in Section 2 can be used to reconstruct an embedded surface from a homogeneous triple arrow. Isotopies are generated by the quadruple arrows in $\mathcal{S}$. This completes the proof. □

Exercise. Give a detailed set of cross-sections for the rest of the example that is depicted in Fig. 2. From these cross-sections, explicitly construct the corresponding movie and glyphs that describe the 3-arrows. Can you apply swallow-tail and critical cancelations to simply the image of the embedding?

The alert reader might be asking, "Where did we need a multi-functor?" Or more precisely, "What is the multi-categorical structure that is associated to an embedded surface?" We address these concerns. Suppose, as usual, that there is an embedded surface $M \subset \mathbb{R}^2 \times [0, 1]$. Imagine space decomposed in a rectilinear fashion. Then think of the components of the complement of the surface M as a collection of objects. An identity 1-arrow is a straight arc that is parallel to a coordinate axis and does not intersect the surface. An identity double arrow that is an identity on an identity 1-arrow similarly does not intersect M. An identity triple arrow on this type of identity double arrow also does not intersect M. A non-identity 1-arrow intersects the surface transversely in at least one point. The identity double arrows on these intersect the surface transversely in arcs. The identity triple arrows intersect the surface in time-vertical sheets. Non-identity double arrows correspond to critical points of curves; identity triple arrows of these are rectilinear neighborhoods of folds. The critical points, cusps, and fold exchanges of $g(M)$ are the generating triple arrows. Composition of triple arrows, then, is tantamount to stacking bricks.

We relate this multi-category to $\mathcal{S}$ by coloring the components of the complement of M alternately black and white.

In brief, the multi-categorical structure of the embedded surface M coincides with decomposing space into a collection of brick-like slabs. We are not only concerned with a particular embedding, but with any embedding. So the bricks that we use are lego-like. There are generators that correspond to births, deaths, saddles, forks, cusps, identity sheets on folds, and identities that have no critical information — walls that separate the black from the white. Any surface can be built from these pieces, and each such piece can be described via the multi-category $\mathcal{S}$.

Remark. In Section 3.4, the collection of 3-arrows in $\mathcal{S}$ were shown to correspond to the arrows in an associated category. In this decategorification, the collection of objects corresponded to red circuit diagrams in a rectangle. The optimal points (Fs and Ⱶs) are at distinct vertical heights. The arrows in the category are generated by exchangers, births, deaths, saddles, forks, and cusps. So the relationships between this category and embedded surfaces includes the possibility that the boundary of the surface also runs vertically up the sides of a rectangular box in a manner analogous to the illustrations in Chapter 4, Figs. 5 or 6. Note that the latter illustration is of a foam rather than a surface, but still pieces of the boundary run up the sides of the box along parallel segments. The theorem and its proof extends to provide a functor between a geometrically defined category and the decategorification of Section 3.4. In this way proper isotopies of "embedded surfaces with corners" can be addressed using the same techniques.

5.4. Why these particular 3- and 4-arrows? From the categorical perspective, it seems arbitrary to define triple and quadruple arrows as we have. Indeed, since the original purpose was to model embedded surfaces, the triple arrows were modeled upon the critical points and singular points of smooth maps from surfaces to the planes while the quadruple arrows were modeled upon the singularities of surface isotopies. On the other hand, starting from the ground up, then the choices of the triple and quadruple arrows can be seen to be less awkward than might have been presumed.

Consider the double arrow $\mathsf{F}(\mathsf{t})$ whose source is the 1-arrow $\mathsf{t} \xrightarrow{\mathsf{b}} \mathsf{f} \xrightarrow{\mathsf{p}} \mathsf{t}$ and whose target is $\mathsf{t} \longrightarrow \mathsf{t}$:

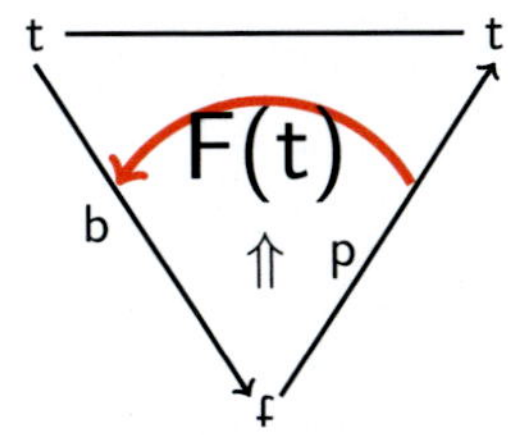

Imagine all the possible ways of composing with any of $\mathsf{F}(\mathsf{t})$, $\mathsf{F}(\mathsf{ɟ})$, $\mathsf{Ⅎ}(\mathsf{t})$, or $\mathsf{Ⅎ}(\mathsf{ɟ})$. Similar consideration, left to the reader, can be applied to the other non-identity double arrows. The two-fold compositions that involve $\mathsf{F}(\mathsf{t})$ are listed.

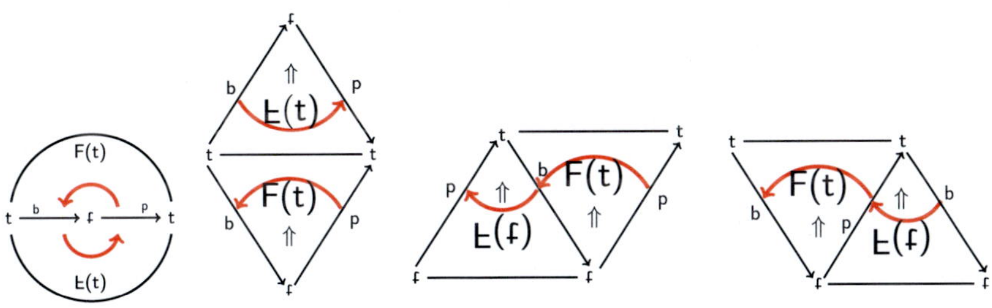

For each of these double arrows, its source 1-arrow agrees with its target 1-arrow. So each can be compared to an identity double arrow. Those comparisons comprise the generating collection of non-identity triple arrows that are formed from $\mathsf{F}(\mathsf{t})$. That is, the comparisons are the triple arrows that were called birth, death, saddle, fork, and cusp.

The relations among these resulting triple arrows are of two types. First, the lips and beak-to-beak relations are the assertions that cusp triple arrows are strongly invertible. To understand the remaining relations consider the possible 3-fold compositions of double arrows. Again concentrate upon $\mathsf{F}(\mathsf{t})$. These are listed here.

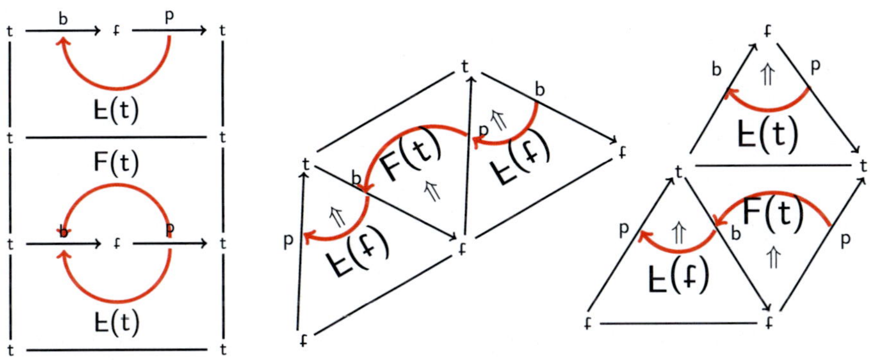

There are two ways to compare any of these double arrows to the composition with an identity. Each composite of three double arrows appears as a source and as a target of the composition of a generating triple arrow and an identity triple arrow. On the left, the given double arrow is the target of a birth and the source of a fork, or the target of a saddle and the source of a death. Either composition of triple arrows can be compared to the identity on Ƚ(t). The trapezoidal double arrow in the middle is the source and target of cusps, and it fits into a swallow-tail configuration. The double arrow on the right is the target of a saddle or the source of an up cusp. It fits into a horizontal cusp sequence.

By considering all possible three-fold compositions of double arrows, we see that the posited quadruple arrows arise because each composite can be resolved by triple arrows in two ways. So it is not coincidence that these correspond to codimension two singularities of surface maps. The situation is forced.

Therefore, the multi-category $\mathcal{S}$ is the most free multi-category generated by a weakly invertible arrow $\mathsf{t} \xrightarrow{\mathsf{b}} ʇ$ with its reverse $ʇ \xrightarrow{\mathsf{p}} \mathsf{t}$ being the weak inverse. That is, double arrows $\mathsf{F}(ʇ) : \mathsf{bp} \Rightarrow \mathsf{I}_{ʇ}$, $\mathsf{F}(\mathsf{t}) : \mathsf{pb} \Rightarrow \mathsf{I}_{\mathsf{t}}$ and their reverses naturally give rise to birth, death, saddle, fork, and cusp triple arrows. These satisfy the quadruple arrow relations critical cancelations (Section 4.1), lip (Section 4.2), beak-to-beak (Section 4.3), horizontal cusps (Section 4.4), and swallow-tail (Section 4.5). Compare this to the cobordism hypothesis of [**6**].

Exercise. Complete the analysis using three-fold compositions that involve $\mathsf{F}(ʇ)$, Ƚ(t), and Ƚ(ʇ).

CHAPTER 7

Beyond surfaces

The multi-category $\mathcal{S}$ is more versatile than isotopy classes of embedded surfaces in 3-dimensional spaces. In Section 1, several multi-categories are examined that have a pair of objects and a pair of non-identity arrows that are reverses. These will be labeled $\mathcal{S}(*)$ where $*$ indicates the positive 1-arrows in the multi-category. Thus Chapter 4, Section 2 is re-examined from the point of view of constructing double, triple, and quadruple arrows. The quadruple arrows will still be named lips, beak-to-beak, and so forth, but their interpretations will depend upon the initial choices of objects t and ɟ.

The approach here is a categorical distillation of Cerf theory [**16**]. Hence, experts will not be surprised that applications here include forms of handle calculus (Section 1.3) since the double arrows F and Ⅎ are modeled upon critical behavior. This chapter will demonstrate the pleasing ubiquity of the methods from Chapter 6.

After providing these examples, an inductive approach is presented to construct weakly invertible multi-arrows in Section 2. In this section, birth, death, saddle, fork, and cuspidal arrows are described in all dimensions, and the critical cancelations, lip, beak-to-beak, swallow-tails, and horizontal cusp relations hold among them. However, these are not explicitly listed since they follow directly from the characterizations of the axioms that have been presented. On the other hand, relationships between critical cancelations and cusps are given.

Then in Section 3, the multi-category FA is revisited and the existence of some new 4-arrows is asserted. Some of these exist for purely algebraic reasons. In particular, an associative multiplication or pairing in the presence of a unit reveals interesting relations. Others exist for topological reasons. The relationships among the topological, the algebraic, and the categorical situations will be further explored in Section 4. Some multiple arrows, in some contexts, should be used only in restricted environments.

1. Different objects and arrows

Exercise. Before proceeding further in this chapter, attempt to catalogue the double arrows F(t), F(ɟ), Ⅎ(t), and Ⅎ(f), the triple arrows of the form birth, death, saddle, fork, and cusp, and the 4-isomorphisms for each choice of t and its reverse ɟ that are presented in the columns of Table 1. After your attempt, whether or not it be successful, continue reading on, and compare your results with those below. Be aware, not all arrows are presented below.

1.1. Units and co-units. Examine the first column of Table 1 and compare it to Chapter 4, Table 13. In this case, the multi-category is labeled $\mathcal{S}($ ⊥ $)$. The double arrow F(t), for example, is an arch. By ignoring the serifs on the glyphs that adorn the double arrows in Table 13 (Ch. 4), these double arrows agree exactly with

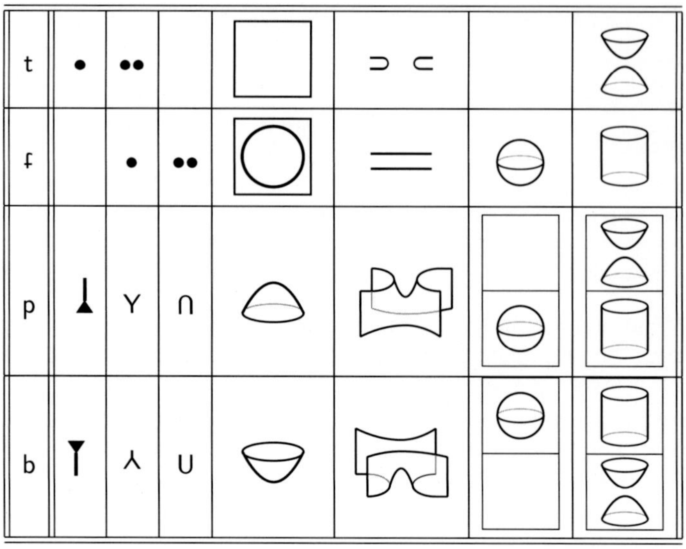

TABLE 1. Some possible choices of objects and the associated 1-arrows

those of Chapter 6. Imagine, for example, that the object ɟ is the natural number 0 and t is 1 where both are depicted in unary notation, then the black region of 3-space is labeled 1 and the white region is labeled 0. The double arrows F and Ⅎ correspond to critical points of curves, and the names of the triple arrows (birth, death, saddle, fork, and cusp) indicate that they are critical events of surface projections.

The multiple arrows of $\mathcal{S}$ are found among the multiple arrows of FA. That is to say that we have not completed the description of FA since we hardly specified any triple or quadruple arrows. Since the double arrows of Ch. 4, Table 13 are among those in FA, the triple and quadruple arrows here should also be included.

1.2. Multiplication and co-multiplication. The double arrows F and Ⅎ that are associated to the second column of Table 1 are found in Ch. 4, Table 12. In this case, the multi-category is named $\mathcal{S}(\mathsf{Y})$.

Exercise. Explicitly concatenate the various zippers that constitute the cuspidal triple arrows. In doing so, it should become apparent that cusps correspond to the cancelation of critical points along the set of seams.

The triple arrows that correspond to births, deaths, saddles, and forks in the current context are depicted below. These all have the property that the topological type of the underlying foams changes under applications of these arrows. These topological differences are a cause for increased caution about meaning. The triple arrows depicted here exist, but since they change the topology of the foam, we may not want to include them among the arsenal of tools to study foams. This topic will be discussed in greater detail in Section 4 of this chapter and Chapter 10. Forks and saddles in the context here are, at best, weak inverses, as are births and deaths (Cf. Chapter 5, Section 3).

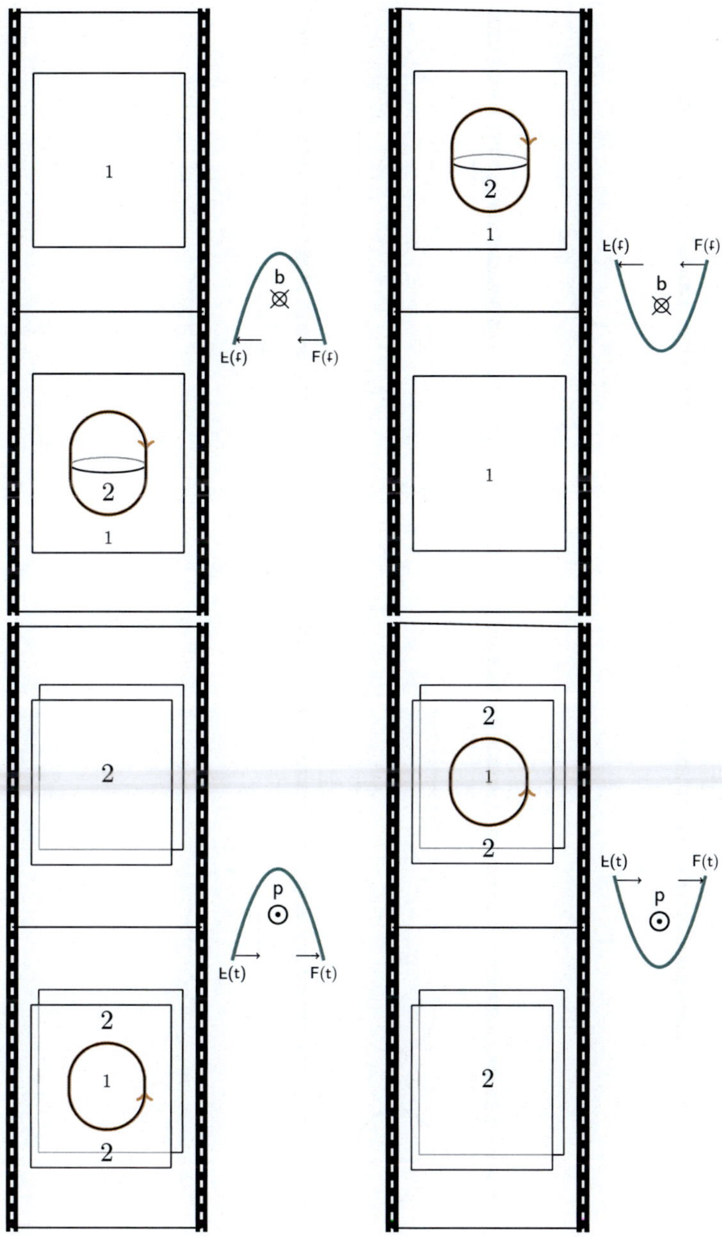
1
2
1
F(f)
b
1
2
1
F(f)
b
2
2
1
2
p
F(t)
2
1
2
2
p
F(t)

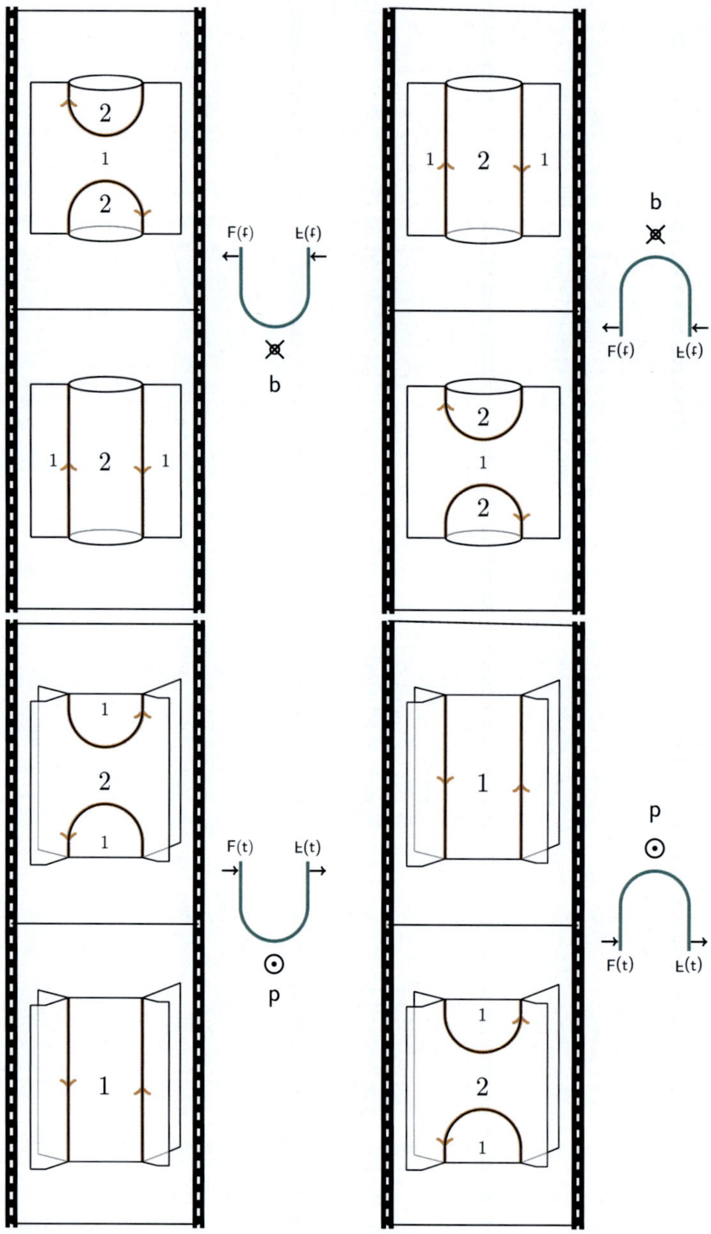

Since cusps are, as expected, cancelations of the critical points of the seams, they define isotopy of foams. In this case, a time-elapsed presentation of the 2-dimensional foam determines a 3-dimensional analogue. Seams evolve into 2-dimensional disks along which three 3-dimensional disks are attached. The critical points (zipper double arrows) evolve to provide local folds in the seams. The lips, beak-to-beak, and swallow-tail quadruple arrows also correspond to phenomena of the same type upon the folds of the evolving seams.

Two examples of critical cancelation quadruple arrows are illustrated. The movies on the left are ambiguous. In the first example, moving from the bottom still to the middle still can be achieved by either a negative birth or a negative fork.

In the second example, the first critical point is either a saddle or a birth. In general, critical cancelations are of the form death/saddle or birth/fork. The glyphs that flank the movies indicate the possible interpretations of the movie. The remaining critical exchanges in these cases are obtained by reflecting each foam in the movies through the horizontal (x, y)-plane.

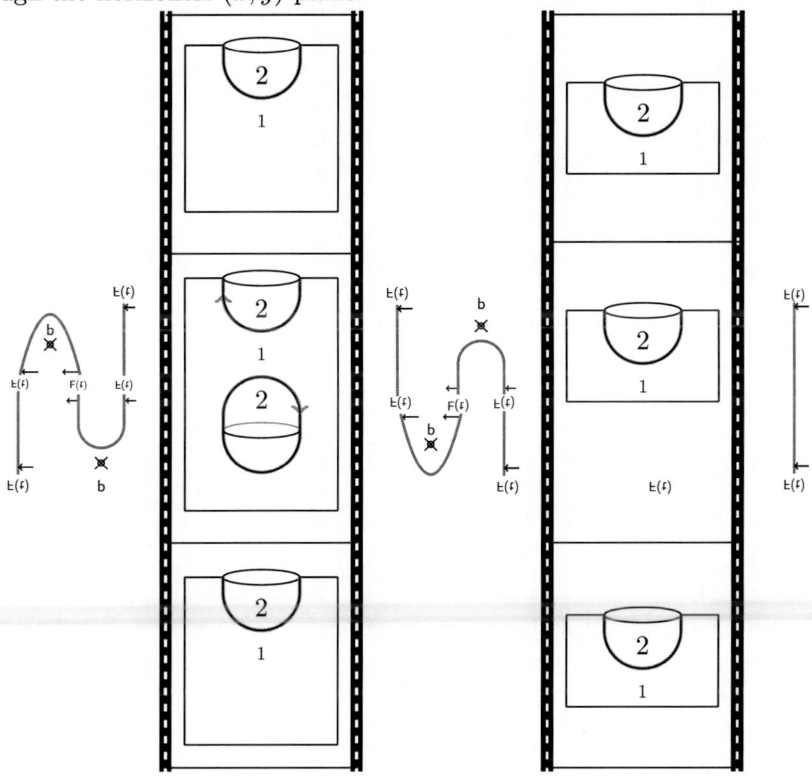

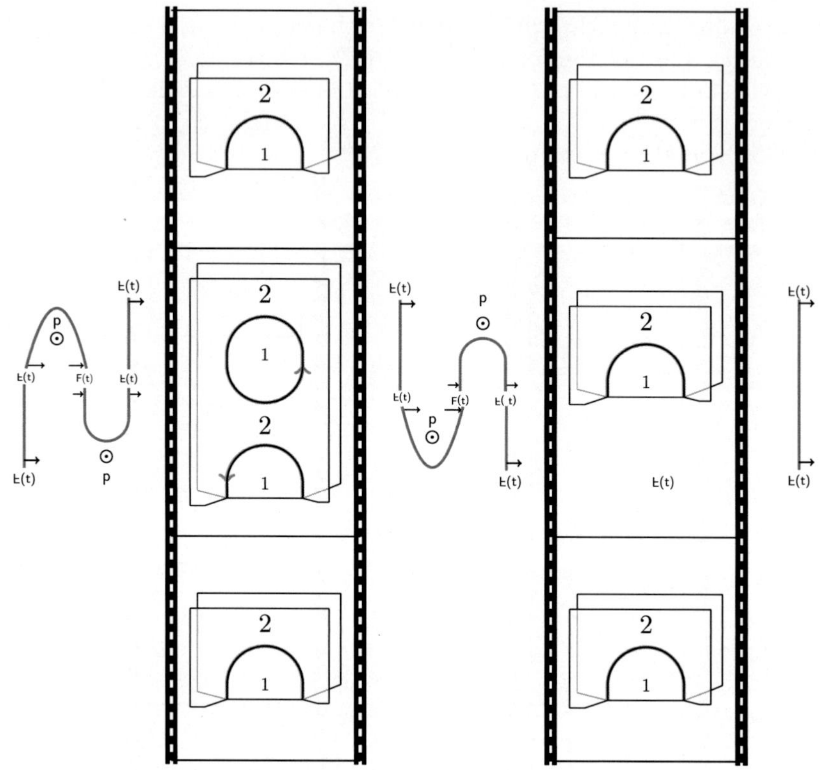

Fig. 1 indicates an example horizontal cusps in the current context. The other horizontal cusps can be obtained from this one by a sequence of reflections and re-orderings of the foams that are depicted.

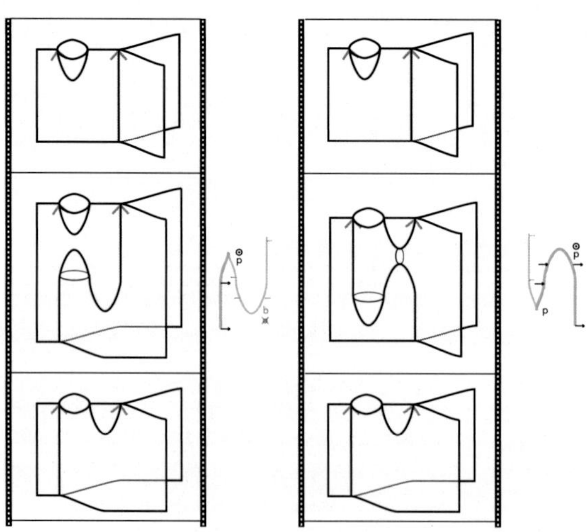

FIGURE 1. A horizontal cusp in the multiplication and co-multiplication multi-category

1.3. Descriptions of handles. The idea of n-dimensional handles that was introduced in Chapter 5, Section 3.4 is formalized. Let $k, n \in \mathbb{N} = \{0, 1, 2, \ldots\}$ denote positive integers. Recall, the *k-dimensional disk* is defined to be the set

$$D^k = \{(x_1, x_2, \ldots, x_k) \in \mathbb{R}^k : \sum_{j=1}^{k} x_j^2 \leq 1\}.$$

Its boundary is the $(k-1)$-dimensional sphere

$$\partial D^k = S^{k-1} = \{(x_1, x_2, \ldots, x_k) \in \mathbb{R}^k : \sum_{j=1}^{k} x_j^2 = 1\}.$$

In general, the n-dimensional disk D^n is homeomorphic to the product $D^n = D^k \times D^{n-k}$ for any $k \in \mathbb{N}$. Such a product is called an n-dimensional *k-handle*. We will abbreviate this to (n-D k-H). There are several important pieces of k-handles that have standard names [**69**].

- The set $S^{k-1} \times \{0\}$ is called the *attaching sphere* or *A-sphere*.
- The set $S^{k-1} \times D^{n-k}$ is called the *attaching region* or *A-region*.
- The set $\{0\} \times S^{n-k-1}$ is called the *belt sphere* or *B-sphere*.
- The set $D^k \times S^{n-k-1}$ is called the *belt region* or *B-region*.
- The set $D^k \times \{0\}$ is called the *core disk*.
- The set $\{0\} \times D^{n-k}$ is called the *co-core disk*.

The n-disk, D^n, can be thought of as an n-dimensional 0-handle or alternatively as an n-dimensional n-handle since D^0 consists of a single point. In these cases, there are some peculiarities in the definition. It is best to consider the boundary sphere S^{-1} as an empty set. Then an (n-D 0-H) has an empty A-region, and an (n-D n-H) has an empty B-region.

A 0-sphere consists of the point set $\{-1, 1\}$. This can be identified with •–•. In the multi-category $\mathcal{S}(\cap)$, the arrow $\mathsf{b} = \cup$ corresponds to a 1-dimensional 0-handle (1-D 0-H) while $\mathsf{p} = \cap$ corresponds to a 1-dimensional 1-handle (1-D 1-H). More generally in Section 1.5 of this chapter, non-identity double arrows in $\mathcal{S}(\cap)$ will be shown to correspond to 2-dimensional k-handles for $k = 0, 1, 2$. Birth, death, fork, and saddle triple arrows correspond to 3-dimensional k-handles, and cusps correspond to handle cancelations.

Consider a compact $(n-1)$-dimensional manifold that does not have a boundary. Such a topological space is defined analogously to a surface (See Chapter 6, Section 5.2) which is a 2-dimensional manifold. Every point in the manifold M has a neighborhood that is homeomorphic to an open subset of $\mathbb{R}^{n-1}$, and the transformations between overlapping neighborhoods is infinitely differentiable. In general, we are interested in two such manifolds M_1 and M_2 that differ by means of *attaching a handle*. In this case, a subset of M_1 that is homeomorphic to the attaching region of the handle is replaced by the belt region. So that M_1 and M_2 are identical except in the attaching or belt regions.

The empty set is, by definition, an $(n-1)$-dimensional manifold for all $n = 1, 2, \ldots$. In this way, the $(n-1)$-sphere is obtained from the empty manifold by means of attaching an n-dimensional 0-handle.

In general, $(n-1)$-dimensional manifolds are thickened by taking the product with an interval $[0, 1]$ and the attaching n-dimensional handle, together with the thickenings, forms an n-dimensional manifold that is bounded by $M_1 \times \{0\}$ at the bottom and $M_2 \times \{1\}$ at the top. This type of thickening was described in some more detail in Chapter 5, Section 3.4.

1.4. Pairing and co-pairing. The multi-category $\mathcal{S}(\cap)$ that corresponds to the third column of Table 1 has its generating double arrows listed in Ch. 4, Table 11. *Thus the double arrows of $\mathcal{S}(\cap)$ coincide with some of the triple arrows of $\mathcal{S}(\perp)$.* But in that case a distinction is made for positive and negative births, deaths, saddles, and forks. In Chapter 5, Section 3.2, the triple arrows that correspond to comparing the compositions $\begin{bmatrix} \mathsf{F} \\ \text{Ⅎ} \end{bmatrix}$ and $\begin{bmatrix} \text{Ⅎ} \\ \mathsf{F} \end{bmatrix}$ were introduced formally. Here these triple arrows will be recalled, an example cusp triple arrow is depicted, and the quadruple arrows will be exemplified.

In the multi-category $\mathcal{S}(\cap)$, a composition of 1-arrows is a sequence of 1-dimensional handle attachments:)○ ⋯ ○;)○ ⋯ (; ○○ ⋯ ○; or ○○ ⋯ (. The non-identity double arrow $\mathsf{F}(\mathsf{t}) = \mathsf{F}(\emptyset)$ is a 2-dimensional 2-handle . The double arrow Ⅎ(t) = Ⅎ$(\emptyset)$ is a 2-dimensional 0-handle , and the remaining non-identity double arrows are both 2-dimensional 1-handles:  or . The illustration below catalogues the birth, death, fork, and saddle triple arrows as 3-dimensional handles. The downward pointing arrows on the left-hand side of the illustrations indicate that the source of the triple arrow is the top still and the target is the bottom.

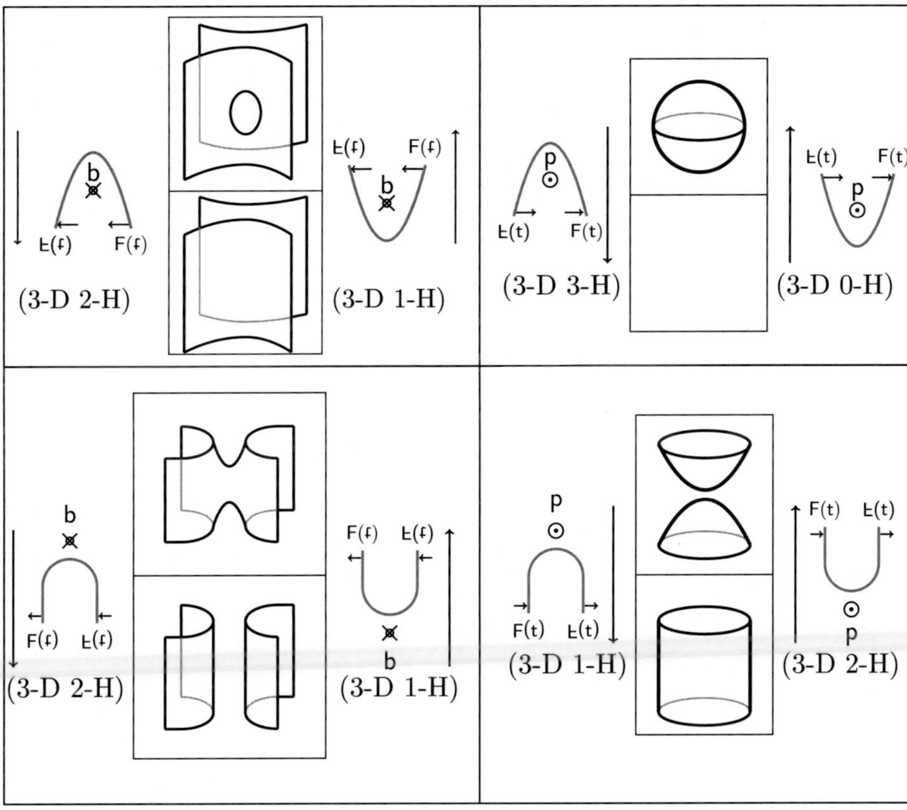

These illustrations should be compared to those in Chapter 5, Section 3.2 in which these triple arrows are described abstractly in terms of the glyphs that represent the geometric phenomena.

Next a negative left down cusp is illustrated below in the multi-category $\mathcal{S}(\cap)$. On the left side of the drawing, a composition of double arrows is given. The fold-lines of the surface and the edges along which the surface pieces are attached are meant to suggest to the reader a method to decompose this surface in terms of its constituent double arrows. For example the vertical fold on the left side indicates the identity upon b. The negative left up cusp is determined by reversing the triple arrow between the two pieces of surface. Other cusps can be obtained via reflections of these figures in various planes.

Exercise. 1. Enumerate the remaining six cusps in the current multi-category. 2. Observe that a cusp triple arrow in $\mathcal{S}(\cap)$ corresponds to a critical cancelation quadruple arrow in $\mathcal{S}(\perp)$. 3. Determine lips, beak-to-beak, and swallow-tail quadruple arrows by examining the cusp triple arrow below.

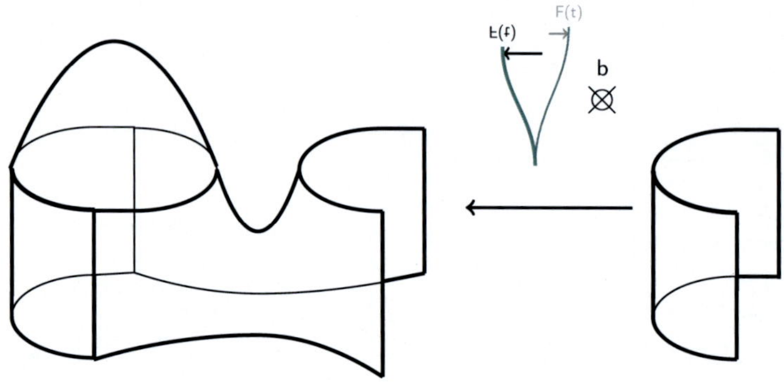

One example of a horizontal cusp is illustrated in Fig. 2. As before, the other variations are obtained by reflecting each of the movie stills.

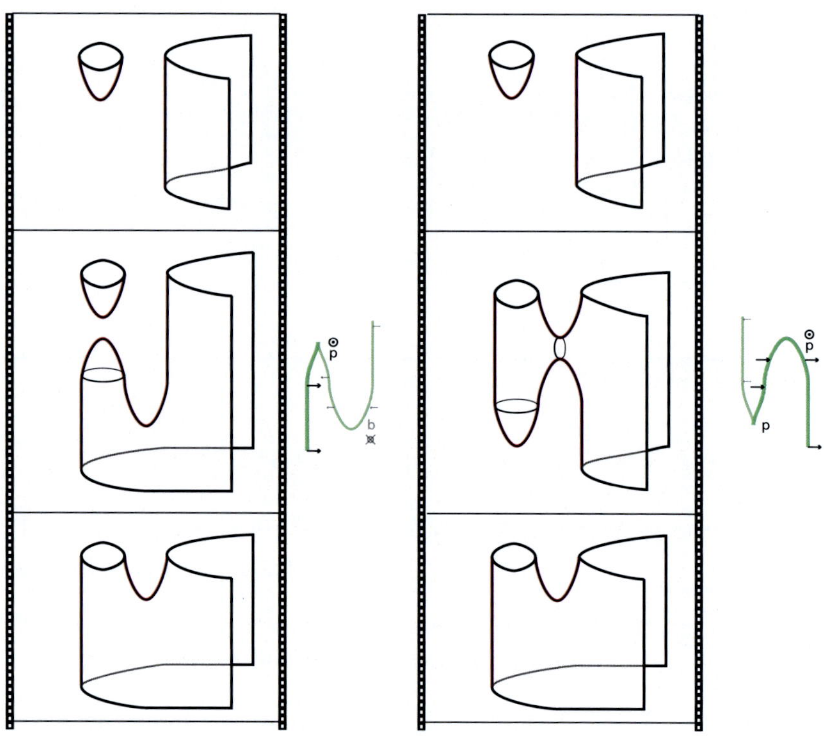

FIGURE 2. A horizontal cusp in the pairing and co-pairing multi-category

1.5. Increasing the dimension: part 1. Positive birth/positive death.

As the fourth and fifth columns of Table 1 suggest, the objects t and ɟ can be 1-dimensional, and then the 1-arrows are represented as surfaces. In this sense, the multi-category $\mathcal{S}(\cap)$ has been de-catgorified since simple closed curves are 1-arrows in $\mathcal{S}(\cap)$, but in the fourth column they are now objects. The fourth and fifth columns will be thought of separately for the time being, but we will amalgamate them subsequently in Section 2.1 of this chapter.

Consider the fourth column of Table 1. The non-identity double arrows have the interpretation indicated below.

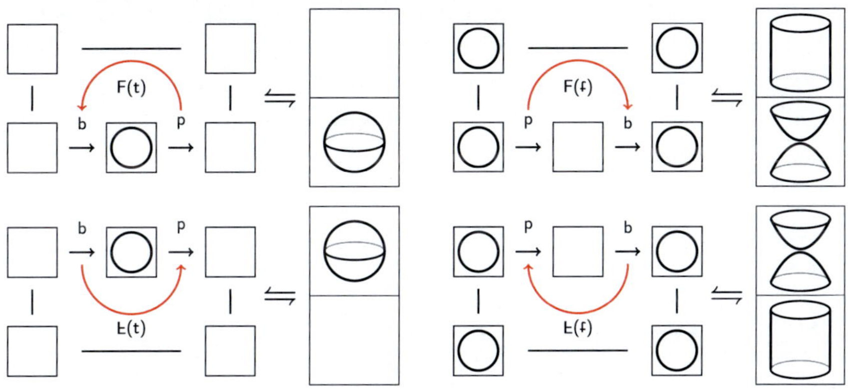

The positive and negative birth and death pairs that are obtained from the composition $\begin{bmatrix} \mathsf{F}(\mathsf{t}) \\ \text{Ⅎ}(\mathsf{t}) \end{bmatrix}$ are indicated below.

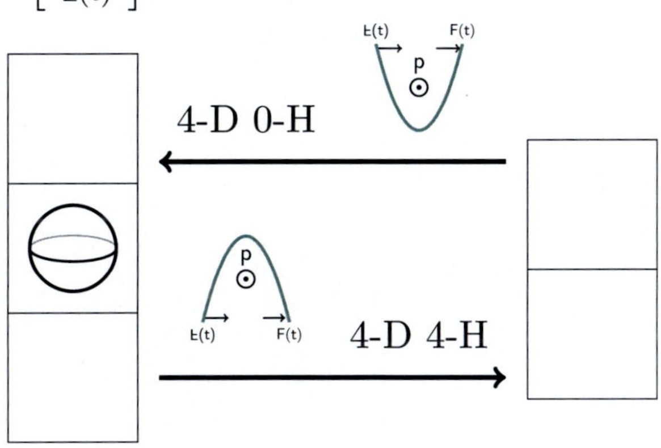

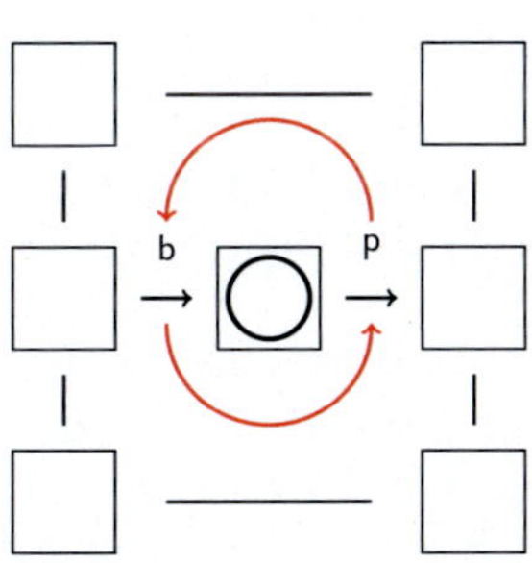

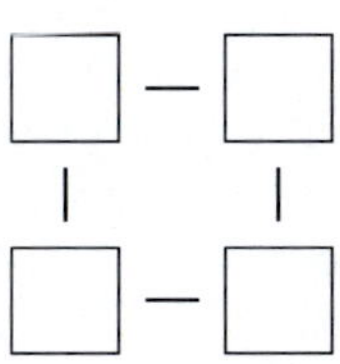

The positive birth triple arrow has an empty movie as its source. The empty two squares that appear along the horizon of the illustration below the movie represent an empty 3-dimensional box — the bottom still of the right-hand movie. Two empty boxes, taken together, represents the top and bottom faces of an empty 4-dimensional box — the right-hand movie. The target movie of the positive birth on the top left of the illustration consists of three stills. The first and third stills are empty. The middle still is the composition of a bowl and a yarmulke or skull cap. This composition is a 2-dimensional sphere. So the double arrow Ⱡ(t), is the birth of a 2-dimensional sphere, and the double arrow F(t) is the death of this sphere. The left-hand movie represents a balloon that inflates and then deflates. The inflation defines a 3-dimensional ball, as does the deflation. The composition, then, is a 3-dimensional sphere —the union of two 3-dimensional balls along a common sphere. The birth of such a 3-sphere, is a 4-ball. Therefore in $\mathcal{S}(\ )$, the positive birth triple arrow is a 4-dimensional 0-handle. And its reflected companion, the positive death arrow, is a 4-dimensional 4-handle.

Negative birth/negative death.

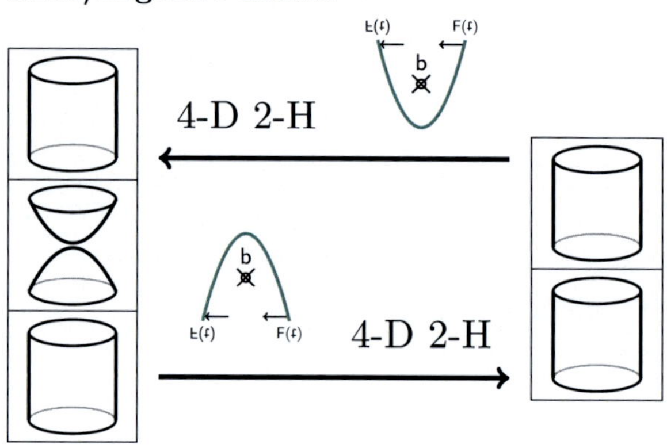

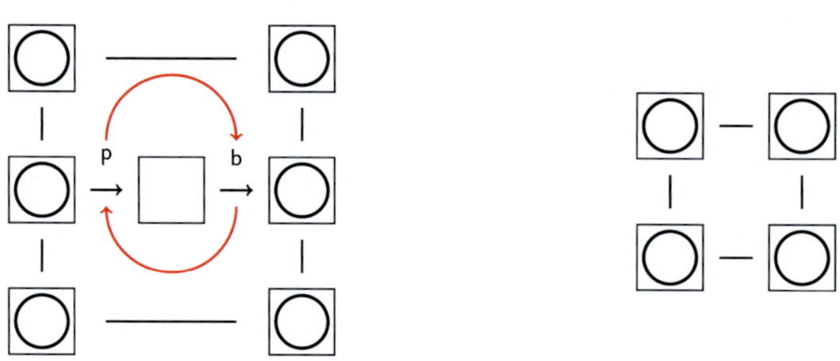

The negative birth and negative death in $\mathcal{S}(\ )$ are both 4-dimensional 2-handles. Look on the right-hand side of the illustration above. The movie depicts a cylinder ($S^1 \times [0,1]$) in either still. Thus the source and target 1-arrows are the identity upon the circle f. The double arrow is the identity upon this. So the source triple arrow of this negative birth is $S^1 \times D^2$.

The left-hand movie is also of the form $D^2 \times S^1$. To see this, observe that the bottom and middle still differ by means of a 3-dimensional 2-handle, $D^2 \times D^1$, and the middle and top differ by means of a 3-dimensional 1-handle, $D^1 \times D^2$. These two solids are glued along a pair of 2-dimensional disks $D^2 \times S^0$. Thus they form a solid torus. Therefore, both of the negative birth and negative death in $\mathcal{S}$ () are 4-dimensional 2-handles.

Positive fork/positive saddle.

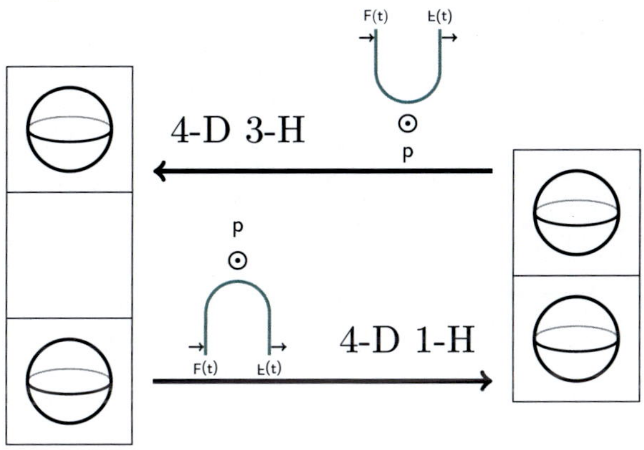

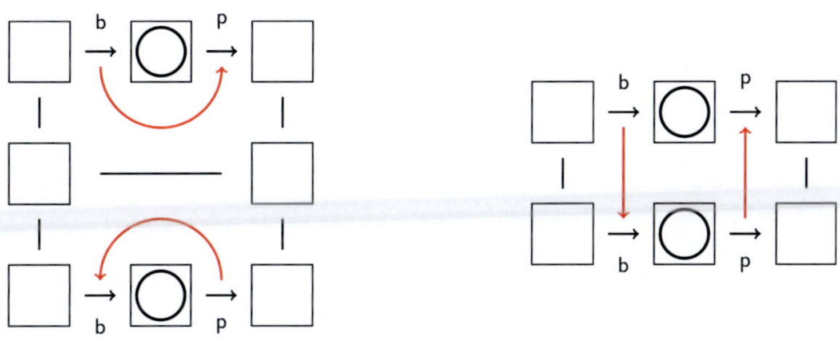

On the right hand side of the movie move two stills appear each of which contains an embedded sphere. The source double arrow of the positive saddle then is of the form $S^2 \times [0,1]$. The target double arrow is the union of two 3 dimensional balls — $D^3 \times \{0,1\}$. So the positive saddle triple arrow in $\mathcal{S}$ () is a 4-dimensional 3-handle while its reflective companion is a 4-dimensional 1-handle.

Negative fork/negative saddle.

The source double arrow for the negative saddle in the multi-category $\mathcal{S}$ () is depicted as the movie on the right-hand side of the illustration below. Each of the two stills consists of a pair of disks, so the movie represents $S^0 \times D^3$. On the left-hand movie, the transition from the bottom still to the middle still is a 3-dimensional 1-handle of the form $D^1 \times D^2$. The transition from the middle to the top is a 2-handle, and these handles are identified upon an annular region.

Thus their union is a $D^1 \times S^2$. The negative saddle triple arrow represents a 4-dimensional 1-handle and its reflective companion, the negative fork, represents a 4-dimensional 3-handle.

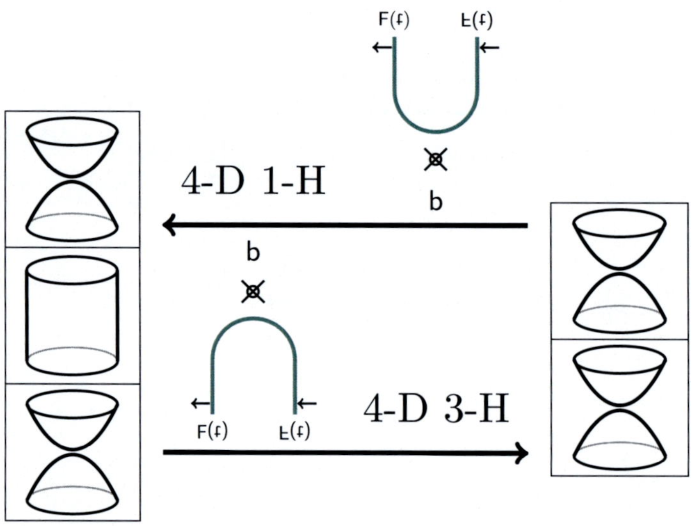

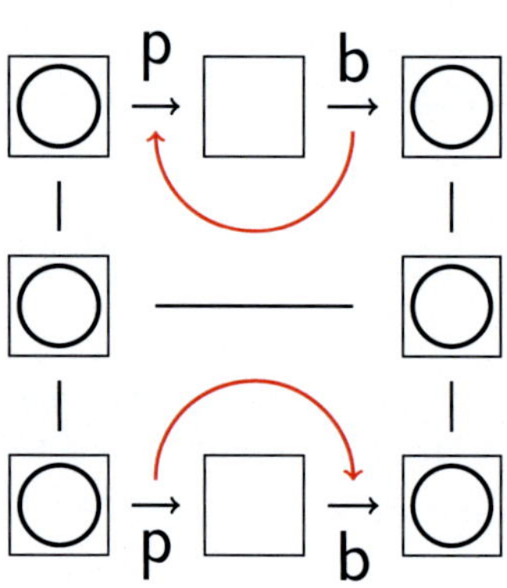

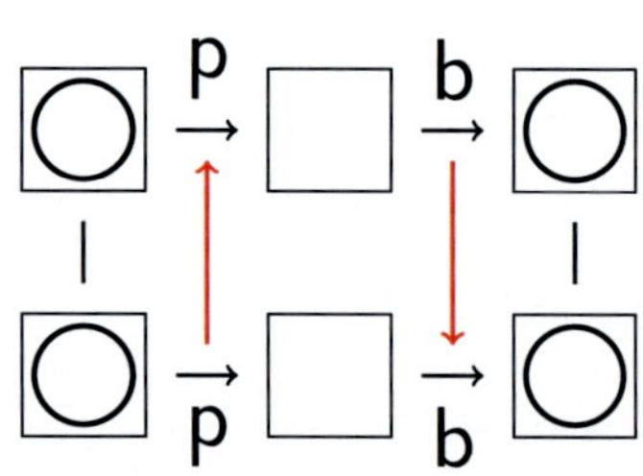

Cusps.

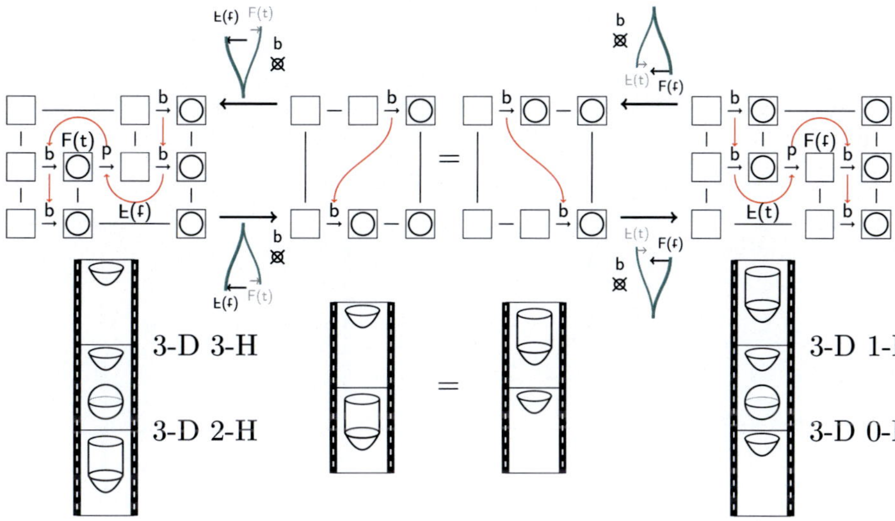

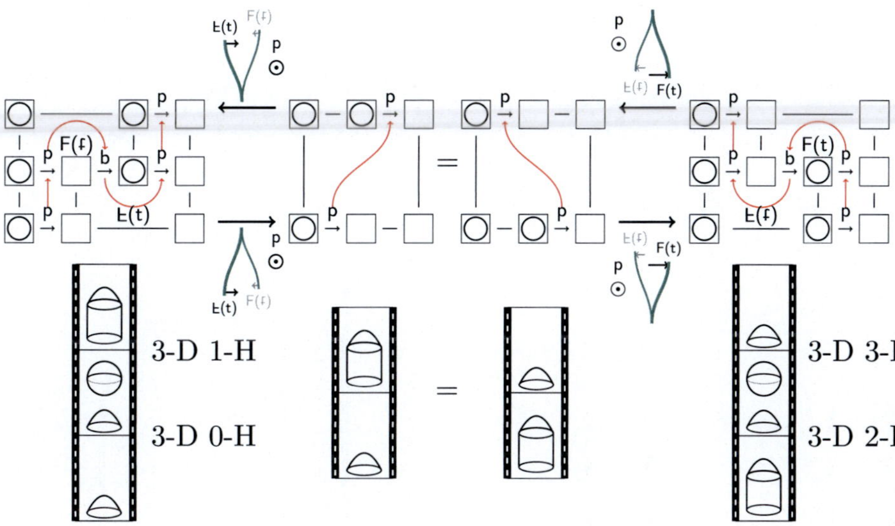

The eight cusp triple arrows in the multi-category $\mathcal{S}$  are indicated above. They correspond to critical cancelations between successive 3-dimensional handles — a 0-handle cancels a 1-handle or a 2-handle cancels a 3-handle. As always, cusps have strong inverses. Up and down cusps of the same type ($\pm$, left/right) are strong inverses in the sense that there are lips and beak-to-beak 4-arrows. The cusps in $\mathcal{S}$  correspond to some of the critical cancelations in $\mathcal{S}(\cap)$. The remaining critical cancelations in $\mathcal{S}(\cap)$ occur as cusps in $\mathcal{S}$ . More will be said about this in Section 2 of this chapter.

Swallow-tails.

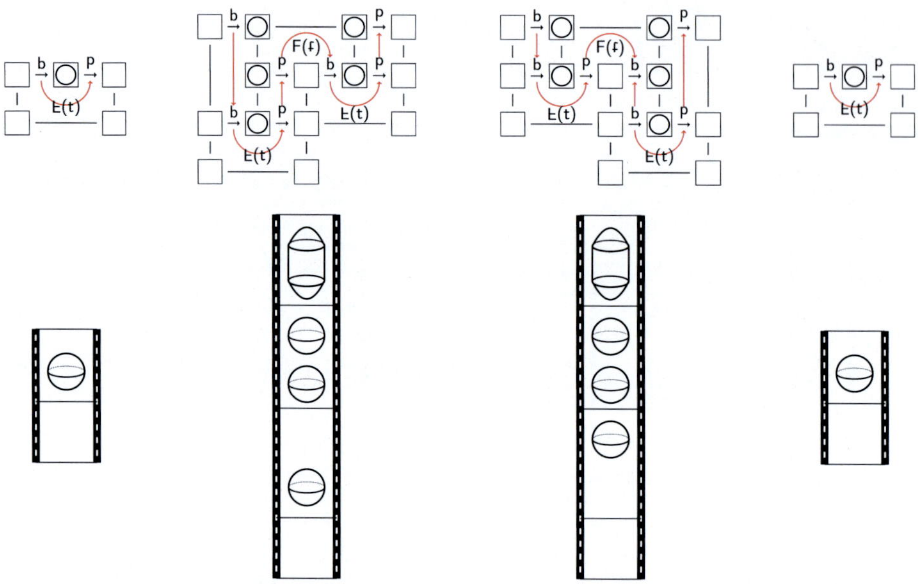

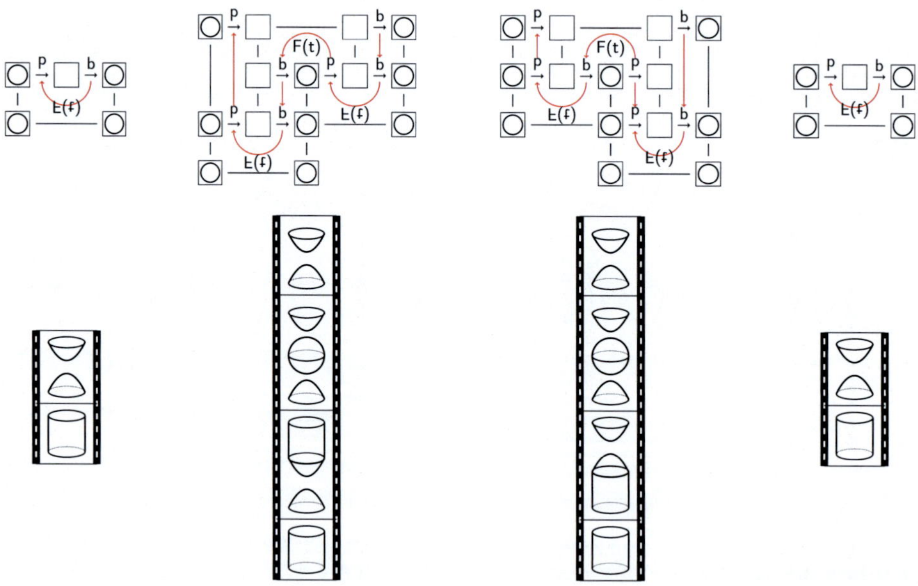

In the illustrations above, the important features of two of the swallow-tail 4-arrows are depicted. In each case, the sequence can either be read from left to right or right to left. And if all the double arrows are reflected through a horizontal line, the other four swallow-tail 4-arrows can be obtained.

The illustrations are described in more detail. The right-hand side (target) of a swallow-tail 4-arrow consists of the identity 3-arrow on one of the double arrows F(t), F(ɟ), Ⅎ(t), or Ⅎ(ɟ).

On the top illustration, Ⅎ(t) flanks the middle two composites of double arrows. This double arrow corresponds to a 3-dimensional 0-handle — the birth of a 2-dimensional sphere. The most left illustration is related to the middle left by a positive left down cusp. From the previous illustrations, this cusp corresponds to the introduction of a canceling pair of 3-dimensional 0- and 1-handles. The middle-left illustration is related to the middle-right by a negative exchange. The middle-right illustration is related to the illustration on the far right by a negative right up cusp — which is the cancelation of a 3-dimensional 0- and 1- handle pair. So in this illustration, the source of the swallow-tail quadruple arrow is the 4-dimensional process: add a canceling pair of handles, exchange heights, and cancel a pair. The target of the swallow-tail is the 4-dimensional process that is a time-elapsed 3-dimensional ball. That is to say it is a 4-dimensional ball.

The bottom illustration has some similarities. The double arrow Ⅎ(f) flanks the middle two composites of double arrows. The source and target double arrows of the 3-arrow are 3-dimensional 1-handles. The triple arrow from left to middle-left is the introduction of a $(2, 3)$ canceling pair of handles. An exchange occurs, and a different pair of 2- and 3-handles cancel.

In both cases, the 3-arrow that results from reversing the sequence of 2-arrows, the description is the same except that the exchanger is positive in this case. In case either is reflected in a horizontal line, then the flanking double arrows correspond to 3-dimensional 3-handles in the top case, and to 3-dimensional 1-handles in the bottom case. The triple arrows that are cusps may correspond to differing $(j, j+1)$ canceling pairs.

Critical Cancelations. The description of the 4-arrows that correspond to critical cancelations in the multi-category $\mathcal{S}$ () require care. Specifically, there is a contrast with the cuspidal 3-arrows that correspond to critical cancelations in $\mathcal{S}(\cap)$. In the current context, $\mathcal{S}$ (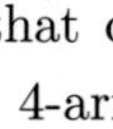), 4-arrows represent 5-dimensional objects, or comparisons between 4-dimensional objects. More specifically, the double arrows correspond to 3-dimensional handles, and so the triple arrows correspond to 4-dimensional things: births, deaths, saddles, and forks are 4-dimensional handles, cusps are cancelations of 3-dimensional handles. The 4-arrows that are critical cancelations, then, will correspond to canceling successive 4-handles.

Recall from Chapter 6, Section 5.4 that the 4-arrows in $\mathcal{S}$ appear by considering all possible composites of three triple arrows. In the case of critical cancelations, three-fold stacks $\begin{bmatrix} \mathsf{F} \\ Ⅎ \\ \mathsf{F} \end{bmatrix}$ or $\begin{bmatrix} Ⅎ \\ \mathsf{F} \\ Ⅎ \end{bmatrix}$ are compared to a single F or Ⅎ. In the next illustrations, two of these compositions of double arrows are illustrated with their representative movies depicted to the right. The other two illustrations are, as before, reflections in a horizontal line. The illustration on the left compares this to Ⅎ(t), and that on the right to Ⅎ(f).

First, consider the situation that is partially illustrated by the configuration on the left below. This critical cancelation 4-arrow compares the identity 3-arrow upon Ⅎ(t) to the composition of a saddle and death or to the composition of a birth and a fork. In the latter case, the 3-dimensional 0-handle that corresponds to the double arrow Ⅎ(t) is transformed to $\begin{bmatrix} Ⅎ(\mathsf{t}) \\ \mathsf{F}(\mathsf{t}) \\ Ⅎ(\mathsf{t}) \end{bmatrix}$ by adding a 4-dimensional

0-handle (positive birth). Here the empty bottom frame in a movie that represents Ⅎ(t) is replaced by three frames. The top and bottom are empty and the middle consists of a sphere. Then a 4-dimensional 1-handle (positive fork) which connects the spheres returns to Ⅎ(t). In the former case, the positive saddle correspond to a 4-dimensional 3-handle. The sphere in the top frame of a movie bifurcates into two. The death 3-arrow corresponds to a 4-dimensional 4-handle. The sphere in the bottom frame is annihilated.

The critical cancelations 4-arrow that are represented via the double arrow illustrated on the right are also comparisons of the identity 3-arrow Ⅎ(ɟ) to the composition of saddle and death 3-arrows or birth and fork 3-arrows. A negative birth transforms the double arrow Ⅎ(ɟ) to this double arrow by adding 4-dimensional 2-handle. The negative fork then transforms it back to Ⅎ(ɟ) by adding a 4-dimensional 3-handle. In the other interpretation, a negative saddle corresponds to a 4-dimensional 1-handle, and the negative death returns the movie representation back to Ⅎ(ɟ) via a 4-dimensional 2-handle.

The remaining four critical cancelations that are obtained by reflecting these double arrows in a horizontal line can be interpreted by the reader. First consider the double arrow F(t) as a 3-dimensional 3-handle and F(ɟ) as a 3-dimensional 1-handle. Then interpret each birth, death, saddle, or fork as a 4-dimensional handle. The critical cancelations occur because the handle indices are successive — of the form $(i, i+1)$.

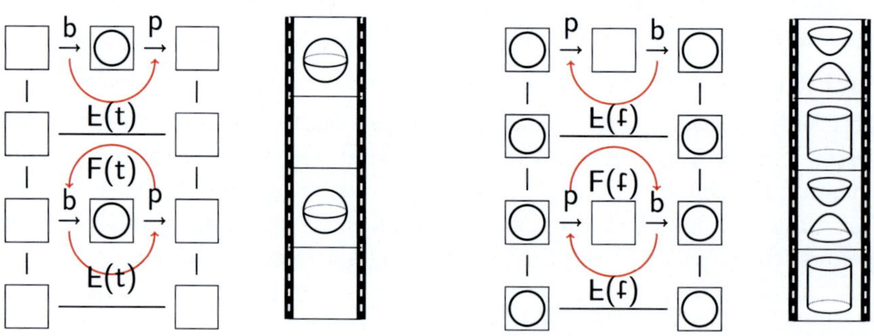

Horzontal cusps.

Recall from the discussion of forks and saddles that in the multi-category $\mathcal{S}$ (⌓) forks and saddles are either 4-dimensional 1-handles or 4-dimensional 3-handles. Both a positive fork and negative saddle are 4-dimensional 1-handles. Yet their movie appearances are different. Now re-examine the illustrations of Chapter 6, Section 4.4 that depict the eight horizontal cusps. Observe that in the presence of a pair of left and right oppositely pointing cusps, a positive saddle is traded for a negative fork, or a negative saddle is traded for a positive fork. And therefore in $\mathcal{S}$ (⌓) these horizontal cusp 4-arrows correspond to reparametrizing the 1- and 3-handles.

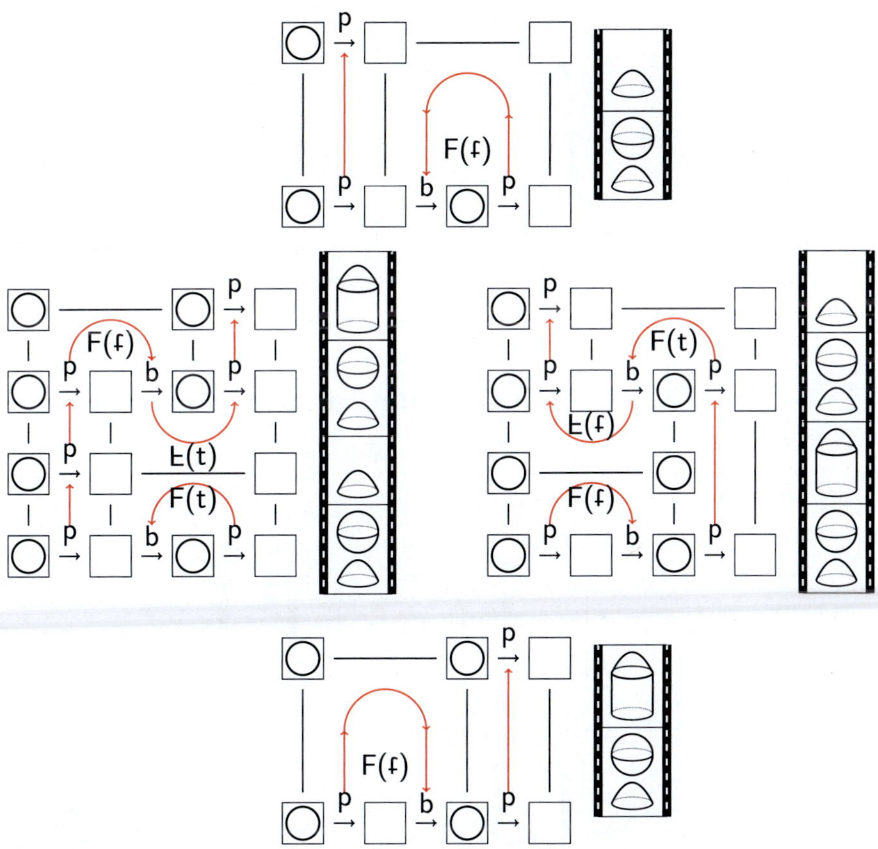

While there are eight total horizontal cusps to consider, only two representative horizontal cusps are depicted by the figures here. Variations can be made by choosing the direction of triple arrows or by reflecting the double arrows and their associated movie representations across a horizontal line. Here we give a description of the horizontal cusp that appears above, and we leave the reader to compile a description of the 4-arrow that is illustrated below.

There are four regions that are associated to the above illustration: south, west, north, and east (S, W, N, and E). The 4-arrow illustrated has as its source 3-arrow the vertical composition $\begin{bmatrix} N \\ W \\ S \end{bmatrix}$. The target is the 3-arrow $\begin{bmatrix} N \\ E \\ S \end{bmatrix}$. Read the 3-arrows going up the page. Then $S \to_3 W$ is a positive saddle and $W \to_3 N$ is a positive left up cusp. As target $S \to_3 E$ is a positive right down cusp, and $E \to_3 N$

is a negative fork. The quadruple arrow represented is

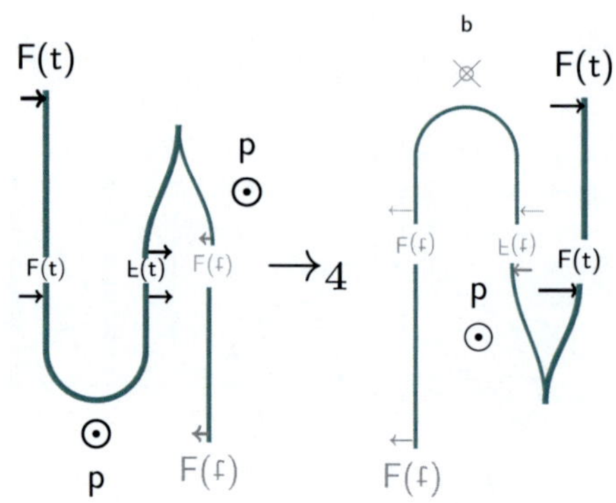

In the case of the horizontal cusp below, $S \to_3 W$ is a negative saddle, $W \to_3 N$ is a negative left up cusp, $S \to_3 E$ is a negative right down cusp, and $E \to_3 N$ is a positive fork.

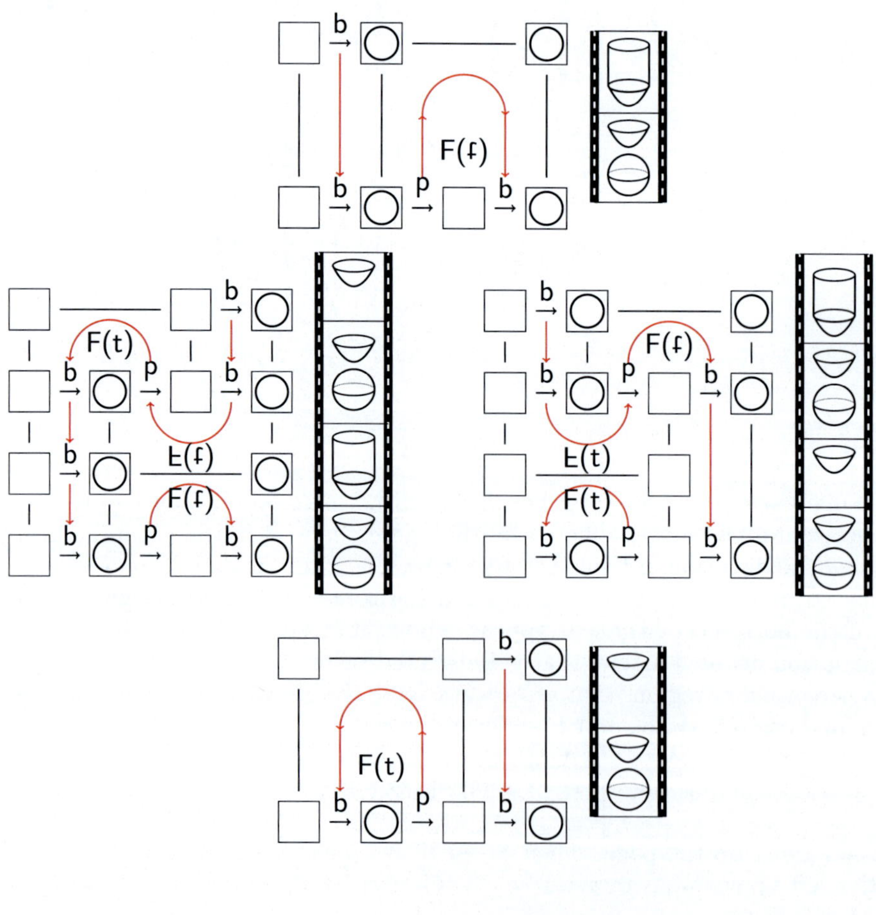

1.6. Increasing the dimension: part 2. In this section, j-arrows, for $j = 2, 3$, and 4, will be illustrated for the multi-category $\mathcal{S}$ () that corresponds to the fifth column of Table 1. In comparison to Section 1.5, the text will be sparse. The point here is that the relations among the higher arrows are all as they were before. The difference is that the 2-arrows represent different types of 3-dimensional handles, and so 3-arrows will also express different 4-dimensional aspects. The 3-dimensional aspects are presented via the associated movies; the 4-dimensional aspects are presented via the associated movie moves; and the 5-dimensional aspects are presented via the 4-arrows in the multi-category.

Generating double arrows.

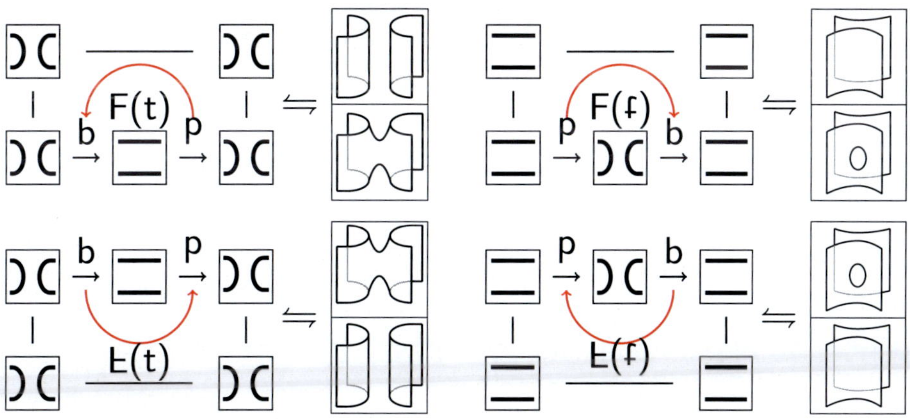

Births and deaths.

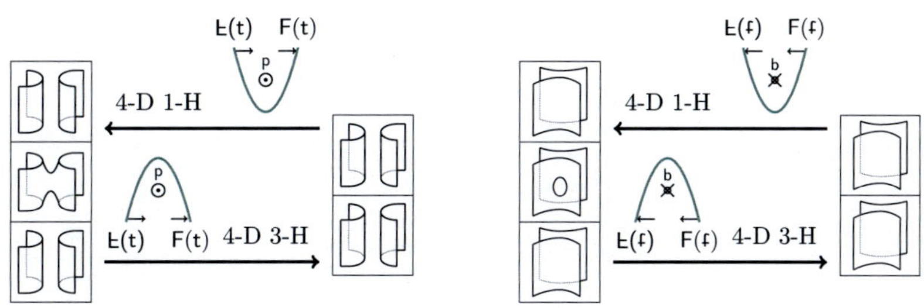

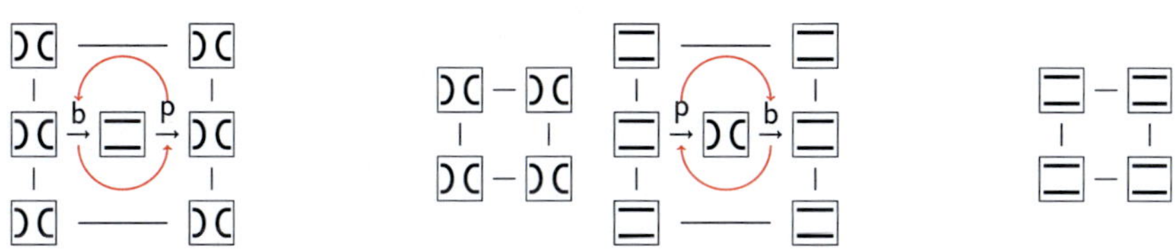

Saddles and forks.

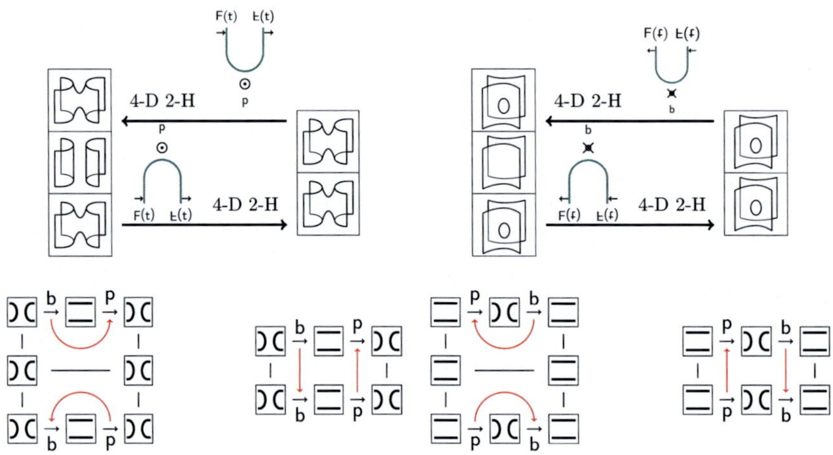

Cusps.

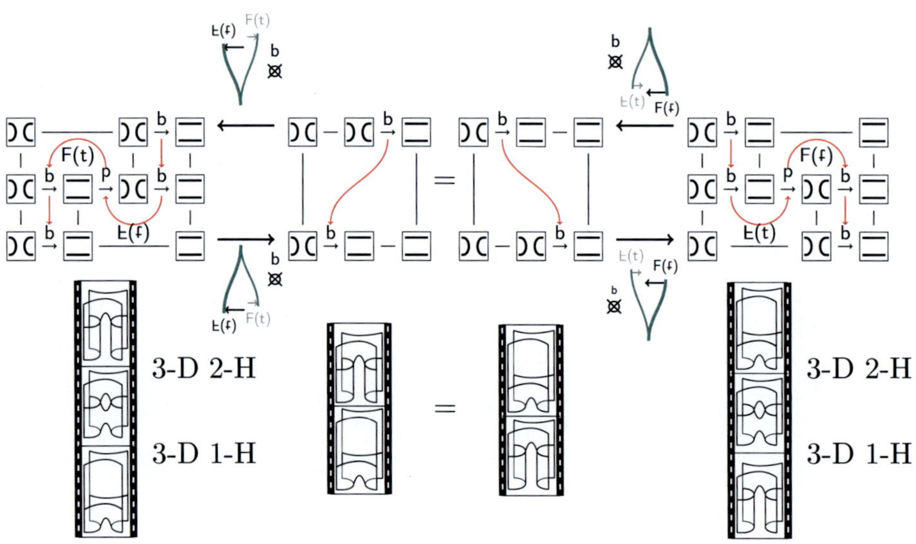

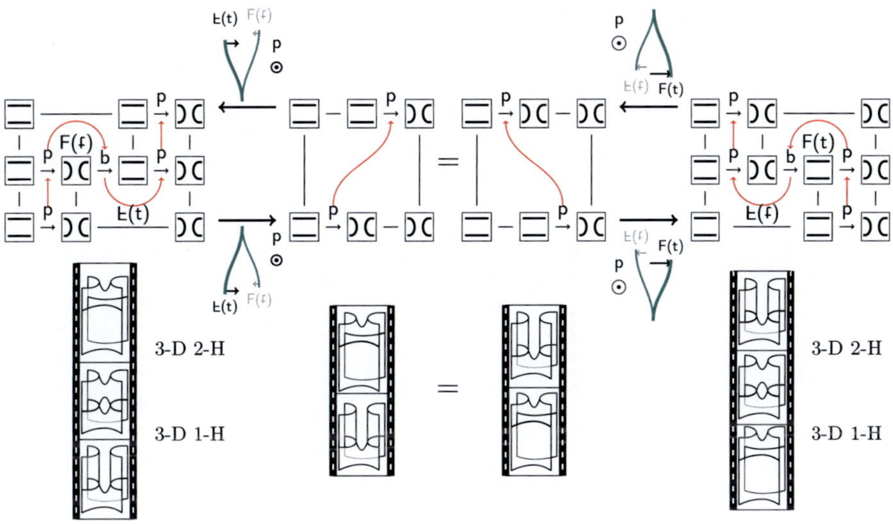

Compare the triple arrows presented in Section 1.4 of this chapter with the double arrows of $\mathcal{S}$ () and $\mathcal{S}$ (). The triple arrows that are negative births, deaths, saddles, and forks in $\mathcal{S}(\cap)$ coincide with the double arrows in $\mathcal{S}$ (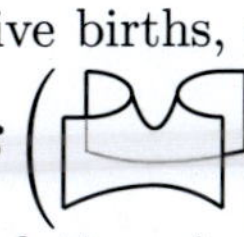). The cusps in $\mathcal{S}$ () and $\mathcal{S}$ (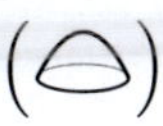), then are the critical cancelations in $\mathcal{S}(\cap)$. By considering arrows of $\mathcal{S}(\cap)$ as objects, a decategorification (Chapter 6, Section 2.2) process occurs, so that k-arrows in $\mathcal{S}(\cap)$ coincide with $(k-1)$-arrows in $\mathcal{S}$ () and $\mathcal{S}$ (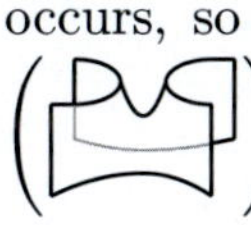).

Swallow-tails.

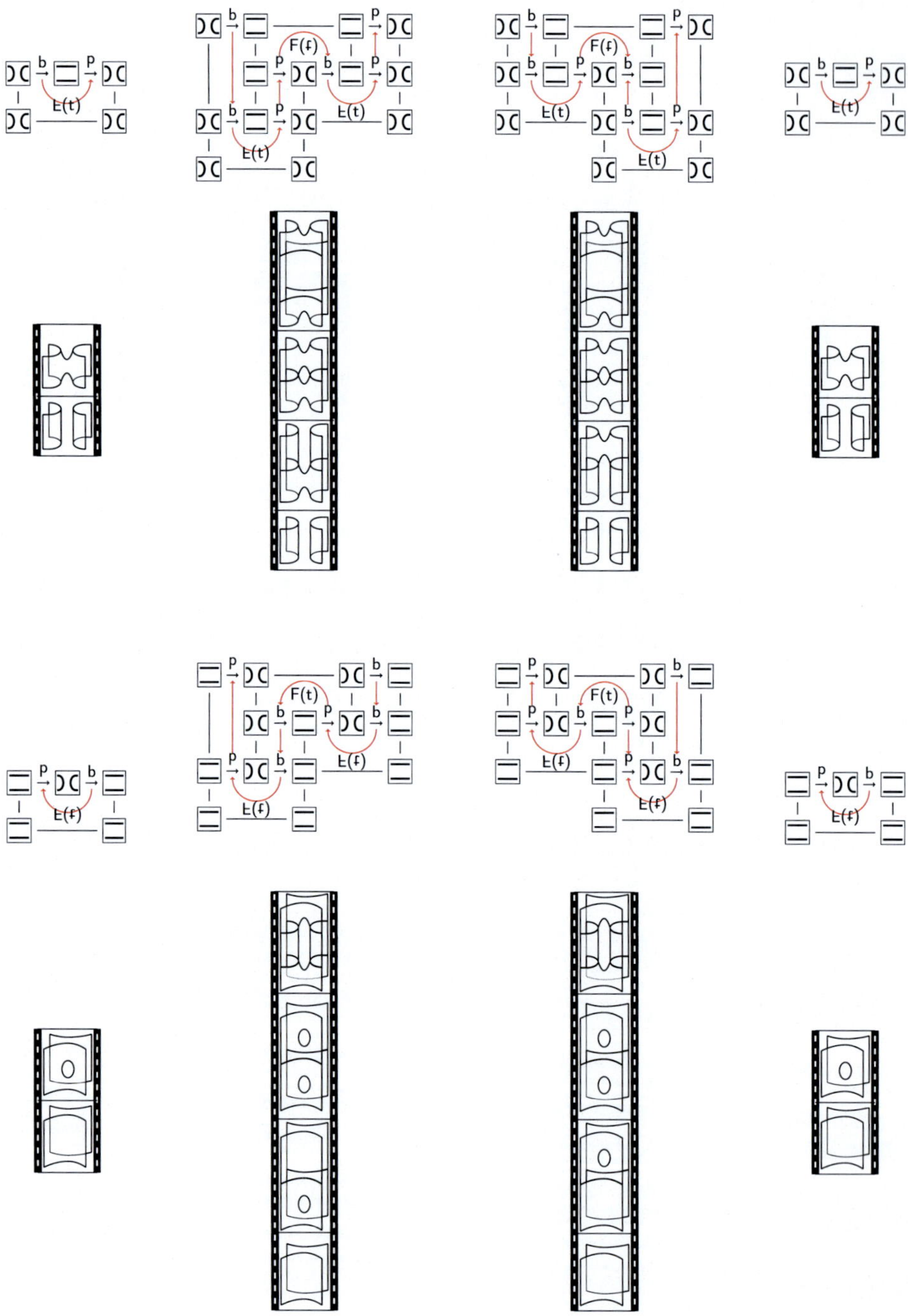

Critical cancelations. As in Section 1.5, only the triple composite of double arrows are illustrated that are involved in the critical cancelation for the current multi-category.

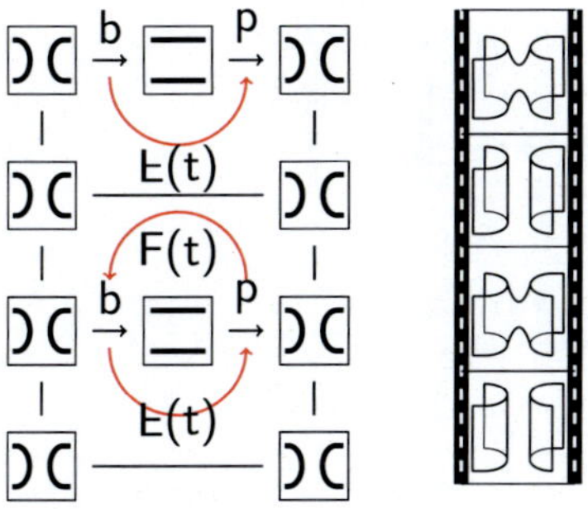

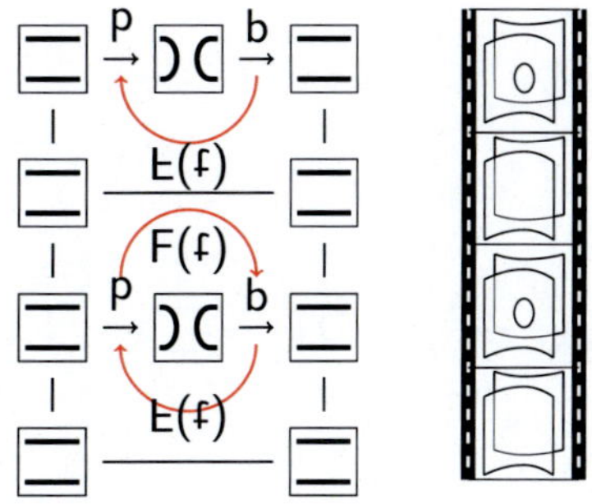

Horizontal cusps.

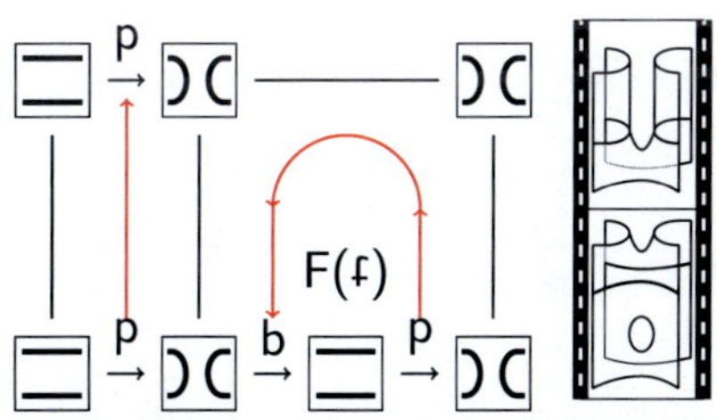

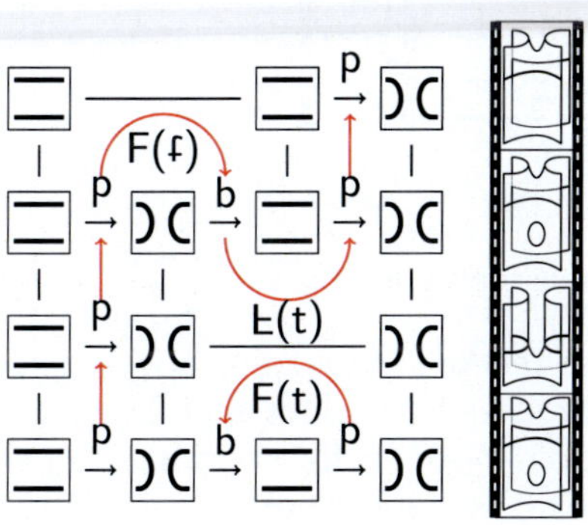

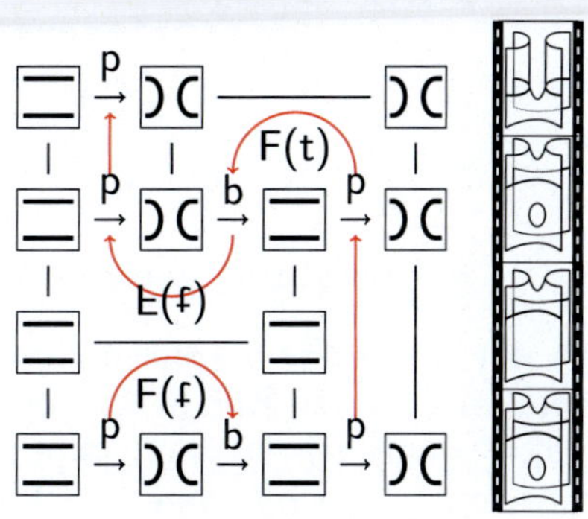

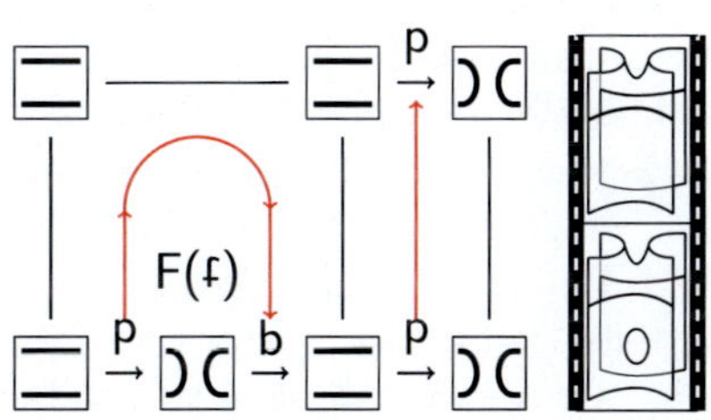

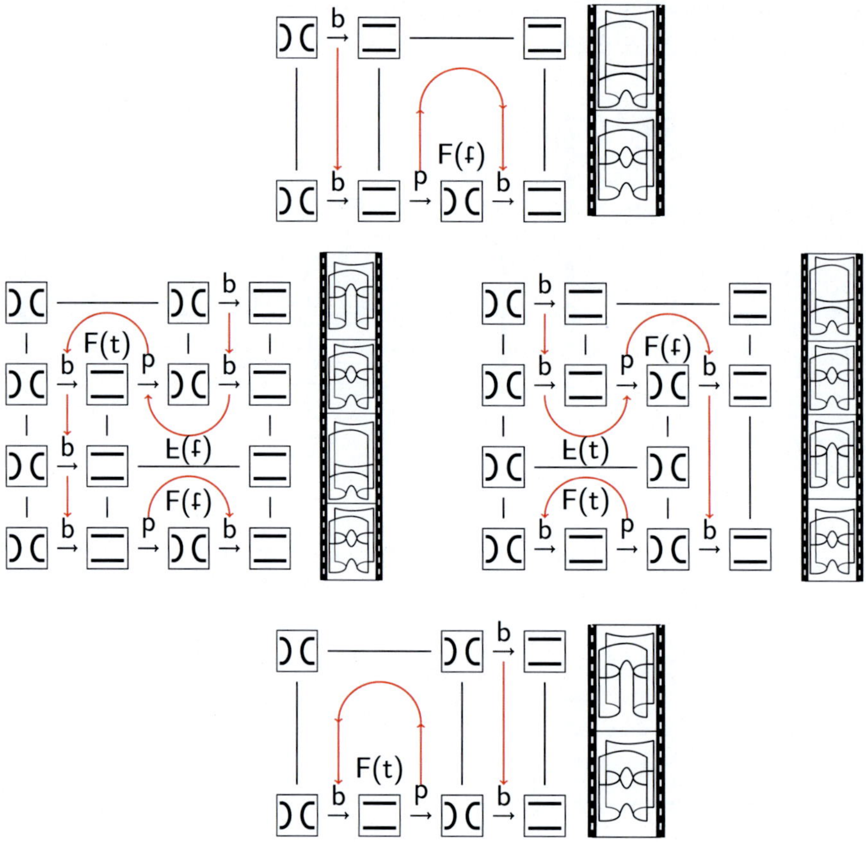

1.7. The remaining examples. The multi-category $\mathcal{S}$  was presented almost entirely graphically so that the reader could enjoy the process of imitating this construction in the context of the last two columns of Table 1 (and beyond) by mimicing the graphics of that section. Specifically, start from a pair of geometric objects such as an empty 3-dimensional box and a sphere contained in a box (as in column 6 of Table 1). Then envision the arrows p and b as a geometric process between them. In the case at hand, the process is the 3-dimensional ball that occurs in a space-time when a perfect balloon is filled or emptied. Then populate the source and target regions of the arrows in the diagrams for F(t) etc. with these geometric objects. Imagine a mythos in which the non-identity double arrows correspond to a higher 4-dimensional process. In the cases the authors are imagining (the sixth and seventh columns of the Table), these are 0-sets of real-valued quadratic forms. Identity arrows are topological products.

So 1-arrows correspond to 3-dimensional objects in 4-space, but of course, not all such objects are represented by a single column. Double arrows correspond to 4-dimensional objects in 5-space, and so forth. The double arrows are 4-dimensional handles while birth, death, saddle, and fork triple arrows are 5-dimensional handles. Cusps correspond to critical cancelations of lower dimensional multi-categories, while critical cancelations are the cancelations of $(i, i+1)$-handle pairs.

Notice also that the double arrows $\mathsf{F}(\mathsf{t})$ and $\mathsf{Ⱶ}(\mathsf{t})$ in $\mathcal{S}$ () could be included as objects in Table 1. Even further, the double arrows $\mathsf{F}(\mathsf{f})$ and $\mathsf{Ⱶ}(\mathsf{f})$ in $\mathcal{S}$ () could also be a t and f pair. The content of the inductively defined k-arrows F of Chapter 5, Section 3 is that these can be used to structure a multi-category of multiplicity $(k+2)$ with birth, death, fork, saddle, and cusp $(k+1)$-arrows that satisfy the $(k+2)$-arrow relations that are imposed by critical cancelation, lips, beak-to-beak, swallow-tail, and horizontal cusps. This idea is examined in detail in the sequel Section 2.

2. Weak inverses revisited

In this section, many of the multi-categories of the form $\mathcal{S}$ will be amalgamated. Since $\mathcal{S}$ () is oriented while $\mathcal{S}(\cap)$ and her daughters are not, notation that will address the orientation issues will be introduced. Again, start from two objects $\mathsf{f} = \emptyset$ and $\mathsf{t} = \bullet$, but because inductive constructions are necessary, these will be denoted as $\mathsf{f} = 0$ and $\mathsf{t} = 1$. In general, an arbitrary pair of objects can be chosen to produce the inductive construction.

Some variation of binary notation for non-negative integers will be employed. Identity k-arrows will be labeled by f_x where x is a binary sequence $x = \epsilon_k \cdots \epsilon_1$, and $\epsilon_j \in \{0,1\}$ for $j = 1, \ldots, k$. As usual, let $\mathsf{p} : \mathsf{f} \to \mathsf{t}$ and $\mathsf{b} : \mathsf{t} \to \mathsf{f}$ denote the generating 1-arrows. To start the induction, let f_0 denote the identity 1-arrow $0 — 0$, and let f_1 denote the identity 1-arrow $1 — 1$. Let t_0 denote the composition pb, and let t_1 denote the composition bp. For $\epsilon \in \{0,1\}$, let $\mathsf{p}_\epsilon : \mathsf{f}_\epsilon \to \mathsf{t}_\epsilon$, and let $\mathsf{b}_\epsilon : \mathsf{t}_\epsilon \to \mathsf{f}_\epsilon$. These are alternative notations: $\mathsf{p}_0 = \mathsf{Ⱶ}(\mathsf{f})$, $\mathsf{p}_1 = \mathsf{Ⱶ}(\mathsf{t})$, $\mathsf{b}_0 = \mathsf{F}(\mathsf{f})$, and $\mathsf{b}_1 = \mathsf{F}(\mathsf{t})$.

Suppose that for $x = \epsilon_{k-1} \cdots \epsilon_1$, arrows of multiplicity $(k-1)$, named f_x, t_x have been defined, with k-arrows, $\mathsf{p}_x : \mathsf{f}_x \to \mathsf{t}_x$ and $\mathsf{b}_x : \mathsf{t}_x \to \mathsf{f}_x$ between them. Let $I[\mathsf{s}_x]$ denote the identity k-arrow upon s for $\mathsf{s} = \mathsf{t}, \mathsf{f}, \mathsf{p}$, or b. Then define k-arrows

- $$\mathsf{f}_{0x} = I[\mathsf{f}_x],$$
- $$\mathsf{f}_{1x} = I[\mathsf{t}_x],$$
- $$\mathsf{t}_{0x} = \mathsf{p}_x \mathsf{b}_x,$$
  and
- $$\mathsf{t}_{1x} = \mathsf{b}_x \mathsf{p}_x.$$

And define non-identity generating $(k+1)$-arrows

- $$\mathsf{p}_{0x} : \mathsf{f}_{0x} \to \mathsf{t}_{0x},$$
  whence $\mathsf{p}_{0x} = \mathsf{Ⱶ}(\mathsf{f}_x)$;
- $$\mathsf{p}_{1x} : \mathsf{f}_{1x} \to \mathsf{t}_{1x},$$
  whence $\mathsf{p}_{1x} = \mathsf{Ⱶ}(\mathsf{t}_x)$;

- $$\mathsf{b}_{0x} : \mathsf{t}_{0x} \to \mathsf{f}_{0x},$$
  whence $\mathsf{b}_{0x} = \mathsf{F}(\mathsf{f}_x)$; and
- $$\mathsf{b}_{1x} : \mathsf{t}_{1x} \to \mathsf{f}_{1x},$$
  whence $\mathsf{b}_{1x} = \mathsf{F}(\mathsf{t}_x)$.

Then define $(k+2)$-arrows of the form $(\pm)$-birth, $(\pm)$-death, $(\pm)$-saddle, $(\pm)$-fork, ($\pm$, left/right, up/down)-cusps, and exchangers. Observe that the birth, death, saddle, and fork $(k+2)$-arrows coincide with $(k+1)$-arrows of type p and b. Finally, construct lips, beak-to-beak, critical cancelations, swallow-tail, horizontal cusp, and exchanger $(k+3)$-arrows.

Inductive Step

$x = \epsilon_k\epsilon_{k-1}\ldots\epsilon_1$; $\epsilon_j \in \{0,1\}$, for $j \in \{1,\ldots,k\}$.

Obj: $\mathsf{t}_x, \mathsf{f}_x$

Generating 1-arrows: $\mathsf{p}_x : \mathsf{f}_x \to \mathsf{t}_x$; $\mathsf{b}_x : \mathsf{t}_x \to \mathsf{f}_x$.

Inductively define

$$\mathsf{f}_{0x} : \mathsf{f}_x \text{ --- } \mathsf{f}_x,\ \mathsf{f}_{1x} : \mathsf{t}_x \text{ --- } \mathsf{t}_x;\ \mathsf{t}_{0x} = \mathsf{p}_x\mathsf{b}_x;\ \mathsf{t}_{1x} = \mathsf{b}_x\mathsf{p}_x.$$

Id. 2-arrows:

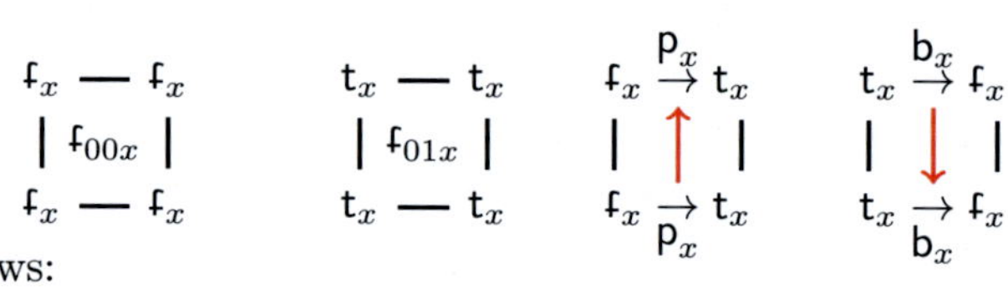

Generating 2-arrows:

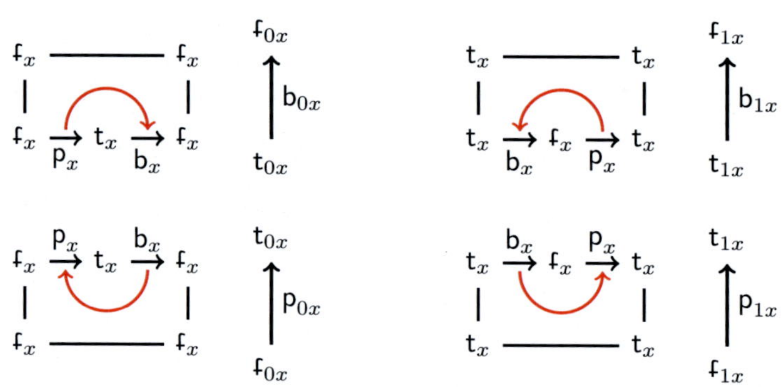

Definition: Weakly Invertible Arrows. We say that the arrows f and t are **weakly invertible** if for all k there are $(\pm)$-birth, $(\pm)$-death, $(\pm)$-saddle, and $(\pm)$-fork inductively defined $(k+2)$-arrows that satisfy the lips, beak-to-beak, critical cancelation, horizontal cusp, swallow-tail, and exchanger $(k+3)$-relations. The arrows f and t depend upon the objects 0 and 1, but it is reasonable to start from an arbitrary pair of objects and create an inductive family of arrows following this construction. The construction can be truncated at any finite level. Finally, the steps in the construction are exponential so it is wise to cut the construction after a very small number of steps.

2.1. Revisiting the diagrams. There are three purposes of this section. First, as an aid to the reader, the k through $(k+2)$ arrows are illustrated in the inductive context. These illustrations were initiated above. Second, critical cancelations are expressed in terms of cusps in the inductive steps. Third, from the objects 0 and 1 the ps and bs are constructed up to a reasonable level.

The inductive step for negative births and deaths.

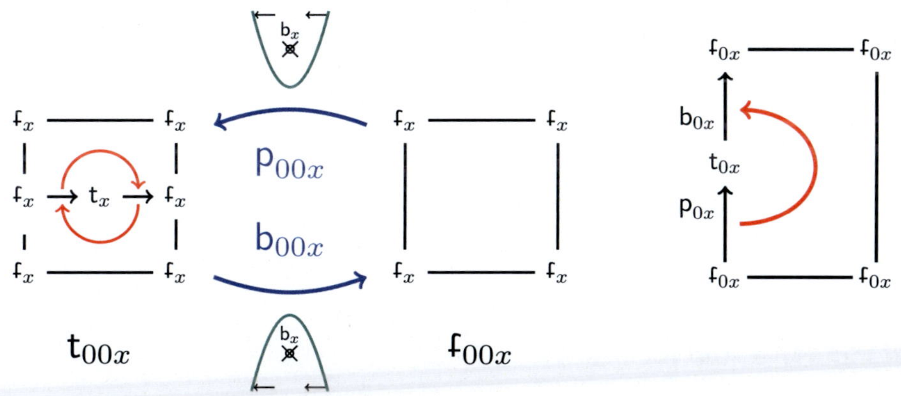

The inductive step for positive births and deaths.

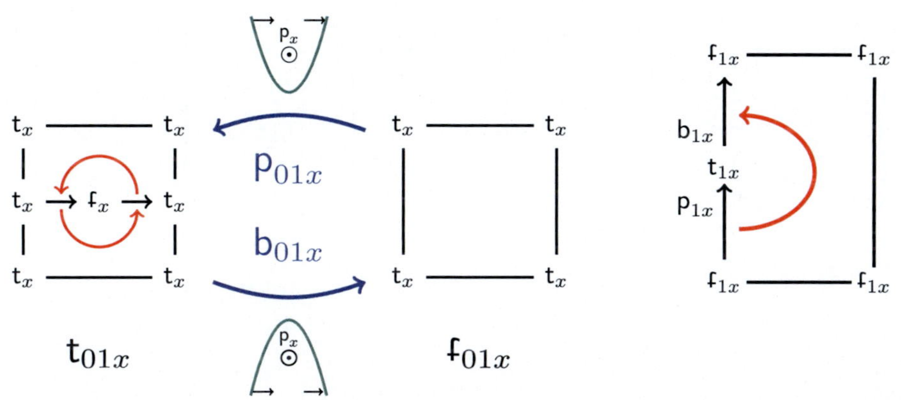

As indicated by these figures, the generating 2-arrows for x are generating 1-arrows for $\epsilon_{k+1}x$ where $\epsilon \in \{0,1\}$. Furthermore, the birth, death, saddle and fork 3-arrows for x are generating 2-arrows for $\epsilon_{k+1}x$. On the right side of the illustration the decategorification (Chapter 6, Section 2.2) is illustrated. These F-type 2-arrows are drawn from the point of view of the arrows $\mathsf{b}_{\epsilon\delta x}$. Furthermore, without any reference to the surface case, their format as arches helps motivate the representative glyphs.

The inductive step for negative saddles and forks.

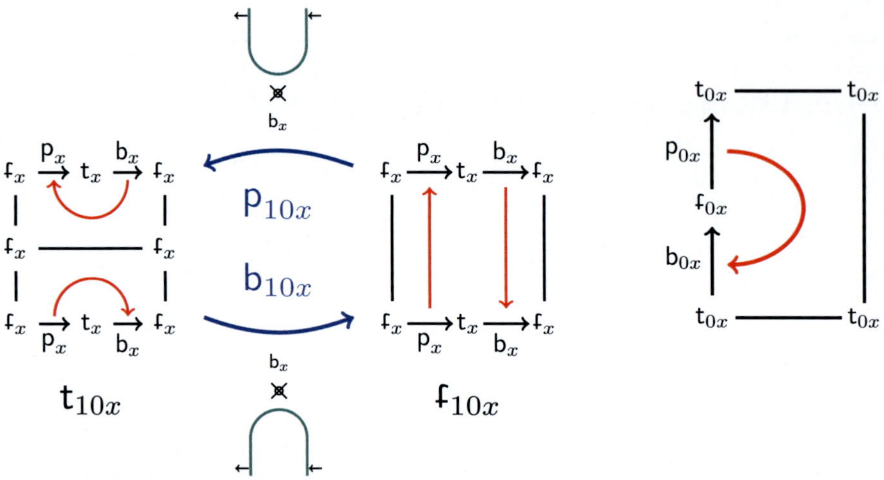

The inductive step for positive saddles and forks.

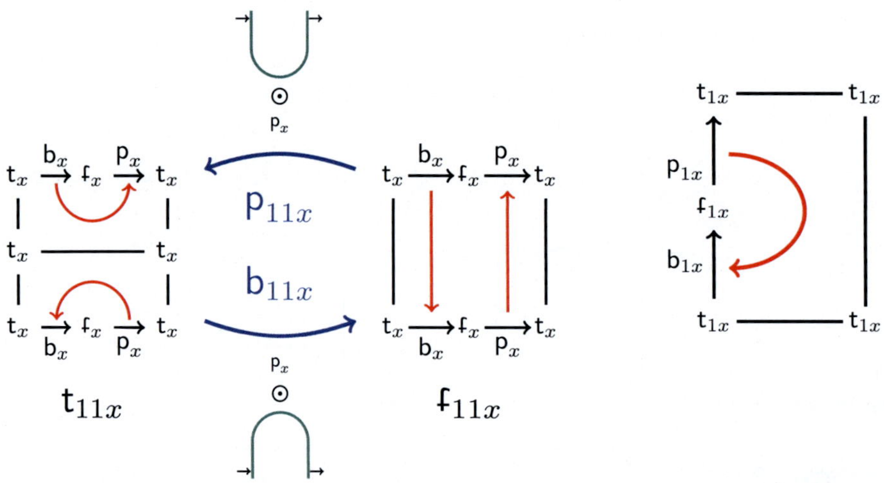

Cusps in the inductive step.

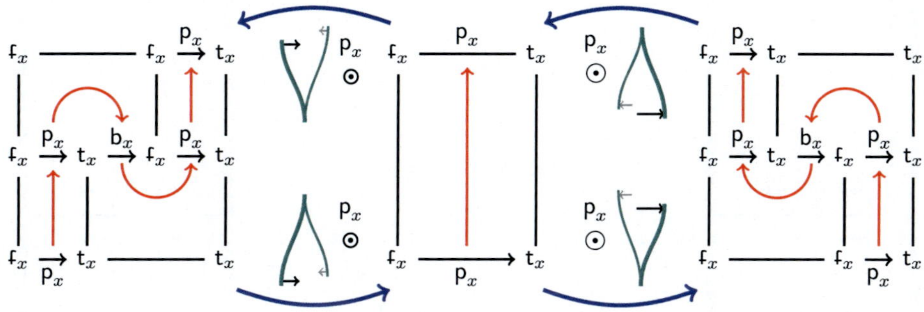

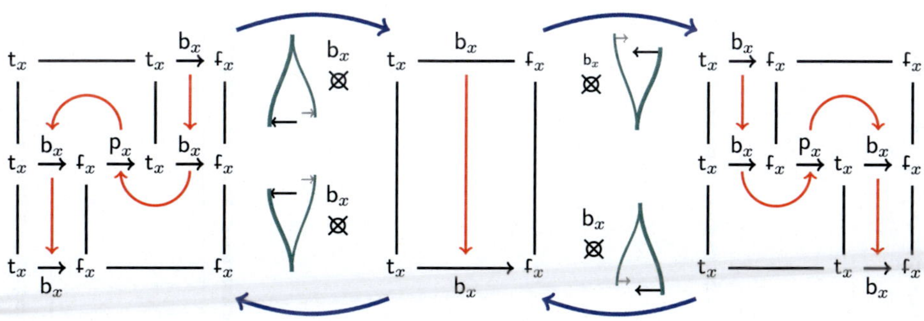

Observe that the lips and beak-to-beak $(k+3)$-arrows can be imagined by following a cycle of down cusp to up cusp or vice versa for any of the $(\pm)$-left/right cusps in the illustrations above. To facilitate the visualization of these, directions of the $(k+3)$-arrows have been drawn inconsistently on the left and right sides and in the positive and negative cusps.

Next two examples of critical cancelations are given. On the left of the illustration is a movie that depicts a positive saddle followed by a positive death. On the right is a positive birth followed by a positive fork. If $x = \epsilon_k \epsilon_{k-1} \dots \epsilon_1$, then each movie represents a $(k+3)$-arrow. The glyph in the center of the figure summarizes that either movie is $(k+4)$-equivalent to the identity arrow on b_{1x}. The $(k+2)$-arrow b_{1x} agrees with $\mathsf{F}(\mathsf{t}_x)$. As was our habit above, we omit the movie that is the identity upon b_{1x}.

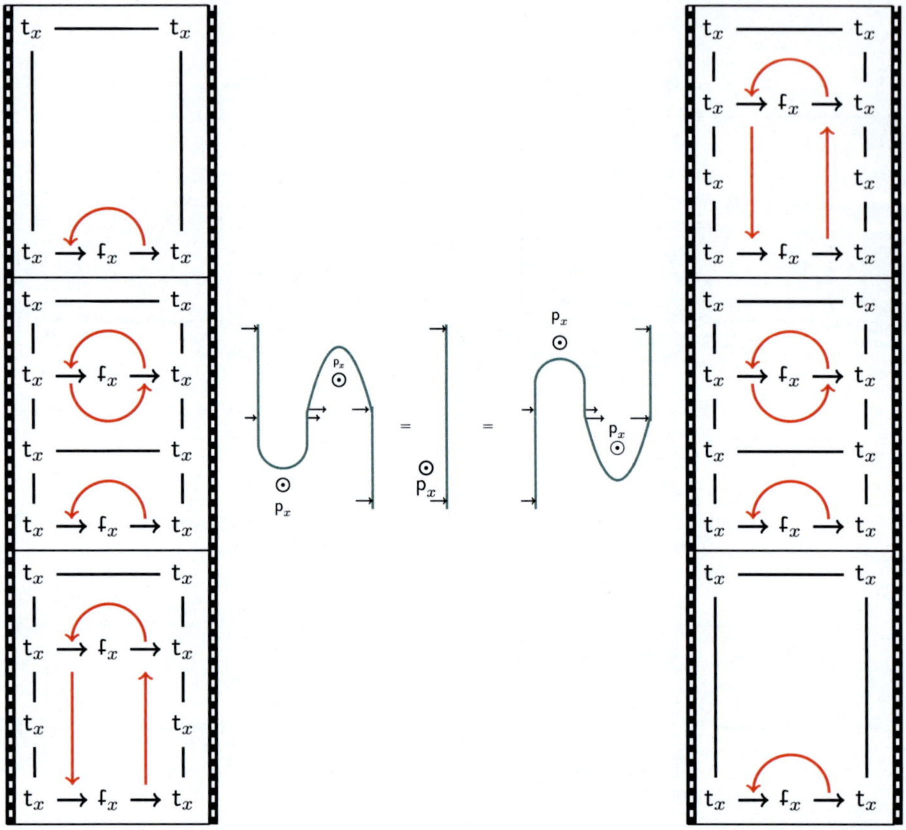

In the next illustration, each multiple arrow is rewritten in one lower dimension. So $(k+1)$-arrows are summarized as objects, $(k+2)$-arrows appear as arrows, and so forth. The subscripts are incremented to reflect their higher nature. In this way, we see that critical cancelations correspond to cusps. The other cases of critical cancelations can be handled in a similar fashion. Explicit arrows that describe these as cusps have been omitted.

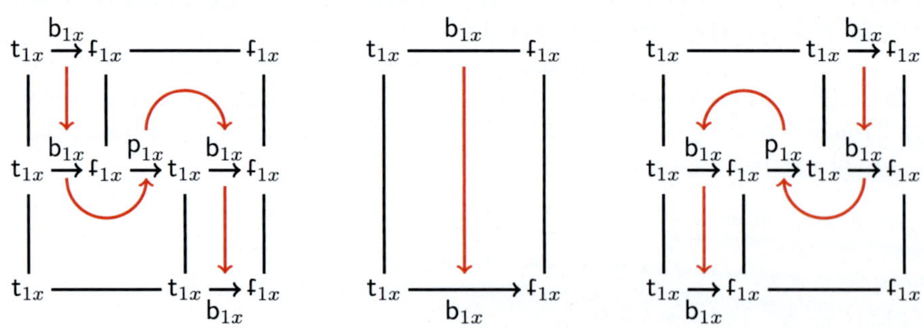

As a last stage in this section, revisit the multi-category $\mathcal{S}\left(\perp\right)$ using the notation that $\emptyset \leftrightarrow 0$ and –•– $\leftrightarrow 1$. Here higher order arrows are included. So the current multi-category turns out to contain oriented versions of $\mathcal{S}(\cap)$, $\mathcal{S}\Big($$\Big)$, $\mathcal{S}\Bigg($$\Bigg)$, and beyond. Orientations appear at the initial stages since the arcs that describe the double arrows point upwards at p. Consequently, 0 is always to the left of an arc that describes a double arrow.

$$\mathsf{t}_0 : 0 \xrightarrow{\mathsf{p}} 1 \xrightarrow{\mathsf{b}} 0 \qquad\qquad \mathsf{t}_1 : 1 \xrightarrow{\mathsf{b}} 0 \xrightarrow{\mathsf{p}} 1$$

$$\mathfrak{f}_0 : 0 \longrightarrow 0 \qquad\qquad \mathfrak{f}_1 : 1 \longrightarrow 1$$

arrows as objects

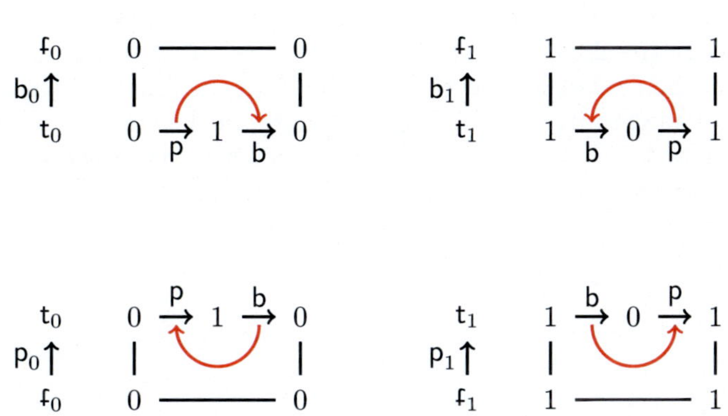

2-arrows as 1-arrows

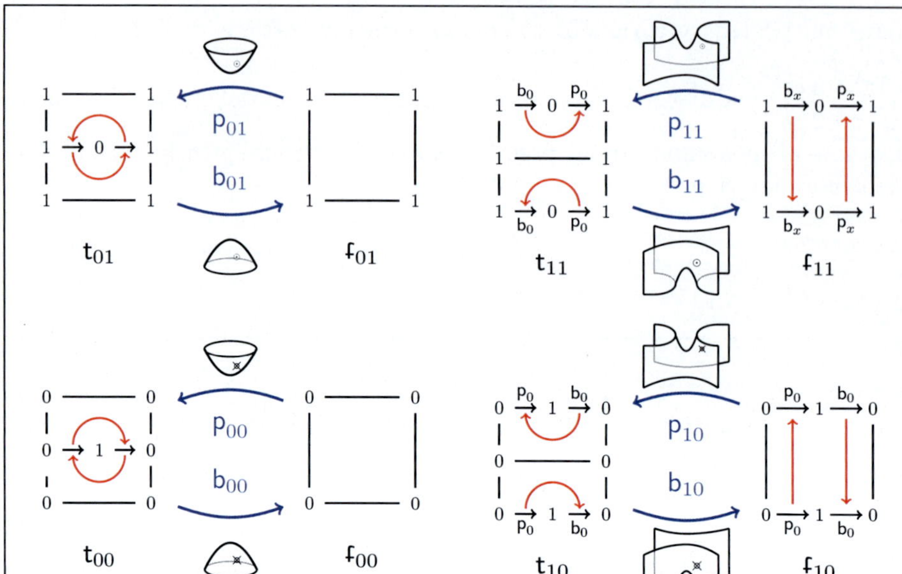

3-arrows as 2-arrows

Roughly speaking, identity k-arrows are $\mathfrak{f}$-type objects, and the compositions pb or bp are t-type objects. In this way, the double arrows $\mathsf{F}(0)$ and $\mathsf{F}(1)$ are interpreted as arrows b_0 and b_1, respectively, and their reflective companions are arrows p_0 and p_1. Above the birth, death, saddle, and fork arrows are illustrated, not via glyphs, but as genuine neighborhoods of critical points of embedded oriented surfaces. The oriented normal vector to a surface points into the 3-dimensional region that is labeled 1.

This perspective is reflected in two 3-dimensional regions that are drawn as sources of the 4-arrows p_{000} and p_{001}. The 2-dimensional sphere that is born as the target of p_{000} is oriented inwardly. Thus the region outside of it is labeled 0, and called "empty" in the illustration that follows. The target of p_{001} is an outwardly oriented sphere. So we labeled the outside "full." In the illustrations thin lines and thick lines are meant to further distinguish the 3-dimensional sources $\mathfrak{f}_{000}$ and $\mathfrak{f}_{001}$ respectively. Here $\mathfrak{f}_{000}$ is the identity upon the identity upon the identity upon 0. That is, the identity upon 0 is a segment $\mathfrak{f}_0$. The identity upon it is an empty square $\mathfrak{f}_{00}$, and the identity upon $\mathfrak{f}_{00}$ is an empty cube.

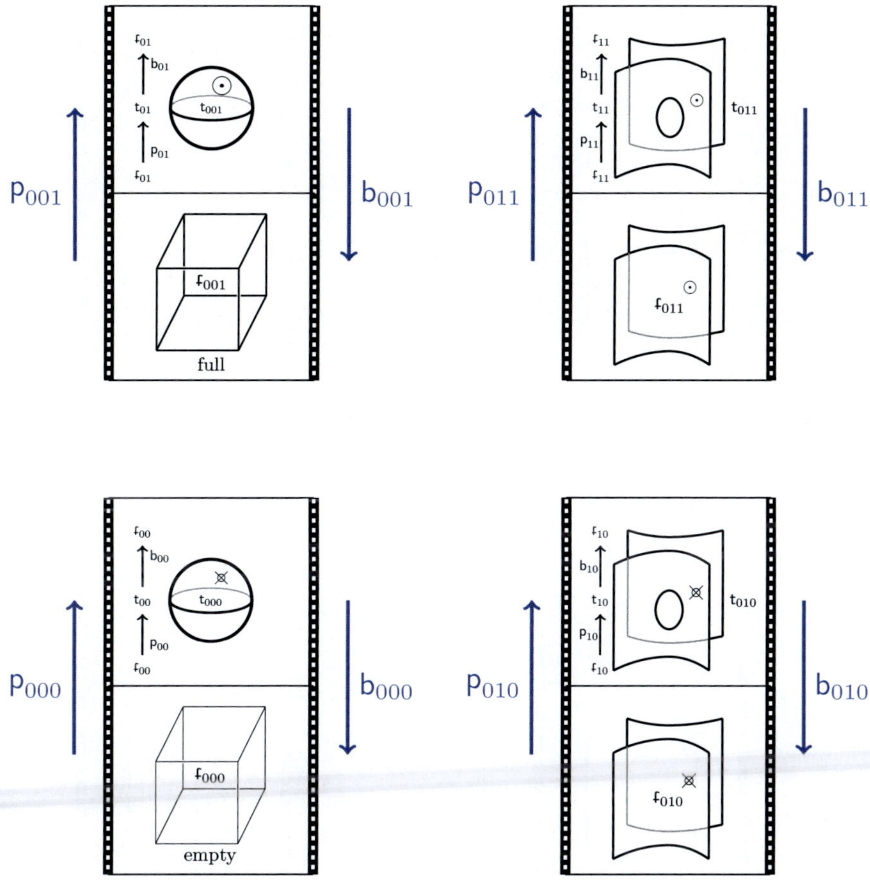

As before, the 4-arrows here correspond to 3-dimensional handles. Now they are oriented. The arrows p_{000} and p_{001} correspond to 3-dimensional 0-handles, the arrows b_{000} and b_{001} correspond to 3-dimensional 3-handles, and the remaining arrows are either 3-dimensional 1- or 2-handles. The p/b notation here differs from that given in Sections 1.5 and 1.6 of this chapter since here we want the notations to be consistent across dimensions, and earlier b was used to suggest "birth." The arrow $\mathsf{p}_{\epsilon_3\epsilon_2\epsilon_1}$ corresponds to a 3-dimensional k-handle if and only if the arrow $\mathsf{b}_{\epsilon_3\epsilon_2\epsilon_1}$ corresponds to a 3-dimensional $(3-k)$-handle.

Within the illustrations, the arrows $\mathsf{p}_{\epsilon_3\epsilon_2\epsilon_1}$ are arranged in two groupings. Those of the form $\mathsf{p}_{0\epsilon_2\epsilon_1}$ appear above this paragraph while those of the form $\mathsf{p}_{1\epsilon_2\epsilon_1}$ appear below this paragraph. The reverses $\mathsf{b}_{\epsilon_3\epsilon_2\epsilon_1}$ appear as downward pointing arrows since their sources are of the form $\mathsf{t}_{\epsilon_3\epsilon_2\epsilon_1}$.

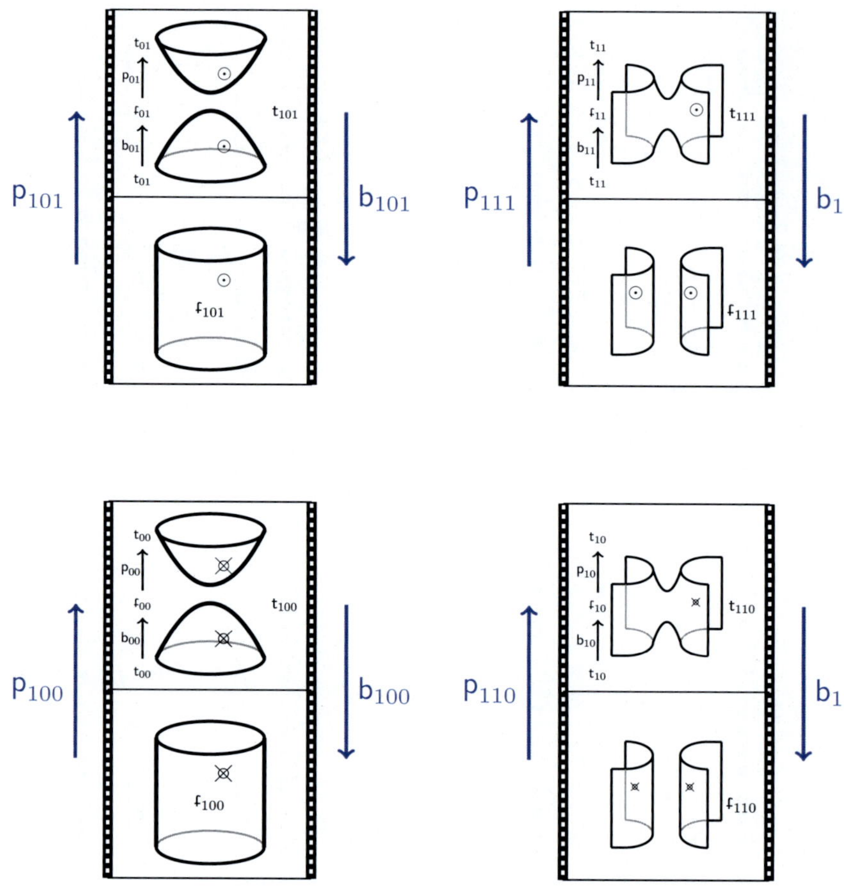

To complete the catalogue of the lower dimensional generating arrows, the sixteen possible arrows $\mathsf{b}_{\epsilon_4\epsilon_3\epsilon_2\epsilon_1}$ and $\mathsf{p}_{\epsilon_4\epsilon_3\epsilon_2\epsilon_1}$ are indicated as relationships between movies, or movie moves. Each movie represents an oriented 3-dimensional manifold. And the arrows $\mathsf{b}_{\epsilon_4\epsilon_3\epsilon_2\epsilon_1}$ or $\mathsf{p}_{\epsilon_4\epsilon_3\epsilon_2\epsilon_1}$ represent 4-dimensional oriented handles that relate them. Here $\mathsf{p}_{\epsilon_4\epsilon_3\epsilon_2\epsilon_1}$ is a k-handle if and only if $\mathsf{b}_{\epsilon_4\epsilon_3\epsilon_2\epsilon_1}$ is a $(4-k)$-handle.

To complete the next stage in the inductive process, compare $\mathsf{b}_{\epsilon_4\epsilon_3\epsilon_2\epsilon_1}\mathsf{p}_{\epsilon_4\epsilon_3\epsilon_2\epsilon_1}$ and $\mathsf{p}_{\epsilon_4\epsilon_3\epsilon_2\epsilon_1}\mathsf{b}_{\epsilon_4\epsilon_3\epsilon_2\epsilon_1}$ to the identity arrows on the source or target.

Recall that by labeling the higher arrows as p and b, we are also imposing cusps as higher level arrows, and these satisfy lips, beak-to-beak, swallow-tail, and horizontal cusp relationships. These follow the same pattern as in the unoriented cases given in Section 1.

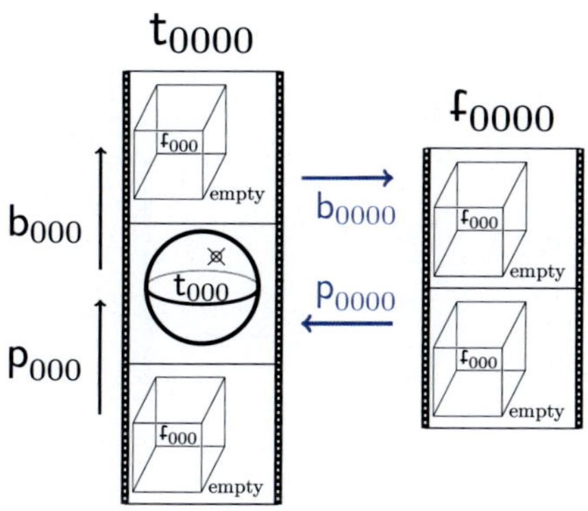
t0000
f000
empty
t000
f000
empty
b000
p000
b0000
p0000
f0000
f000
empty
f000
empty

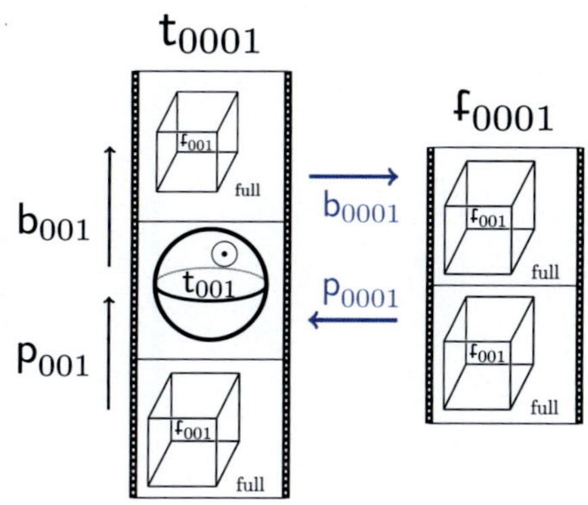
t0001
f001
full
t001
f001
full
b001
p001
b0001
p0001
f0001
f001
full
f001
full

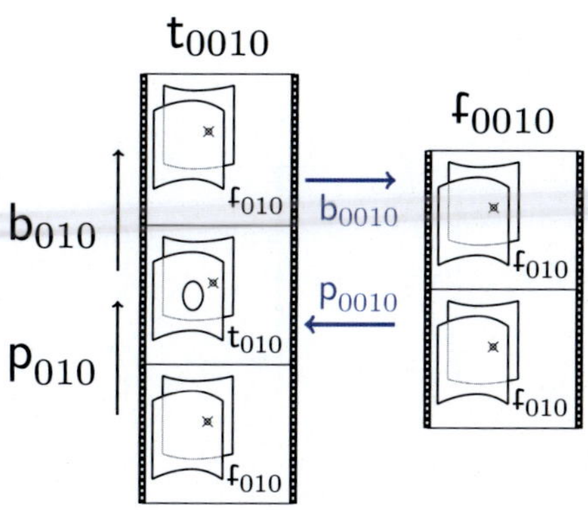
t0010
f010
t010
f010
b010
p010
b0010
p0010
f0010
f010
f010

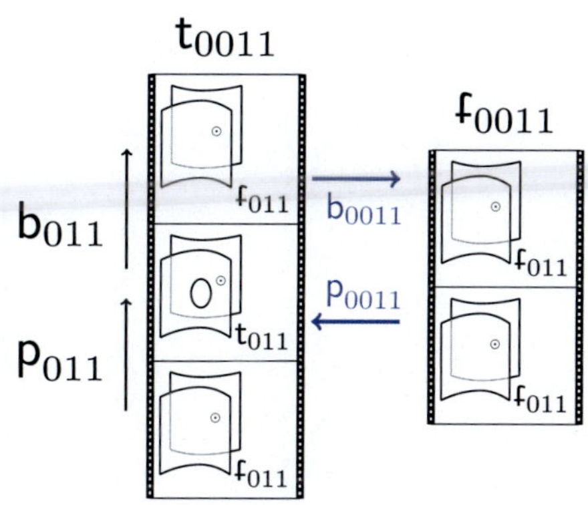
t0011
f011
t011
f011
b011
p011
b0011
p0011
f0011
f011
f011

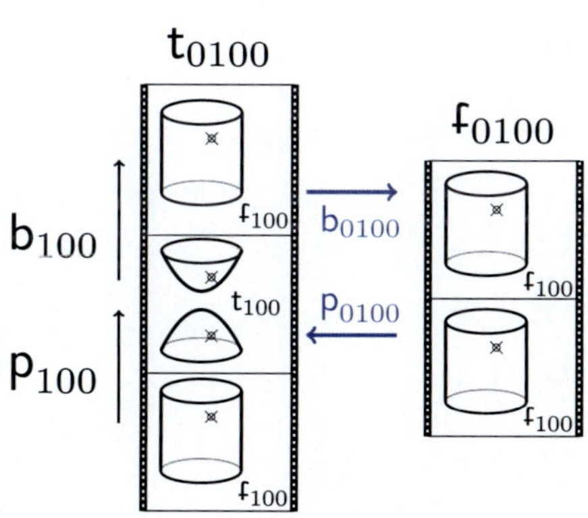
t0100
f100
t100
f100
b100
p100
b0100
p0100
f0100
f100
f100

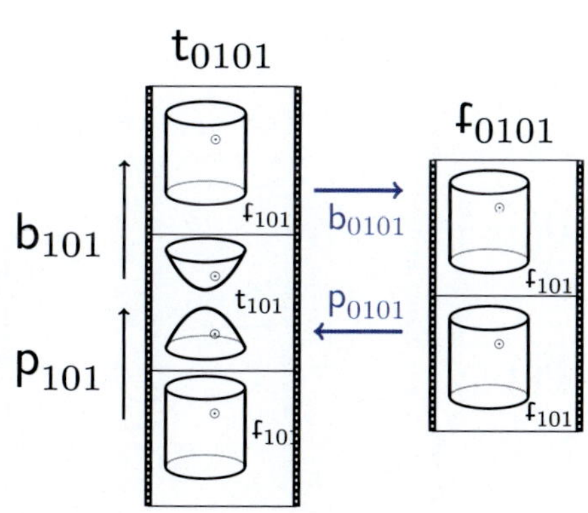
t0101
f101
t101
b101
p101
b0101
p0101
f0101
f101
f101

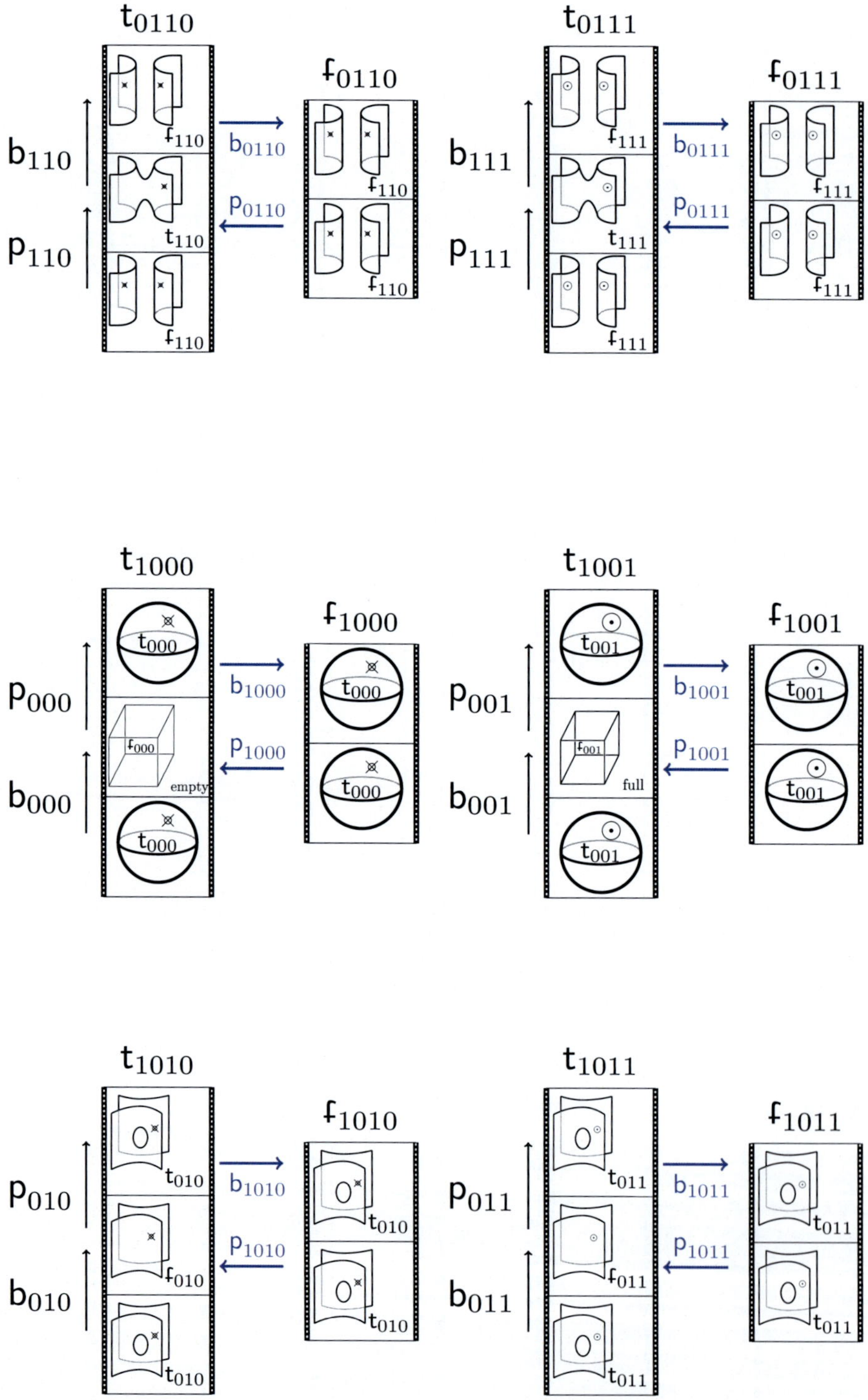
t0110
f0110
b110
p110
b0110
p0110
t0111
f0111
b111
p111
b0111
p0111
t1000
f1000
p000
b000
b1000
p1000
empty
t1001
f1001
p001
b001
b1001
p1001
full
t1010
f1010
p010
b010
b1010
p1010
t1011
f1011
p011
b011
b1011
p1011

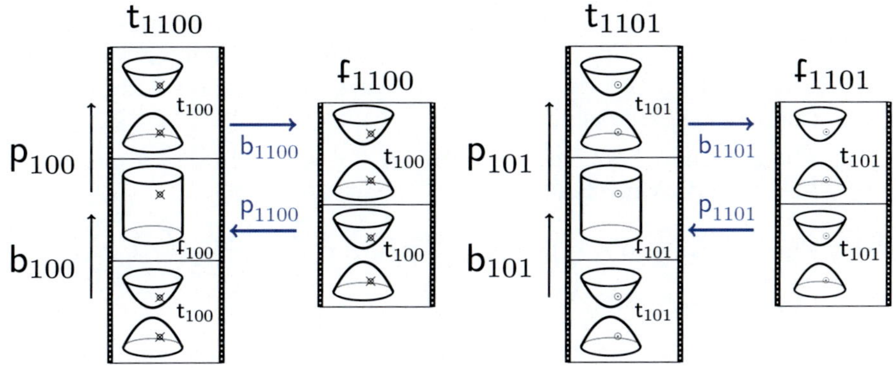

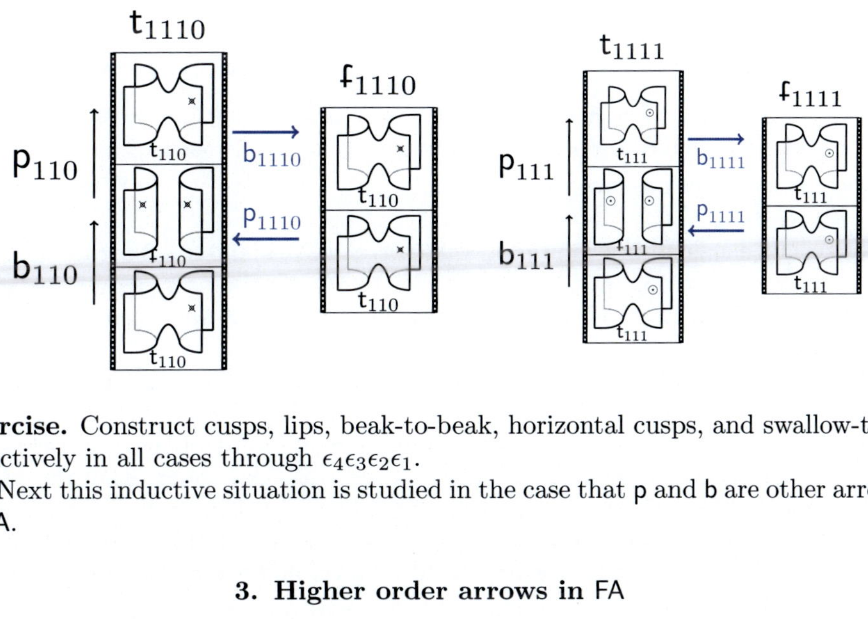

Exercise. Construct cusps, lips, beak-to-beak, horizontal cusps, and swallow-tails inductively in all cases through $\epsilon_4\epsilon_3\epsilon_2\epsilon_1$.

Next this inductive situation is studied in the case that p and b are other arrows in FA.

3. Higher order arrows in FA

In this section, the multi-category FA is revisited from the points of view developed above. In the original description the collection of objects corresponded to the natural numbers, $\mathbb{N} = \{0, 1, 2, \ldots\}$. This monoid also is a category that has one object; its morphisms are generated by a non-identity generator. Since the things that were originally called objects also are arrows, the arrows will be promoted to being 2-arrows, and so on. Thus the definition of the multi-category FA is recapitulated.

(1) The multi-category FA has 1-object. It will exist as an un-named entity.
(2) The 1-arrows of FA are freely generated by a non-identity generator. The collection of 1-arrows correspond to the set $\mathbb{N} = \{0, 1, 2, \ldots\}$. The 1-arrows will be denoted by Arabic numerals and also be denoted in unary notation. Thus $1 \leftrightarrow$ -•-, $2 \leftrightarrow$ -•-•-, etc. while 0 may be denoted by the empty character.

(3) The 2-arrows are generated by $\mathsf{U}, \mathsf{\cap}, \mathsf{Y}, \lambda$, ⊥, ⊤ , and of course I which is the identity upon -•-.
(4) The 2-arrows are composed globularly and via skew compositions. These compositions are presented mutatis mutandis in Chapter 3, Section 2.2. Compare this with the discussion in Chapter 4, Section 1.1 in which skew compositions are defined.
(5) The grouping of double arrows $\{\mathsf{U}, \mathsf{\cap}\}$, $\{\mathsf{Y}, \lambda\}$, and { ⊥, ⊤ } indicates that these are weak inverses.
(6) Thus the 3-arrows of the form 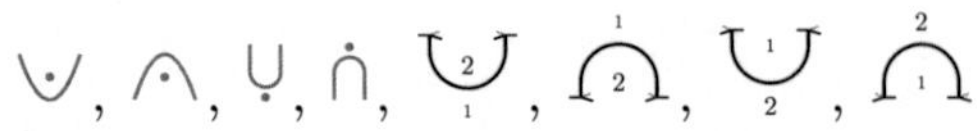

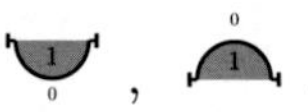 , 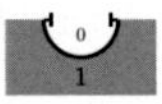, and  exist.
(7) These 3-arrows above appear in weak inverse pairs. This implies that they can be expressed as ps and bs, and higher order weak inverse relations hold. There are cuspidal relations, critical cancelations, lips, beak-to-beak, swallow-tail, and horizontal cusp relations among them and their higher order inductively defined multiple arrows.
(8) Associated to the 2-arrows U and ∩ are the cuspidal 3-arrows: V, Λ, V, and Λ. These are strong inverse pairs (so that lip and beak-to-beak identities hold), and they satisfy the swallow-tail relations. Furthermore, these cusps interact with saddles U and forks ∩ to satisfy horizontal cusp relations.
(9) The elements in Chapter 4, Tables 1, 2, 6, 7, and 9 are 3-arrows that are listed as strong inverse pairs.
(10) The elements in Chapter 4, Tables 3, 4, 5, and 8 are 3-arrows that are listed as weak inverse pairs.
(11) The diagrammatic proof of the coalgebra axioms given in Chapter 5, Section 4 are among the 4-arrows.
(12) An additional 4-arrow is formulated in Section 3.1 of this chapter as the (3, 2)-relation. Some other higher order relations follow from the higher order Stasheff relations that are articulated in Chapter 8.
(13) Additional 4-arrows articulated in Section 3.2 of this chapter that encode the unit axioms as expressed in Joyal and Street [**36**] and analogues thereof.
(14) Additional 4-arrows are formulated in Section 3.3 in this chapter that allow for units, co-units, and folds to move past optima on the seam set.

A 2-dimensional foam that is embedded in 3-dimensional space is analogous to a surface that is embedded in 3-space. In general, the boundary of a foam consists of embedded trivalent graphs in the plane, and edges that are formed as identities of units or co-units. Such edges can originate or terminate at seams. For example, any of the relations that describe the left or right unit or co-unit axiom such as  will also be called a *shark's fin*. This colloquialism is an abuse of notation, but it simplifies the text below. We can describe the foam as a finite sequence of compositions of triple arrows. The foam diagrams of Chapter 4, Section 1.3 are among those 3-arrows.

Definition. To keep track of the extra compositional dimension of triple arrows, a pair of non-negative integers is assigned to each seam or fold. The foam F is embedded in 3-space and a visualization (and therefore a glyphographic description) of the foam is tantamount to a generic projection onto a 2-dimensional plane. Assume that the foam F appears in front of the plane upon which the projection occurs. Consider a ray that is perpendicular to the plane, that usually pierces F transversely, and that is defined by the pair of points (point on the seam or fold, its projected image). Say the ray is $\vec{SQ}$ with $p(Q) = S$ — the projected image of the point Q that lies on the fold or seam. Then the pair of non-negative integers $(\#([S,Q)\cap F), \#((Q,\infty)\cap F))$ is associated to the component of the fold or seam. To say that the intersection is "usually transverse" means that it is transverse except at the point Q. The *bookmark* is the function that assigns such a pair to the components of the fold and seam sets. Here a component of the fold is either a closed curve or it has its end points on an exchanger, cusp, or seam. Similarly, a component of a seam can form a simple closed curve, end at a vertex, and (for the purpose of describing the bookmark) can end at a terminal point of a fold.

The glyphographic description of the foam then is enhanced by assigning bookmarks to the edges of the glyphs. The corresponding triple arrows can be composed by gluing together segments of the glyphs along their serifs provided the bookmarks and the serifs that are associated to the edges agree. In Sections 3.1 and 3.2 of this chapter, the bookmarks will be omitted from the legs of the glyph so the figures are more easily read. However, the notion of bookmark was used to insure that the compositions are possible.

Exercise. Supply bookmarks when drawing compositions of triple arrows.

3.1. The $(3,2)$-move. First, the $(3,2)$-move is depicted as a quadruple arrow using the glyph that describes the associator. This move is a weakly invertible 4-arrow. Choose a direction for the move, and the weak inverse follows the other direction.

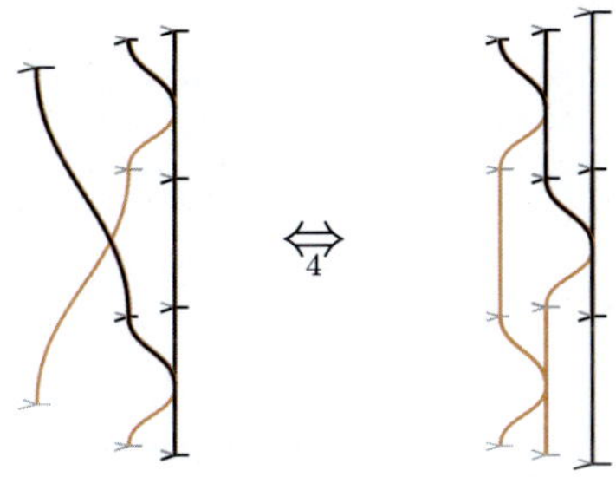

Next to facilitate further understanding, the sources and targets of the triple arrows are expressed as binary trees. The $(3,2)$-move that is expressed by this quadruple arrow is realized as if it is a commutative diagram. From this point of view, the $(3,2)$-move has been decategorified (Chapter 6, Section 2.2). The diagram appears to indicate that the binary trees are objects, the associators are arrows, and the $(3,2)$-move (or its weak inverse, the $(2,3)$-move) is a double arrow. However, recall that in this section that each binary tree is a double arrow that is formed from the generating double arrow ⅄.

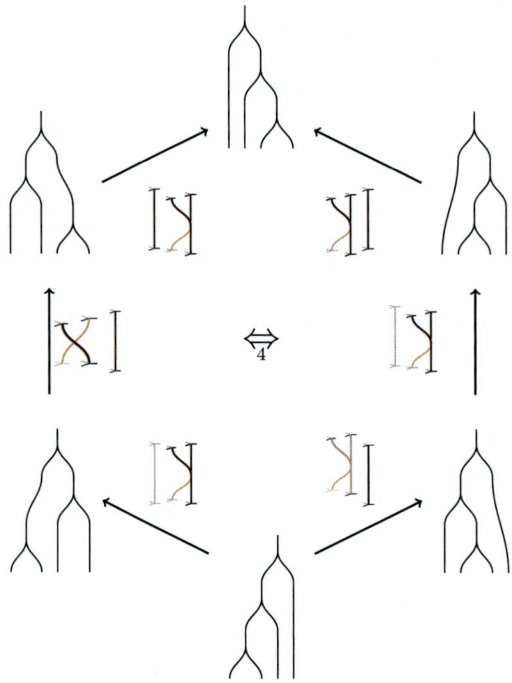

In Fig. 3, the $(3,2)$-move has been illustrated as a move to foams. The two foams

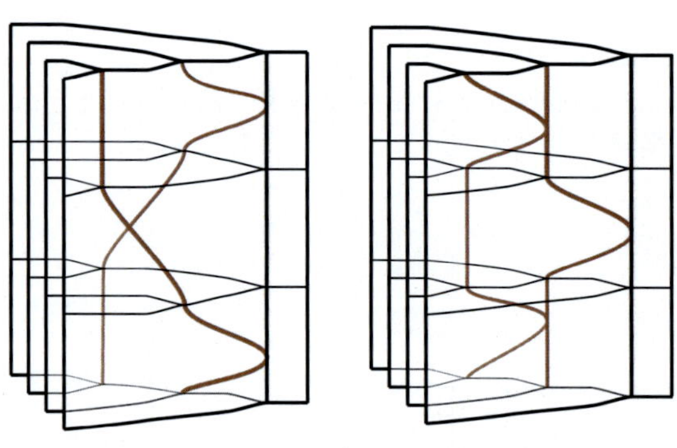

FIGURE 3. The $(3,2)$-move as a relationship between foams

that are represented are not homeomorphic as topological spaces. On the left-hand side of the figure, there are two vertices. On the right-hand side there are three vertices. Each vertex is the junction of four seams and six faces. The $(3,2)$-move seems quite unnatural from both the point of view of foams and from the point of view of the weak invertibility of the associator . On the other hand, in Chapter 8, the move is a natural consequence of a desire to weaken the associativity axiom, and the move is topologically significant when foams (or their cousins: special spines) are used to represent 3-dimensional manifolds.

We also assert the corresponding 4-arrows for the co-associator. The figures are similar.

Exercise. Develop diagrams that describe the $(3, 2)$-move for co-associators.

3.2. Unit 3-arrows, the associator, and pairing. In their seminal work [**36**], Joyal and Street describe tensor categories in terms of associators and unit maps. The book you are reading includes generalizations of these concepts in higher categorical contexts. Even though our multi-categories are small (the objects form a set), they form models for other categorical generalizations.

The $(3, 2)$-move that is given in Section 3.1 is an expression of the pentagon relation that is equation (a) of [**36**]. It appears as a hexagon here because the exchanger is included as a triple arrow. Chapter 8 discusses its topological significance. Joyal and Street also codify relationships between units and the associator. Here these 4-arrows are formulated in terms of the glyphographic notation, as commutative diagrams, and as moves to foams.

The triple arrow  is called the *associator*.

Once we establish the corresponding 4-arrows, we will revisit these ideas in a different context, that of the triple arrow  which corresponds to the associativity of the pairing *double arrow* ∩.

For all of these relations, we will only describe these 4-arrows in the contexts of multiplication ⅄, associativity, and the pairing. However, there are corresponding 4-arrows in the co-multiplicative contexts. Since the corresponding glyphs are reflections in a vertical axis, the formulation of those are left to the reader as an exercise.

To help organize the material, consider the products $(1b)c$, $(a1)c$, and $(ab)1$. In these cases the unit is upon the left, in the middle, or on the right, respectively. The formulations of the corresponding 4-arrows will be called *unit left*, *unit center*, and *unit right*.

Unit left. When the unit appears as the left most multiplicand, there are two possible resolutions. On one side, a shark's fin appears. On the other, first an associator is applied, the unit exchanges height with the other multiplication, and then a shark's fin appears. This is illustrated in glyphographic notation here. Recall that the glyphs are read from bottom to top.

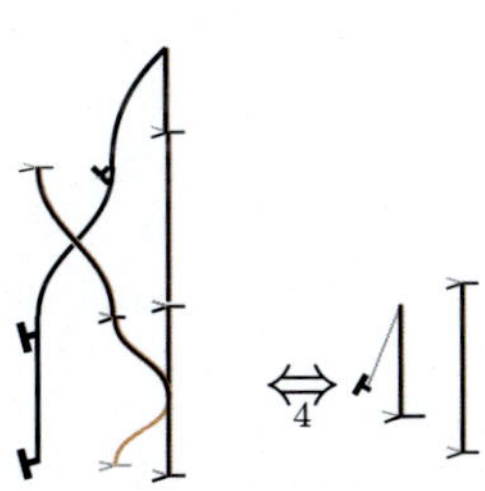

Next this quadruple arrow is illustrated as a commutative diagram.

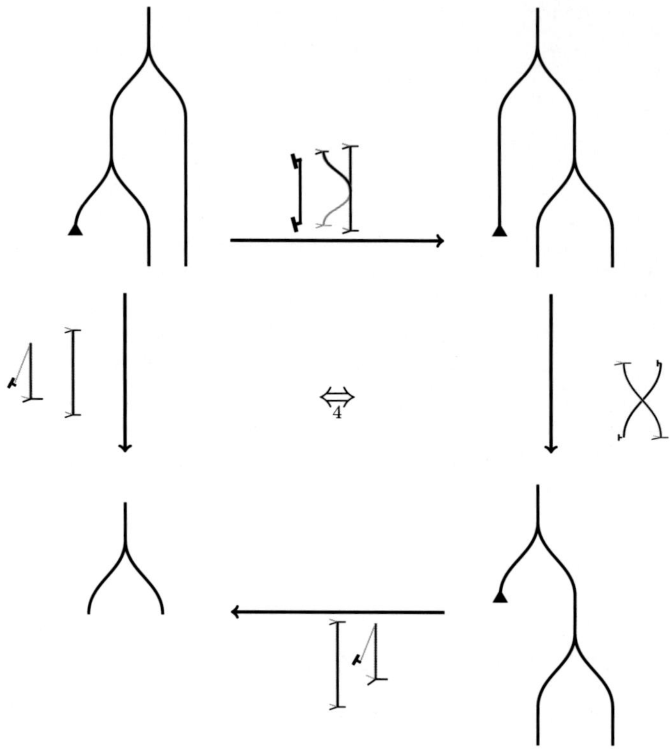

Fig. 4 indicates this quadruple arrow as a move to foams.

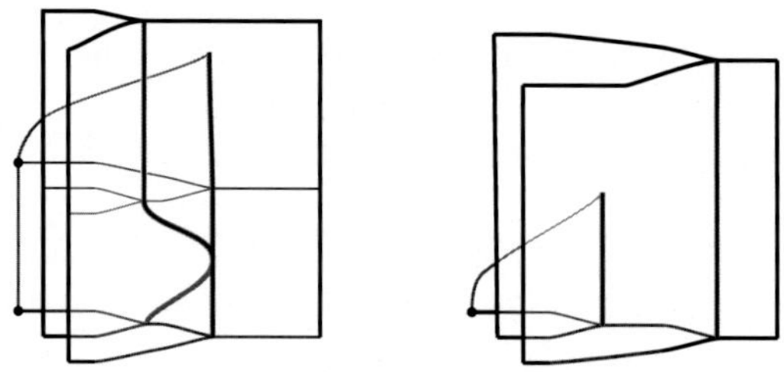

FIGURE 4. The shark's fin on the back can slide over the other seam to create a foam vertex

Unit center. The weakly invertible quadruple arrow describing the interaction between the unit and associator when the unit starts on the center is given in glyphographic form here.

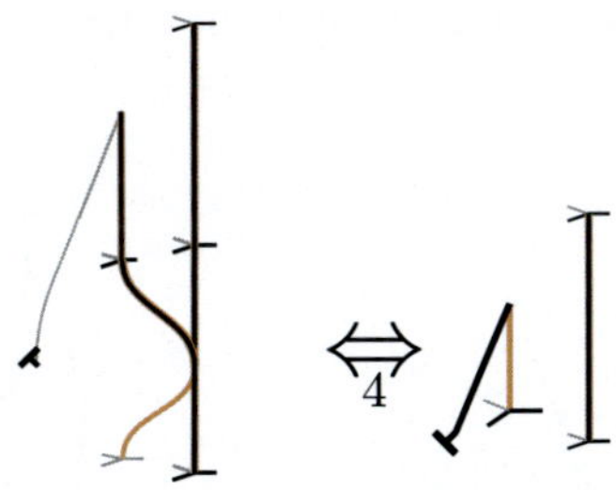

To facilitate the understanding of this 4-arrow, it is displayed as a commutative diagram here and as a foam in Fig. 5.

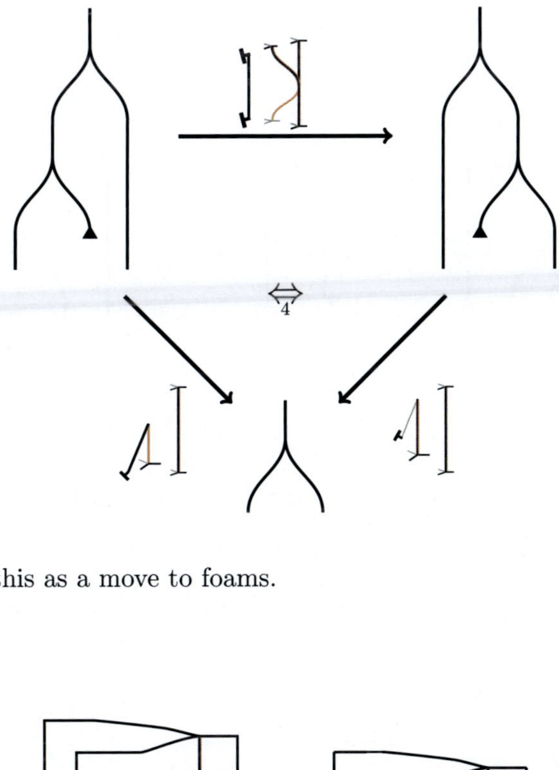

Fig. 5 indicates this as a move to foams.

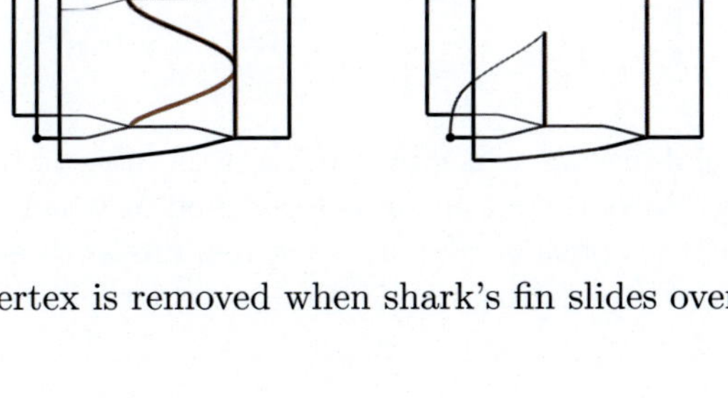

FIGURE 5. A vertex is removed when shark's fin slides over a major seam

Unit right. As in the previous two cases, the weakly invertible 4-arrow that expresses the relationship between associativity and a unit as the right most multiplicand is illustrated first in glyphographic notation as follows.

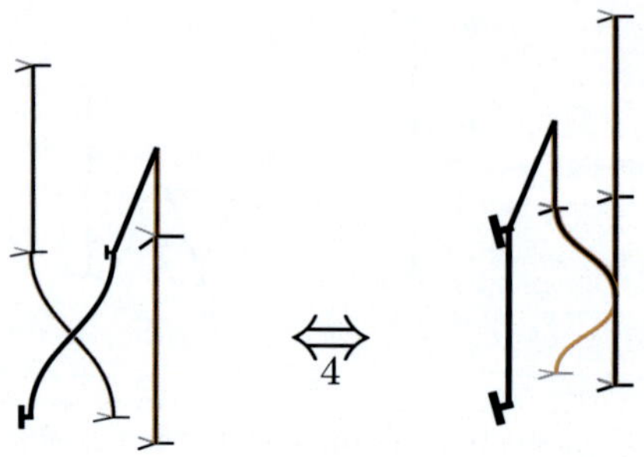

Next the same move is expressed as a commutative diagram.

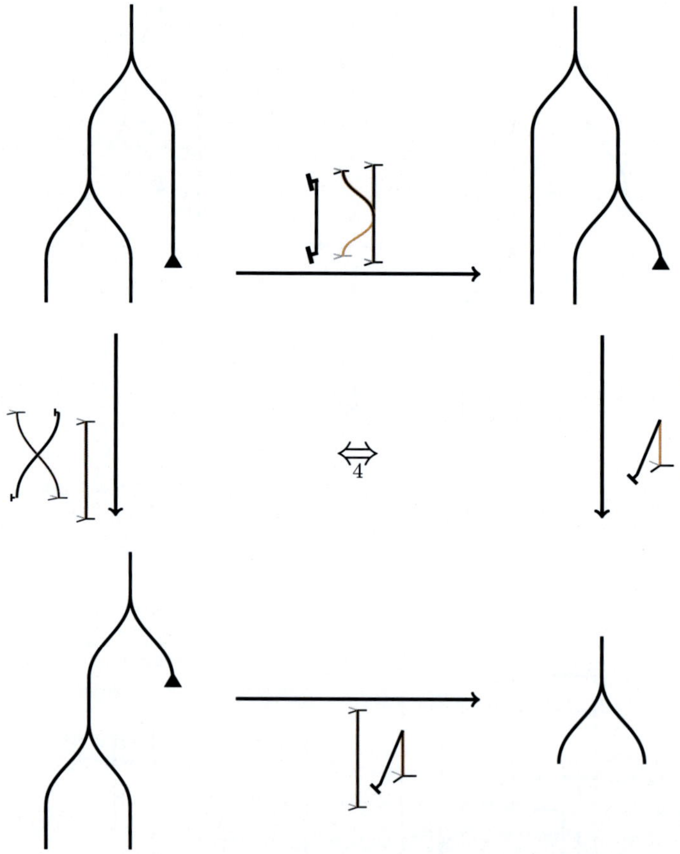

An illustration of the 4-arrow as a move to foams is presented in Fig. 6.

Units and pairing. Next the situation is examined in which a unit is a multiplicand in a pairing. First, consider algebraic expressions such as $\langle 1a, b\rangle$ and $\langle a, b1\rangle$. These two inputs are written categorically on the upper left-hand node of either of

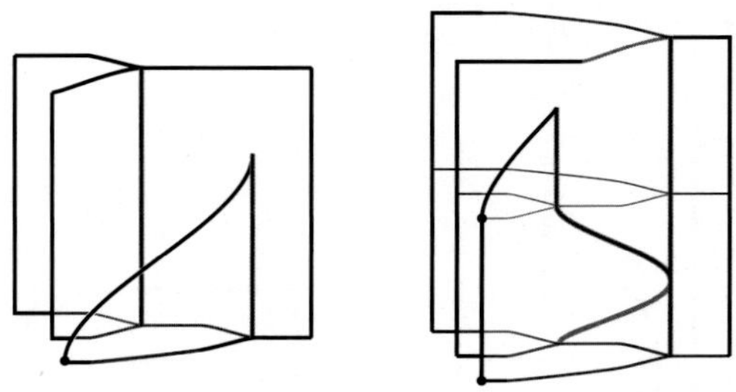

FIGURE 6. The shark's fin of the unit axiom can slide over the other seam to create a foam vertex

the quadruple arrow expressions below. Then either the associator for the pairing or its inverse triple arrow is applied. The shark's fin triple arrow from Chapter 4, Table 5 completes the commutative diagram. At the level of Frobenius algebras these diagrams commute. One can think of the shark's fins as a theorem, with the other four sides of the pentagonal diagrams constituting the proof. When considering the non-degenerate pairing in the matrix algebra $M(n,n)$, the pairing is given by $\langle A, B\rangle = \mathrm{tr}(A \cdot B)$, and the co-unit applied to a matrix C is $(C) = \mathrm{tr}(C)$. A similar computation for the unit and the co-pairing is not difficult to give in this case.

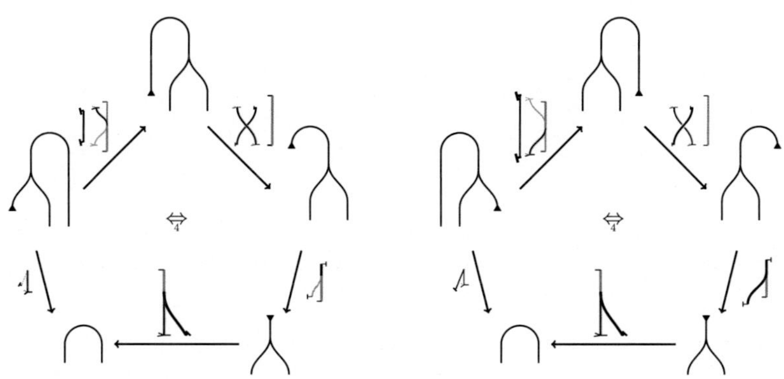

Fig. 7 illustrates the 4-arrow that is expressed by the right-hand commutative diagram above. The foam in the case of the left-hand diagram is a reflection in the plane of the paper, and will not be illustrated now. The glyphographic expression of the right-hand 4-arrow can be read from the foam move.

Unit center. The unit appears in the center of the expression $\langle A \cdot 1, B\rangle$. The products $\langle A \cdot 1, B\rangle$, $\langle A, 1 \cdot B\rangle$, and $\langle A, B\rangle$ are compared via the 4-arrow that

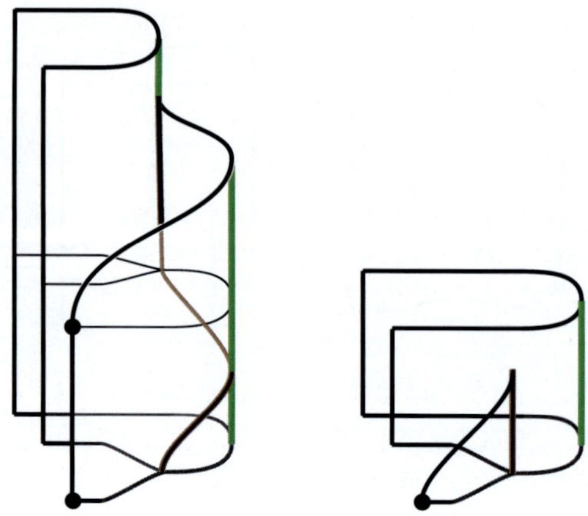

FIGURE 7. The shark's fin can roll over the fold

appears below.

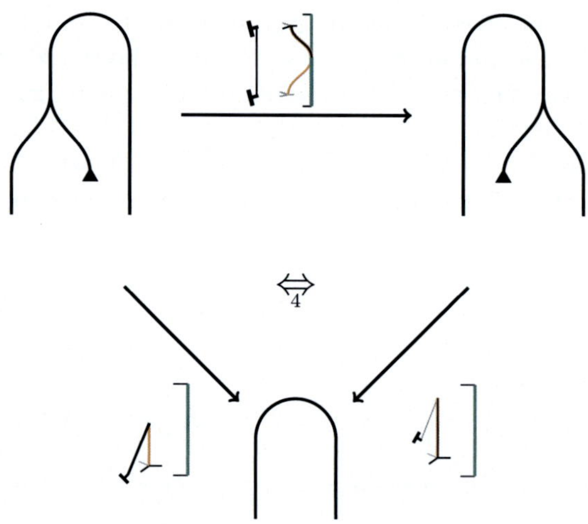

The move-to-foam representation of this 4 arrow is depicted in Fig. 8. On the left of the illustration, the shark's fin passes over the fold. On the right, it remains on the far side of the surface.

3.3. Zippers, units, co-units, and folds. Recall from Chapter 5, Section 3.3, that the triple arrows associated to the cancelation or creation of weak-inverse pair {Y, ⅄} of double arrows are the triple arrows 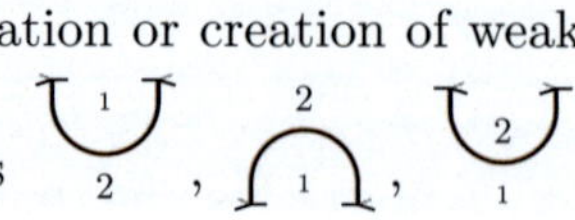

and 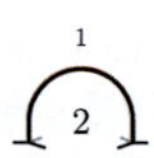. There are other optima that are associated to weak inverse pairs {⊥, ⊤} and {U, ∩}. In this section, the interactions of these and other optima with the units or co-unit axiom are examined, e.g. $\sqrt{}_{(+)}$, the definition of multiplication in

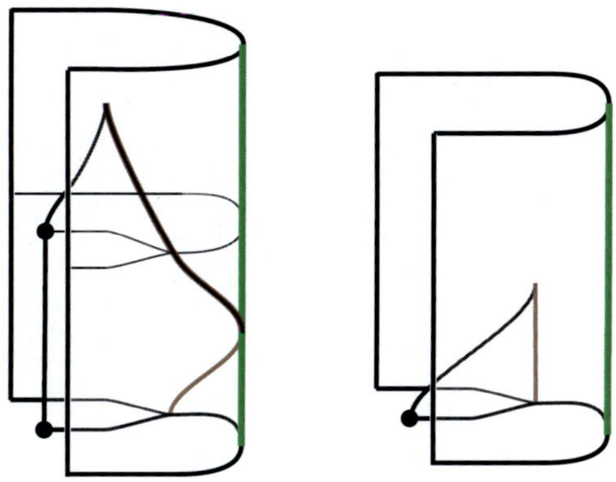

FIGURE 8. The shark's fin can roll over the fold

terms of co-multiplication, e.g. , the definition of co-units in terms of units, e.g. , and the shark's fin, e.g.  .

There are many instances of these phenomena. All of these follow a pattern that is called *a fish hook*. Some fish hook quadruple arrows that involve births or deaths can be re-written in terms of fish hook quadruple arrows for saddles or forks provided that critical cancelation and exchanger 4-arrows are available. The fish hook quadruple arrows in only the four examples below will be presented. See also item 3 of Chapter 10, Section 2.2.

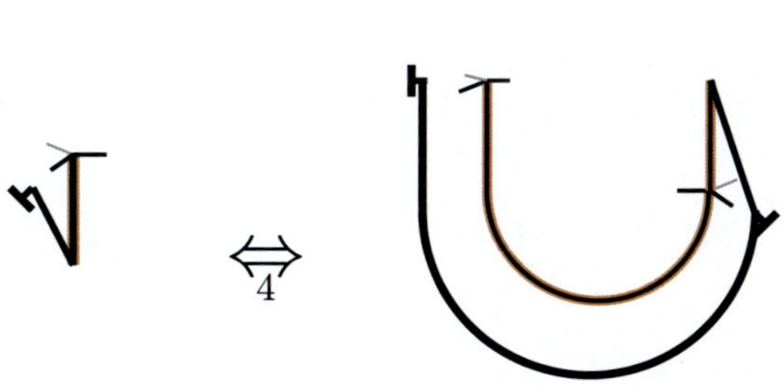

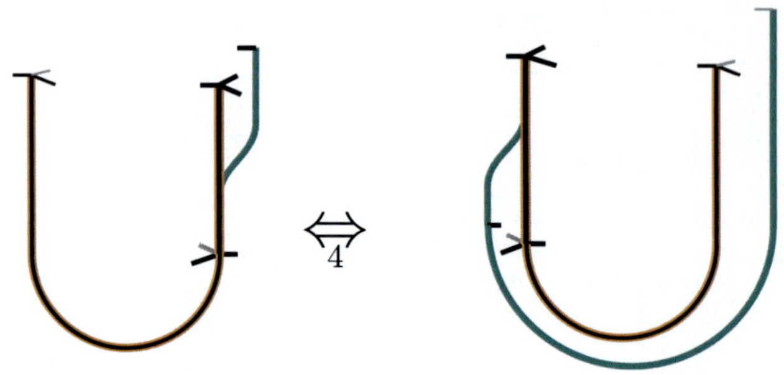

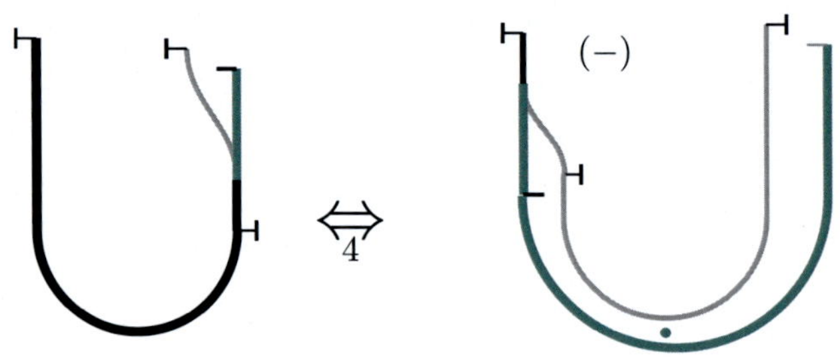

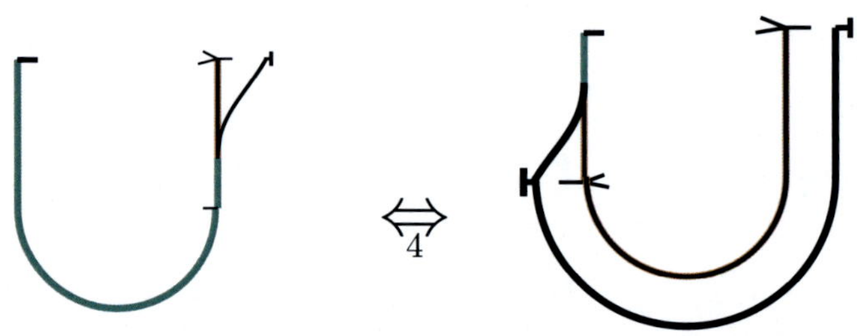

4. Restricting the collections of arrows

4.1. No seams. First, consider foams that have no seams. In this case the double arrows Y and ⅄ and any higher arrows that have these among sources or targets are excluded. Such a foam is a surface with boundary that is embedded in 3-space, but the boundary is not necessarily embedded properly. Here are two examples:

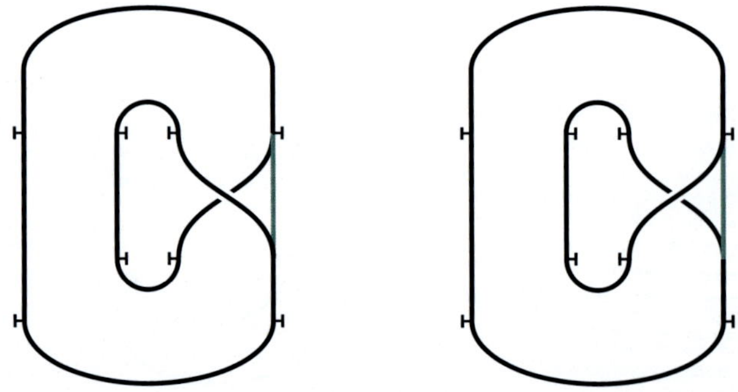

Exercise. Are these two triple arrows related by a finite sequence of applications of quadruple arrows? Give a proof.

More generally one can consider a knot or a link and an oriented or non-orientable Seifert surface that is bounded by the link. Putting the link into a nice position with respect to a height function allows the surface to be considered as a triple arrow in FA that does not involve seams. It may be interesting to determine if such surfaces are equivalent (that is related using quadruple arrows) when seams are introduced or removed by means of foam bubble or saddle moves (Section 3.3).

4.2. No seams and no edges. Second, exclude units ⊥, co-units ⊤, multiplication ⅄, and co-multiplication Y. Then this is the multi-category $\mathcal{S}$ of Chapter 6. In addition, exclude the arrows that correspond to 3-dimensional handles (births/deaths of spheres, and 1- and 2-handles) and any higher arrows that are derived from them. In this way, we are only considering isotopy classes of embedded surfaces and not cobordism relations between them.

The following two diagrams indicate a composition of triple arrows.

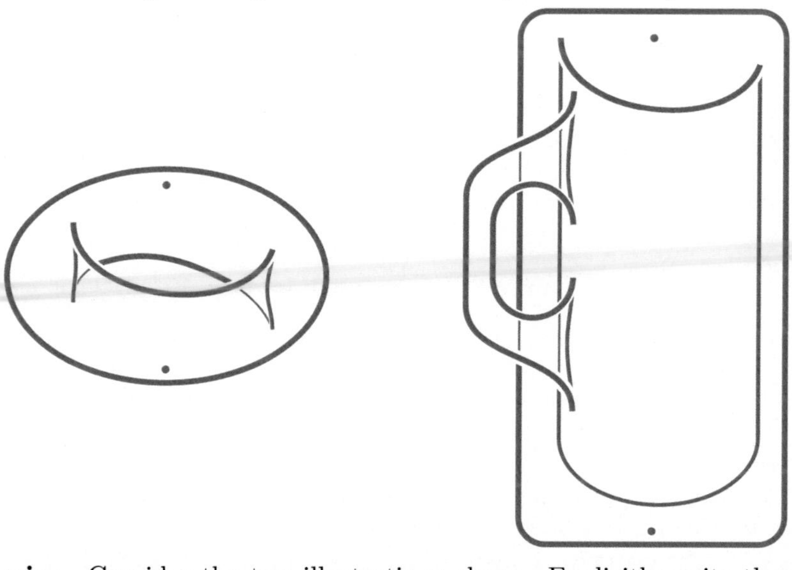

Exercise. Consider the two illustrations above. Explicitly write them as triple arrows. Include bookmarks. Demonstrate a finite application of quadruple arrows that relates the two diagrams.

4.3. Seams but no edges. Third, consider the situation in which units ⊥ and co-units ⊤ are excluded. Then multiple arrows that involve these 1-arrows are also non-existent. Before considering this thoroughly, observe that there are situations in which edges that bound faces ending upon seams can be thought of as vestigial. Metaphorically, such an edge is analogous to a rain flap on a purse or pack. Of course, from the real world perspective, a rain flap protects a zipper and the contents of a purse. And from the strictly topological perspective, such flaps affect the topology of the underlying foam. On the other hand, the weak invertibility of the unit and co-unit axioms, allows shark's fin type cuts to occur along a flap. These shark's fins can move across transverse seams and folds using the quadruple arrows from Section 3.2, and they can twist around optima using the quadruple arrows form Section 3.3. So it is possible, using these quadruple arrows, and weak invertibility that the sheets bounded by edges of foams may be removed. Investigate.

Keep in mind that surfaces with boundary are among the most interesting and useful topological feature of 3-dimensional space. They include Seifert surfaces for knots, and their various embeddings in 3-space explains the richness, and consequently the topological limitations, of 3-space. Surfaces with boundary easily get entangled in space. That is why a box full of unattended ribbon usually is ensnarled in a thrackle.

Still foams that have edges that cobound seams seem to be somewhat less interesting than foams that do not have them.

So in this section, we consider foams that do not have edges. Be aware. Such foams may have *boundaries* that are planar trivalent graphs. There are two contexts in which such foams are considered. These distinct aspects also deserve care.

Case 1: 4-arrows that preserve the topology of the foam. In this case, the double arrows are generated by U, ∩, Y, ⅄, and I. A double arrow corresponds to a planar trivalent graph. Since units  and co-units  are **not** among the double arrows, **none of the triple arrows** that are listed in Chapter 4, Tables 1, 2, 3, 4, or 5 are included. Neither are the 3-arrows that appear in Chapter 4, Table 13 or any consequences thereof (such as those given in Chapter 5, Section 3.1).

Since the topological type of the foam is to remain unchanged, the associator  **no longer is assumed to be a weak inverse** of . Similarly, the co-associator  **no longer is assumed to be a weak inverse** of . That is there are no quadruple arrows that connect the compositions of these to the identity upon the source or target. The identity depicted in Ch. 4, Fig. 18 does not hold. We also disallow the birth and deaths of spheres or the attachment of handles to surfaces that are listed in Chapter 5, Section 3.2. In a similar fashion, the quadruple arrows that are described in Chapter 5, Section 3.3 are excluded, again because these change the topological type of the foam. Please also notice that in Chapter 5, these arrows were described as triple arrows, but here we consider them as quadruple arrows because of the monoidal structure ($\mathbb{N}$) on the object set. Finally, when the topology of the foam is meant to be preserved, then the $(3, 2)$-move of Section 3.1 (Fig. 3) is excluded.

On the other hand, folds that end on seams, to relate multiplication and co-multiplication, will be included since these do not change the topology of the foam. Cusps, births and deaths of simple closed curves, saddles, forks, and the various zippers all are among the triple arrows that are used to construct foams.

The quadruple arrows that remain (lips, beak-to-beak, swallow-tails, horizontal cusps, fish hooks for rewriting multiplication in terms of co-mulitplication and vice versa, and exchangers) should be sufficient for performing isotopies of foams in space while leaving their properly embedded boundaries fixed.

Exercise. As an experiment, create an embedded closed foam in 3-dimensional space as a composite of triple arrows in FA. Find a sequence of quadruple arrows that achieve the rotation of the foam 180 degrees around a vertical or horizontal axis. Let us give some additional context to this exercise.

We start from a lower dimensional analogy. Consider vertical line segment that is rotated 180 top to bottom. The segment is horizontal at an instant. Then the projection of the segment is a single point. This phenomena is illustrated below and a generic perturbation is indicated.

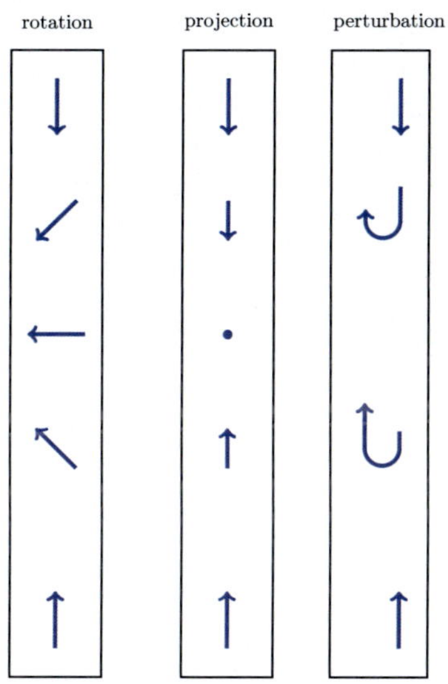

Suppose that a rotation is parameterized by time, and a fixed 2-dimensional plane is chosen onto which the foam is projected. A rigid foam will not project generically at various points of time during the rotation. Moreover, there may be instances at which the change in projection is not governed by a single generating quadruple arrow. Thus when a rigid foam is rotated, vertical segments of seams and folds will be coincident at certain times. Also, that which is the fold set changes as the foam is rotated. By definition, the folds are the portions of the foam upon which the planar projection drops in rank. These change as the foam is rotated. As in the cases of surfaces, lips, exchanges of fold lines, beak-to-beak, swallow-tails, and horizontal cusp quadruple arrows achieve some of the changes that occur during the rotation. The point of the exercise is to find an explicit collection of quadruple arrows that can achieve the rotation.

We propose that GLOUBULAR.SCIENCE (see Chapter 12) can be used to this end. Furthermore, the glyphography that is developed within this text is an enhanced version of the iconography that is present within GLOBULAR. In fact, some of the mouse hovers in GLOBULAR indicate one component of the bookmark. The enhanced notation that we have given here includes serifs and bookmarks that indicate necessary conditions for composability.

Case 2. 4-arrows that change the topology of the foam. Foams are closely related to special spines that were introduced by S. Matveev and R. Piergallini [**56, 63**] as a method of studying 3-dimensional manifolds. In such a setting, the intrinsic topology of the foam is not as important as the way that it is embedded in the ambient 3-dimensional space. So even though the quadruple arrows of Chapter 5, Section 3.3 and Section 3.1 of this chapter, and the interactions between units and the associator given in Section 3.2 (this chapter) change the topology of the foam, the way that the foam represents the surrounding 3-dimensional space remains consistent. However, the triple arrows that we have in this situation can only describe foams that are properly embedded in $\mathbb{R}^3$.

Nevertheless, there are several standard foam embeddings that are obtained from the usual handle decompositions of 3-dimensional space. Start from a standardly embedded genus-2 handlebody. Attach four 2-dimensional disks to the boundary surface. Two of these disks are the cores of the inside handlebody and two are the cores of the outside handlebody. A foam that is embedded in 3-space is the union of the genus-2 surface and these four core disks. The resulting foam is the *standard foam associated to a genus* 2 *splitting of* $\mathbb{R}^3$.

The construction generalizes easily to any standard genus $g > 0$ Heegaard splitting of $\mathbb{R}^3$.

Exercise. 1. Give a description of this foam as a composition of 3-arrows. 2. Generalize to give a description of the genus g foams. 3. Determine if the resulting foams are related by quadruple arrows.

CHAPTER 8

Parentheses and so forth

In this chapter, the Temperley-Lieb algebra, higher associativities, and the higher order Pachner moves [**61**] are discussed. These topics are inter-related, in part, because Catalan numbers are so frequently encountered.

One of the first examples of diagrammatic algebra given was Kauffman's description of the Temperley-Lieb algebra [**42**]. Section 1 will present the Temperley-Lieb algebra from Kauffman's point of view, and it will develop the representations that describe the Jones polynomial (Chapter 9, Section 4). In Section 3, the higher dimensional Stasheff polytopes are presented to develop higher order relations that are analogous to the $(3,2)$-move. In Section 3.1, the relationships between these higher order associativities and moves to triangulated manifolds are described.

1. The Temperley-Lieb algebra

As V. F. R. Jones mentioned in [**34**], the Temperley-Lieb algebra plays a prominent role in the algebraic theory of his celebrated polynomial invariant of classical knots. In [**42**], L. H. Kauffman gave the diagrammatic description of the Temperley-Lieb algebra that is presented in this section. Kauffman's approach to this algebra derives from Penrose's much earlier notion of abstract tensors [**62**]. Thus the current section continues the first thread that was introduced in Chapter 2, Section 2.

While the manipulation of knot diagrams according to certain rules and indeed skein theory [**19, 22**] in general are primary examples of diagrammatic algebra in the wild, the diagrammatic description of the Temperley-Lieb algebra, as presented by Kauffman and influenced by Penrose, is one of the earliest expressions of diagrammatic algebra in which the idea that diagrams encode algebra is presented.

The Temperley-Lieb algebra $TL_n(\delta)$ is a vector space over the complex numbers $\mathbb{C}$ that has dimension $\frac{1}{n+1}\binom{2n}{n}$. Here $\delta \in \mathbb{C}$ is called the *loop value.* A parameter $A \in \mathbb{C}$ is related to δ (the loop value) by the formula $\delta = -A^2 - A^{-2}$. A basis for $TL_n(\delta)$ is constructed from the monoid generated by elements $1, h_1, ..., h_{n-1}$ that satisfy the relations given below.

$$\left\langle 1, h_1, \ldots, h_{n-1} : \begin{array}{ll} h_i h_{i+1} h_i = h_i, & \text{for } i = 1, \ldots, n-2, \\ h_i h_{i-1} h_i = h_i, & \text{for } i = 2, \ldots, n-1, \\ h_i h_j = h_j h_i, & \text{for } |i-j| > 1, \\ h_i^2 = \delta h_i & \text{for } i = 1, \ldots n-1 \end{array} \right\rangle.$$

Let a non-negative integer $n > 1$ be given. For $j = 1, \ldots, n-1$, define a double arrow

$h_j =$ 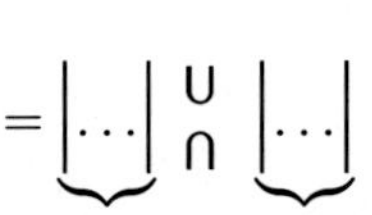 .

For example, when $n = 3$, the generators are the arrows: $1 =$ 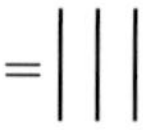, $h_1 =$ 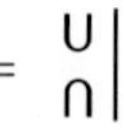, and $h_2 =$ 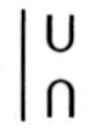. These can be composed globularly. Then the resulting diagrams can be reduced using the cuspidal triple arrows in Chapter 4, Table 10. The basis for $TL_3(\delta)$ is listed here.

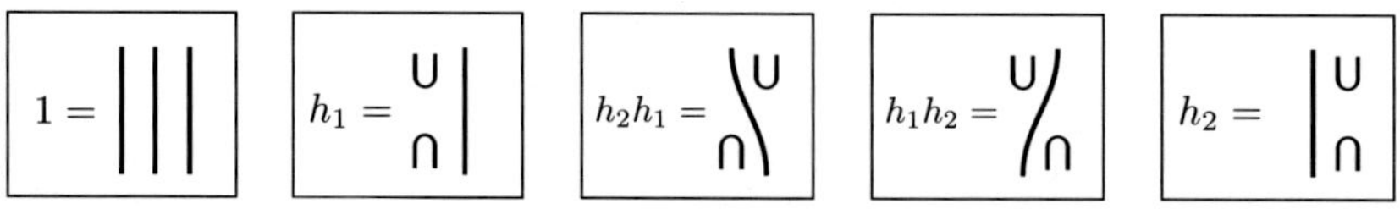

The value of h_{j+1}^2 is computed diagrammatically as indicated below, where the vertical line subscripted by j (resp. $n-j-2$) indicates j (resp. $n-j-2$) parallel vertical lines.

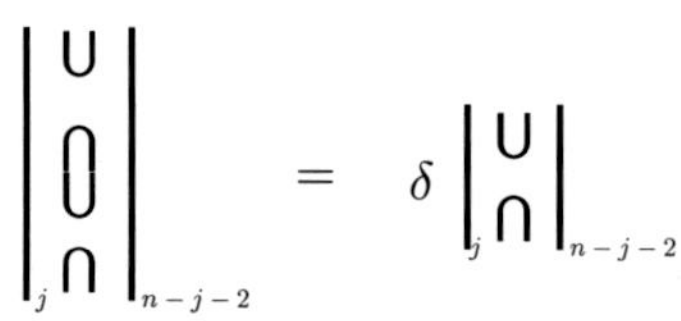

Each time a circle (simple closed curve) appears in a diagram, it is removed from the diagram no matter where it might lie, and the constant $\delta \in \mathbb{C}$ replaces it. In the diagrammatic reasoning, such constants commute with any other parts of the diagram.

Since the generators are double arrows, they can be combined in skew compositions as well. In the case of $TL_3(\delta)$, there are four types of skew compositions. The resulting diagrams are (up to a multiple of δ) basis elements for $TL_4(\delta)$ or $TL_5(\delta)$. The formats of the skew composition follow.

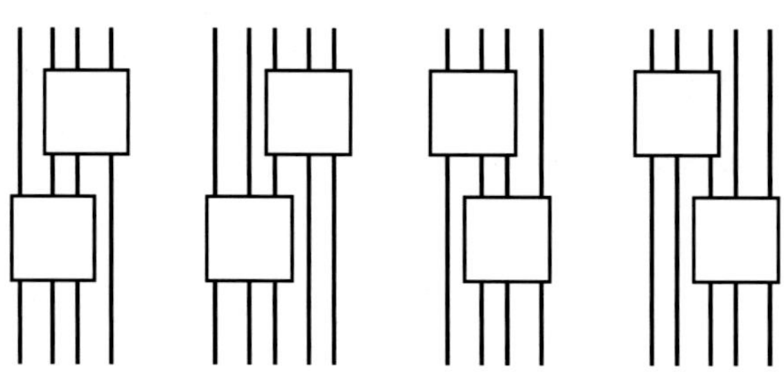

From left to right these are called *offset* 1 *right, offset* 2 *right, offset* 1 *left, offset* 2 *left* with the top box representing the item being offset. In general, in TL_n there are $2(n-1)$ possible offsets for skew composition.

Exercise. Consider the bottom box to be labeled A and the top box to labeled B in the illustration immediately above. 1. Compute each of the offset by 1 products for $A, B \in \{1, h_1, h_2, h_1h_2, h_2h_1\}$, and write each result in terms of products of the generators for TL_4. 2. Compute the offset by 2 products for A, B as above and write each product in terms of the generators for TL_5.

1.1. Counting parentheses. In this section, the dimension of TL_n as a vector space over $\mathbb{C}$ is deomonstrated to be $\mathcal{C}_n = \frac{1}{n+1}\binom{2n}{n}$. Using TL_3 as an example, a *fold over map* takes the basis $\{1, h_1, h_2, h_1h_2, h_2h_1\}$ into strings of matched parentheses. A pair of matched parentheses is represented as ∩, and strings of such are various nestings thereof. Such nested arches of strings are counted by relating each to a corresponding *Dyck path* that consists of a collection of segments in the plane that have slope ± 1, that lie in the first quadrant, and that join $(0,0)$ to $(2n,0)$. The proof presented is well-known since it is commonly taught in a combinatorics course, and is easy to find on the internet, for example. It is a nice argument that is worth repeating. First we illustrate the fold-over map.

Starting from the generating set $\{1, h_1, \ldots, h_{n-1}\}$, form all possible reduced diagrams that represent a product. A diagram is *reduced* if all simple closed curves have been replaced by δ — the loop value, and adjacent critical point have been cancelled via cusp triple arrows. In the latter case, products of the form $h_i h_{i\pm 1}$ are reduced to having two critical points, and $h_i h_{i\pm 1} h_i$ is reduced to h_i. Furthermore, we are only interested in the diagram that represents the product, not the number of δ factors.

Among all possible finite products of the generating set $\{1, h_1, h_2, \ldots, h_{n-1}\}$, there are finitely many reduced expressions. Let a word $h_{i_1} h_{i_2} \cdots h_{i_k}$ be given. Consider the number of critical points (∪s or ∩s) in the expression. If $|i_k - i_{k+1}| > 2$ (sufficiently distant), then the expression can be reordered so that the subscripts are ascending. In this case, there are exactly $(2k)$ critical points. On the other hand, if $h_{i_{j+1}} = h_{i_j}$, then a factor of δ can be removed, and the word can be reduced. Similarly, if $i_{j+1} = i_j + 1$ (that is there is a pair of successive subscripts), then at least one pair of critical points can cancel using a cusp. Two pairs of critical points can cancel in a replacement $h_i h_{i\pm 1} h_i \mapsto h_i$. Therefore, for any given unreduced diagram that represents the word $h_{i_1} h_{i_2} \cdots h_{i_k}$, there are at most $(2k)$ critical points, and a reduced representative diagram may have fewer.

Presume that the reduced diagram (as a double arrow from n to n) is represented in a rectangle of height 2 units. To the right of the diagram, at height 1, insert a marker, and fold the top of the diagram down to the right. In the illustration below, the five basis elements of TL_3 are indicated. The marker is a dot to the right of the diagram. Immediately to the right of the diagram-with-marker is the result of folding the top of the strings down and anchoring them to the right of the bottom (left) strings. The five parentheses strings that occur are, in order, ((())), ()()(), ()(()), (())(), and (()()).

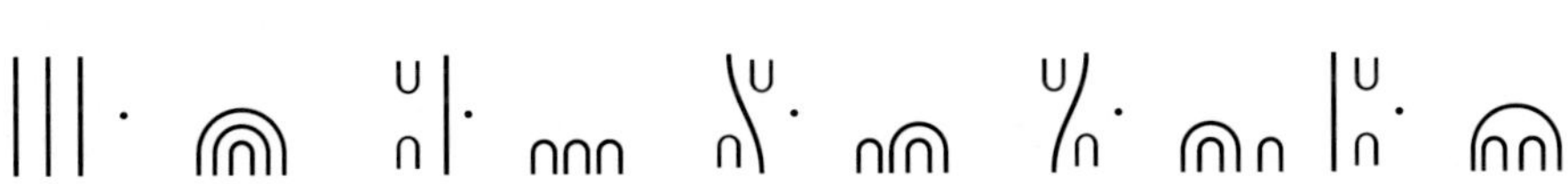

This process generalizes to demonstrates how an element in the vector space basis for $\mathrm{TL}_n(\delta)$ folds into a unique string of n pairs of matched parentheses.

The number of such strings is known to be the Catalan number $\mathcal{C}_n = \frac{1}{n+1}\binom{2n}{n}$. There are a zillion thing[1] that are counted by Catalan numbers [**73**] including binary trees, associated products, and triangulation of an (n+2)-gon in which the vertices

[1] Stanley [**73**] lists at least 214.

of the triangles are all on the boundary. Dyck paths are particularly easy to count and to put into one-to-one correspondence with matched strings of parentheses.

A Dyck path consists of a collection of connected segments that lie in the first quadrant of the plane, that have slope ± 1, and that join $(0,0)$ to $(2n,0)$. A given matched parentheses string corresponds to such a path. Each left parenthesis, (, corresponds to a segment of slope 1, and each right parenthesis,), corresponds to a segment of slope -1. Counting the number of such paths seems difficult, so instead count the total number of paths with segments of slope ± 1 from $(0,0)$ to $(2n,0)$. Such a path is a binary sequence with the same number of +s and −s. There are $\binom{2n}{n}$ such. Now we will count the bad paths — those that dip below the x-axis. For any bad path there is a first integer k at which the path intersects $(k,-1)$. Then reflect the segments that lie to the left through the line $y=-1$. This gives a path from $(0,-2)$ to $(2n,0)$. There are $\binom{2n}{n-1}$ of these. The difference between all paths and bad paths is

$$\binom{2n}{n} - \binom{2n}{n-1} = \frac{1}{n+1}\binom{2n}{n}.$$

The values run 1, 2, 5, 14, 42, and so on.

1.2. Matrix representation. It is time now to reinterpret the ideas that were introduced in Chapter 2, Section 2 and give an alternative matrix representation for $\cap$ and $\cup$. Specifically, imagine that $\mathbb{C} \xleftarrow{\cap} V \otimes V$ where V is a 2-dimensional vector space over the complex number field $\mathbb{C}$. Analogously, $V \otimes V \xleftarrow{\cup} \mathbb{C}$. Let $A \in \mathbb{C}$, or alternatively, consider A as an abstract variable.

In [**13**], the vector space V is chosen to be the set of linear expressions $V = \{\alpha x + \beta y : \alpha, \beta \in \mathbb{C}\}$ and it is also thought of as the 2-dimensional irreducible representation space for the Lie group $\mathrm{SL}_2(\mathbb{C})$ or its quantum group analogue. These interpretations of V were used to define, interpret, and simplify the presentation of Turaev-Viro [**76**] and Witten-Reshetikhin-Turaev invariants [**78**] [**65**] for 3-dimensional manifolds. The Kauffman-Lins book [**44**] indicates the diagrammatic foundation of this approach, and [**13**] is a detailed presentation of the underlying representation theory.

When $V = \{\alpha x + \beta y : \alpha, \beta \in \mathbb{C}\}$, an ordered basis for $V \otimes V$ is $(x \otimes x, x \otimes y, y \otimes x, y \otimes y)$. The matrix representation of $\cap$ in this basis is the row vector $[0, \boldsymbol{i}A, -\boldsymbol{i}A^{-1}, 0]$, and $\cup \leftrightarrow [0, \boldsymbol{i}A, -\boldsymbol{i}A^{-1}, 0]^T$ where T denotes the transpose. The loop value δ is the composition $\cap \circ \cup = -A^2 - A^{-2}$. In this matrix representation,

$$\cap \otimes \mathsf{I} \leftrightarrow [0I, (\boldsymbol{i}A)I, (-\boldsymbol{i}A^{-1})I, 0I],$$

and

$$\mathsf{I} \otimes \cup \leftrightarrow \begin{bmatrix} \cup & 0 \\ 0 & \cup \end{bmatrix}$$

where $I = \begin{bmatrix} 1 & 0 \\ 0 & 1 \end{bmatrix}$ is the (2×2) identity matrix. Here we are abusing notation and using the double arrow $\cup$ to indicate its matrix representation. The matrix product $(\cap \otimes \mathsf{I}) \circ (\mathsf{I} \otimes \cup) = I$ — the (2×2)-identity matrix. A similar computation shows that $(\cap \otimes \mathsf{I}) \circ (\mathsf{I} \otimes \cup) = I$. Meanwhile,

$$\begin{matrix}\cup \\ \cap\end{matrix} = \begin{bmatrix} 0 & 0 & 0 & 0 \\ 0 & -A^2 & 1 & 0 \\ 0 & 1 & -A^{-2} & 0 \\ 0 & 0 & 0 & 0 \end{bmatrix}.$$

Exercise. 1. Verify the claims about the various matrix products. 2. Compute that the matrix product $\begin{bmatrix} \cup \\ \cap \end{bmatrix}^2 = \delta \begin{bmatrix} \cup \\ \cap \end{bmatrix}$ for $\delta = -A^2 - A^{-2}$.

These identities demonstrate that the Temperley-Lieb algebra can be represented as a matrix algebra. The representation that has been given here is constitutive for the Kauffman bracket definition of the Jones polynomial given in Chapter 9, Section 4.

2. Other Catalan-like things

This section surveys one-to-one correspondences among associated products, based binary trees, triangulated polygons, and the set of n pairs of matched parentheses. Each is counted by the Catalan number $\mathcal{C}_n = \frac{1}{n+1}\binom{2n}{n}$. The triangulations of a regular $(n+2)$-gons have no interior vertices.

To establish the correspondences, start from a string of parentheses, for example, ()(()), and intersperse variables a, b, c, d from the left. The natural number n refers to the number of matched parentheses. The variable a, goes immediately to the left of the first "(." Subsequent variables are inserted so that the result is a grammatically correct expression that indicates the expression enclosed in parentheses acts upon the express that lies to the left. For example, $a(b)$ indicates that a is acted upon by b. Thus $a(b)(c(d))$ indicates that the result of d acting upon c acts upon the result of the action of b upon a. In case the action is a binary operation, $a(b) = a \cdot b = ab$, then the expression $a(b)(c(d))$ translates into $(ab)(cd)$. That parenthesized string corresponds to a binary tree.

In the current framework, the two vertices of the binary tree can occur at the same level, but in any computation, one or the other products (ab) or (cd) takes priority. So within this section, the corresponding $(3,2)$-move is encoded by a pentagon, whereas earlier product priorities were chosen and possibly exchanged.

The binary tree encoding the parenthesized product is dualized to give a triangulation of an $(n+2)$-gon. When three variables are multiplied, the polygon is a square; with four, the polygon is a pentagon. Given a regular $(n+2)$-gon, label the vertices $0, 1, \dots, n+1$ sequentially and clockwise around the boundary. For $i = 1, \dots, n+1$ associate a variable a_i to the edge $[i-1, i]$. When a triangulation is given and there is an internal edge $[j-1, j+1]$, the product $(a_j \cdot a_{j+1})$ is associated to this edge. In this way, an inductive process determines an associative product to a triangulation of the $(n+2)$-gon where the parenthesized product that corresponds to the triangulation appears along the edge $[0, n+1]$. The three charts below indicate the correspondences.

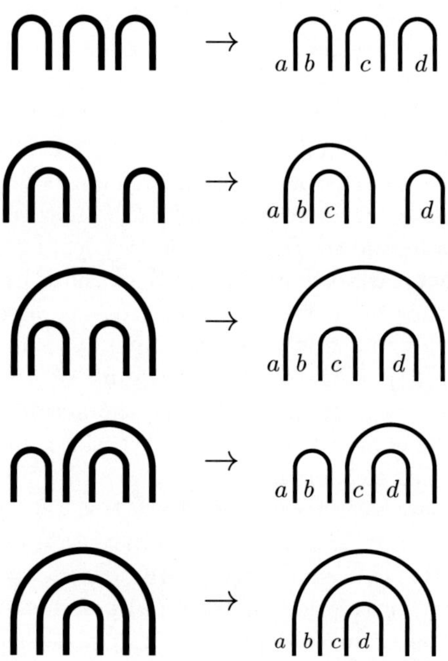

$\leftrightarrow a(b)(c)(d) \leftrightarrow ((ab)c)d$

$\leftrightarrow a(b(c))(d) \leftrightarrow (a(bc))d$

$\leftrightarrow a(b(c)(d)) \leftrightarrow a((bc)d)$

$\leftrightarrow a(b)(c(d)) \leftrightarrow (ab)(cd)$

$\leftrightarrow a(b(c(d))) \leftrightarrow a(b(cd))$

 $\leftrightarrow a(b)(c)(d) \leftrightarrow ((ab)c)d \leftrightarrow$ 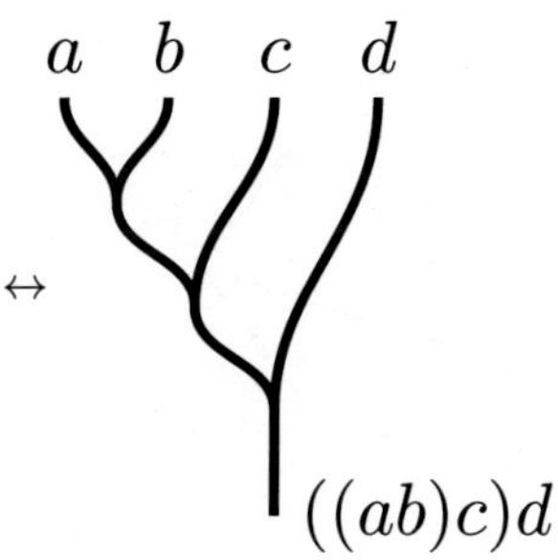

 $\leftrightarrow a(b(c))(d) \leftrightarrow (a(bc))d \leftrightarrow$ 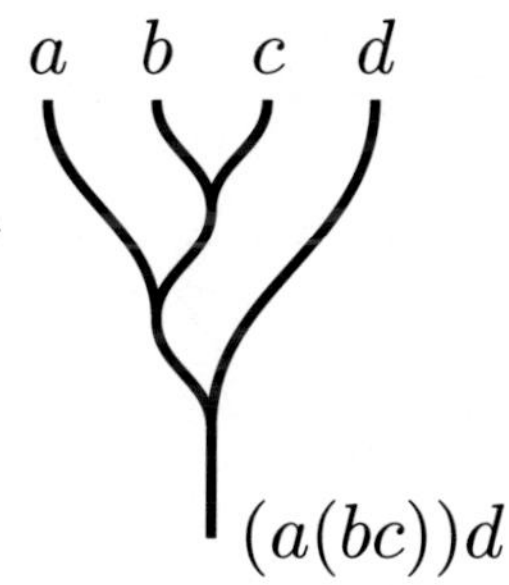

 $\leftrightarrow a(b(c)(d)) \leftrightarrow a((bc)d) \leftrightarrow$ 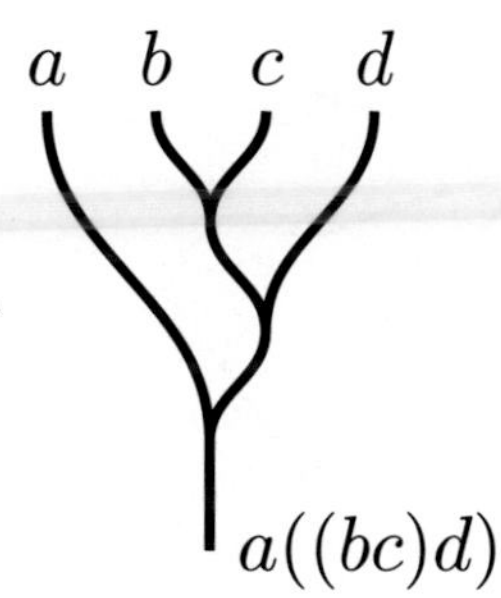

 $\leftrightarrow a(b)(c(d)) \leftrightarrow (ab)(cd) \leftrightarrow$ 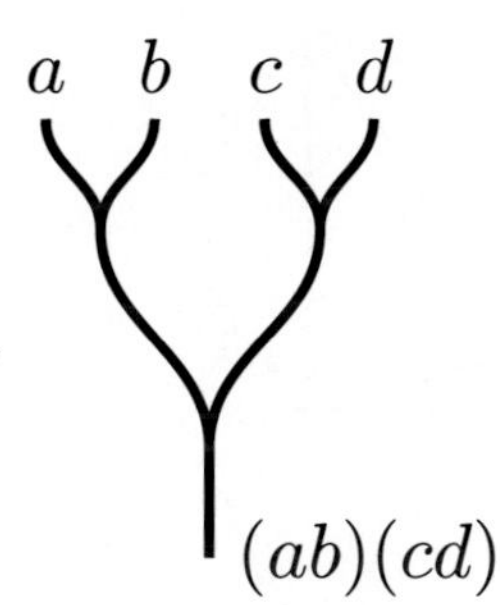

 $\leftrightarrow a(b(c(d))) \leftrightarrow a(b(cd)) \leftrightarrow$ 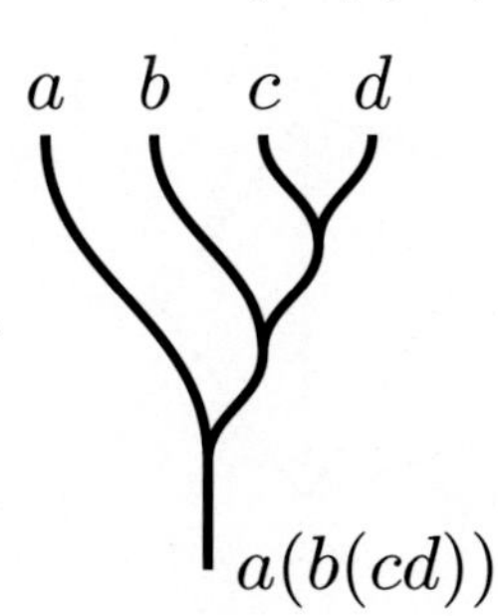

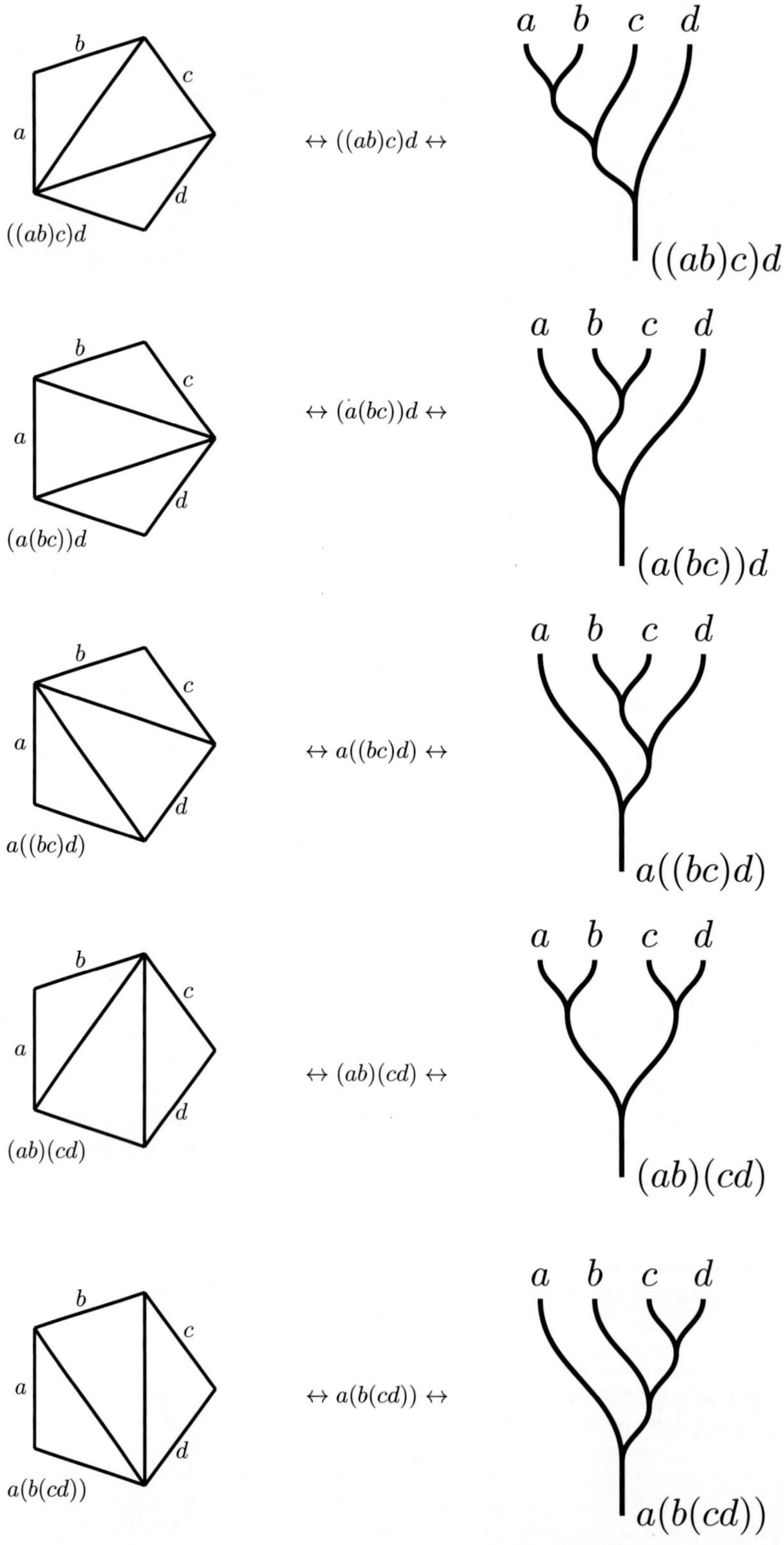
↔ $((ab)c)d$ ↔
$((ab)c)d$
↔ $(a(bc))d$ ↔
$(a(bc))d$
↔ $a((bc)d)$ ↔
$a((bc)d)$
↔ $(ab)(cd)$ ↔
$(ab)(cd)$
↔ $a(b(cd))$ ↔
$a(b(cd))$

2.1. Triangulations of the pentagon. The duality between the binary trees and triangulated polygons allows an alternative description of the $(3,2)$-move. First, a neighborhood of a vertex, as depicted on the right-hand column of Chapter 4, Table 8, is dual to a tetrahedron in the sense that the faces of the foam correspond to edges of the tetrahedron, the edges of the foam correspond to the faces of the tetrahedron, and the vertex of the foam corresponds to the 3-cell. The $(3,2)$ relation as interpreted on the triangulations of the pentagon consists of three tetrahedra glued along an edge being replaced by two tetrahedra that are glued along a face. In the illustration below, the glued edge is the edge that is labeled with the product bc. On the right, the two tetrahedra are glued along the triangle that has edges labeled (ab), (ac) and $(abcd)$. These five tetrahedra form the boundary of a 4-dimensional simplex Δ^4.

The standard *n-dimensional simplex* is the set

$$\Delta^n = \{(x_1, \dots, x_{n+1}) \in \mathbb{R}^{n+1} : \\ \sum_{j=1}^{n+1} x_j = 1 \ \&\ 0 \le x_j \text{ for } j = 1, \dots, n+1\}.$$

It is the convex hull of the set of basis vectors $\{e_1, e_2, \dots, e_{n+1}\}$.

The $(2,3)$-move, as a replacement between sets of tetrahedra, is one of the primary moves that allow a change in triangulations upon a 3-dimensional manifold. (A $(2,3)$-move is the inverse move of a $(3,2)$-move). Specifically, imagine that two tetrahedra within the 3-manifold share a face in a triangulation. Attach a 4-dimensional simplex Δ^4 to the manifold along the shared boundary that consists of these two tetrahedra. Then replace the two tetrahedra by the other side of the boundary of the 4-simplex Δ^4.

In addition, there are similar $(4,1)$ and $(1,4)$-moves to triangulations. These, too, come from the pentagon condition that is illustrated below. The additional moves depend upon the associator being an invertible object. In the geometric situation, two tetrahedra that are glued along two faces, can be collapsed. In the diagrammatic situation, one of the two triple arrows on the right of the pentagon that is depicted below is reversed. That pentagon represents the quadruple arrow of the $(3,2)$-move, but now is written in terms of triangulations.

Following this illustration, additional interpretations of the associators and co-associators triple arrows , , , and  are discussed in the context of arches that represent parentheses.

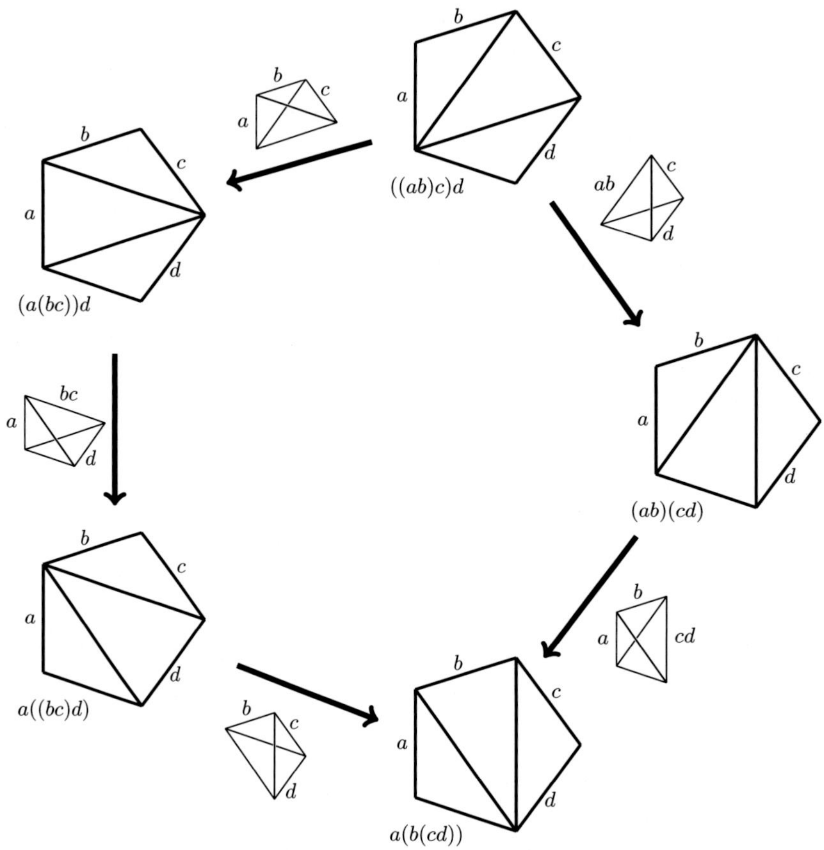

The pentagon relation that is depicted above appears to be a commutative diagram in which the objects are triangulations of a pentagon, the arrows are tetrahedra, and the double arrow is the 4-simplex Δ^4 whose boundary is the union of these tetrahedra. It is a translation of the $(3, 2)$ relation that is expressed in terms of binary trees in Chapter 7, Section 3.1. In the earlier context, the trees represented double arrows whose source is -•- $\leftrightarrow 1$ and whose target corresponds to 4. So in that earlier context, the commutative diagram was thought of as a quadruple arrow. The process of "decategorification" occurs when the conceptual multiplicities of arrows is reduced, that is, when a double arrow is thought of as a single arrow or as an object (Chapter 6, Section 2.2).

Exercise. 1. It is amusing and worthwhile to explicitly construct a multi-category whose objects are triangulations of polygons, whose arrows are tetrahedra, and whose double arrows are 4-simplices. In case such an exercise appears to be difficult, try an alternative approach and promote the triangulations to be double arrows.

2. More generally, take any of the 214 sets that are listed in Stanley's book as a set that is counted by the Catalan number $C_n = \frac{1}{n+1}\binom{2n}{n}$ find an explicit correspondence between it a binary trees or parentheses, and define corresponding Temperley-Lieb multiplication, skew compositions, and categorical descriptions.

On the boundary of the pentagon that describes the $(3, 2)$-move, the binary tree that corresponds to $((ab)c)d$ (fully left associated) is a source and $a(b(cd))$ (fully right associated) is the target. Under the correspondence that we have chosen, the string ()()() is the source and ((())) is the target. More generally, the un-nested string $()()\cdots()$ is a source and the fully nested string $((\cdots()\cdots))$ is the target. In the next section, the intermediate targets are found by putting a partial order on the strings of parentheses.

2.2. Arches and handle slides. A *partial order* on a set X is a reflexive, anti-symmetric, and transitive relation $\leq$. Specifically,

(1) *reflexive:* For all $x \in X$, $x \leq x$.
(2) *anti-symmetric:* If $x \leq y$ and $y \leq x$, then $x = y$.
(3) *transitive:* If $x \leq y$ and $y \leq z$, then $x \leq z$.

Associated to a partially ordered set X is its *Hasse diagram*[2] which is a graph with directed edges. The elements in X correspond to the vertices. There is a directed edge $x \to y$ if (1) $x \leq y$, $x \neq y$ and (2) there is no $z \neq x, y$ such $x \leq z \leq y$. That is, the edges connect two elements of X provided that there is no intermediate element between them. A partially ordered set, X, is a type of category. There is an arrow between x and y if and only if $x \leq y$. The arrows correspond to directed paths in the corresponding Hasse diagrams.

Let n be fixed and consider the set of valid strings of n pairs of parentheses. A partial order on this set is induced by the relation $()() \to (())$. Here the arrow indicates the ordering. More generally, if a, b, and c are smaller, possibly empty, and grammatically correct strings, then $a(b)(c) \to a(b(c))$ is an edge in the Hasse diagram. Imagine that the right element of a matched pair can engulf and slide over the pair of parentheses that enclose c. The structure, in case $n = 4$, is illustrated in Fig. 2.

In the arch-like depiction, each arrow indicates that the right foot of the left arch slides over the arch towards the right, which remains stationary, while the left foot of the sliding arch also remains stationary. This sequence of handle slides is depicted in the $n = 3$ case in Fig. 1 and in the $n = 4$ case in Fig. 2.

In the $n = 3$ case, generators for the first homology group of a disk that has three handles attached to it (which is $\mathbb{Z}^3$ where $\mathbb{Z} = \{0, \pm 1, \pm 2, \ldots\}$ denotes the set of integers) are depicted as x, y, and z at the top of the figure. Each handle slide is a transformation of the bases of the form $a \mapsto a+b$, $b \mapsto b$, and $c \mapsto c$. The resulting action on the homology group is represented by a matrix that is represented along the arrows. In the center of the figure, the two possible matrix multiplications are indicated. The resulting transformation sends x to the sum $x + y + z$, y is sent to $y + z$, and the third generator z is stationary. While we have not spent any effort in describing homology within this paragraph, it is our belief that the diagrammatics communicate these transformations clearly.

3. Higher associativities

An associated product, for example $(((ab)c)d)e$, involving $(n + 1)$-variables corresponds to a binary tree. Such a tree is a 2-arrow in the multi-category $\mathcal{S}(\mathsf{Y})$.

[2]Hasse is said to have applied for membership to the Nazi party in 1933, but his application was denied because of his Jewish ancestry. The idea of a Hasse diagram seems to have been found earlier by Vogt in 1985.

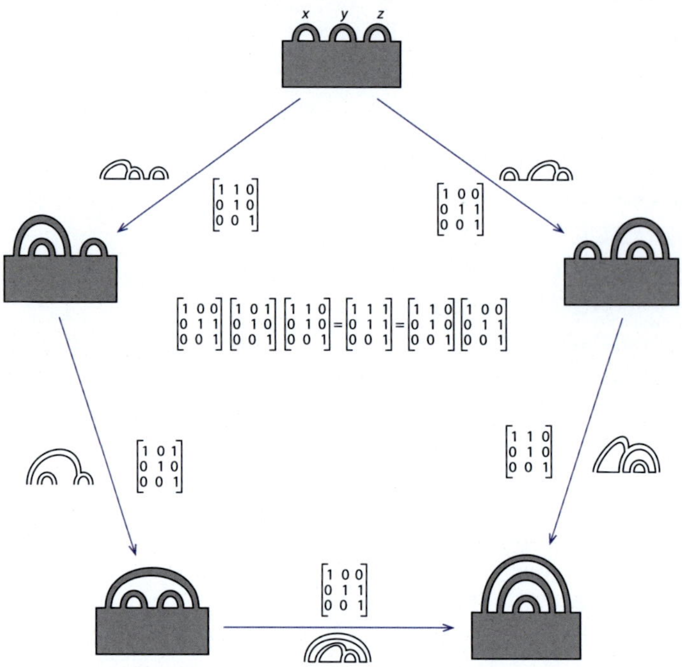

FIGURE 1. Parentheses and handle slides

The associator  is a triple arrow, and the pentagon relation is a quadruple arrow. A 5-arrow is associated to the Stasheff 3-polytope.

In Fig. 4, the 3-polytope is depicted in a 3-dimensional view. The central projection is depicted in Fig. 3 as a Hasse diagram for the partially order set of parenthesized strings or equivalently the triangulations of the hexagon.

Exercise. Describe the vertices of the 3-dimensional Stasheff polytope as binary trees.

The arrows between triangulated hexagons are labeled by tetrahedra, and the pentagonal faces ((3, 2)-moves) are labeled by the 4-simplices that these bound. These 4-simplices form the boundary of a 5-simplex. The square faces are not labeled since each corresponds to a pair of tetrahedra glued along one of the three diagonals to the hexagon. In the dual picture, they correspond to exchangers between pairs of associator triple arrows.

The general *associahedron* or *Stasheff polytope* [**74**] is an $(n-1)$-dimensional polytope — the vertices of which are labeled by associated products, strings of n matched pairs of parentheses, or equivalently triangulations of a regular polygon with $(n+3)$ vertices. The oriented edges correspond to the edges of the Hasse diagram of the underlying partially ordered set. The 2-dimensional faces are pentagons and quadrilaterals. The latter faces correspond to exchangers.

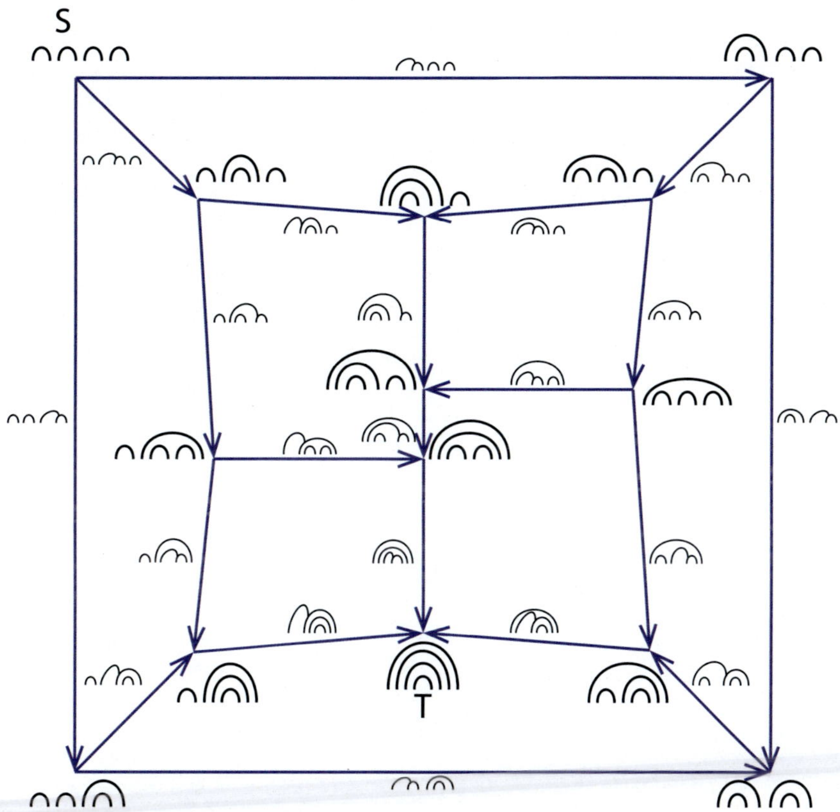

FIGURE 2. The 3-dimensional Stasheff polytope, first version

When the vertices are labeled by triangulations of polygons, the edges are labeled with tetrahedra, and the pentagonal faces are labeled by 4-simplices. The 3-dimensional enneahedral[3] cells can then be labeled by 5-simplices. Other 3-dimensional faces are cubical or pentagonal prisms. Again these involve exchangers.

THEOREM 8.1. *Suppose that an $(n-1)$-dimensional associahedron has its vertices labeled by triangulations of a polygon that has $(n+2)$ edges. Then there are $(n+2)$ facets that are $(n-2)$-dimensional associahedra labeled as n-simplices. The union of these form the boundary of an $(n+1)$-dimensional simplex.*

PROOF. The proof by induction follows. When $n = 1$, there is a unique triangulation of a triangle. The two triangulations of a square cobound a tetrahedron. Similarly, the five triangulations of a pentagon are connected edgewise by such tetrahedra, and the Stasheff pentagon corresponds to a 4-dimensional simplex.

In general, the triangulations of a regular polygon that has $(n+2)$-vertices are being considered. Label the vertices in cyclic order, clockwise, $0, \ldots, n+1$. For $i = 1, \ldots, n$, label the edge $[i, i+1]$ with a variable a_{i+1}. The label on the final edge $[n+1, 0]$ corresponds to an association of the product $a_1 \cdot a_2 \cdot \cdots \cdot a_n$ where

[3]having nine faces

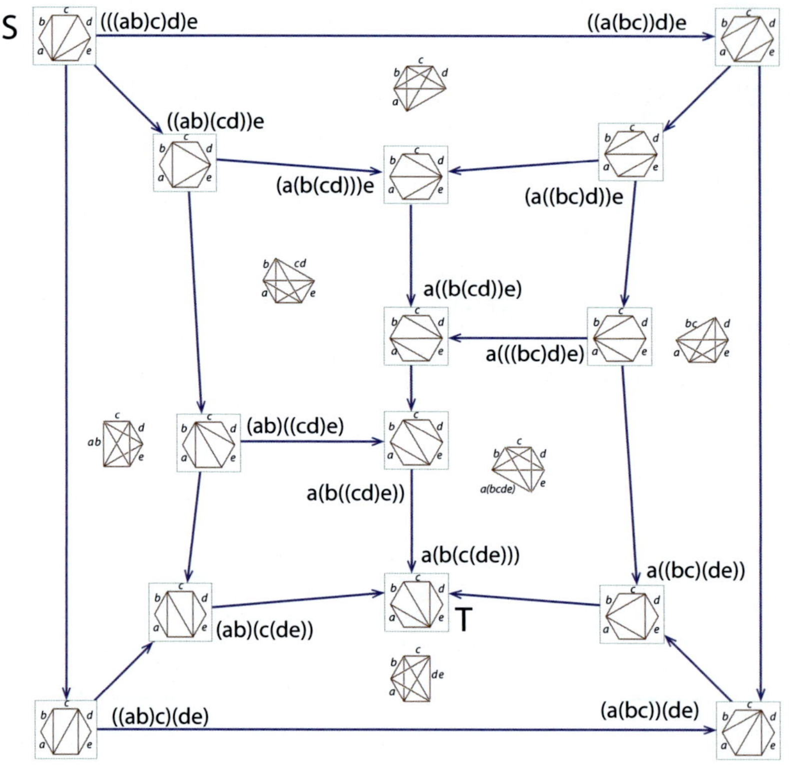

FIGURE 3. The 3-dimensional Stasheff polytope, second version

the association is induced by the triangulation. This association is also inductively defined. For $i = 1, 2, \ldots, n$ consider the set of those triangulations that contain an edge $[i, i+2]$. When a triangulation corresponds to an associated product, such an edge can be labeled $a_{i+1} \cdot a_{i+2}$. The set of the triangulations that includes the edge $[i, i+2]$ correspond to a triangulation of a polygon with $(n+1)$ vertices and edges that have labels, $a_1, \ldots, a_i, a_{i+1} \cdot a_{i+2}, a_{i+3}, \ldots, a_n$. The final edge is labeled as an associated product of the form $a_1 \cdot \cdots \cdot (a_{i+1} \cdot a_{i+2}) \cdot \cdots a_{n+1}$. Those triangulations that contain the edge $[n+1, 1]$ have products of the form $a_1 \cdot (a_2 \cdot \cdots \cdot a_n)$ along the edge $[n+1, 0]$. The triangulations that contain the edge $[0, n]$ have products of the form $(a_1 \cdot \cdots \cdot a_n) \cdot a_{n+1}$ along the bottom edge $[n+1, 0]$. Thus within the $(n-1)$-dimensional associahedron, there are $(n+2)$ facets that are $(n-2)$-dimensional associahedra. Each of these, therefore, is labeled by an n-dimensional simplex.

In fact, the triangulations that contain the edge $[n+1, 1]$ are the vertices of the associahedron that corresponds to the face $(0, x_2, \ldots, x_{n+2})$ of the simplex. The triangulations that contain the edge $[i, i+2]$ are the vertices of the associahedron that corresponds to the face $(x_1, \ldots, x_{i-1}, 0, x_{i+1}, \ldots, x_{n+2})$. Finally, the triangulations that contain the edge $[0, n]$ are the vertices of the associahedron that corresponds to the face $(x_1, \ldots, x_{n+1}, 0)$. This completes the proof $\square$.

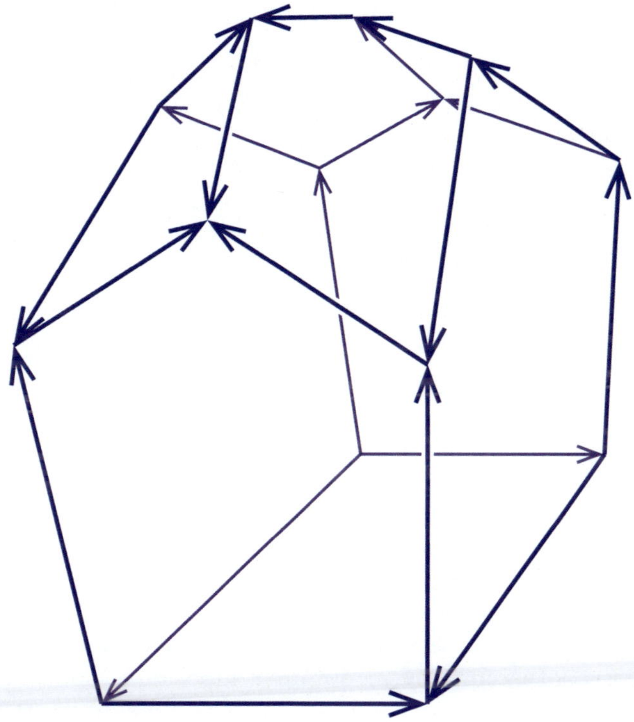

FIGURE 4. The 3-dimensional Stasheff polytope in 3-dimensions

The proof above bears a kinship with the ideas of group homology, and indeed Chapter 10, Section 3 will discuss this further.

The Stasheff 3-polytope can also be understood in terms of the associator and the $(3,2)$-move as a quadruple arrow. That point of view is dual to the picture that involves simplices. It will be described in terms higher dimensional foams in Section 4 in this chapter. First, the Pachner moves are introduced.

3.1. Pachner moves. There are two types of "nice"[4] manifolds. A *smooth manifold* is one upon which the general notions of calculus apply. A *triangulated manifold* is one that is constructed from n-dimensional simplices and certain additional conditions hold. Pachner's Theorem lets us determine when two different triangulations of an n-dimensional triangulated manifold are related. In order to discuss it, we will look closely at the boundary of an $(n+1)$-dimensional simplex.

[4]Of course, being "nice" is a subjective property. We have colleagues who prefer topological manifolds.

Recall that the standard $(n+1)$-dimensional simplex is a subset of $\mathbb{R}^{n+2}$ consisting of the points that have non-negative coordinates ($0 \le x_j$ for $j = 1, \dots, n+2$) and $\sum_{j=1}^{n+2} x_j = 1$. The boundary of the simplex Δ^{n+1} consists of its intersection with the coordinate planes. For $j = 1, \dots, n+2$, let

$$\Delta_j^n = \{(x_1, \dots, x_{n+2}) \in \Delta^{n+1} : x_j = 0\}$$

denote the intersection of Δ^{n+1} with the jth coordinate plane. For each $k = 1, \dots, n+1$, the union

$$L_k = \bigcup_{j=1}^{k} \Delta_j^n$$

can be identified with an n-dimensional ball. Let

$$R_k = \bigcup_{j=k+1}^{n+2} \Delta_j^n;$$

this, too, is an n-dimensional ball. For an n-dimensional manifold M, if there is a configuration of simplices that can be identified with L_k it can be replaced by R_k without changing the topological type of the manifold. This is the *Pachner move* of type $(k, n+2-k)$. Pachner [**61**] showed that two triangulations of an n-dimensional manifold are related by a finite sequence of moves of type $(k, n+2-k)$. Of course, differing values of k are used when the triangulations are related.

The $(n-2)$-dimensional facets of the $(n-1)$-dimensional associahedron can be thought of as multi-arrows in the multi-category whose 1-arrows are the natural numbers and whose 2-arrows are rooted binary trees, generated by Y. That multi-category is dual to another in which arrows are compositions of directed edges (with consistent directions), and whose double arrows correspond to inserting an internal vertex. That is Y is dual to $\Delta = \Delta^2$ — a triangle. In this sense, the Pachner moves correspond to assuming that arrows at the penultimate level are invertible, and the highest level arrows relate two possible compositions thereof. In the next section, we explicate the duality by constructing higher dimensional foams that are analogues to 2-foams.

4. Higher dimensional foams

The model for a higher dimensional foam is the $(n-1)$-dimensional complex that is constructed from the barycenters of an n-dimensional simplex.

Recall, that Δ^1 consists of the line segment in the plane $\{(x_1, x_2) : 0 \le x_1, x_2 \ \& \ x_1 + x_2 = 1\}$, and Δ^2 is the convex hull of $(1,0,0)$,$(0,1,0)$, and $(0,0,1)$. Let $Y^0 = (\frac{1}{2}, \frac{1}{2}) \subset \Delta^1$. We inductively define a subset of Δ^j for $j \ge 1$. Suppose that Y^{j-2} in Δ^{j-1} has been defined. For $k = 1, \dots, j+1$, let Y_k^{j-2} be a copy of Y^{j-2} embedded in the *kth face* $\Delta_k^{j-1} = \{\vec{x} \in \Delta^j : x_k = 0\}$ of the simplex Δ^j. Cone the union $\cup_{k=1}^{j+1} Y_k^{j-2}$ to the barycenter $\frac{1}{j+1}(1, 1, \dots, 1)$ of the j-simplex. Define $Y^{j-1} \subset \Delta^j$ to be this cone. The space Y^1 is homeomorphic to the alpha-numeric character Y. The space Y^2 is depicted in Fig. 5. Of course, this is an alternative view of the associator. In general, the space Y^{j-1} has $\binom{j+1}{\ell}$ faces of dimension ℓ for $\ell = 0, \dots, j-1$. The ℓ-dimensional faces of Y^{j-1} are obtained as the union of the cones on the $(\ell-1)$-dimensional faces found in the copies of Y^{j-2} that appear in the boundary of Δ^j. The space Y^{j-1} is considered both as an abstract topological space and in terms of this fixed embedding in Δ^j.

Let $n = j - 1$. An *n-dimensional foam* is a compact space X such that each point $x \in X$ has a neighborhood that is homeomorphic to a neighborhood of a point in Y^n. If the point in x maps, via such a homeomorphism, to $\frac{1}{n+1}(1, 1, \ldots, 1)$, then x is said to be a *vertex of the foam.* The points in X that have neighborhoods that map to neighborhoods of points in $\cup_{k=1}^{j+1} Y_k^{j-2}$ are called *boundary points.*

A triangulated $(n + 1)$-dimensional manifold has a foam naturally embedded within the duals to the simplices. Moreover, the Pachner moves can be treated as moves to foams. Foams can be constructed as compositions of higher dimensional arrows in $\mathcal{S}(\mathsf{Y})$. These higher order arrows are conceived via the associahedron. Invertibility conditions can be imposed, and these will allow the full range of moves that correspond to the Pachner moves to be consequences of the next order associahedron. Here the correspondence between Pachner moves on the one hand, and foam moves on the other is given directly via relabeling the vertices of the Stasheff polytope by triangulations of polygons or by the corresponding binary trees that encode multiplications.

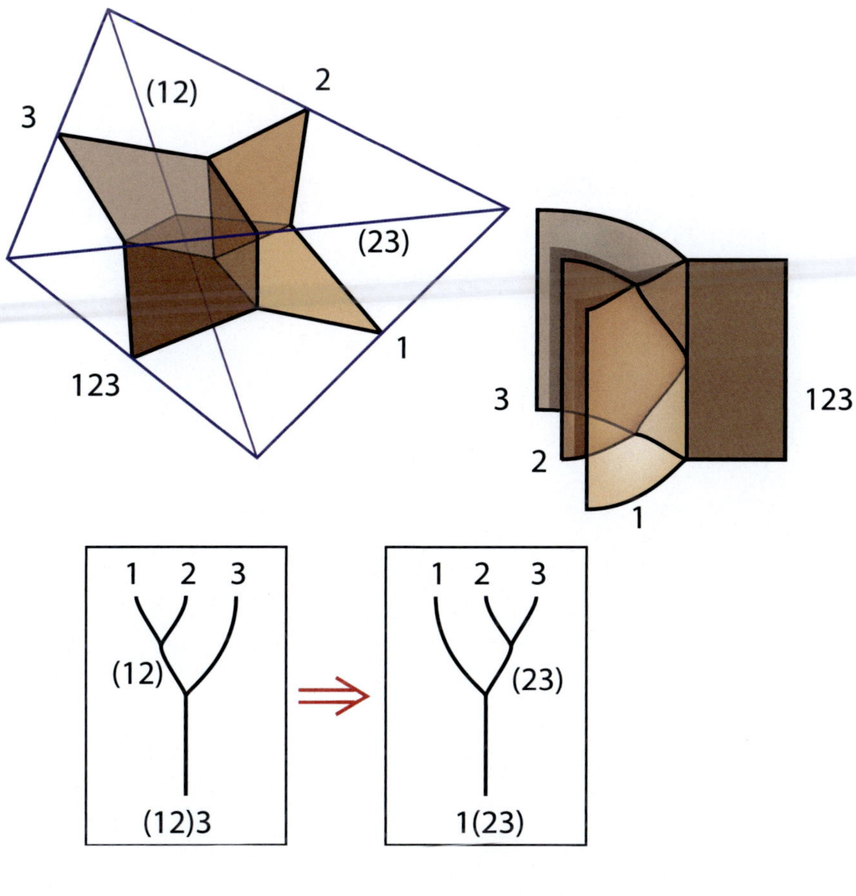

FIGURE 5. Another view of the associator

CHAPTER 9

Knots in space

Contemporary manifestations of diagrammatic algebra originated in the study of knots, links, and braids. The highly influential works of Kauffman, Reshethikin, Turaev, Viro, et al.[**41, 43, 75, 65, 76**] indicate that quantities associated to knot and link diagrams that are unchanged under the moves relating the diagrams are, in fact, analogies[1] between a diagrammatic algebraic system and a traditional algebraic system. In some formulations, Joyal and Street [**36**] and Turaev [**75**] described tangles as a category with properties that reflect the knotting phenomenon, and they demonstrated that the Jones polynomial [**34**] and its generalizations were functors between categories. Kauffman took this analogy along with Penrose's abstract tensors and used diagrams to simplify algebraic computations. Feynmann diagrams of elementary particle interactions also manifest these techniques.

This chapter will be a small sample of these ideas. It will start by further generalizing the categories constructed thus far and demonstrate some standard constructions in knot theory. It is not a comprehensive treatise on knot theory since there are many magnificent texts in that regard [**2, 41, 47, 53, 60, 67**].

1. Oriented knots and higher categories

There are several categorical approaches to the categorical characterization of tangles [**36, 72, 75, 80**]. Each has its advantages. Roughly, the category of tangles is the free braided monoidal category with a single object that generates it. Tangles can be oriented and/or framed, and consequently, there are several ways of approaching the Reidemeister type I move. These are described in the literature as rigid, ribbon, tortile, and pivotal. The paper [**71**] helps distinguish these structures. Sometimes, when a reader is hungry for information, reading the subtle differences in the menu items becomes a distraction. The approach herein will be to enhance some aspects of the multi-categories that have been constructed within the text thus far. Axioms will be articulated carefully, but a proliferation of adjectives will be avoided.

A multi-category $\mathcal{OFT}$ is to be defined. The name is meant to suggest oriented framed tangles, but the category is, in fact, an abstract entity that has one unnamed object. The only 1-arrow in $\mathcal{OFT}$ is an identity arrow upon that object. The identity 1-arrow is both source and target of any of the three kinds of generating 2-arrows : (1) a negative 2-arrow, sometimes called "pointing down;" (2) the identity 2-arrow; (3) a positive 2-arrow called "pointing up." These are illustrated.

[1]or functors

Compositions are globular, but the identity 2-arrow can be attached via a horizontal composition, and a variety of tensorial compositions agree. Specifically, skew compositions of double arrows alway can contract to a globular composition. So double arrows are represented as a phalanx of dots along a line segment. The situations are indicated below with an unlabeled gray dot used to indicate either a positive or a negative double arrow. Within the illustration, the vertical $\updownarrow$ is meant to indicate that globular compositions with the identity double arrow can be contracted or expanded at will. Thus by contracting all the identities upon the identities of the unique object, every double arrow is uniquely and unambiguously expressed as a word w in (+)s and (−)s. While these dots are drawn vertically here, the word w will be expressed horizontally by using the convention given in Ch. 4, Fig. 1.

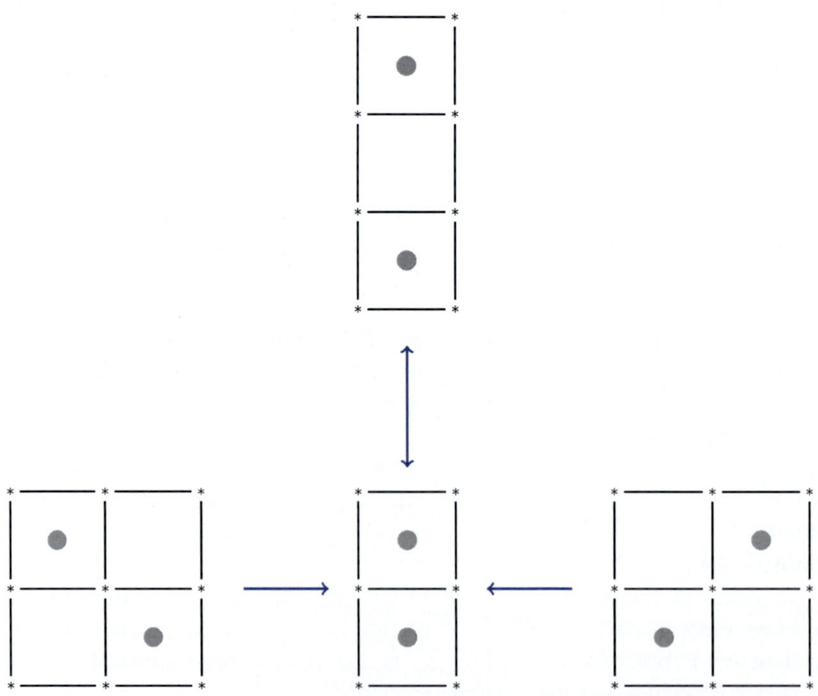

The possible 3-arrows that exist are the main points of interest. The identities upon (+) and (−) are, respectively, I_+ and I_- or, alternatively, $\uparrow$ and $\downarrow$. The reader should imagine these as upward or respectively downward oriented strings in the center of a 3-dimensional cube. There are eight possible generating exchangers Xs, two types of Us, and two types of ∩s. These also occupy 3-dimensional space. They are indicated below.

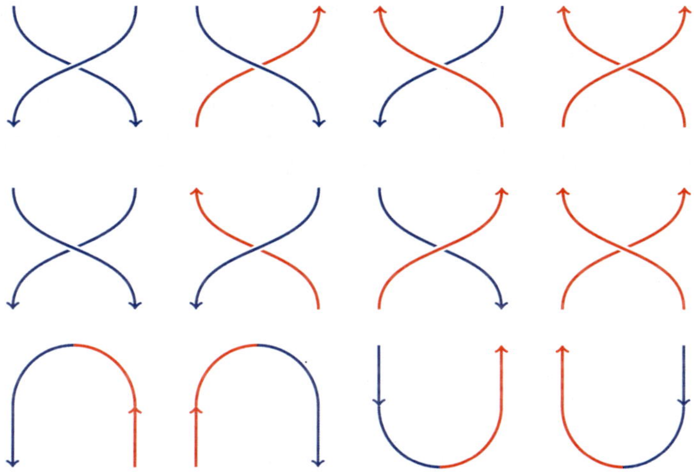

Both globular and skew compositions are allowed for triple arrows. *In this way, composites of triple arrows correspond to oriented tangles in space.* This correspondence is functorial between the diagrammatically defined multi-category $\mathcal{OFT}$ and a geometrically defined category of tangles. The composites of triple arrows also are formal expressions in $\mathcal{OFT}$. At this stage, the embedding of the tangle is considered to be quite rigid, and a viewer's perception of the tangle is exactly described as a composite of triple arrows. Quadruple arrows will be used to define an equivalence relation upon these, and that relation will correspond to isotopy of the corresponding tangles, but we are not that far along in the construction.

Let w_0 and w_1 denote finite words in the characters $\pm$. Each of these words is a juxtaposition of signs with no further rhyme nor reason. Define the *signature* $s(w_i)$ to be the difference in the non-negative integers $s(w_i) = \#(+)\text{s} - \#(-)\text{s}$ — that is the difference in values between the number of $(+)$s and the number of $(-)$s in the word w_i for $i = 0, 1$. If $s(w_0) \neq s(w_1)$, then the collection of triple arrows between them is empty. When $s(w_0) = s(w_1)$, the collection of triple arrows between them consist of all possible finite globular and skew compositions of Xs, $\overline{\mathsf{X}}$s, Is, ∩s, and Us. These correspond to finite tangle diagrams.

Since these are triple arrows, there is a bookmark associated to each critical point or crossing in the diagram that represents the triple arrow. By tradition, the composition of triple arrows is arranged on the page in a rectangular box. The bookmark is a pair of non-negative integers that simply counts the number of strings to the left and right of the crossing point of an X or the critical point of a U or ∩. If necessary, the bookmark can be enhanced to describe a given non-identity 3-arrow as a triple uAv where u and v are finite words in $\{-, +\}$ and A is of type X, $\overline{X}$, U, or ∩. The triple arrow X can be sub-, and/or super-scripted with signs as well in order to describe its type.

These typographic choices are somewhat optional that depend upon the need to disambiguate the notation. An explicit and composable sequence of triple arrows gives rise to an unambiguous tangle diagram. For example, write

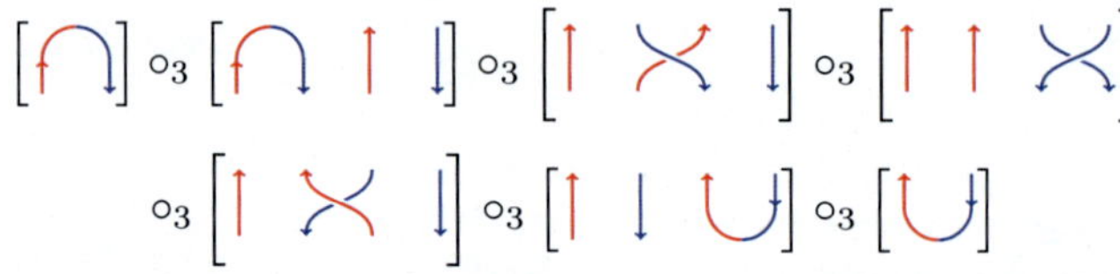

to indicate a composition of triple arrows where $\circ_3$ indicates that the item on the left is stacked above the item on the right sequentially. By vertically stacking these triple arrows, and placing them above each other so that their serifs match[2], the corresponding tangle diagram is apparent.

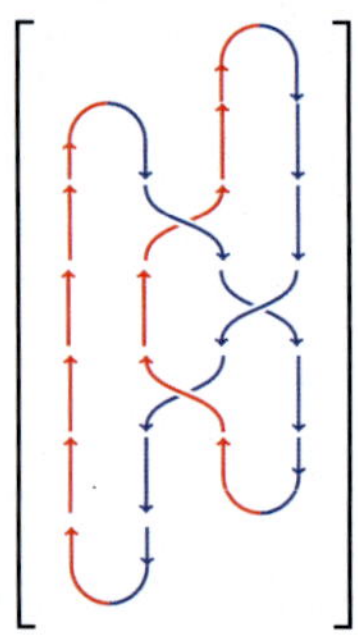

There are two aspects of diagrammatic knot theory. So far, a categorical system has been developed that allows for descriptions of tangle diagrams. In the next Section 2, quadruple arrows will be given that define an equivalence relation upon the set of diagrams. That relation will correspond to the modulation of geometric classes of diagrams by mean of isotopy. To summarize, a system of glyphs allows tangle diagrams to be named and represented, and relations among these will determine when two diagrams represent the same isotopy class of tangles.

In general an *oriented tangle* is a proper smooth embedding of a compact 1-dimensional manifold into $\mathbb{R}^2 \times [0,1]$. The boundary points are required to lie disjointly in the intervals $[0,k] \times \{0\} \times \{0,1\}$ for some value $0 < k \in \mathbb{R}$. Tangent lines to arcs are vertical in a neighborhood of these points. By small perturbations, the projection onto $\mathbb{R} \times \{0\} \times [0,1]$ is assumed to consist of a collection of transverse double points, and critical points; double points each occur at separate times $t \in [0,1]$. This last condition indicates that the interval $[0,1]$ can be used to distinguish critical and crossing points.

In somewhat loose terminology, it is often said that projection onto the interval $[0,1]$ provides a *Morse function* for the tangle. That is, this projection has only non-degenerate critical points. But more importantly, those points that correspond to crossing points are at distinct heights and these heights are distinct from the (also distinct) critical values. It is worth noting that contemporary understanding of critical behavior of real valued functions that are defined upon manifolds is derived from analyses due to Marston Morse as articulated in the treatise by Morse and Cairns [**59**].

---

[2]Here the serif is merely an arrow.

2. Reidemeister moves

Consider an oriented tangle $T : (L, \partial L) \hookrightarrow (\mathbb{R}^2 \times [0,1], [0,k] \times \{0\} \times \{0,1\})$ where L is a compact 1-dimension manifold (a union of a finite number of intervals and circles). Its *tangle diagram* is obtained from its generic projection into $\mathbb{R} \times \{0\} \times [0,1]$ by a trompe l'oeil that indicates over and under crossing information. The *under-arc* is broken into two segments at a crossing, and the *over-arc* is the unbroken segment near the crossing. Of course, an over-arc at a given crossing may pass under a subsequent crossing. In this section, a finite set of moves are given, such that two diagrams representing isotopic tangles are related by means of a finite application of a sequence of these moves. This classical result was first published by Kurt Reidemeister [**64**]. An alternative proof was given by J. W. Alexander and G. B. Briggs a little later [**3**]. These moves to diagrams are commonly called the *Reidemeister moves.* Apparently they were proposed by Maxwell. They occur in three types.

We will abuse notation since the glyphographic representation of the variety of Xs, ∩s and Us that were introduced in Section 1 of this chapter coincides with the visual representation of crossings, maxima, and minima.

First, some variations of the Reidmeister type II move are listed. These are presented in all orientations and all variations of over/under crossing information. These type II moves are indistinguishable from the strong invertibility of the exchangers when these are represented as the X-like glyphs. Thus at the level of knot theory, doing followed by undoing is the same as not doing. Here "doing" means the crossing of strands, one in front of the other, and in "undoing" the front strand remains in front as the strands are uncrossed. Two paragraphs from here, other variations of type II moves that are tantamount to the naturality of the exchanger with respect to ∩ and U will be presented.

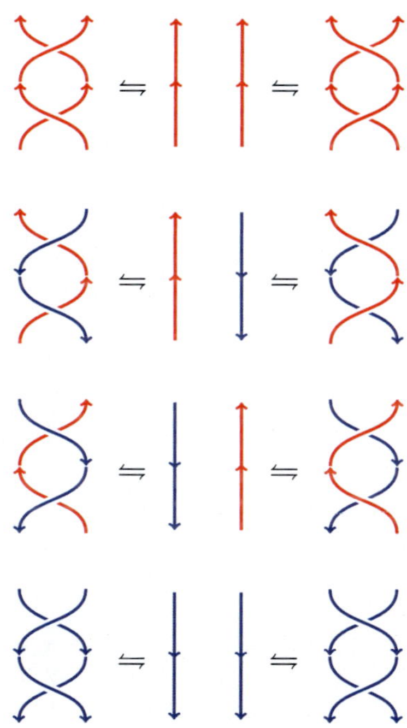

The Reidemeister type III move is depicted using black strings devoid of direction in order to indicate that it holds with any variation of upward or downward pointing strings. See also Chapter 11, Section 2 where variations are listed for the move in one direction. Furthermore, in the presence of all variations of the type II move that are given above, only one type III move is needed. Otherwise, considering all possible orientations and crossings, there are 48 variations of the type III move. The type III move is a consequence (or a restatement) of the naturality axiom of the exchanger. In the language of change and exchange, the change is an exchange between the positions of two strands, and the exchange repositions one strand with respect to a pair of them.

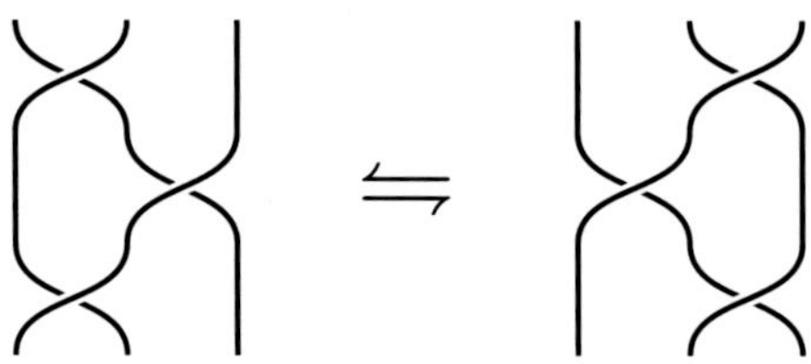

Below, another sort of type II move also follows from the naturality of the exchanger. In these cases, change (joining two oppositely oriented strands) followed by exchange (one strand exchanging places with two) is the same as exchange followed by change. However, after the strands are joined, the exchange in the latter instance is with an identity that appears to be empty.

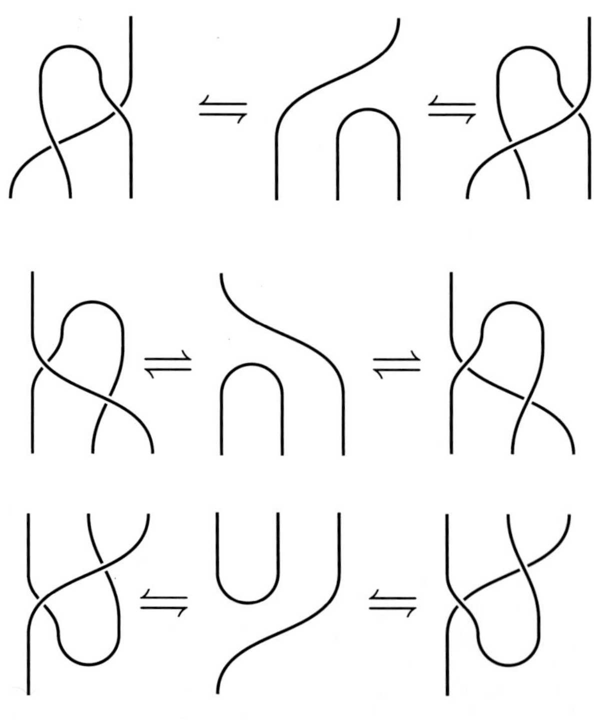

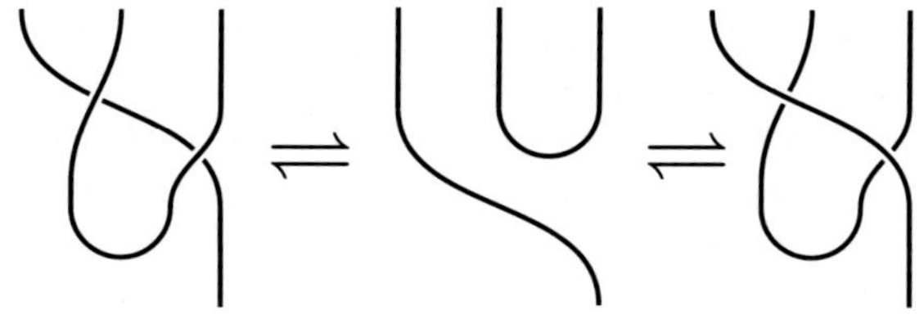

In the tangle context, critical points and crossings are required to lie at distinct horizontal levels. Thus each X, U, and ∩ occurs at a distinct height. Tangle diagrams also contain skew compositions such as $(\cap \otimes \mathsf{I}) \circ_3 (\mathsf{I} \otimes \mathsf{U})$. These can be straightened out using the geometric analogue of the cusp axiom that follows from weak invertibility of the pair U and ∩ — cusps. In addition, distant critical or crossing points can have their heights exchanged by the analogue of the quadruple arrow exchanger.

Isotopy classes of oriented framed tangles correspond to tangle diagrams modulo Reidemeister type II and type III moves, zig-zag moves, and the exchange of distant critical points and/or crossings. Similarly, the multi-category $\mathcal{OFT}$, that is generated by the double arrows  and  that are weak inverses, has exchanger X, cancelation ∩, and creation U triple arrows. The exchanger is asserted to be strongly invertible and natural with respect to any other triple arrow. Skew compositions of U and ∩ are identities because U and ∩ are a weakly invertible pair of reverses. Finally, the quadruple arrow exchanger gives that distant triple arrows commute.

There is a categorical structure on oriented framed tangles. Objects are finite words in (+) and (−), arrows are oriented tangle diagrams. These are generated by crossings and critical points. Two tangle diagrams represent isotopic framed tangles provided they are related by type II and type III Reidemeister moves, zig-zags, ψ-moves and commutativity of distant critical points and crossing points. **This category is equivalent to the decatgorification of $\mathcal{OFT}$ in which double arrows are considered as objects, and exchangers and creation and annihilation operators are considered to be arrows.**

2.1. Framing and the type I move. An *oriented framed tangle* is a proper embedding of a 1-dimensional manifold in space $(\mathbb{R}^2 \times [0,1])$ such that a direction, in $\{0\} \times \mathbb{R} \times [0,1]$ perpendicular to the tangent of each arc, is chosen. When a 1-manifold is described by a diagram in $\{0\} \times \mathbb{R} \times [0,1]$, a special framing, called *blackboard framing*, is described as follows. The arcs of a diagram project non-singularly into the plane $\{0\} \times \mathbb{R} \times [0,1]$, so that the tangent lies in that plane. Thus the driver of a vehicle riding along a curve could hold out one arm to indicate the normal, this normal lies in the plane, and its projection is not 0. In a tangle diagram, this normal induces the blackboard framing. The diagram lives on the plane of the blackboard or the page of the paper with the interval factor $[0,1]$ corresponding to the vertical direction. At each point on a component, a short direction arrow is chosen to point perpendicularly to the left of the tangent direction that is specified by the $\uparrow$ or $\downarrow$ of the arc.

A blackboard framed curve gives rise to a curve that is pushed off from the original arc, and the curve and its pushed off companion typically link. The amount of linking can be determined by defining signs for self-crossings and counting the accumulation (positive versus negative) of signs over all of the self crossings.

The type I Reidemeister moves affect the framing. This is why they were excluded in the discussion of $\mathcal{OFT}$. Here the upward oriented type I moves are presented. Section 2.2 demonstrates that the downward type I move follow from zig-zig, naturality of the exchanger, the upward type I, and the exchanger quadruple arrows.

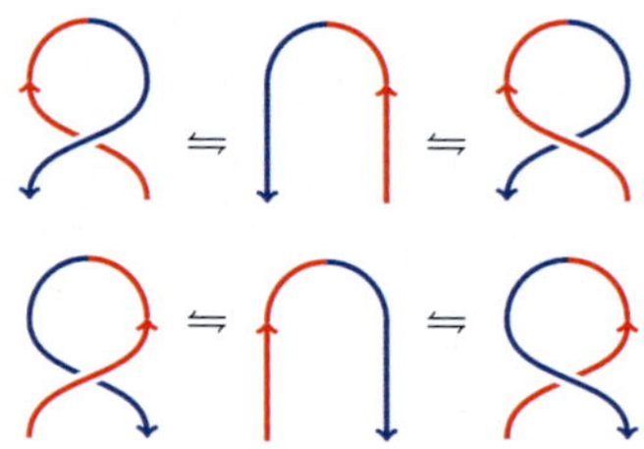

2.2. Moves that are consequences of others. When a tangle diagram is given in the plane, the vertical direction serves as a height function. By convention, critical points and crossings occur at distinct heights. More precisely, if two critical events or crossings occur at the same height with respect to the vertical, then by tilting the diagram slightly, the coincidence can be avoided. Such tilts are part and parcel of the exchangers. Moreover, singularity theory informs knot theory. When a height function is chosen for a knot diagram, the Reidemeister moves and changes in height correspond to codimension one singularities in the time direction of the isotopy. For example, the type III move corresponds to the coincidence of the crossing of a pair of strings with a transverse sheet. The type II move corresponds to a tangency between curves. The type I move corresponds to the end point of an arc of double points. The zig-zag moves correspond to cusps in the fold set of the isotopy.

Another codimension one singularity corresponds to a double point passing over a fold. This is called a ψ-move if there is a minimum involved or a $\rotatebox[origin=c]{180}{ψ}$-move when a maximum is involved. Four types of these moves are depicted at the extremes of Fig. 1. In this figure, the ψ-moves are demonstrated to be a consequence of naturality of the exchanger with respect to optimal points on the curves. In a similar way, this naturality can be seen to be a consequence of ψ-move.

Fig. 2 demonstrates that one of the ψ-moves is a consequence of one of the $\rotatebox[origin=c]{180}{ψ}$-moves, the zig-zag (or cusp) move, and the exchange of distant critical points or crossings.

Fig. 3 demonstrates that one of the down versions of the type I move follows in a similar manner from the type I up move. These proof are similar to those given in Ch. 2, Fig. 4, and they follow the same methodology: to prove the upside-down version of a theorem use the zig-zig rules to dualize the computation, and apply commutativity between items that are distant from the center of calculation.

Add the type I relations as invertible quadruple arrows to the multi-category $\mathcal{OFT}$ to create a multi-category $\mathcal{OT}$ that has:

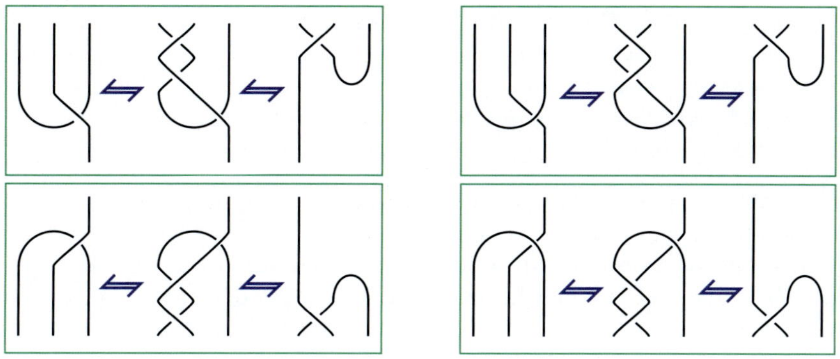

FIGURE 1. Sliding a crossing over an optimum follows from naturality

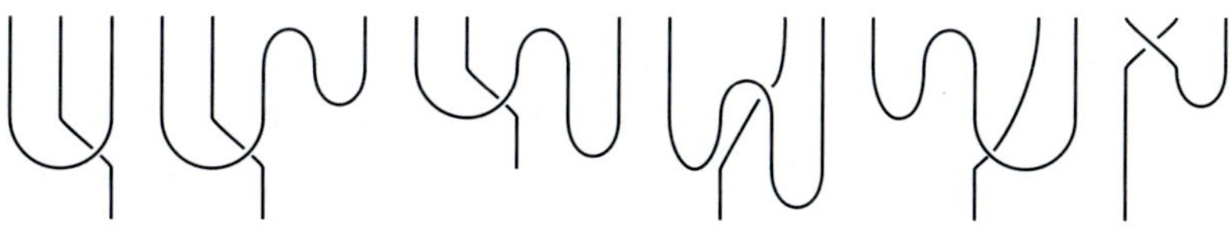

FIGURE 2. An up move gives a down move

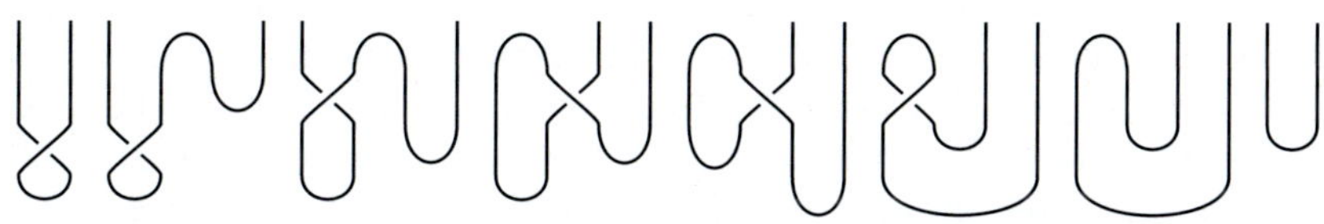

FIGURE 3. The type I down move follows from the up move

(1) one unnamed object,
(2) an identity 1-arrow,
(3) generating 2-arrows  and  that are weak inverses,
(4) triple arrows formed by exchangers Xs, $\overline{\mathsf{X}}$, creations: Us, and annihilations: $\cap$s that express that $(+)$ and $(-)$ are weak inverses,
(5) quadruple arrows formed from the invertibility and naturality of the exchanger that are isomorphisms (and hence thought of as equalities),
(6) cuspidal quadruple arrows, i.e. expressions of zig-zag identities that relate as skew compositions of U and $\cap$ to identities (these follow from our definition of weak inverses),
(7) quadruple arrows of the form of the type I Reidemeister move that are also isomorphisms, and

(8) exchanger quadruple arrows that are isomorphisms.

THEOREM 9.1. [**36, 72, 75, 80**] *The decategorification of $\mathcal{OT}$ that collapses the double arrows into objects, exchangers* X*s,* $\overline{\mathsf{X}}$*s, creations* U*s, and annihilations* ∩*s into arrows, and that expresses the quadruple arrows as equalities between arrows is equivalent to the category of oriented tangles. This is also known as the free braided monoidal category*[3] *on an object generator that has a dual.*

The duality has been built into the axioms by requiring that (+) and (−) are weak inverses, and the braiding follows from the exchanger axioms and the requirement that horizontal composition is forbidden unless one double arrow is an identity. The statement is somewhat inaccurate since Reidemeister type I moves are included in the tangle case, but not accounted for among the axioms of a braided monoidal category. As far as we know, including a type I move among the axioms does not have a standard name [**71**].

Finally in case the double arrow is unsigned, then this multi-category agrees with the category of unoriented tangles in an analogous manner.

3. The fundamental group and related invariants

In this section and in Section 4 algebraic quantities that are associated to tangle diagrams will be defined. These quantities do not depend upon the diagram that represents the tangle. That is, the quantities remain unchanged or *invariant* when a diagram is changed by one of the *generalized Reidemeister moves.* In addition to the Reidemeister moves of type I, type II, and type III, zig-zag moves that cancel successive critical points along arcs, the variations on the ψ-move that is a consequence of the naturality of the exchanger, and the exchange of distant critical points and/or crossings are all considered.

The categorical axioms which characterize equivalent diagrams also hold in traditional algebraic contexts. This is a major success of diagrammatic algebra. Geometric features are abstracted to an axiomatic context, and other algebraic systems satisfy these axioms. On the other hand, the geometric context allows a bit of latitude with conventions. Within this section, labels on diagrams may flow from the top down or from the bottom up of a diagram so the vertically up direction that is the direction of categorical considerations need not be respected.

In the discussion above, we wrote of tangle diagrams. A *knot or link* diagram is a tangle diagram in which each component curve is a closed curve. So the tangle represented is a union of circles that are embedded into 3-dimensional space. The vertical axis is a height function, and the direction perpendicular to the plane of the paper is the direction in which exchangers pass in front or in back.

3.1. Quandles.

A *quandle* is a set Q together with a binary operation $\triangleleft$ that is presented by in-fix notation: for every $a, b \in Q$, there is an element $a \triangleleft b \in Q$ (read a acted upon by b, or b acts on a) such that

I. for all $a \in Q$, $a \triangleleft a = a$;
II. for all $a, b \in Q$, there is a unique $c \in Q$ such that $c \triangleleft b = a$; and
III. for all $a, b, c \in Q$, $(a \triangleleft b) \triangleleft c = (a \triangleleft c) \triangleleft (b \triangleleft c)$.

[3] An additional adjective such as "pivotal" or "tortile" is used to restrict the use of the type I Reidemeister move.

The last condition is called *(right) self-distributivity.* The axioms for a quandle correspond to a diagrammatic method of assigning variables to the arcs near a crossing in a tangle diagram. To demonstrate this correspondence, the arcs of a tangle diagram are oriented downward, For a negative crossing, label the bottom ends of the arcs from left to right a and b, respectively. At a positive crossing label the top ends of arcs in the same way.[4] Both ends of the over-arc at a crossing are labeled b. If you imagine standing on the over-arc and holding out your left arm, then your left hand will point towards the under-arc that receives the quandle product $a \triangleleft b$. The arc to the right is labeled a.

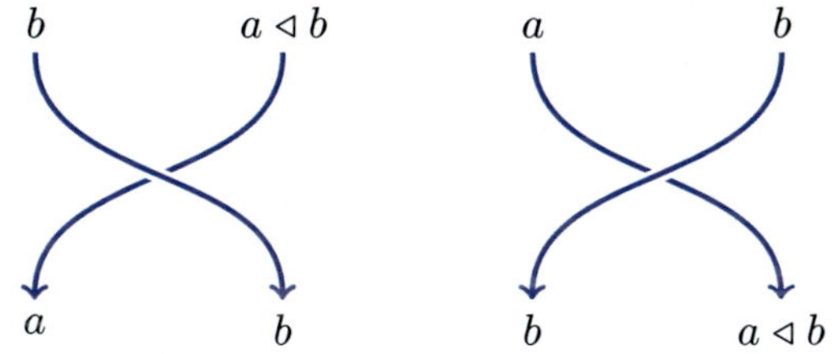

Often, the element c that is defined by the relationship $c \triangleleft b = a$ in the second axiom is labeled $c = a \triangleleft^{-1} b$, and the operation $\triangleleft^{-1}$ also forms a quandle. Below the quandle axioms are illustrated in the context of the Reidemeister moves. In the illustration of a type I move, the downward pointing over-arc is labeled with an a, after it passes over the maximal point, the label remains a. As a self crossing, $a \triangleleft a$ is the label on the lower right under-arc. This label must be a so that it agrees with the lower right label of the right-hand side of the illustration. In the type II move, the bottom labels on the under-arcs must agree with the labels at the top of the diagram. In the type III move, the labels on the right-hand lower under arc on either side of the move should agree. This gives self-distributivity.

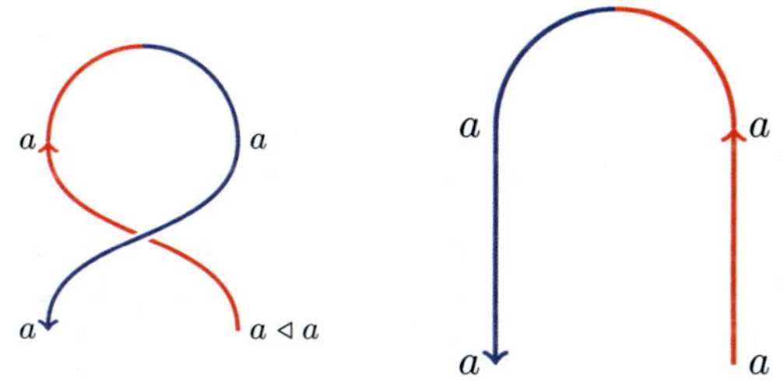

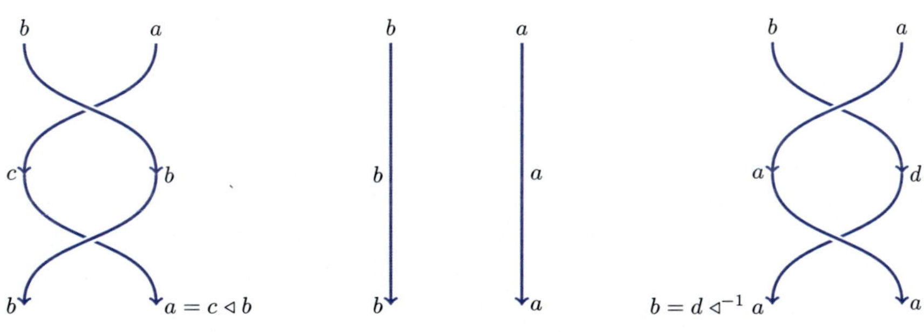

[4]Signs of crossings will be defined below in Section 4

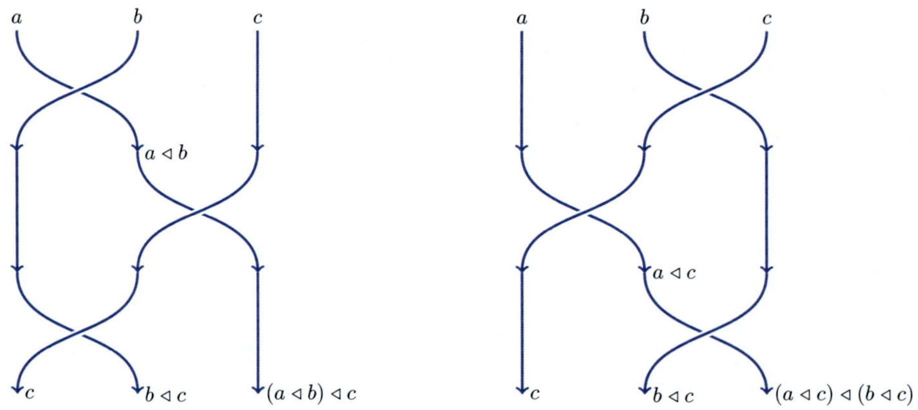

Conjugation. A typical first example of a quandle is a group for which conjugation in the group is the quandle operation. We explain in detail. A *group* is a set G together with an associative, unital, binary operation such that each element has an inverse. In short, a group is a category that has one element and for which each arrow is invertible. The operation $(\cdot)$ is written by in-fix notation so that

(1) for any $a, b \in G$, the element $a \cdot b = ab \in G$;
(2) for any $a, b, c \in G$, the associative law $(ab)c = a(bc)$ holds;
(3) there is an element $1 \in G$ such that $1 \cdot a = a = a \cdot 1$;
(4) for each $a \in G$, there is a unique element $a^{-1} \in G$ such that $aa^{-1} = 1 = a^{-1}a$.

Given a group G, for any $n \in \mathbb{Z} = \{0, \pm 1, \pm 2, \ldots\}$, the operation $a \triangleleft^n b = b^{-n}ab$ defines a quandle structure on G. When $n = 1$, we call this operation *conjugation.* If S is a subset of G that is closed under conjugation among its elements, then S is also a quandle.

A favorite example of such a closed subset is provided by considering the *dihedral group* that coincides with the symmetries of a regular polygon with n-vertices. The group D_{2n} is given by the presentation

$$D_{2n} = \langle s, r : s^n = r^2 = (sr)^2 = 1 \rangle.$$

Exercise. From this presentation, demonstrate the following claims. (1) The set of elements in D_{2n} can be represented as $\{1, s, s^2, \ldots, s^{n-1}, r, rs, rs^2, \ldots, rs^{n-1}\}$. The set $R_n = \{r, rs, rs^2, \ldots, rs^{n-1}\}$ is called the set of *reflections.* (2) Since conjugating a reflection by a reflection gives another reflection, then R_n is a quandle that is called the *dihedral quandle.* (3) Show that $rs^a \triangleleft rs^b = rs^{2b-a}$. Also see Chapter 12.

While we do not demonstrate this here, every quandle can be obtained as a variation upon the following construction. Let G denote a group with elements a, b, c etc. Let ϕ be an automorphism of G that acts upon the right. As a function, ϕ is injective and surjective, and it is a *group homomorphism* from G to itself. So $(ab)\phi = (a)\phi(b)\phi$. Define a quandle structure upon G, by $a \triangleleft_\phi b = (ab^{-1})\phi\ b$. It is a straight-forward traditional algebraic calculation to demonstrate that for any automorphism ϕ of the group G, the binary operation $\triangleleft_\phi$ is a quandle. Both of the papers by Joyce [**37**] and Matveev [**57**] demonstrate how to identify a given quandle with a family of group cosets whose quandle operations are obtained from this construction.

Specialize to the case in which G is an abelian group that is written in additive notation. So that $a + b = b + a$, and the automorphism ϕ is written as T. In this case, write $a \triangleleft_T b = (a - b)T + b = aT + b(1 - T)$ where 1 indicates the identity automorphism. Quandles of this form are called *(generalized) Alexander quandles.* Quotients of the polynomial ring $\mathbb{Z}[T, T^{-1}]$ are often considered among Alexander quandles. The quotient polynomial ring $(\mathbb{Z}/n)[T, T^{-1}]/(T + 1)$ has the property that $a \triangleleft b = 2b - a$, and so it agrees with the dihedral quandle above.

A *quandle coloring* of an oriented knot or link diagram by a quandle Q is an assignment of a quandle element in Q to each over-arc in a diagram such that the quandle rule holds at each crossing. Thus when standing upon an over-arc facing in the direction of the orientation, the under-arc on your left is colored $a \triangleleft b$ where a is the color on the right-hand under-arc and b is the color on the over-arc. This condition holds at all the crossings.

Colorings by a conjugation quandle: $a \triangleleft b = b^{-1}ab$ are among the most studied. In this case, the quandle coloring gives rise to a homomorphism from the fundamental group into the underlying group G. For our current purposes, the fundamental group is the group that is generated by the arcs in the diagram subject to Wirtinger-like relations at the crossings. At a positive crossing, the top ends of the arcs are labeled a and b from left to right, and the lower arcs are labeled left to right b and c where $c = b^{-1}ab$. In a moment, a more topological definition will be presented.

When a knot or link diagram is colored by the Alexander quandle $\mathbb{Z}[T, T^{-1}]$, relations induced from the crossings gives a presentation for the *Alexander module* which is the homology of the infinite cyclic covering space of the knot complement. This is discussed below, but only from the point of view of giving a presentation of this module. A discussion of the infinite cyclic cover is beyond the scope of the text.

There are topological definitions for the fundamental group and the fundamental quandle. We start with the latter. A knot in space is an embedding of a circle into 3-dimensional space. Customarily the embedding is required to be smooth. In particular, the knot has a tubular neighborhood N, so that the embedding extends to an embedding of a solid torus $N = S^1 \times D^2$ into three dimensional space $\mathbb{R}^3$. The knot exterior E is the 3-dimensional manifold that has a torus as its boundary and that is obtained by removing an open tubular neighborhood of the knot. Choose a point p in the exterior E. Often p is selected to be on the boundary of the tubular neighborhood, but for intuition, it is more convenient to assume it is a little bit further away from the knot. If $z \in S^1$ is a point on the circle S^1 that is knotted, then consider a pair (D_z, α) where D_z is a disk contained in the tubular neighborhood N that is transverse to the knot at the point z, and α is a path from a point on the boundary of D_z to the base point $p \in E$. The *fundamental quandle* $\pi_Q(K)$ of the knot is constructed as homotopy classes of such pairs.

A *homotopy* of such pairs consists of a collection $\{(D_t, \alpha_t) : t \in [0, 1]\}$ of such pairs that is parametrized by the unit interval. The arc α_t lies in the exterior, has its terminal point at the base point p for all time $t \in [0, 1]$, and has its initial point

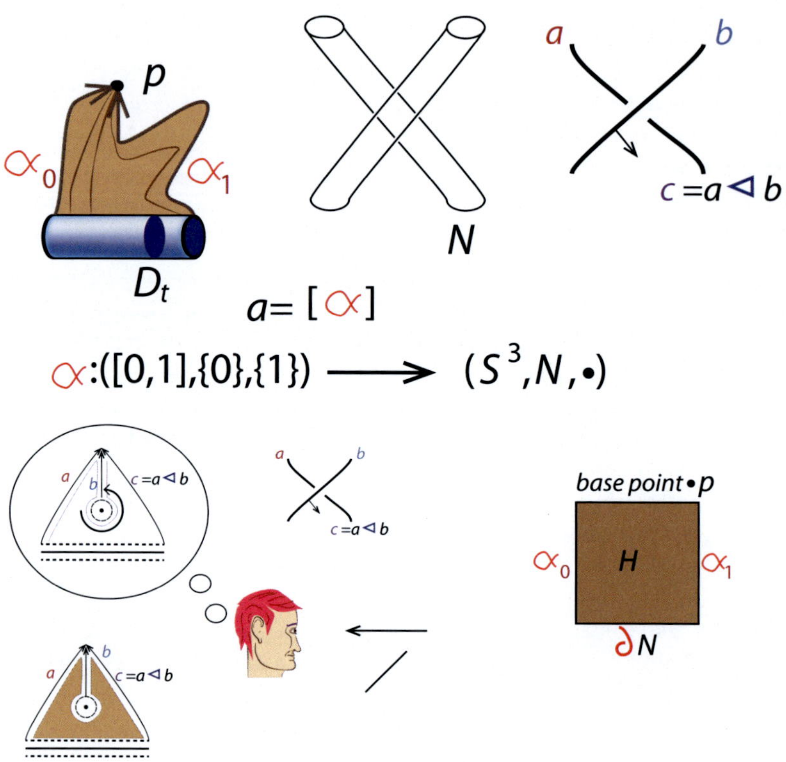

FIGURE 4. The homotopy relation in the fundamental quandle

at a transverse disk D_t that lies within the tubular neighborhood N. In this way, each path α_t starts somewhere along the tubular neighborhood, and always ends at p. Assume that the disks D in the pairs of elements are oriented so that their boundaries wind anti-clockwise with respect to the orientation tangents to the knot.

The quandle operation between pairs (D_z, α) and (D_w, β) when z and w are points along the knot is the pair $(D_z, \alpha\cdot\beta^{-1}\cdot\partial D_w\cdot\beta)$ where the composition in the paths is to follow the path α up to p, move backwards along β, run anti-clockwise along the boundary of D_w and return to p along β. Such a path is well-defined up to the homotopy relation, and the collection of homotopy classes satisfy the quandle axioms. This path is written as the path composition $\alpha \cdot \beta^{-1} \cdot \partial D_w \cdot \beta$. But notice that the path $(\partial D_z)^{-1} \cdot \alpha \cdot \beta^{-1} \cdot \partial D_w \cdot \beta$ is homotopic since homotopies are only required to remain along the boundary of a tubular neighborhood. Thus the meridional elements act via conjugation. Let ϕ denote the automorphism that acts upon the right $(x)\phi = \mu^{-1}x\mu$ where μ is a meridian. Then the path quandle agrees with $a \triangleleft_\phi b = (ab^{-1})\phi b$ that was defined above.

Along any arc in a knot diagram, one can imagine that the *meridional disk* D_z and its tail α can slide freely, but as the arc approaches an over-crossing of a transverse arc in the diagram, the tail will be come increasingly entangled. In this way, each arc in a diagram corresponds to a unique quandle element. The relations in the fundamental quandle are precisely the relations indicated above: the under-arc towards which the left hand normal of an over-arc points receives a quandle product. These matters are illustrated in Fig. 4.

Keeping the base point p as is, we can also consider loops of the form $(\alpha)^{-1} \cdot (\partial D_z) \cdot \alpha$. The loop ∂D_z that encircles the knot once positively is called a *meridian* or a *meridional loop.* Homotopy classes of paths that start and end at the base point p are said to form the elements of the *fundamental group,* $\pi_1(E)$ *of the knot exterior* E. The idea of homotopy here is that it is sort of floppy. A loop is specified by an explicitly parametrized curve, but as long as the loop doesn't break or pass back through the tubular neighborhood N, it is in the same homotopy class. That $\pi_1(E)$ is a group follows because path multiplication forms the group operation when paths are considered up to homotopy. Given loops $(\alpha)^{-1} \cdot (\partial D_z) \cdot \alpha$ and $(\beta)^{-1} \cdot (\partial D_w) \cdot \beta$, multiply them as paths: $(\alpha)^{-1} \cdot (\partial D_z) \cdot \alpha \cdot (\beta)^{-1} \cdot (\partial D_w) \cdot \beta$. Then use the homotopy relations to establish the associative law and inverses. The constant path at p serves as the identity.

Given a crossing in a knot diagram, place a square handkerchief below the crossing so that the tangents to the arcs of the knot are perpendicular to the edges of the handkerchiefs. Look down at the crossing, and imagine your nose to be the base point p. Hold four threads between your teeth, and loop each along one of the four edges of the handkerchief. Then return the other end of each thread and bite down on the other end. Now gather the threads and tape them to the end of your nose. The handkerchief provides the homotopy from the edge labeled $b^{-1}ab$ to the path product of the other three threads. And you look pretty silly, so un-tape the threads from your nose. The construction is depicted for a positive crossing in Fig. 5.

Colorings. Both the fundamental group and the fundamental quandle are obtained by thinking of the arcs in an oriented knot diagram as generators, and establishing relations at the crossings. In the quandle case, the under-arc that is the target of a left pointing arm along an over-arc receives a product $a \triangleleft b$. In the group case, it receives a product $b^{-1}ab$. In both the quandle and group cases, quantities are found that are more easily computable by choosing a test quandle or group, and labeling the arcs with elements from the test quandle. The labels satisfy the quandle conditions at the crossing, and the result is the homomorphic image of the fundamental quandle or group into the test quandle or group.

As a necessary aside, the notion of a *homomorphism* is a function that preserves the type of structure (binary operation subject to some axioms) which two algebraic objects may share. A *group homomorphism* is a function $f : G \to H$ between groups G and H such that $f(xy) = f(x)f(y)$ where $x, y \in G$. A *quandle homomorphism* is a function g between quandles Q and R such that $g(a \triangleleft b) = g(a) \triangleleft g(b)$.

It is a remarkable fact that among the first 1.7 million prime knots of sixteen crossings or fewer [**30**], can be distinguished by means of a relatively small number of groups, quandles, or quandle cocycles. The number of groups that are needed are fewer than 200.

These ideas will be exemplified by making a few computations on the knot 6_1 that is known as the stevedore's knot. It is illustrated below as an oriented knot

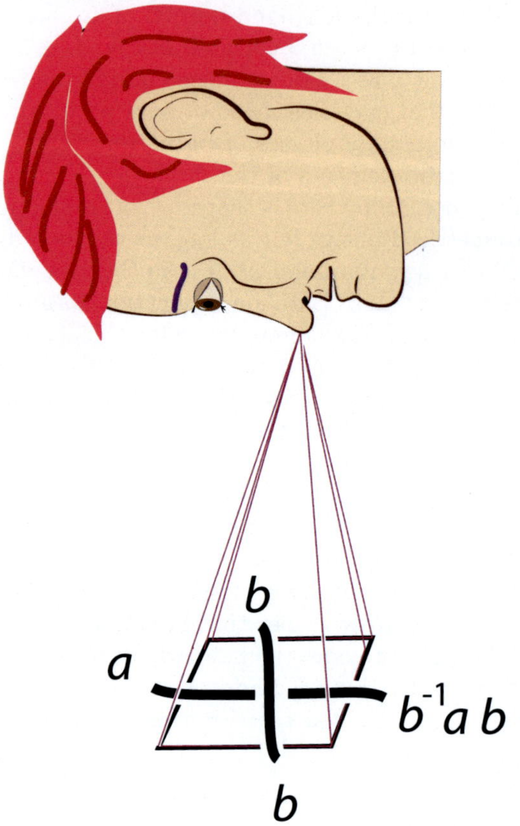

FIGURE 5. The Wirtinger relation at a positive crossing

with labels upon each of its over-arcs. Presentations for the fundamental group and the fundamental quandle are obtained as indicated.

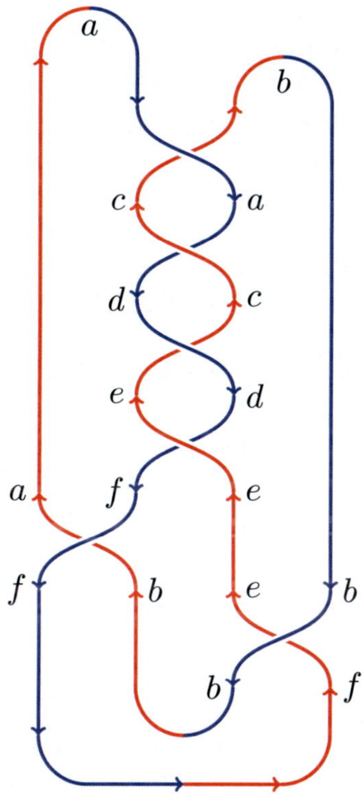

$$b = c \triangleleft a$$

$$d = a \triangleleft c$$

$$c = e \triangleleft d$$

$$f = d \triangleleft e$$

$$b = a \triangleleft f$$

$$f = e \triangleleft b$$

$$\pi_Q(6_1) = \left\langle a, b, c, d, e, f : \begin{array}{lll} b = c \triangleleft a, & d = a \triangleleft c, & c = e \triangleleft d, \\ f = d \triangleleft e, & b = a \triangleleft f, & f = e \triangleleft b \end{array} \right\rangle.$$

$$\pi_1(E(6_1)) = \left\langle a, b, c, d, e, f : \begin{array}{lll} b = a^{-1}ca, & d = c^{-1}ac, & c = d^{-1}ed, \\ f = e^{-1}de, & b = f^{-1}af, & f = b^{-1}eb \end{array} \right\rangle.$$

As mentioned earlier, labelings by Alexander quandles are used to obtain a presentation for the first homology of the infinite cyclic cover. This is a module over $\mathbb{Z}[T, T^{-1}]$ where T corresponds to a deck transformation of the cover. As such it is given as a ring quotient $\mathbb{Z}[T, T^{-1}]/(\Delta(T))$. The polynomial $\Delta(T)$ is defined up to multiples of units of $\mathbb{Z}[T, T^{-1}]$, i.e., $\pm T^{\pm 1}$. Colorings give the Alexander polynomial. First, the quandle operation is $x \triangleleft y = Tx + (1-T)y$. Then for each relation, for example $b = c \triangleleft a = Tc + (1-T)a$, write the rows in the matrix below to indicate the coefficient of the variables a through f in each relation:

$$\begin{bmatrix} 1-T & -1 & T & 0 & 0 & 0 \\ T & 0 & 1-T & -1 & 0 & 0 \\ 0 & 0 & -1 & 1-T & T & 0 \\ 0 & 0 & 0 & T & 1-T & -1 \\ T & -1 & 0 & 0 & 0 & 1-T \\ 0 & 1-T & 0 & 0 & T & -1 \end{bmatrix}.$$

The standard procedure for obtaining the Alexander polynomial is to compute any (5×5) minor to obtain the common factor $-2T^2 + 5T - 2$.

Alternatively, solve the various relations in terms of the variables b and f. So since $f = Te + (1-T)b$, solve for $e = T^{-1}[f + (T-1)b]$. Similarly, the relation

$b = Ta + (1-T)f$ gives $a = T^{-1}[b + (T-1)f]$. Substitute the value of e into $f = Td + (1-T)e$ and solve for $d = T^{-2}[(1-2T+T^2)b + (2T-1)f]$. Then since $c = Te + (1-T)d$, we obtain $c = (-f(1-3T+T^2) + b(1-3T+2T^2))/T^2$. Finally, substitute expressions into the equation $d = Ta + (1-T)c$ to obtain

$$T^{-2}[(1-2T+T^2)b + (2T-1)f]$$

$$= [b + (T-1)f] + (1-T)(-f(1-3T+T^2) + b(1-3T+2T^2))/T^2.$$

This expression is trivial when $2-5T+2T^2$ is trivial. The relation $b = Tc+(1-T)a$ follows similarly.

The method of back substitution works in this case because the knot is a 2-bridge knot. So all the relations reduce to a relationship between b and f. In general, the technique of giving a presentation matrix and exhibiting the minors will work no matter how many generators and relations in the presentation of the knot group.

Consider the permutations of four elements $\{1,2,3,4\}$, and write these in cyclic notation. So $a = (1234)$ represents the permutation $1 \mapsto 2 \mapsto 3 \mapsto 4 \mapsto 1$. The collection of 4-cycles $a = (1234)$, $A = (1432)$, $b = (1243)$, $B = (1342)$, $c = (1423)$, and $C = (1324)$ is a subset of the permutation group. This set of cycles is closed under conjugation, and it can be used to color the knot 6_1 as depicted. Parentheses and commas are omitted from the expressions of the permutations to keep font sizes larger.

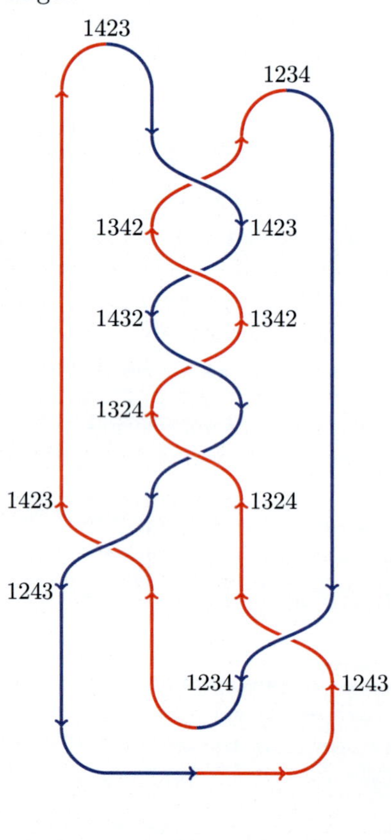

$1234 = 1342 \triangleleft 1423$

$1432 = 1423 \triangleleft 1342$

$1342 = 1324 \triangleleft 1432$

$1243 = 1432 \triangleleft 1324$

$1234 = 1423 \triangleleft 1243$

$1243 = 1324 \triangleleft 1234$

4. The Jones polynomial

The theory of knots underwent a historic change upon V. F. R. Jones's discovery of a new polynomial invariant for knots and links [**35**]. In this section, the definition is outlined and expanded upon with the purpose of further exemplifying the notion of diagrammatic algebra.

To state the definition, start with an unoriented knot diagram, and orient it arbitrarily. At each crossing a sign is defined via a right-hand rule: point the index finger of your right hand along the over-arc; point your thumb in the same direction as the under-arc; if your middle finger points downward, the crossing is positive. If not, the crossing is negative. For convenience the eight possibilities are listed below. Observe that if the orientation of knot is reversed, then both directions for the under-arcs and over-arc reverse. The sign, then, remains the same. The *writhe of a knot diagram, D* is the difference

$$w(D) = \#(\text{positive crossings}) - \#(\text{negative crossings}).$$

The writhe depends the diagram, but it remains unchanged under type II, type III, and ψ-moves. A type I Reidemeister move changes the writhe of a diagram by ± 1.

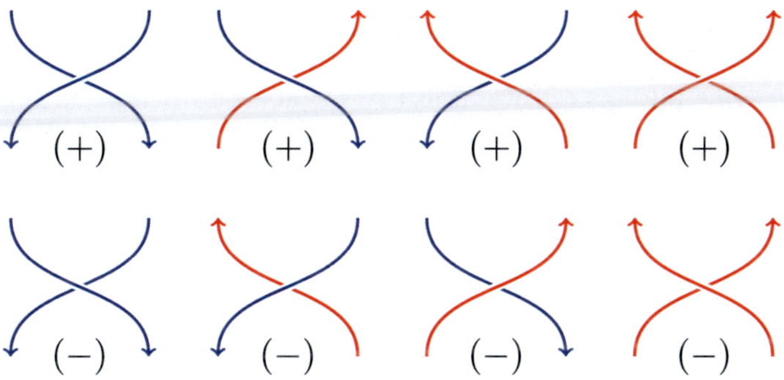

The *Kauffman bracket* model will be used to define the Jones polynomial. It remains unchanged when a type II, type III, or ψ-move is applied. And the type I move has a predictable effect. It is defined by the three relations that are given below. The interpretation is that a crossing in a diagram is chosen, and the crossing is replaced by a linear combination of the two smoothings on the right-hand side. Once a diagram with no crossings is obtained, then each disjoint closed loop is replaced by the value $\delta = -A^2 - A^{-2}$. Thus the third relation indicates that a disjointly embedded loop is replaced by δ. Many authors also find it convenient to divide the result of the "skein calculation" by δ. When each crossing is replaced, two daughter curves appear. Because of this, the bottom level of the calculation is called a "skein tree." In other approaches (specifically Khovanov homology [**8, 9, 48**]; see Section 6.2 of this chapter), the states of the calculation at which all crossings have been resolved, are arranged in an n-dimensional cube, where n is the crossing number of the diagram.

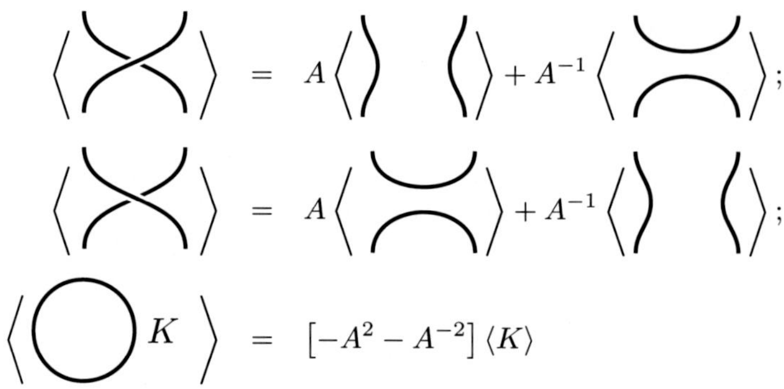

The origins of skein theory appeared in Conway's seminal paper [**19**] in which he found a method of computing the Alexander polynomial via a recursive relation. Jones's original approach included a skein relation that involved the same terms that Conway's did, but different coefficients. Conway coined the term "skein relation" to indicate such a recursively computable function in which properties of the original knot diagram were derived by computing daughters, granddaughters, and so forth. He has described these as a form of diagrammatic algebra in the sense that one does computations in a geometric context for as long as possible and then examines the results through an algebraic lens. The authors here are recalling, perhaps incompletely, some phrases of several lectures that Conway gave in the late 1980s.

Next the Kauffman bracket is demonstrated to be invariant under type II and type III moves. Kauffman's observation is that invariance under type II gives invariance under type III.

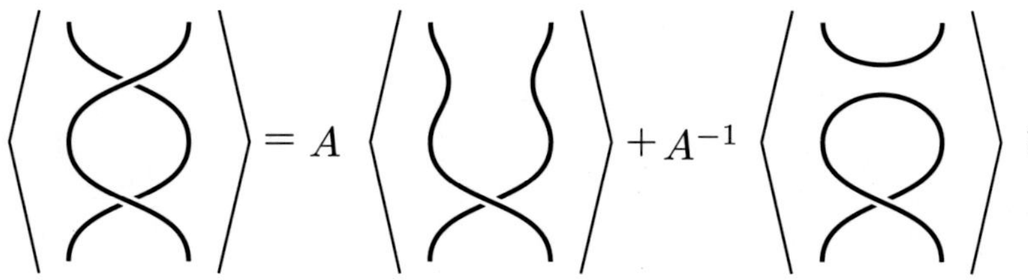

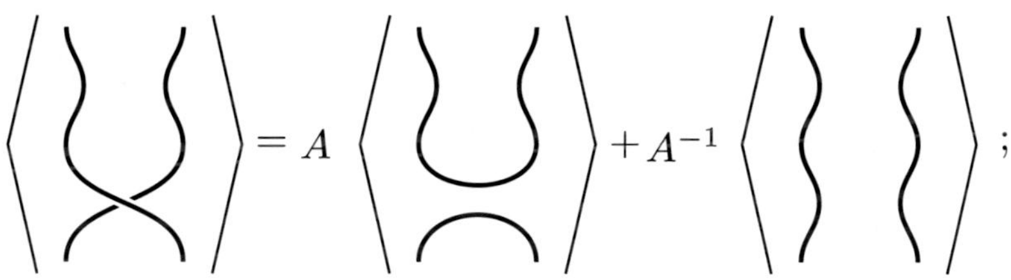

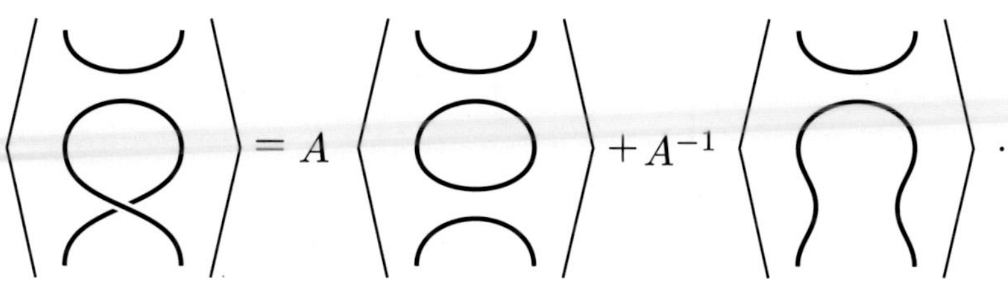

Therefore,

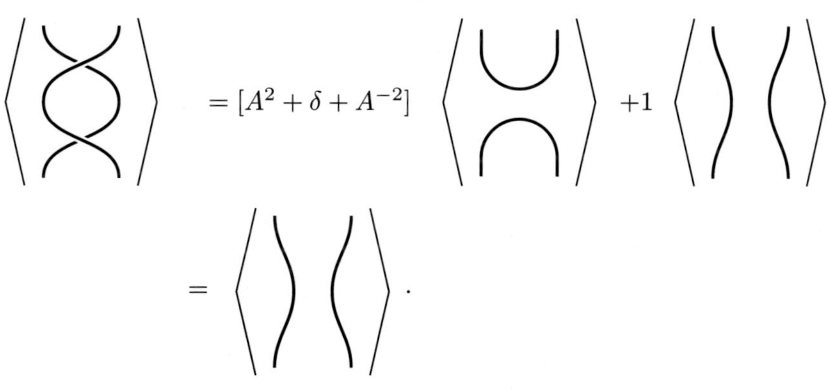

Invariance under the type III move follow from the calculation:

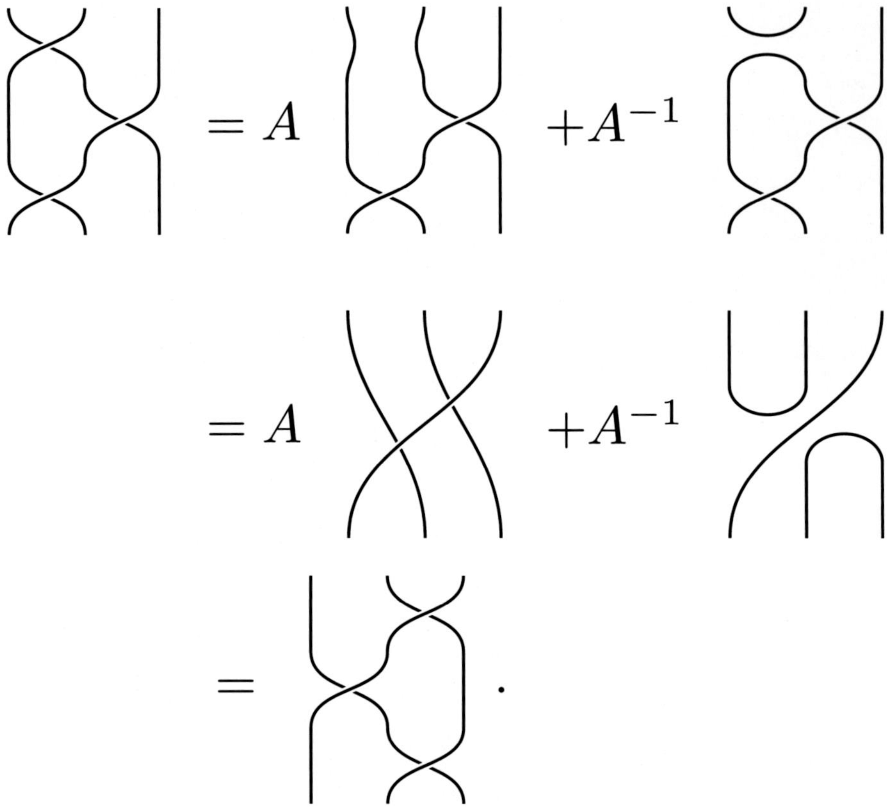

Therefore, the Kauffman bracket is an invariant of framed tangles. Its behavior under the type I move, and consequently changes of framing is computed as follows.

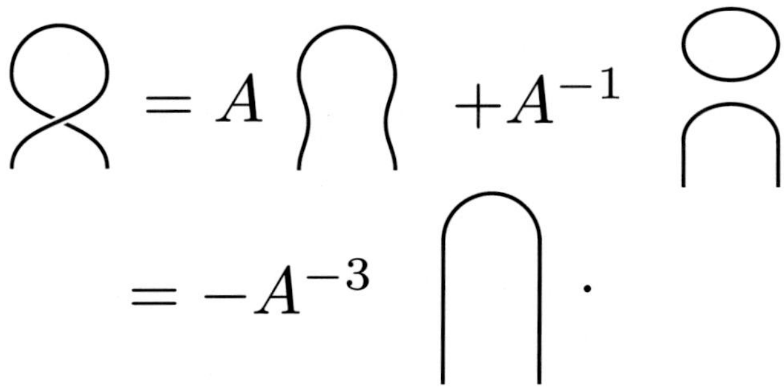

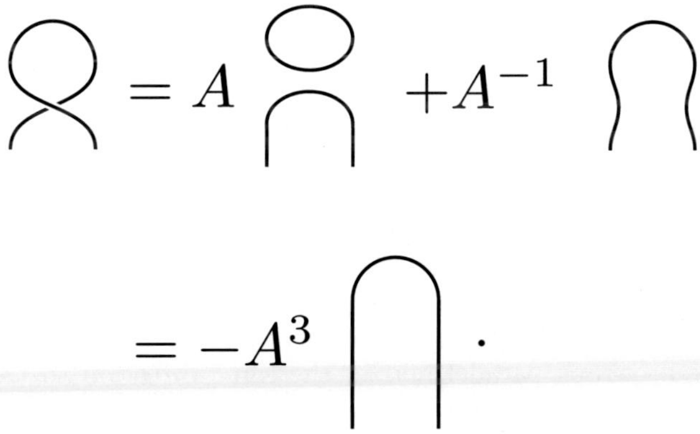

For a diagram $D(K)$ of a knot K, define $J(K)[A] = (-A)^{-3w(D)}\langle K\rangle$ where $w(D)$ indicates the writhe of the diagram $D(K)$ and $\langle K\rangle$ indicates the result of the skein computation that arrises by smoothing each crossing in an A or A^{-1} manner. Then $J(K)[A]$ is invariant under all of the generalized Reidemeister moves. So it depends only on the knot K and not the diagram $D(K)$ that represents K. Under the substitution $A = t^{1/4}$, Jones's original polynomial is obtained.

Of course, one should show that $J(K)$ is not the same for all knots. Interestingly, it can be used to distinguish the left- and right-handed trefoil.

4.1. Abstract tensors revisited. The matrix representations of $\cap$ and $\cup$ that were presented in Chapter 8, Section 1.2 are revisited. Each is expressed in terms of $\left[0, iA, -iA^{-1}, 0\right]$ with $\mathbb{C} \xleftarrow{\cap} V\otimes V$ and $V\otimes V \xleftarrow{\cup} \mathbb{C}$. where V is a 2-dimensional vector space over $\mathbb{C}$. Thus $\cap$ is a row vector and $\cup$ is a column vector. Their outer product is

$$\begin{matrix}\cup\\ \cap\end{matrix} = \begin{bmatrix} 0 & 0 & 0 & 0 \\ 0 & -A^2 & 1 & 0 \\ 0 & 1 & -A^{-2} & 0 \\ 0 & 0 & 0 & 0 \end{bmatrix}.$$

The bracket relation in this matrix representation is written as

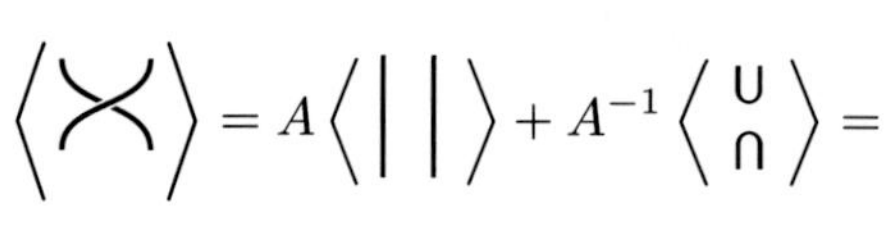

$$= \begin{bmatrix} A & 0 & 0 & 0 \\ 0 & 0 & A^{-1} & 0 \\ 0 & A^{-1} & A - A^{-3} & 0 \\ 0 & 0 & 0 & A \end{bmatrix}.$$

Let V denote the 2-dimensional vector space $\{\alpha x + \beta y : \alpha, \beta \in \mathbb{C}\}$, and consider $V \otimes V \xleftarrow{X} V \otimes V$ — the linear transformation that is given via the matrix above with respect to the ordered basis $\{x \otimes x, x \otimes y, y \otimes x, y \otimes y\}$ of $V \otimes V$. Then X satisfies the Yang-Baxter equation $(X \otimes I)(I \otimes X)(X \otimes I) = (I \otimes X)(X \otimes I)(X \otimes I)$ as an operator from $V \otimes V \otimes V$ back to itself. The Yang-Baxter relation is diagrammatically represented by the Reidemeister type III move. Moreover, X is invertible with X^{-1} represented using the bracket expansion of .

5. The braid group

In this section, the bracket polynomial is used to define a representation of the braid group into the Temperley-Lieb algebra. The Alexander quandle is used to define a representation of the braid group into a matrix group. This representation is called the Burau representation. Braid theory is important to knot theory because of two major results. Alexander's theorem states that any knot or link is isotopic to one that is a closure of a braid. That is any knot or link is isotopic in 3-dimensional space to one that is embedded into a solid torus $S^1 \times D^2$ in such a way that every strand of the knot wraps parallel and in the same direction to the circle factor S^1. Markov's theorem indicates when two braids represent the same knot. The corresponding words in the braid group can represent the same braid element, or a strand is added with an extra braid generator in the subsequent braid group attached.

The *braid group*, B_n, of degree n is the group of (the equivalence classes) of braids of degree n. This group is generated by

$$\sigma_1, \ldots, \sigma_{n-1},$$

which are called the *standard generators* or the *Artin's braid generators*. The braid group has a group presentation below.

$$\left\langle \sigma_1, \ldots, \sigma_{n-1} : \begin{array}{ll} \sigma_i \sigma_{i+1} \sigma_i = \sigma_{i+1} \sigma_i \sigma_{i+1}, & \text{for } i = 1, \ldots, n-2, \\ \sigma_i \sigma_j = \sigma_j \sigma_i, & \text{for } |i-j| > 1 \end{array} \right\rangle.$$

The relations above are called *braid relations*. Chapter 11 discusses the braid group in greater detail. Here a geometric representation of the generator σ_i is illustrated.

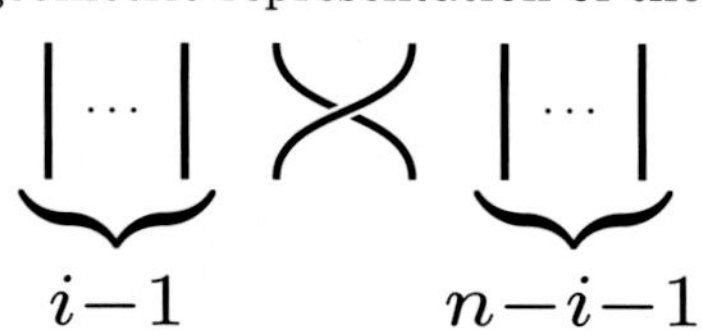

Recall that a group is a category with one object in which each arrow is invertible. In the case of the braid group B_n, the object is the natural number n. The arrows are generated by the exchangers, $\mathsf{X}(i-1, n-i-1)$ with inverses $\overline{\mathsf{X}}(i-1, n-i-1)$

which are depicted using their bookmarked notation. So $\sigma_i = \mathsf{X}(i-1, n-i-1)$. The exchangers, that above were thought of as triple arrows, have been decategorified. In this way, the exchangers generate the set of arrows, and the naturality of the exchanger (which was previously thought of as a quadruple arrow) and the quadruple arrow exchangers are now considered to be the double arrows that are the relations in the group.

From Chapter 8, Section 1, the bracket relation

gives a representation of the braid group into the Temperley-Lieb algebra. One of victories of Jones's work is the existence of this representation.

Furthermore, consider $V = \{\alpha x + \beta y : \alpha, \beta \in \mathbb{C}\}$ — the 2-dimensional vector space of linear expressions in unknowns x and y. Then the matrix representation of Section 4.1 of this chapter gives an explicit representation of B_n into the group of linear transformations upon $V^{\otimes n}$.

The Burau representation is described via the illustration:

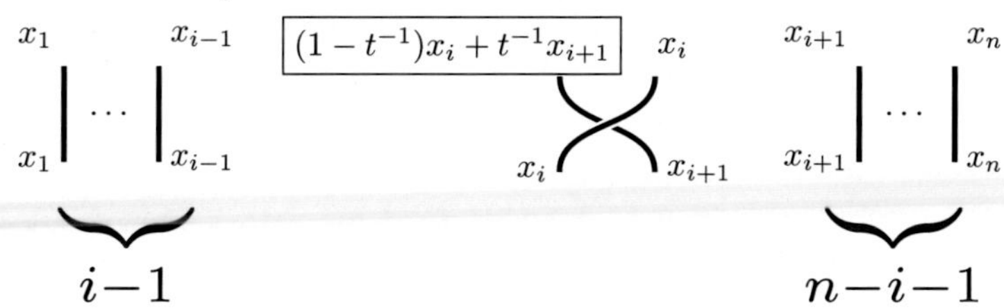

So as a matrix

$$\sigma_i \mapsto \left[\begin{array}{c|cc|c} I_{i-1} & & 0 & 0 \\ \hline 0 & 1-t^{-1} & t^{-1} & 0 \\ & 1 & 0 & \\ \hline 0 & & 0 & I_{n-i-1} \end{array}\right].$$

The version of the Burau representation given here is the inverse of the traditional one because we are using the quandle representation, inputing a column vector on the right, and reading from the bottom to the top of a positive crossing.

The positive trefoil is the braid closure of σ_1^3. If variables a and b are input at the bottom of the diagram, then after computing the Burau representation of σ_1^3 at the top of the diagram the output is $[(a(-1+t-t^2+t^3)+b(1-t+t^2)/t^3, (a(1-t+t^2)+(t-1)b)/t^2]$. That the Alexander polynomial is (up to factors of $\pm t^{\pm n}$) is t^2-t+1 follows since these top colors should equal the bottom colors (a,b) upon braid closure.

6. More algebraic structures

In this section, three more diagrammatically defined algebraic systems will be presented. Firstly, projections from $V^{\otimes n}$ to a subspace will be defined that are important in generalizing the Jones polynomial to invariants of 3-dimensional manifolds. Secondly, Frobenius algebras will be re-examined from the point of view of topological quantum field theories and briefly in relation to Khovanov homology. Thirdly, the axioms of a Hopf algebra are presented in a diagrammatic form. Invariants of 3-manifolds can also be obtained from Hopf algebras [**28, 45, 51**].

The discussions introduced in this section are far from complete. They are just a sampling of how diagrammatic techniques grew out of the bracket model for the Jones polynomial. Furthermore, skein modules are neglected even though they are used to form the foundational AJ conjecture and the volume conjecture that relate diagrammatically defined invariants to invariants that arise from hyperbolic geometry.

6.1. The Jones-Wentzl idempotents. To describe the Jones-Wentzl projectors, we take a step back and consider the group of permutations on n letters. This *symmetric group* Σ_n is given by the presentation:

$$\left\langle t_1,\dots,t_{n-1} : \begin{array}{ll} t_it_{i+1}t_i = t_{i+1}t_it_{i+1}, & \text{for } i=1,\dots,n-2, \\ t_it_j = t_jt_i, & \text{for } |i-j|>1 \\ t_i^2 = 1 & \text{for } i=1,\dots,n-2. \end{array} \right\rangle .$$

There are $n! = n(n-1)!$ permutations in Σ_n (where $0! = 1$). In cyclic notation the generator t_i corresponds to the permutation $t_i \leftrightarrow (i, i+1)$ that transposes the ith and $(i+1)$st elements and leaves others fixed.

If the variable $A = 1$ in the Kauffman bracket model, the values associated to positive and negative crossings are indistinguishable. In this case, we obtain a representation of the permutation group. A generator t maps to $V \otimes V \xleftarrow{X} V \otimes V$ where $X(x \otimes y) = (y \otimes x)$ and $X(y \otimes x) = (x \otimes y)$. Let a permutation $s \in \Sigma_n$ be given, and consider $[s]$ to be the linear map $V^{\otimes n} \xleftarrow{[s]} V^{\otimes n}$ that is defined via the bracket at $A = 1$, e.g.

$$\begin{bmatrix} 1 & 0 & 0 & 0 \\ 0 & 0 & 1 & 0 \\ 0 & 1 & 0 & 0 \\ 0 & 0 & 0 & 1 \end{bmatrix}$$

corresponds to X. In this way the braid generator and its inverse $\sigma_i^{\pm 1}$ are both projected to the transposition t_i.

The vector space $V^{\otimes n}$ has as its basis the 2^n vectors that correspond to monomials in non-commuting variables x and y. That is, for each binary sequence in $\{0,1\}^n$, write a sequence of tensor products of x and y where, say, x appears as the ith tensorand if and only if $f(i) = 0$ for the function $\{0,1\} \xleftarrow{f} \{1,2,\dots,n\}$. There is a subspace that corresponds to the set of symmetric polynomials, It is defined to be the image of the projection $P = (\sum_{s\in\Sigma_n}[s])/n!$. The subspace of symmetric polynomials forms an irreducible representation for the lie algebra $\mathfrak{sl}_2$.

The Jones-Wentzl projectors are known to be the quantum analogues of the projections P. They are defined in a few steps. First, given a permutation s, write it in terms of the standard generators $t_1,\dots,t_{n-1}$ of the permutation group Σ_n. Second, lift each generator t_i to the positive braid generator σ_i. Third, evaluate the resulting braid element via the bracket relation. Let $\langle \hat{s} \rangle$ denote the result of this lifting and evaluation. Finally, the factor $n!$ is replaced by with a so-called quantum integer $[n]! = [n][n-1]! = \prod_{k=1}^{n-1}\left[\sum_{j=0}^{k} q^j\right]$ where $q = -1/A^2$. Then define

$$ _n = \frac{1}{[n]!}\sum_{s\in\Sigma_n}\langle\hat{s}\rangle. $$

One can also show that  $_n$ is a projection onto the subspace of symmetry polynomials in $V^{\otimes n}$. It turns out that these correspond to the irreducible representations of the quantum analogue of the lie algebra $\mathfrak{sl}_2$.

6.2. 2-D TQFT. A *topological quantum field theory* (TQFT) is a functor between a category of cobordisms and a category of vector spaces, or more generally a category of modules over a fixed ring. Other features in the cobordism category will be reflected in the category of vector spaces. For 2-dimensional TQFTs (2-D TQFTs), vector spaces are assigned to circles, and linear maps are assigned to cobordisms. We consider the oriented case. Surface cobordisms can be generated by a few fundamental pieces. There are cylinders that serve as identities on circles, trinians (or pants) that connect two circles to one, and disks that bound the circles. These are given representationally, via an associated composition of glyphs, and then distilled to their essence as the arrows in a Frobenius algebra context.

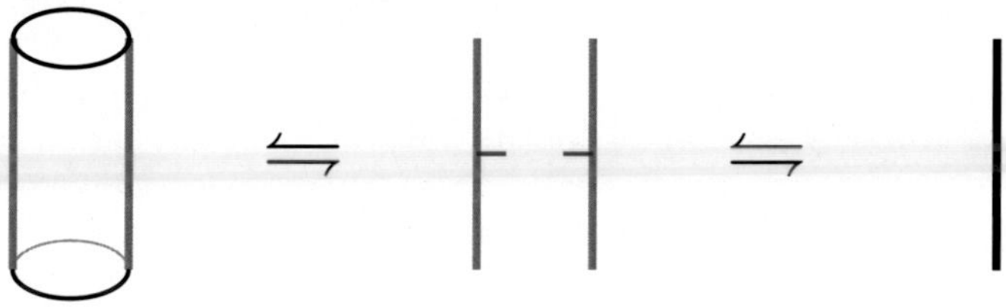

The birth and death of circles are represented as

,

and

.

Importantly, the cobordisms that connect one circle to two and vice versa are among the generating morphisms.

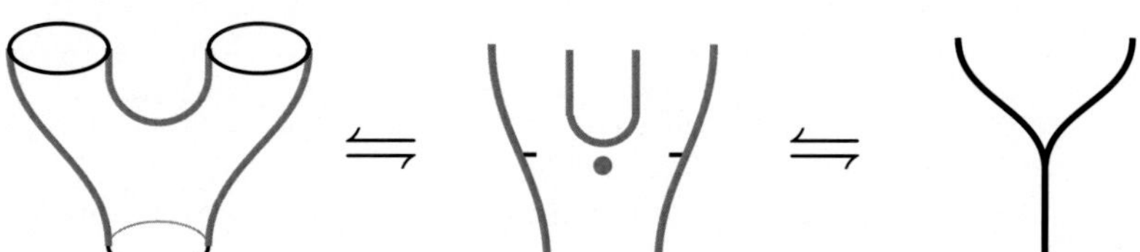

and

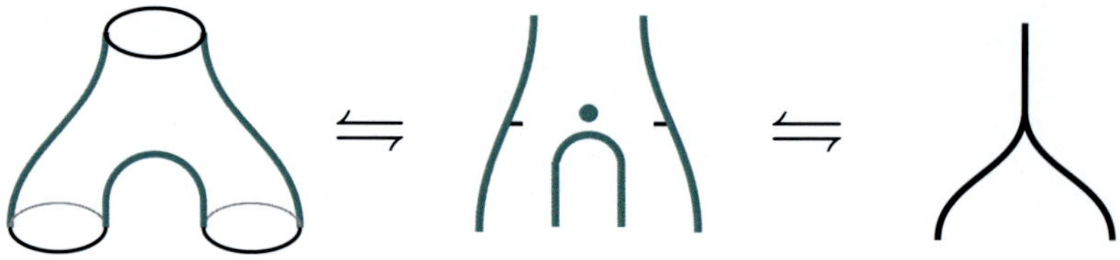

Now translate the Frobenius algebra axioms given in Ch. 2, Fig. 1, that there is an associative unital multiplication and a non-degenerate associative pairing, back to the glyphographic context. The pairing is given as the composition

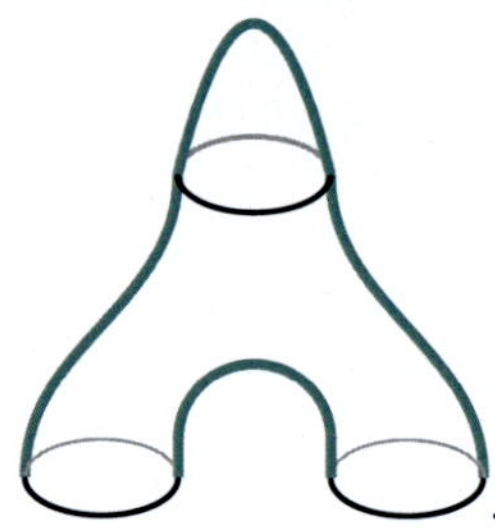

The geometric manifestation of the Frobenius algebra axioms correspond to altering the critical decomposition of the surface. Specifically, associativity corresponds to exchanging the heights of critical points. The unit and co-unit axioms correspond to critical cancelations birth/fork and death/saddle. Writing co-multiplication in terms of multiplication and co-units in terms of units, the co-associativity, and the non-degeneracy of the co-pairing all translate directly to the context of cobordisms between circles.

Therefore in a 2D-TQFT, a finite dimensional vector space (or module) V is assigned to each circle, and tensor powers are assigned to the disjoint union of circles. A multiplication or co-multiplication is assigned to a trinion (pair of pants). These operations are associative or co-associative and they have units and co-units that are assigned to disks. The paring is associative; the co-pairing is co-associative. So the image of any functor from this cobordism category to the category of vector spaces must be a Frobenius algebra.

Often, the image of this functor need not be a vector space, but only should be a module over a ring in which the module has a multiplication, unit, and pairing. Khovanov [**48**] uses $R[X]/(X^2)$ to describe the Frobenius algebra that gives the categorification of the Jones polynomial. Multiplication is ordinary polynomial multiplication with the ordinary unit, and the co-multiplication is

$$\mathsf{Y}(1) = 1 \otimes X + X \otimes 1 + cX \otimes X, \quad \mathsf{Y}(X) = X \otimes X,$$

with co-unit

$$\top(1) = -c, \quad \top(X) = 1,$$

where $c \in R$ is some element.

Exercise. Verify that $R[X]/(X^2)$ is co-associative. Verify that $\top \circ \curlywedge$ defines a non-degenerate pairing. Define the co-pairing.

On first pass, it may be advantageous to let $c = 0$. For this value of c, a homology theory still is defined.

Consider a knot diagram. For example, in the figure below the trefoil knot is written in closed braid form on the lower right of the illustration. There are two smoothings in the bracket relation associated to each crossing. If the coefficient in the bracket is A, that smoothing is called an A smoothing, and otherwise it is a B smoothing. A *state* of the knot diagram consists of smoothing all the crossings. The eight states that are associated are arranged in a cubical array.

There are a number of book keeping matters towards which attention is drawn.

The crossings in this diagram are ordered from top to bottom. The cube is leveled by the number of B states. So the bottom $0B$ level corresponds to the all A state AAA. It is depicted at the bottom back of the cube at coordinate $(0,0,0)$. The $1B$ level contains the states that correspond to (BAA, ABA, AAB). These are respectively placed at the coordinates $(1,0,0)$ (forward), $(0,1,0)$ (right), and $(0,0,1)$ (up) loci of the smoothing cube. The $2B$ level is (BBA, BAB, ABB). Respectively these states are at $(1,1,0)$, $(1,0,1)$, and $(0,1,1)$. The $3B$ level is BBB located at $(1,1,1)$.

In full generality, crossings are ordered, and an element in $\{A,B\}^n$ is associated to each state. The correspondences $A \leftrightarrow 0$ and $B \leftrightarrow 1$ maps the states to the vertices of $[0,1]^n$. A directed edge in the cube moves from an A smoothing to a B smoothing. A change from A to B is indicated by $(\cdot)$ in the coordinate of the crossing for which the smoothing is changed.

A state at a vertex of the cube contains a number of circles. These too will be ordered. A 2D-TQFT associates a module to each circle, and a tensor product of modules to the disjoint union of circles. In the case at hand, each circle is assigned the module $R[X]/(X^2)$. A basis for this module is $\{1, X\}$. In the AAA state, the circle on the left corresponds to the left tensor factor; the circle on the right corresponds to the right tensor factor. In the $(2B)$ states, the circles are nested. The inner circles will be associated to the right tensor factor. In the BBB state, the outer-most circle corresponds to the left most factor, the top inner circle corresponds to the middle tensor factor, and the bottom inner circle corresponds to the right most tensor factor.

The correspondences that are given above will be used in order to compute differentials that are induced from the underlying 2D-TQFT. These differentials are built from the edges in the cube. Each arrow corresponds to a trinion (pair of pants) because an edge corresponds to either joining two circles or separating them. Thus the edges in the cube of smoothings should be labeled by a Y or a $\curlywedge$. The table below indicates the correspondences, and it augments the arrows $(\cdot)BB$, $B(\cdot)B$, and $BB(\cdot)$ with identities. Going from a $2B$ state to the $3B$ state splits one of the circles, and the ordering that was specified above indicates which circle is left alone. The third column of the table indicates an auxiliary sign that is attached to the arrows of the cube. The sign is determined by counting the number of As that precede the given change.

$(\cdot)AA$	⅄	$(+)$
$(\cdot)AB$	Y	$(+)$
$(\cdot)BA$	Y	$(+)$
$(\cdot)BB$	YI	$(+)$
$A(\cdot)A$	⅄	$(-)$
$A(\cdot)B$	Y	$(-)$
$AA(\cdot)$	⅄	$(+)$
$B(\cdot)A$	Y	$(+)$
$B(\cdot)B$	IY	$(+)$
$AB(\cdot)$	Y	$(-)$
$BA(\cdot)$	Y	$(-)$
$BB(\cdot)$	YI	$(+)$

The multiplication ⅄ and the co-multiplication Y correspond to critical points for trinians (pairs of pants). The choice of $(\pm)$-signs that are distributed among edges of the smoothing cube induce differentials[5], d, and the signs are chosen in a way that forces $d^2 = 0$.

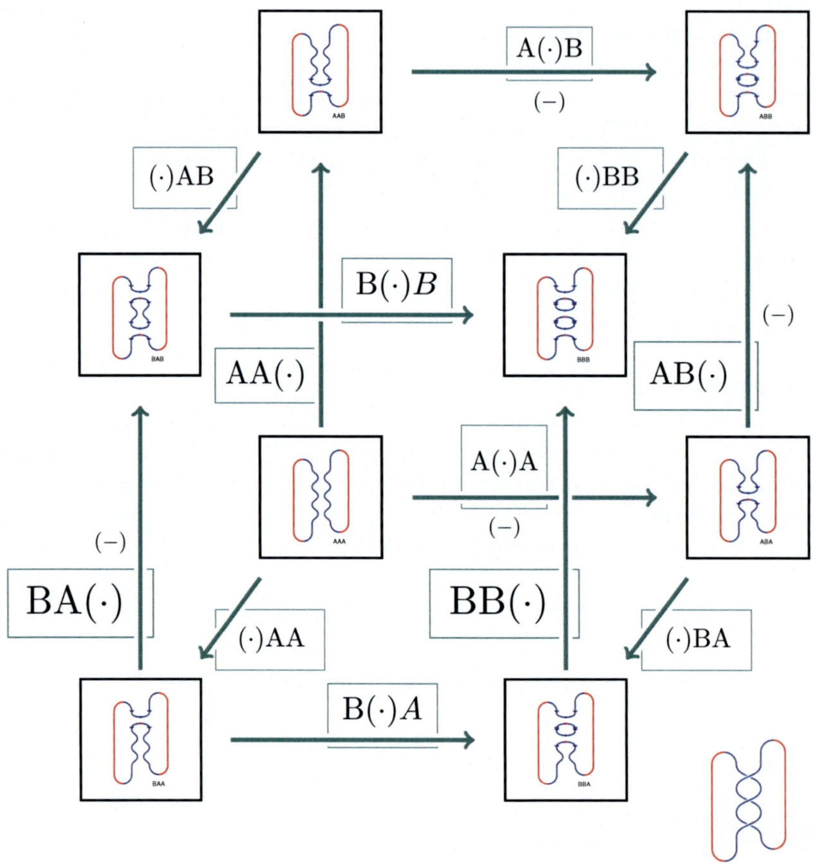

[5]The idea of homology and differentials will be briefly sketched in an side comment below.

Let us examine one of the squares of this cube. Stating from the AAA-state, compare the composition[6] $[A(\cdot)A]\circ[AB(\cdot)] = [-\curlywedge]\circ[-\mathsf{Y}]$ to $[AA(\cdot)]\circ[A(\cdot)B] = [\curlywedge]\circ[-\mathsf{Y}]$. They differ by a sign, so the back square of the cube anti-commutes. In each composition, two circles fuse and then one splits into two.

Next compare the compositions $[BA(\cdot)]\circ[B(\cdot)B] = [-\mathsf{Y}]\circ[\mathsf{I}\otimes\mathsf{Y}]$ and $[B(\cdot)A]\circ[BB(\cdot)] = [\mathsf{Y}]\circ[\mathsf{Y}\otimes\mathsf{I}]$. Because of co-associativity, these also differ by a sign. Therefore the front square anti-commutes. In each composition, a circle splits twice into three circles.

On the left face of the cube, $[(\cdot)AA]\circ[BA(\cdot)] = [\curlywedge]\circ[-\mathsf{Y}]$ while $[AA(\cdot)]\circ[(\cdot)AB] = [\curlywedge]\circ[\mathsf{Y}]$. The two compositions differ by a sign. In each composition, two circles fuse and then one splits into two.

On the right face of the cube, $[(\cdot)BA]\circ[BB(\cdot)] = [\mathsf{Y}]\circ[\mathsf{Y}\otimes\mathsf{I}]$ while $[AB(\cdot)]\circ[(\cdot)BB] = [-\mathsf{Y}]\circ[\mathsf{Y}\otimes\mathsf{I}]$. The two compositions differ by a sign because Y is co-associative. The situation is analogous to the front square.

On the bottom face of the cube, $[(\cdot)AA]\circ[B(\cdot)A] = [\curlywedge]\circ[\mathsf{Y}]$ while $[A(\cdot)A]\circ[(\cdot)BA] = [-\curlywedge]\circ[\mathsf{Y}]$. The two compositions differ by a sign. Each is a fusion followed by a fission.

On the top face of the cube, $[(\cdot)AB]\circ[B(\cdot)B] = [\mathsf{Y}]\circ[\mathsf{I}\otimes\mathsf{Y}]$ while $[A(\cdot)B]\circ[(\cdot)BB] = [-\mathsf{Y}]\circ[\mathsf{Y}\otimes\mathsf{I}]$. The two compositions differ by a sign because Y is co-associative.

Side Comment. The idea of homology is that it measures something called "in-exactness" of a chain complex. When homology is not trivial, there are cycles that are not boundaries. Topologically, there are holes in a space that correspond to such cycles. A short overview follows.

A *chain complex* $\mathcal{C}$ consists of a sequence C_n of $\mathbb{Z}$-modules[7] for $n = 0,1,2,\ldots$, and module homomorphisms $C_{n-1} \xleftarrow{d_n} C_n$, that satisfy $d_n \circ d_{n+1} = 0$ for all $n = 0,1,2,\ldots$. The d_ns are called *boundary operators*. Colloquially, we will write $d^2 = 0$. The *kernel* of the module homomorphism d_n is the set

$$Z_n = \{v \in C_n : d_n(v) = 0\}.$$

Elements in the kernel are called *n-cycles*. The set of *boundaries* is the submodule

$$B_n = \{w \in C_n : w = d_{n+1}(u) \text{ for some } u \in C_{n+1}\}.$$

Since $d_n \circ d_{n+1} = 0$, the set of boundaries is a subset of the set of cycles: $B_n \subset Z_n$. The homology group $H_n = Z_n/B_n$ is the quotient group of the set of cycles modulo the set of boundaries. A cycle $c \in Z_n$ represents the homology class $[c] = c + B_n$. Thus cycles c and c' represent equivalent homology classes if $c - c' = d_{n+1}u$ for some $u \in C_{n+1}$. That is $c - c' \in B_n$.

The boundary operators are also called differentials. The differentials may also increase the indices in the chain groups. So it is possible that $C_n \xleftarrow{d^n} C_{n-1}$. Purists

---

[6]On the left-hand expressions, the source is fed to the operators from the left. In the right-hand expressions, the source is at the bottom of the diagram, and the left operator is adjoined below the right.

[7]A $\mathbb{Z}$-module is a commutative group written additively. A positive integer k acts upon an element a as $k \cdot a = \underbrace{a + a + \cdots + a}_{k}$.

would call the chains "co-chains," and discuss cohomology rather than homology. Herein, we will abuse the traditional vocabulary.

In Khovanov homology theory for the trefoil, there are three chain groups: the 0-chains are found in the module $C_0 = R[X]/(X^2) \otimes R[X]/(X^2)$ that is associated to the AAA state. The 1-chains are found in direct sum $C_1 = R[X]/(X^2) \oplus R[X]/(X^2) \oplus R[X]/(X^2)$ in which the summands correspond respectively to the AAB, ABA, and BAA states. The 2-chains are found in the direct sum $C_2 = \left[R[X]/(X^2) \otimes R[X]/(X^2)\right] \oplus \left[R[X]/(X^2) \otimes R[X]/(X^2)\right] \oplus \left[R[X]/(X^2) \otimes R[X]/(X^2)\right]$. The summands correspond respectively to the ABB, BAB, and the BBA states. The 3-chains are found in the three fold tensor product $C_3 = R[X]/(X^2) \otimes R[X]/(X^2) \otimes R[X]/(X^2)$ which corresponds to the BBB state.

The chain complex is of the form

$$0 \leftarrow C_3 \stackrel{d_3}{\leftarrow} C_2 \stackrel{d_2}{\leftarrow} C_1 \stackrel{d_1}{\leftarrow} C_0 \leftarrow 0$$

where the differentials are given as $d_1 = (⅄, -⅄, ⅄)$, $d_2 = ((-\mathsf{Y}, \mathsf{Y}, 0), (-\mathsf{Y}, 0, \mathsf{Y}), (0, -\mathsf{Y}, \mathsf{Y}))$, and $d_3 = (\mathsf{Y}\mathsf{I}, \mathsf{I}\mathsf{Y}, \mathsf{Y}\mathsf{I})$.

Exercise. Choose bases that are made of direct sums of tensors of 1 and X for the chain groups, and express the chain maps d_1,d_2, and d_3 as matrices with respect to these bases. Compute the image and the kernel of each chain map.

A detail that will not be discussed herein is the bi-grading that is associated to the chain groups. The gradings are chosen so that the differentials respect one of these gradings. In this way, a homology theory is constructed whose graded Euler characteristic is the Jones polynomial.

Consult Bar-Natan [**8, 9**] or the original Khovanov paper [**49**] to complete the description of Khovanov homology as a categorification of the Jones polynomial.

6.3. Hopf algebras. In this section, the axioms of a finite dimensional Hopf algebra are presented in diagrammatic form. Axiomatically, a Hopf algebra consists of a vector space, H, together with a unital, associative multiplication $H \stackrel{⅄}{\longleftarrow} H \otimes H$, a co-unital, co-associative co-multiplication $H \otimes H \stackrel{\mathsf{Y}}{\longleftarrow} H$, and an antipode $H \stackrel{S}{\longleftarrow} H$. These satisfy the rules that are described in Fig. 6. The co-multiplication Y is an algebra homomorphism so that $\mathsf{Y}(⅄(a,b)) = ⅄(\mathsf{Y}(a), \mathsf{Y}(b))$, and the antipode S satisfies $⅄ \circ (S \otimes \mathsf{I}) \circ \mathsf{Y} = \perp \circ \top = ⅄ \circ (\mathsf{I} \otimes S) \circ \mathsf{Y})$.

When the elements $\mathsf{Y}(a), \mathsf{Y}(b) \in H \otimes H$ are multiplied, it is customary to use the Sweedler notation to indicate the images under co-multiplication $\mathsf{Y}(a) = \sum_i a_{(i)} \otimes a'_{(i)}$ where $a_{(i)}, a'_{(i)} \in H$. Then the product becomes $\sum_i ⅄(a_{(i)}b_{(i)}, a'_{(i)}b'_{(i)})$. The exchanger in the middle of the relation $\mathsf{Y} \circ ⅄ = (⅄ \otimes ⅄) \circ (\mathsf{I}\ \mathsf{X}\ \mathsf{I}) \circ (\mathsf{Y} \otimes \mathsf{Y})$ that expresses this axiom represents the fact that the $b_{(i)} \otimes b'_{(i)}$ is split, the $b_{(i)}$ goes to the left, and the $a'_{(i)}$ goes to the right.

There is a category with object set $\mathbb{N} = \{0, 1, 2, \ldots\}$ in which there are arrows Y, ⅄, $\top$, $\perp$, and S that satisfy relations that are analogous to the Hopf algebra axioms as depicted in Fig. 6.

Since $\mathbb{N} = \{0, 1, 2, \ldots\}$ corresponds to the category that has one object and a non-identity arrow $\bullet$, the arrows Y, ⅄, $\top$, $\perp$, and S are promoted to being double arrows. In that situation, the relations depicted in Fig. 6 become triple arrows.

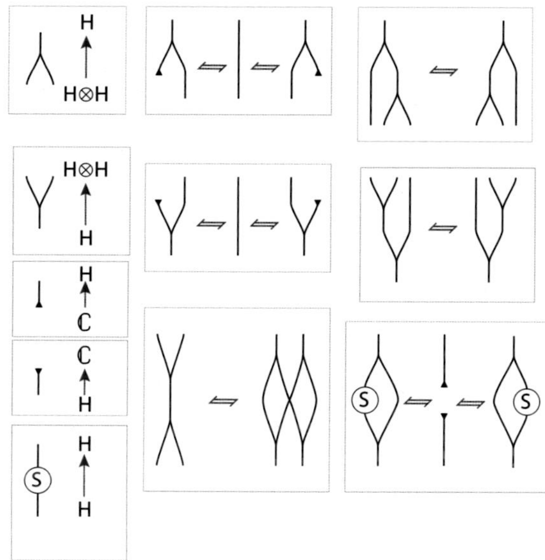

FIGURE 6. The Hopf algebra axioms

A typical example of a Hopf algebra is given by considering a vector space whose basis elements consists of elements of a finite group, G. This example is known as a group algebra $\mathbb{C}[G]$. Multiplication is induced by group multiplication and a distributive law. The unit map assigns to $1 \in \mathbb{C}$ the identity element in G. The co-unit gives the value $\top(g) = 1 \in \mathbb{C}$ for all $g \in G$. The co-multiplication is given as $\mathsf{Y}(g) = g \otimes g$, and the antipode is $S(g) = g^{-1}$. Notice that in this case, $S(\curlywedge(g \otimes h)) = \curlywedge(h^{-1} \otimes g^{-1}) = \curlywedge(S(h) \otimes S(g))$. Thus S is an anti-homorphism.

In [**51**], G. Kuperberg diagrammatically proved a pair of lemmas to demonstrate that $\overset{\top}{\curlywedge} = \top\,\top$. Occasionally, authors include this identity as an axioms. We will do so here to demonstrate the diagrammatic proof that S is an anti-homomorphism. This proof is recreated from that in [**54**] and is presented in Fig. 7.

There are several diagrammatic approaches to finding 3-manifold invariants via diagrammatic methods. Kauffman and Radford [**45**] use a method of decorating knot diagrams with Hopf algebra elements, and they use a calculus of knot diagrams to represent a general 3-dimensional manifold. They show that the decorated diagrams are invariant under the moves that give equivalent 3-manifold. Kuperberg [**51**] decorates handle decompositions with Hopf algebra operations. Hennings's [**28**] approach is very close in spirit to Kauffman and Radford, and Lawrence [**52**] develops the invariant from the point of view of the universal R-matrix.

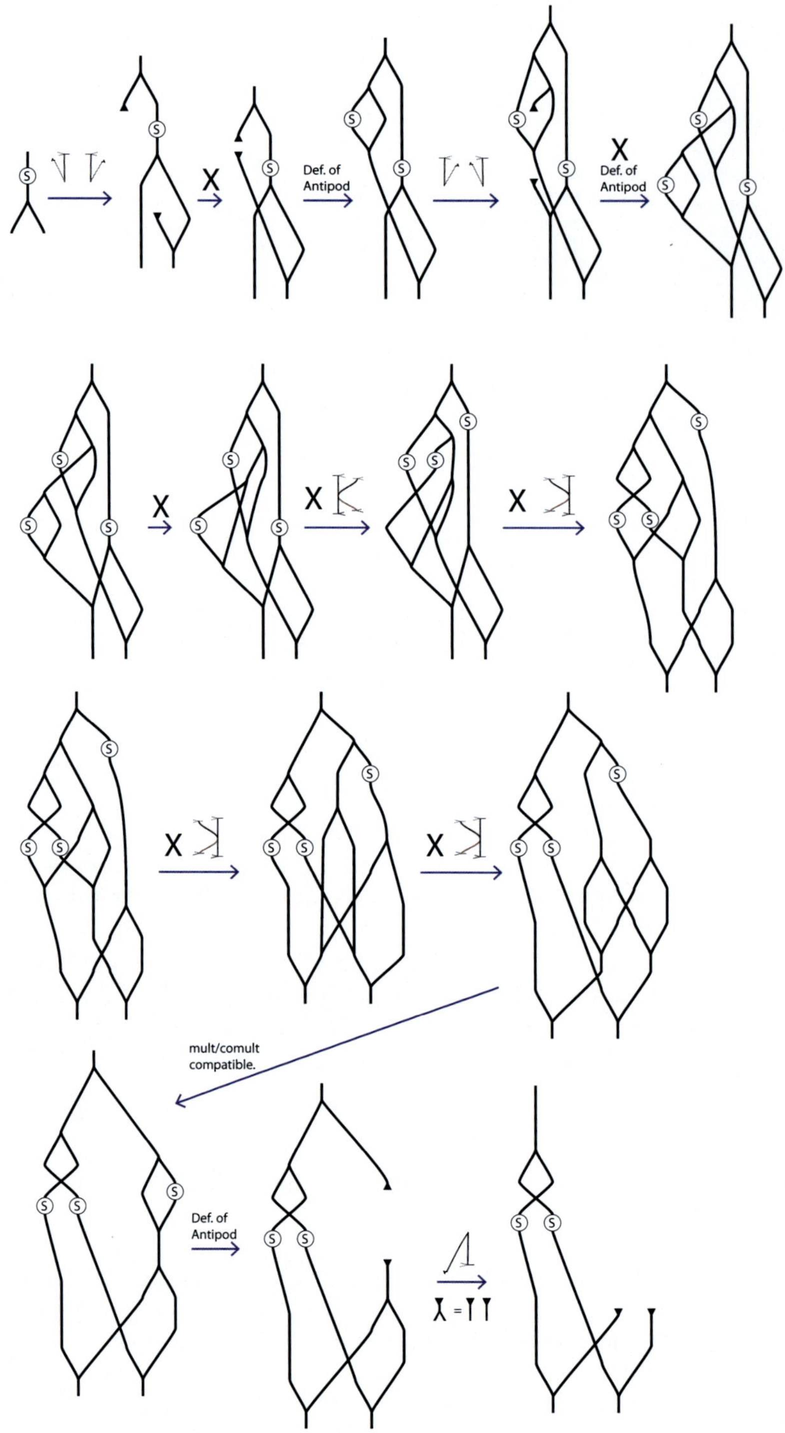

FIGURE 7. The antipod is an anti-homomorphism

7. Trivalent graphs

Trivalent graphs consist of a finite collection of vertices and edges for which each vertex has valence three. A given edge may return to the same vertex as in the case of the handcuff graph that is indicated below. If a trivalent graph is embedded in the plane, then it is said to be un-knotted. There are trivalent graphs embedded in $\mathbb{R}^3$ that are knotted in the sense that they are not isotopic to any embedding in $\mathbb{R}^2$ of the underlying graph. To model these in the diagrammatic context, assume that there are generating double arrows of the form $\boxed{\bullet}$ with triple arrows, Y, ⅄, U, ∩, X, and $\overline{\mathsf{X}}$ where these last two are the exchanger and its inverse.

Two important trivalent graphs are the *handcuff graph*, 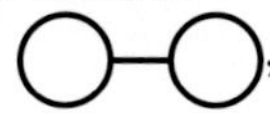, and the *θ-graph*, . Both can be embedded in 3-dimensional space as knotted objects.

Knotted embeddings of the θ-graph and the handcuff graph are illustrated in Fig. 8. The Kinoshita θ-curve, on the right, has the *Brunnian property*: if any of the three arcs are removed, the resulting circle is unknotted. Knotted trivalent graphs also can be used to represent embeddings of handlebodies[8]. The knotting of the handcuff graph on the left of the Figure represents a trivial embedding of a handlebody.

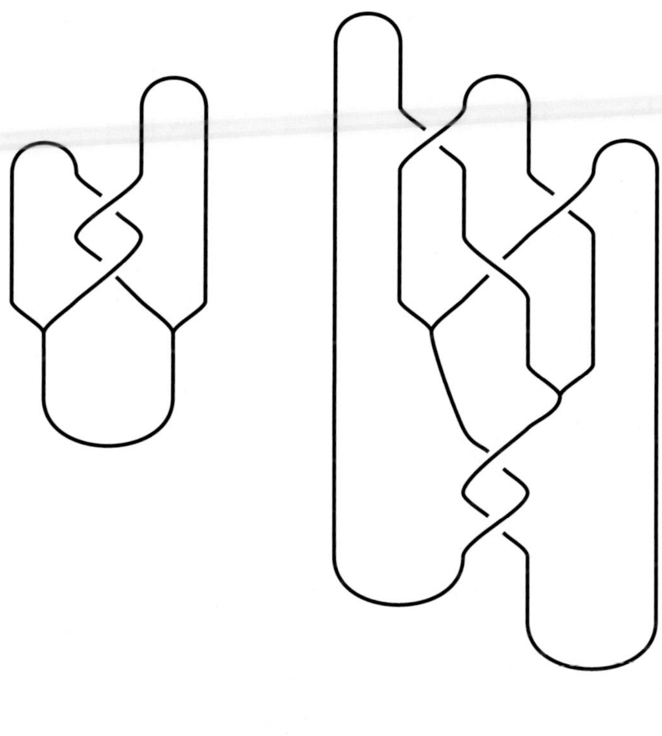

FIGURE 8. Two knottings of trivalent graphs

[8]to be defined in Section 7.3 of this chapter

Exercise. Show that the Kinoshita θ-curve is Brunnian by removing each of the arcs, one by one, at the trivalent vertices. The remaining two arcs form a circle. Determine isotopies that demonstrate each of these three circles is unknotted.

It is a bit more complicated to see that the entire graph is knotted. An argument is given in [**58**]. To see that the handcuff graph is knotted, remove the lower arc in the diagram that is depicted on the left of Fig. 8. The remaining diagram is a Hopf link, and an unknotted embedding of a graph should not contain a knotted or linked sub-graph.

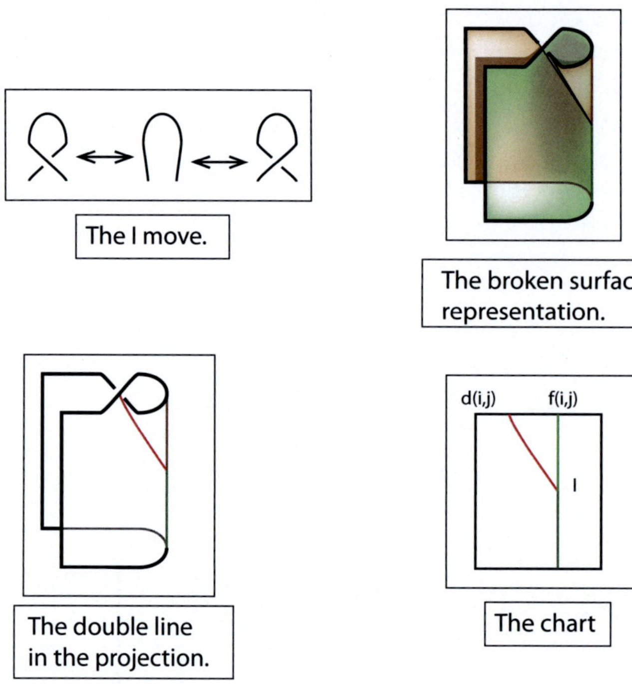

FIGURE 9. The type I move and its chart

Diagrams of knotted trivalent graphs correspond to compositions of triple arrows of the form Y, ⅄, U, ∩, X, and $\overline{\text{X}}$ where the non-identity double arrows are generated by [•]. The quadruple arrows that define equivalences of diagrams are:

(1) the Reidemeister moves of type I, type II, and type III;
(2) cusps:

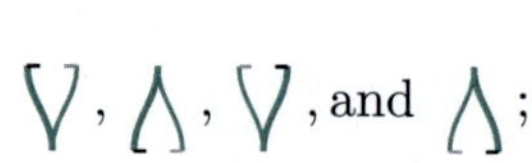

(3) associativity and co-associativity (resp.) of the pairing and co-pairing:

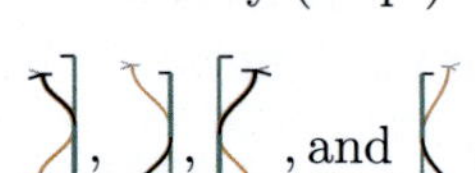

(4) writing multiplication in terms of co-multiplication and writing co-multiplication in terms of multiplication:

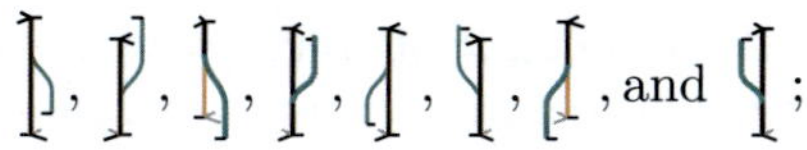

(5) naturality of X and $\overline{\mathsf{X}}$ with respect to U, ∩, Y and ⅄; and
(6) twisting a trivalent vertex (also illustrated below).

Even though the quadruple arrows in items 2, 3, and 4 above were defined in a lower categorical setting, those definitions extend to the current situation. Trivalent vertices can move past optima. Thus, for example, in the algebraic context, (∩) ∘ (⅄I) = (∩) ∘ (I⅄) (associativity of a pairing), and (YI) ∘ (U) = (IY) ∘ (U) (co-associativity of the co-pairing). Furthermore, Y and ⅄ can be written terms of each other (writing multiplication in terms of co-multiplication or vice-versa); (I⅄) ∘ (UI) = Y = (⅄I) ∘ (IU) and (∩I) ∘ (IY) = ⅄ = (I∩) ∘ (YI). We repeat: the glyphography that describes the relations (with the exceptions of associativity and co-associativity) in FA applies to embedded trivalent graphs. In addition, the Reidemeister moves hold. In the next section, we establish glyphs and fonts that express the relations for these additional moves.

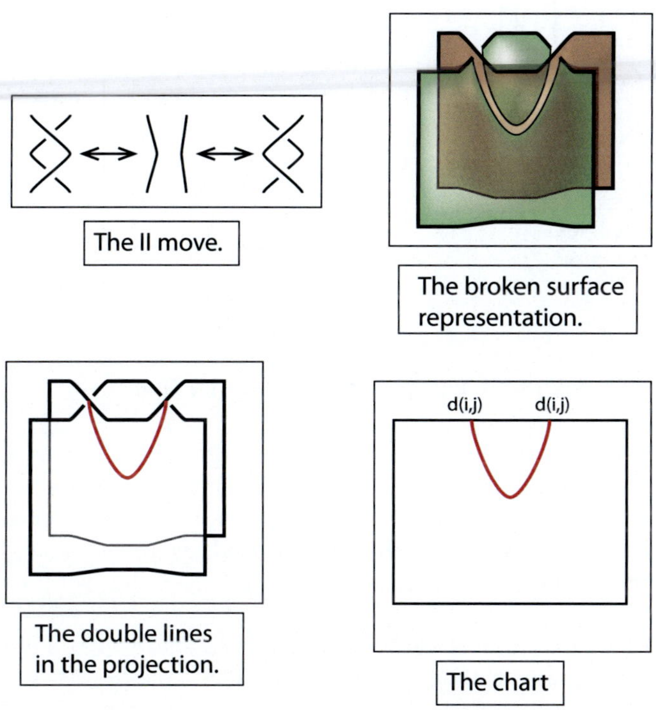

FIGURE 10. The type II move and its chart

7.1. Using charts to construct glyphs. In this section, the ideas introduced in Chapter 4, Section 1.4 are mimicked to develop glyphs that reflect the Reidemeister moves and the ψ-moves. The notion of a chart will be introduced, and the Reidemeister moves will be thought of kinematically. Be aware, in Figs. 11, 13, and 14 the sources of the quadruple arrows are written at the top of the illustrations on the upper right side. All of these quadruple arrows are strongly invertible. The bottom-to-top arrow direction describes the strong inverse.

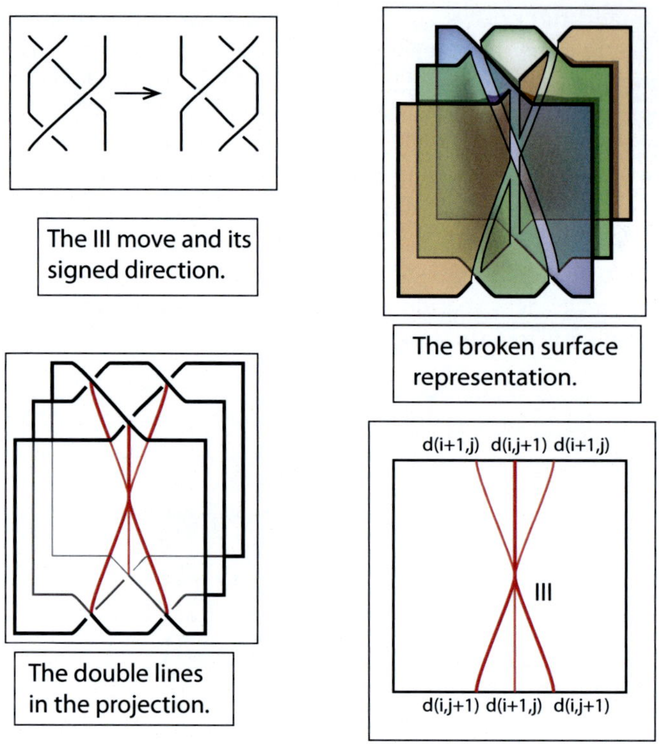

FIGURE 11. The type III move and its chart

Each move can be thought of as a quadruple arrow, or a motion of strings in a space-time; as a string changes (or remains static), it engenders a surface. The surfaces that are the result of Reidemeister moves are, then, embedded in a 4-dimensional space. The trompe l'oeil that is used to draw knot crossings extends to create broken surface diagrams. Only representative diagrams are illustrated in the figures.

The broken surface diagram of each move occupies space in a 3-dimensional cube. The regions along which the breaks occur are called *double curves*. These are projected further into a plane and labeled with the letter d that is enhanced with a bookmark. Thus we write $d(i,j)$ or $d_{i,j}$, for example. The first index of a bookmark records the number of sheets between the edge and the viewer; the second indicates the number of sheets on the other side of the edge. In the cases of the ψ-moves and the type I moves, there is a *fold line* that is similarly projected onto the plane and

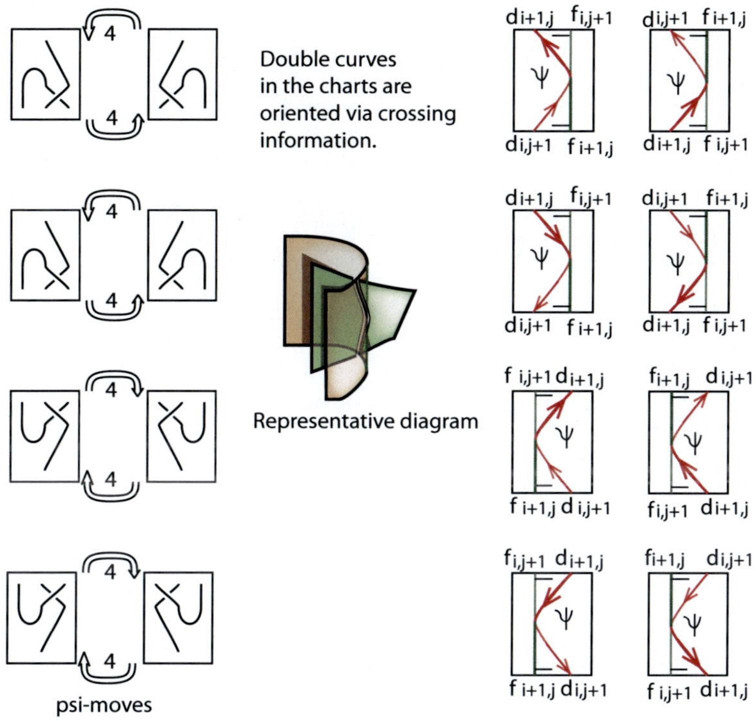

FIGURE 12. The ψ-moves and their charts

labeled with a bookmarked f. The fold line is the region of the surface at which the derivative of the projection drops rank by 1. Each such projection defines a portion of a *chart* which is a labeled planar graph that has bookmarked edges labeled d (double curves), s (seams) or f (folds), vertices and critical points that correspond to incidences of the double curves, seams, and/or folds.

It is possible that sheets in the broken surface have oriented normals. In this case, a bookmark can also consist of a pair of sequences of signs that indicate if the normal to a sheet is parallel or anti-parallel to the upward oriented normal in the plane of the chart.

These situations are depicted in Figs. 9, 10, 11, 12, 13, 14, and 15. In Fig. 12, the double curves are oriented, and the orientation is indicated by an arrow. Orientations for the double curves for the type I and II moves and the ψ moves will be presented within their glyphographic descriptions. In the cases of the type III, the type IY, the type YI, and the twist moves orientations will not be depicted.

The charts for each move is the graph that lies in the rectangle for which the edges are labeled d, f, or s and bookmarks are indicated.

The figures given here serve several purposes. From the standpoint of the current section, they will be used to construct glyphs that are analogous to those given in Chapter 4, Section 1.4. These glyphs name the generating quadruple arrows (and consequently the equivalences between representative diagrams) in the multi-category that describes knotted trivalent graphs. In Chapter 10, the glyphs and charts will be used to describe 2-foams that are embedded in 4-space.

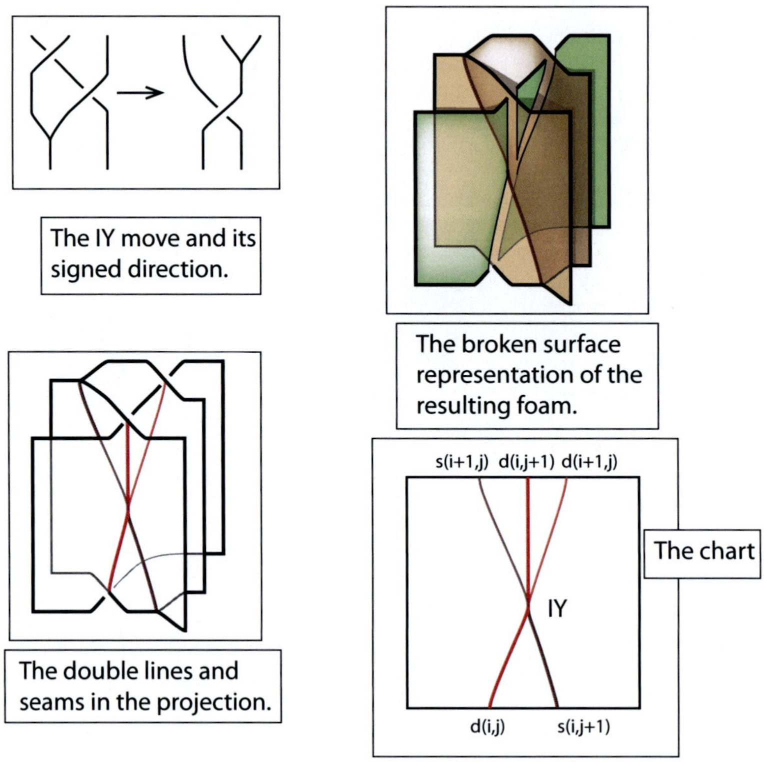

FIGURE 13. The IY move and its chart

Therein important moves among the foams are illustrated. As has been the case throughout, there are coincidences between the categorical and the geometric.

7.2. Describing glyphs. In this section, an explicit methodology for constructing glyphs is presented. A glyph has three parts: edges, serifs, and vertices.
Edges. The edges in a glyph either represent (1) a fold (f in the chart) that is drawn as a green interval with a segment as a serif, (2) a seam (s in the chart) that is drawn as a thinner black line juxtaposed over a thick brown line. Sometimes a seam is only indicated by a brown line. The serif upon a seam is a Y or ⅄. (3) A double curve (d in the chart), drawn in red with a serif of the form X or $\overline{\mathsf{X}}$. The thicknesses and relative transparencies of the edges are meant to suggest their visibility or the number of sheets (the first index of the bookmark) that occlude them from the viewer. Thinner, more transparent, edges have more sheets in front of them. Bookmarks are rarely specified within the depictions because at the level of describing the next-order relations (5-arrows), the number of sheets that are involved is small and the line thicknesses are suggestive.

Serifs. (1) The short segment serifs that lies along a fold] are meant to suggest that the fold (as a 4-arrow) is an identity upon U or ∩. The segment lies in the region of the diagram along which two surfaces overlap. In the case drawn here the surfaces are to the left of the fold. (2) The ≻ or ≺ shaped serif that lies at the top

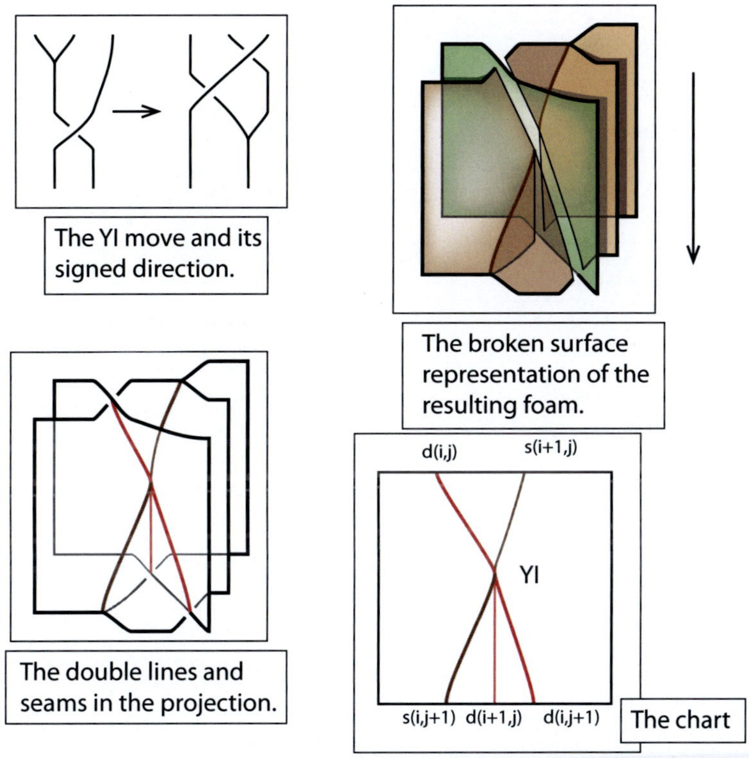

FIGURE 14. The YI move and its chart

or bottom of a seam ⊺ suggests that the seam is the identity upon Y or ⅄, and the seam is the junction of three sheets in the foam model. When the serif appears as ≻, two sheets are to the left and one is to the right. When the serif appears as ≺, two sheets are to the right and one is to the left. (3) The $\overline{\mathsf{X}}$ and X shaped serifs lie on the top or bottom of a red arc $\overline{\mathsf{X}} \leftrightarrow$ ↧ or X $\leftrightarrow$ ↥ that represents a double curve. Here arrows along the double arcs indicate orientations. The orientations indicated in the text are dependent upon the two intersecting sheets being normally oriented in the same direction — either away from or towards the reader. When a double curve passes over a fold via a ψ-move, the fold changes the normal orientation of one sheet. Therefore, the glyphs that are associated with the ψ-moves have consistent orientations along the double curves, but at the bottom of the glyphs, the serifs may differ from one's expectation given the orientation. That is, the choice of $\overline{\mathsf{X}}$ or X serifs coincide with the crossing information and is independent of the orientation of the double curve within the ψ-move glyphs.

Vertices. (1) In the glyph that represents a type I move, a red double arc emanates from a green fold. Equivalently, three edges are coincident — two folds and one double arc. Within the text, a type I move may be represented by a font of the type ⟨, ⟩, ⟩, or ⟨. (2) In the glyph that represents a type II move, a red arc achieves an optimal value with respect to the vertical direction. The vertex, then, is a critical point. Within the text, fonts such as ∪ or ∩ may be used to represent type

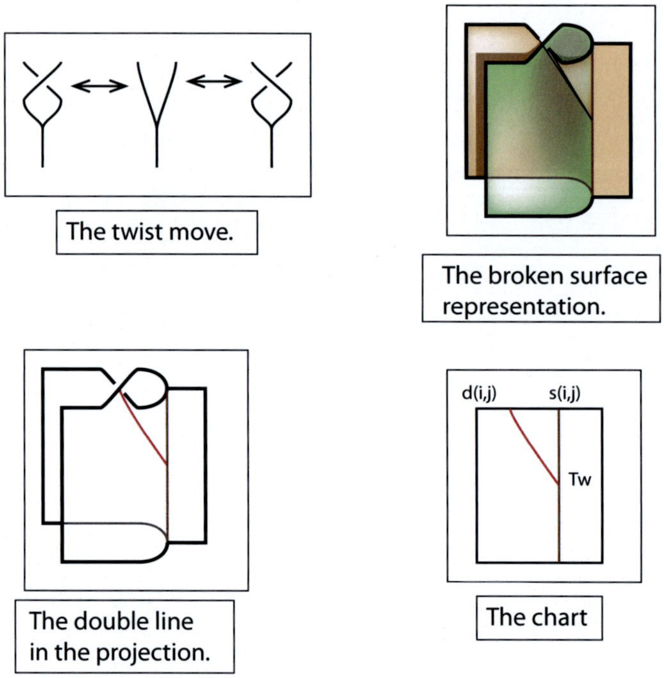

FIGURE 15. The twist move (T-move) and its chart

II moves when they won't be confused with union or intersection. (3) In the glyph that represents a type III move six edges are coincident at a vertex. The vertex is represented by a small white region enclosed in a black circle. Three consecutive edges are oriented away from the vertex, and the other three are oriented towards it, but in many of our depictions the orientations will be absent. The bookmarks at serifs alternate as in Fig. 11. For convenience, a font such as X will also be used to indicate a type III move. (4) In the glyph that represents the YI-move, as illustrated in Fig. 14, a vertex is the junction of three red double curves and two brown seams. Two red double curves and a one brown seam leave the vertex from the top while one red double curve and one brown seam enters from the bottom. The chart indicated in Fig. 14 is the chart of the strong inverse. Sometimes a font W will be used to represent the YI move. The glyph of the inverse is essentially a reflection in the horizon of this. The representative font will be M which should not be confused with an M. The YI move is the expression of naturality of the exchanger X with respect to Y. (5) In the glyph that represents the IY move, as illustrated in Fig. 13, a vertex is the junction of three red double curves and two brown seams. Two red double curves and a one brown seam enter the vertex from the bottom while one red double curve and one brown seam leave toward the top. The chart indicated in Fig. 13 is the chart of the strong inverse. Sometimes a font M will be used to represent the IY move. The glyph of the inverse is essentially a reflection in the horizon of this. The representative font will be W which should not be confused with a W. The IY move is another expression of naturality of the exchanger X with respect to Y. (6) In the glyphs that represents the twist moves, as illustrated in Fig. 15, a red double curve and two brown seams emanate from a vertex. (7) The remaining glyphs have been depicted in Chapter 4, Section 1.4.

The glyphs below are arranged so that those which represent the strong inverses appear vertically above and below each other. Be aware, the serifs differ from left to right. The strong inverse to an exchanger X may be denoted by $\overline{\mathsf{X}}$ or $\underline{\mathsf{X}}$ since these are turned sideways and the figure seems easier to read in one form or the other.

Type I glyphs. The orientations on the double curves of the type I glyphs can be determined by choosing an arbitrary orientation of the twisted α-shaped arc and computing the sign of the induced crossing. This sign is independent of the chosen orientation.

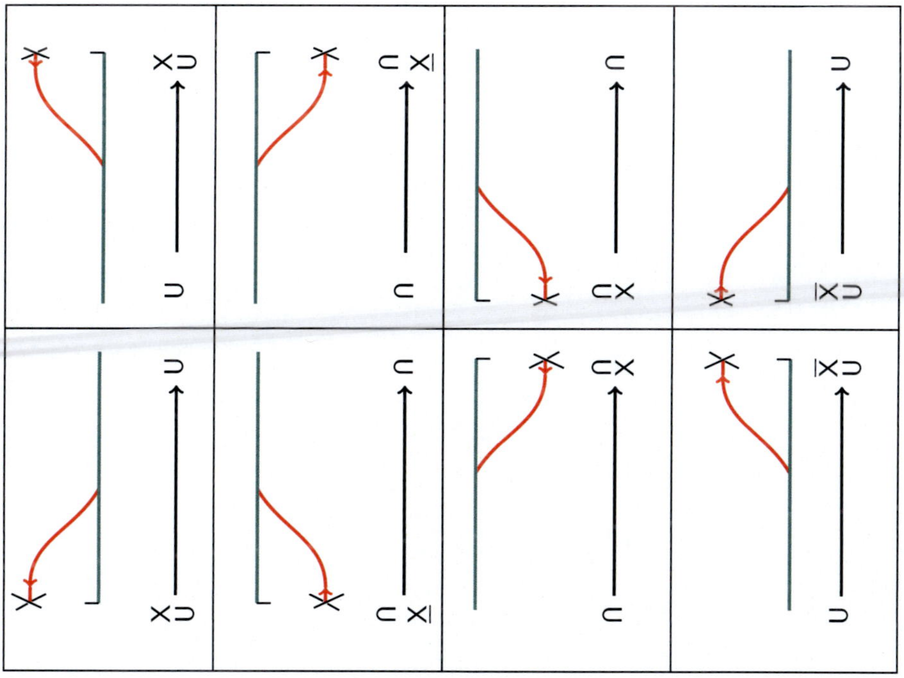

Type II glyphs. There are four type II glyphs.

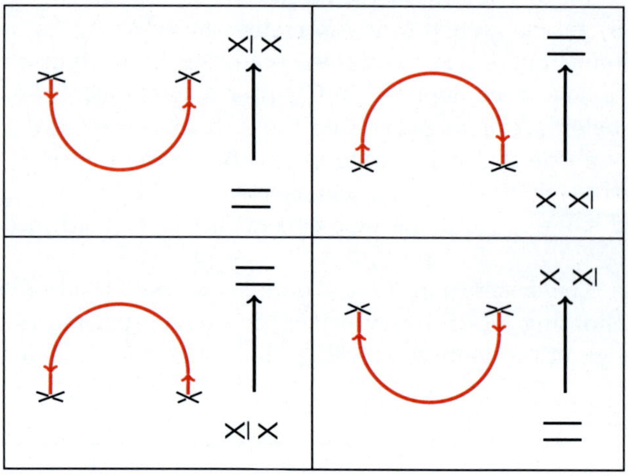

Type III glyphs. Representative type III glyphs are illustrated. Alternative arrangements of serifs are possible, and since normal orientations of the sheets have not been included, the double curves remain unoriented.

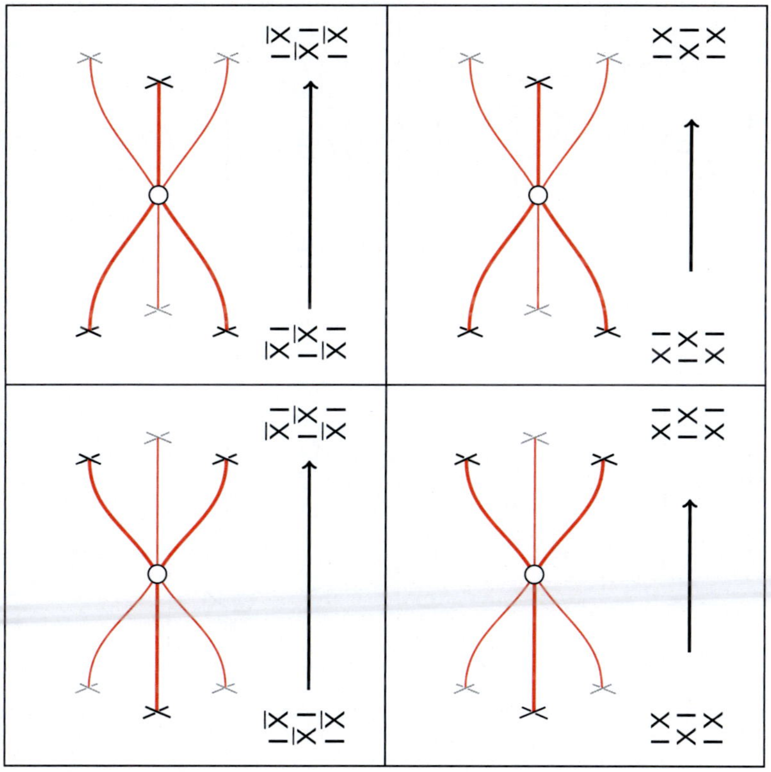

ψ-glyphs. Each of the ψ-glyphs that are depicted below can have the red double curves oriented upwards or downwards dependent upon the normal orientations of two sheets that intersect in the 3-dimensional projections.

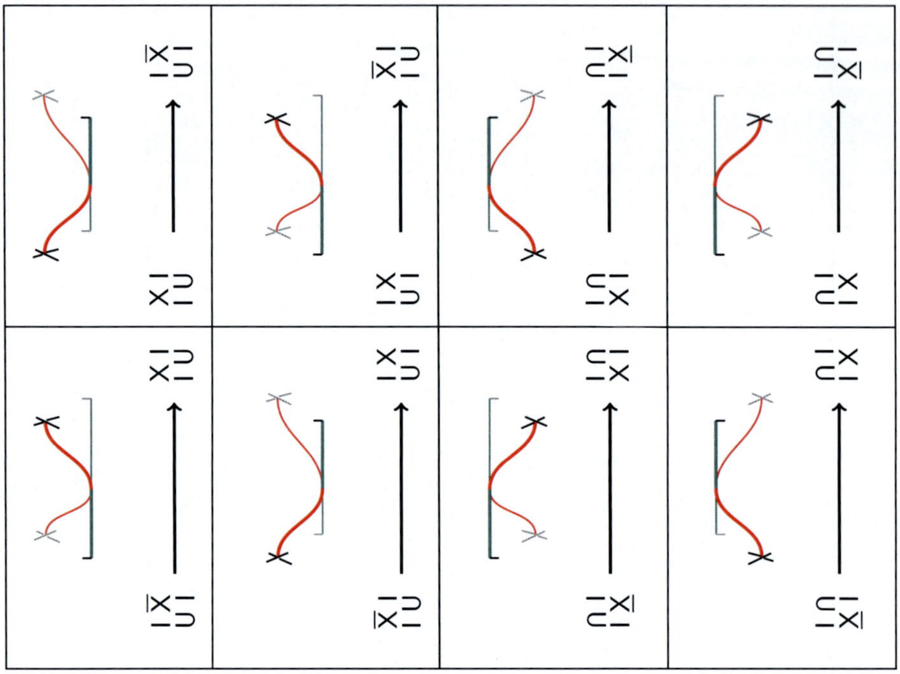

YI and IY glyphs. The naturality of the exchanger X with respect to Y is drawn in Figs. 13 and 14. The glyphs are depicted below.

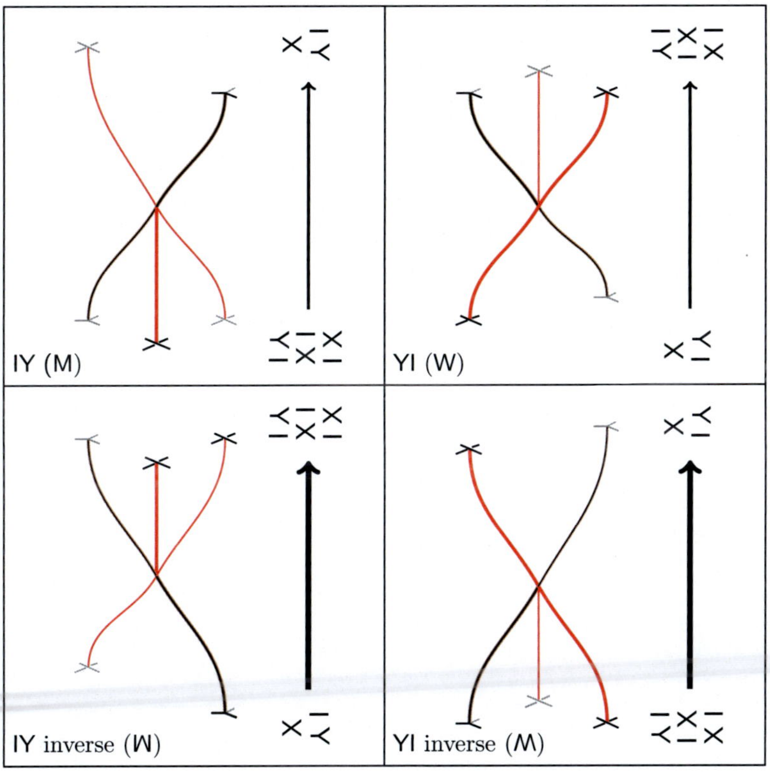

Twist glyphs. The glyphs below represent the variations upon the twist vertex T that is depicted in Fig. 15.

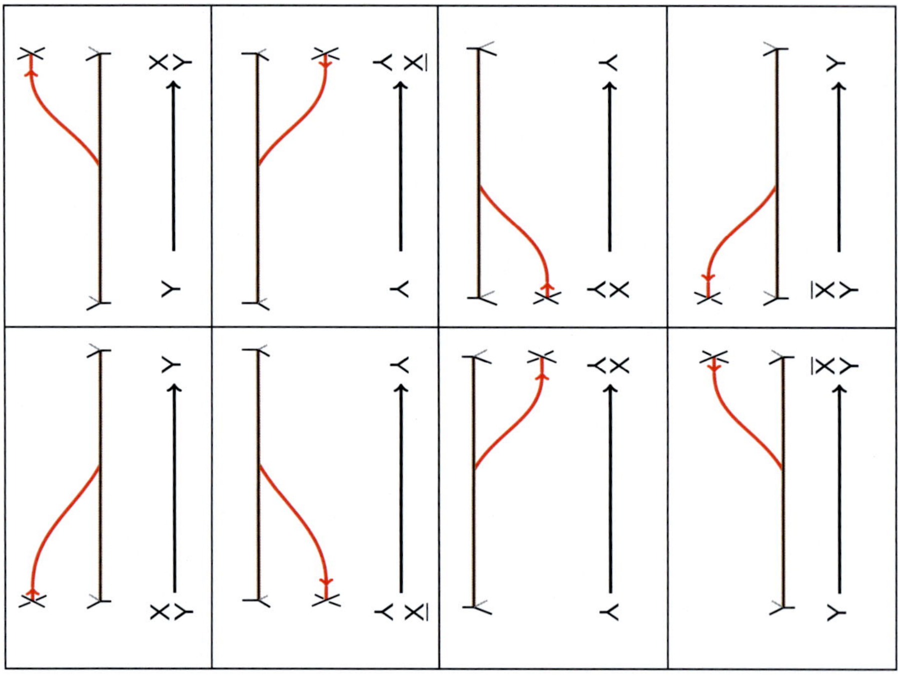

To recap, the moves to knotted trivalent graphs are encoded by a family of glyphs. These encapsulate the Reidemeister moves for classical knots, methods of bending and rotating trivalent vertices, and naturalities of the exchanger. For convenience, the naturality of the exchanger with respect to optima is rewritten in the equivalent formulation in terms of ψ-moves. Each of the moves here is an atomic piece of a foam that is embedded in 4-space. These glyphs together with those for associativity, co-associativity, births, deaths, saddles, and forks constitute generating quadruple arrows that completely describe those foams. Finally, a contrast between knotted trivalent graphs and handlebody knots occurs where associativity and co-associativity moves are included.

7.3. Handlebody knots. A *handlebody* is a 3-dimensional manifold that is constructed in the abstract by attaching 3-dimensional 1-handles ($D^1 \times D^2$ attached along $S^0 \times D^2$) to a solid ball D^3. It has as its boundary an orientable genus g surface, and the fundamental group of the 3-manifold is free of rank g. This section focuses upon handlebodies that are embedded in the 3-sphere, S^3. In this case, one may think that the handles have knots tied in them and that they interlink. When a handlebody is knotted, its complement in S^3 has a non-free fundamental group. Detecting whether or not a handlebody is knotted is a subtle problem. Some diagrammatic tools can be used to address it.

A small tubular neighborhood of a knotted trivalent graph in the 3-sphere, S^3, is an embedding of a handlebody. A given handlebody knot retracts to its spine which is such a knotted trivalent graph. On the other hand, inequivalent knotted trivalent graphs may represent the same handlebody knot. The point of the discussion is that the associativity and co-associativity moves (quadruple arrows) when applied to knotted trivalent graphs do not affect the topology of their tubular neighborhoods. Thus two knotted trivalent graphs that are related by associators and co-associators:  represent equivalent handlebody knots.

Let two knotted trivalent graph be presented as compositions of the triple arrows U,∩,Y,⅄,X, and $\overline{\text{X}}$. Consider the resulting planar diagrams up to topological equivalence. The topological equivalences are generated by the quadruple arrows :

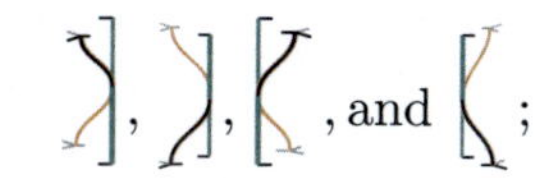

and quadruple arrow exchangers.

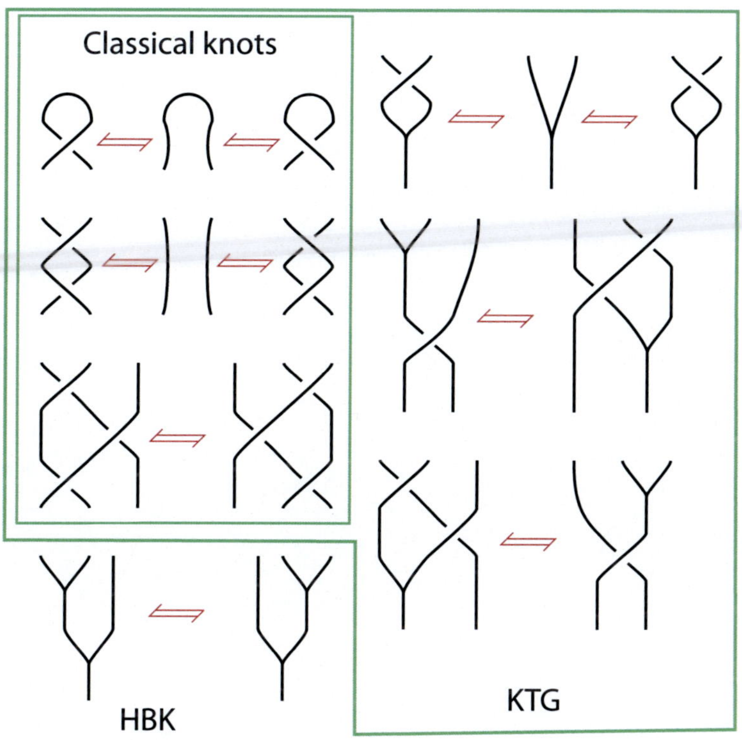

FIGURE 16. The Reidemeister type moves for KTGs that represent HBKs

A theorem of Ishii [**31**] formulates that two diagrams represent the same knotted handlebody if and only if they are related by the moves that are depicted in Fig. 16 and topological equivalences. Within the illustration, the classical Reidemeister moves are grouped on the upper left, the additional moves for knotted trivalent graphs (KTGs) are on the right, and the bottom left of the illustration indicates the associator/associator inverse.

Exercise. Determine a sequence of quadruple arrows that transform the KTG on the left of Fig, 8 to an unknotted handcuff graph. Observe that the topological types of the embeddings depicted here: 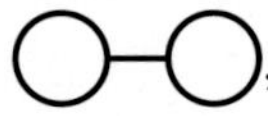, and  are related via associators.

CHAPTER 10

Foams and surfaces in 4-space

This chapter addresses several ideas. (1) Knotted surfaces and knotted foams will be described as compositions of quadruple arrows. (2) The moves among them are quintuple arrows. In the case of knotted surfaces, a complete set of these will be presented as relationships between two compositions of glyphs. In the case of foams, many known relations will be presented. But as of this writing, a complete set of quintuple arrows is not known. (3) An algebraic structure called a shalgebra is defined via the quadruple arrows that describe naturality of the triple arrow that is the exchanger. (4) A homology theory is presented for shalgebras that encompasses both group and quandle homology. (5) From the homological study and from considerations about the quintuple arrows, systems of abstract tensor equations will be formulated. These equations return us to the ideas of Chapter 2, and they bear some resemblances to other algebraic structures. The homology theory gives a set of solutions.

1. Knotted surfaces

The multi-category $\mathcal{T}$ has:

(1) one unnamed object;
(2) an identity 1-arrow;
(3) a generating 2-arrow $\boxed{\bullet}$ that is its own weak inverse;
(4) triple arrows formed by the identity I, exchangers X and $\overline{\mathsf{X}}$, a creation U and an annihilation $\cap$ that express that $\boxed{\bullet}$ is its own weak inverse;
(5) in particular there are birth, death, saddle, and fork weakly invertible quadruple arrows (since U and $\cap$ are weak inverses), and strongly invertible cusps;
(6) additional quadruple arrows formed from the ψ-moves, and the type I, type II, and type III moves that are strong isomorphisms in a manner to be described below;
(7) exchanger quadruple arrows that occur in natural families and are isomorphisms; and
(8) the additional quintuple arrows that are described below.

An oriented multi-category that includes double arrows of the form $(-)$ and $(+)$ as in the oriented tangle category could be considered. But in that case, certain lists of isomorphisms at the 5-arrow level would become un-manageable.

A composition of quadruple arrows corresponds to a properly embedded surface in $\mathbb{R}^3 \times [0,1]$. The situation is analogous to Theorem 6.1 which is illustrated via Fig. 2 (Chapter 6).

1.1. Quintuple arrows. Many quintuple arrows will be described. These are deemed to be, in the current context, isomorphisms. The term 5-*isomorphism* will be used to refer to them. In addition to five of the Roseman moves, other quintuple arrows are constructed via promoting the categorical dimension. For example, all of the quadruple arrows from Chapter 6, Section 4 are considered to be 5-isomorphisms since here -•- has been promoted to |•|. Therefore, critical cancelations, lips, beak-to-beak, horizontal cusps, and swallow-tail isomorphisms are among the 5-isomorphisms.

Consider the type II quadruple arrow following the technique of Chapter 7. There is a multicategory $\mathcal{S}$(II). Both t and ƒ are the composition |:|, the arrow b corresponds to X, and the arrow p corresponds to $\overline{\mathsf{X}}$. Then F(t), F(ƒ), Ⅎ(t), and Ⅎ(ƒ) are the type II quadruple arrows that are represented by the glyphs

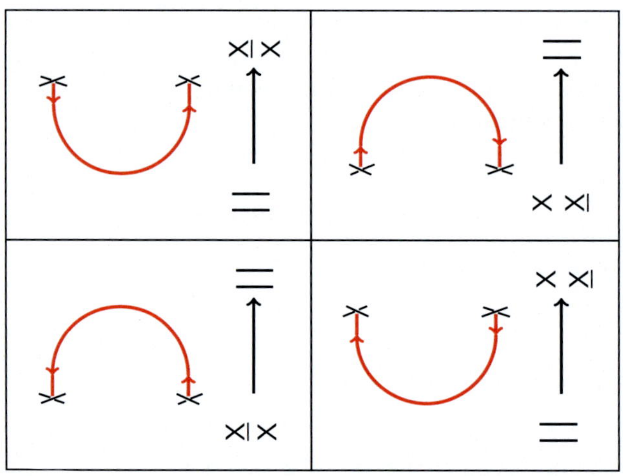

Here, we won't be too fussy about which item is which, and in the future orientations along double curves will usually be omitted. Choose a correspondence and maintain consistency. Then under this correspondence, the birth, death, saddle, and fork arrows are 5-isomorphisms with the two pairs (birth and death) and (saddle and fork) grouped as strong inverses. These moves constitute two of the Roseman moves that are called the type II bubble (birth or death) and the type II saddle (saddle or fork). One version of the bubble move identifies the composition

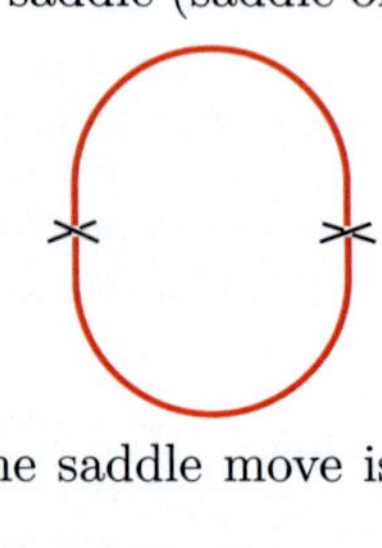

with the identity upon the identity on two dots. A version of the saddle move is the isomorphism depicted here:

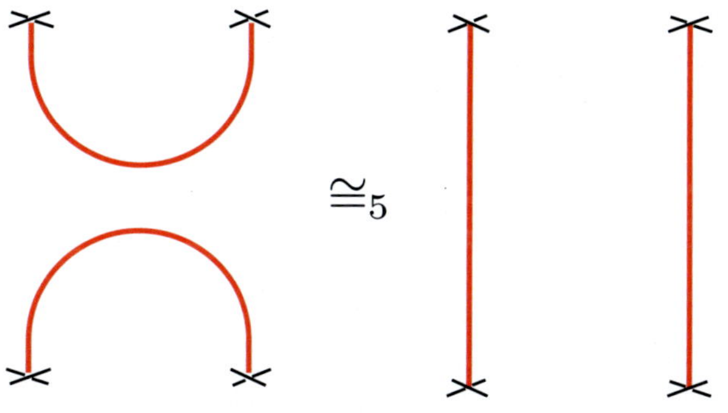

Furthermore, the cusp arrows in the context of skew compositions also are strong isomorphisms that correspond to a zig-zag being straightened or introduced in an arc of double curves. In this case, the red curves that describe the double arrows in $\mathcal{S}(\text{II})$ coincide with the red curves in the glyphs that indicate double arcs since the glyphs used to describe type II moves are enhancements of the glyphographic representations that were used in the expressions for $\mathsf{F}(\mathsf{t})$, etc.

The type I quadruple arrows are strongly invertible. One pair of the isomorphisms is illustrated. The 5-isomorphism on the left is called a *type I saddle* move; that on the right is the *type I bubble move.* The other 5-isomorphisms are reflections, have their serifs reflected, or both.

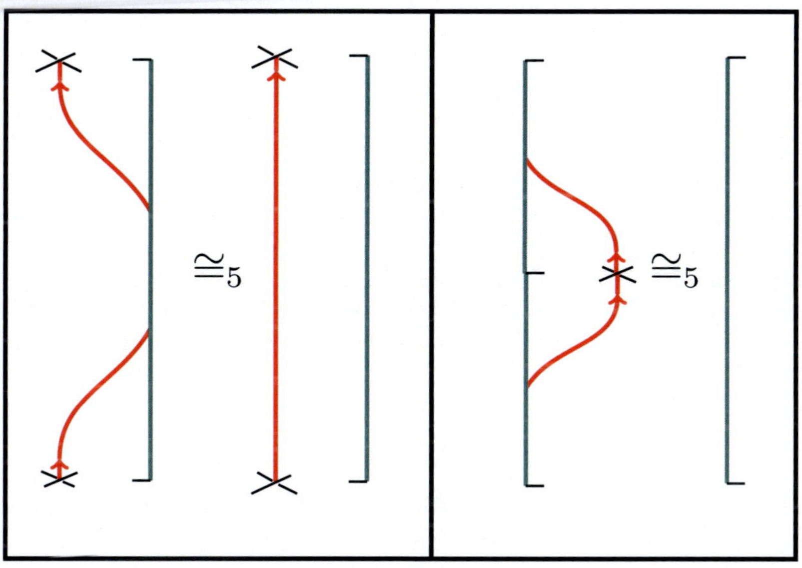

The type III quadruple arrows are strongly invertible. The format of the invertibility is common among the various compositions of glyphs. Only one 5-isomorphism is illustrated.

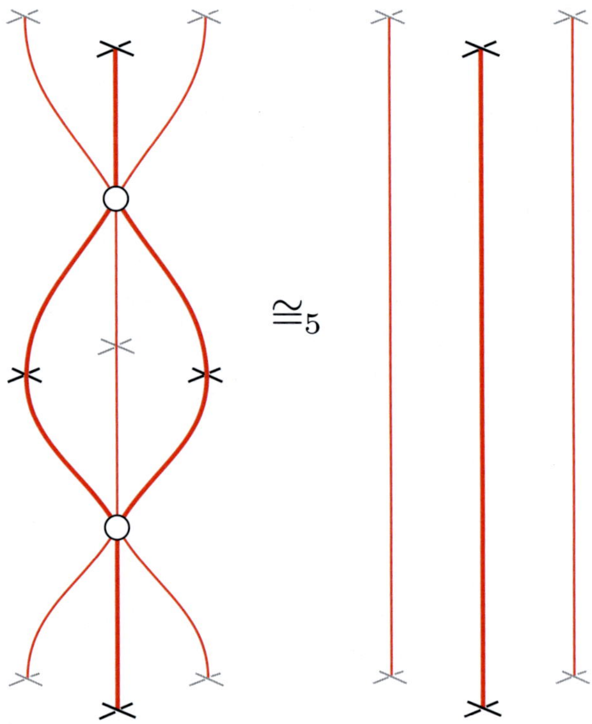

The strong invertibility of the type I, II, and type III moves correspond to five of the Roseman moves that are the analogues for knotted surfaces in 4-space of the Reidemeister moves for knotted curves in 3-space. These are illustrated in Fig. 1.[1] The strong invertibility of these quadruple arrows is an expression of categorical principle 3 (See Chapter 2, Section 5). In this case, doing followed by undoing *is the same* as not-doing.

The Roseman moves were described in a preprint of Dennis Roseman that appeared in the early 1980s and was subsequently published in [**68**]. In the decade of the 1980s diagrammatic approaches to knotted surfaces, such as Roseman's work and that of Giller [**23**], built upon the initial work of Yajima [**79**].

The ψ-moves are strongly invertible. Exemplary 5-isomorphisms are indicated.

[1]This figure originally appeared in this form in [**21**]

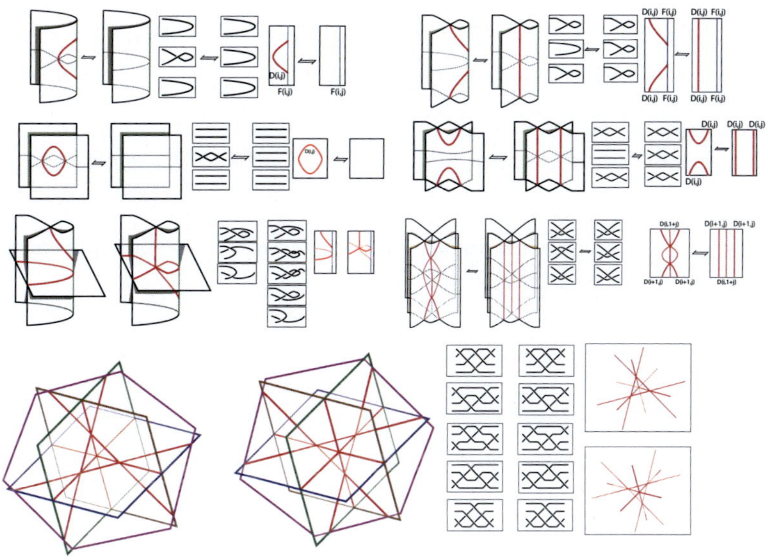

FIGURE 1. The Roseman moves for knotted surface isotopies

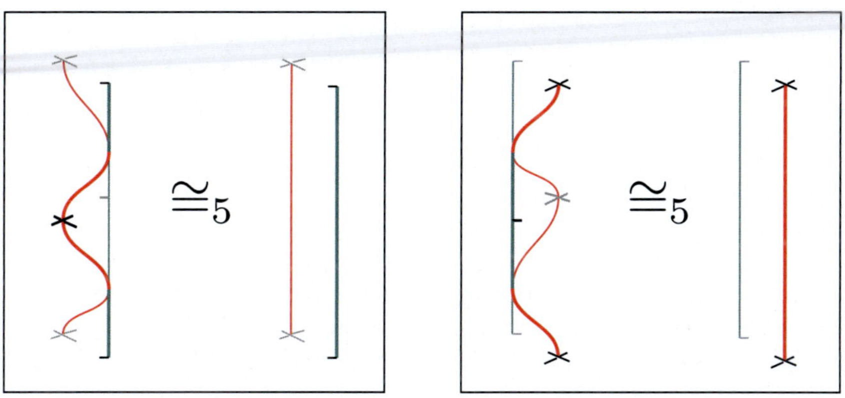

1.2. Exchanger quadruple arrows. The remaining two Roseman moves (pulling a branch point through a transverse sheet and the quadruple point move) are described in terms of the naturality of the family of exchanger quadruple arrows. Recall from Chapter 2, Section 5, that naturality is codified via the maxim "change followed by exchange is comparable to exchange followed by change." Here the comparison will be made by means of 5-isomorphisms. Next the naturality of the exchanger quaduple arrows is examined more closely. Additional relations among quadruple arrows will become apparent. We will begin with a metaphor that is convenient to describe one of the exchangers.

Consider a red chopstick (hashi) that lies on a table to the left of a blue chopstick which is parallel and a few centimeters to the right. Both sticks point away from you. With your right hand, reach over the blue chopstick and lift the red stick while using your left hand to slide the blue stick to its left and into the original position of the red stick[2]. Now place the red stick on the right of the table where the blue stick once was. The initial configuration of the sticks represents the identity  . The final represents the identity[3] 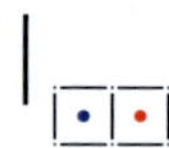 . The space-time process corresponds to the quadruple arrow that is the exchange from  to 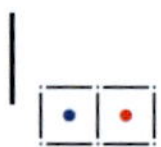 . That is the process of moving one stick over the other is the quadruple arrow exchanger 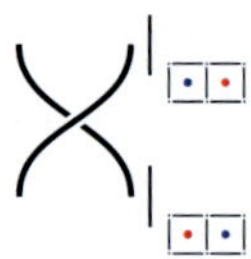 .

Alternatively, consider the quadruple arrow that is the identity upon the exchanger 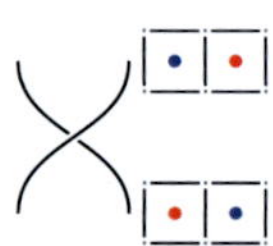 . This quadruple arrow 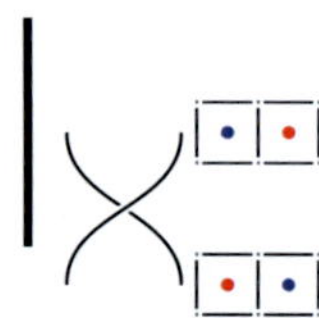 is presented by the glyph  . Both of the quadruple arrows 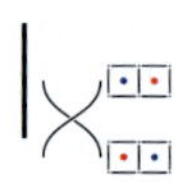 and 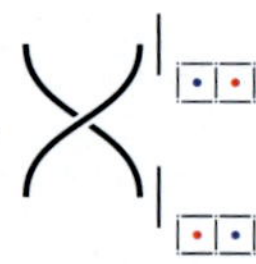 are represented by topologically equivalent embedded surfaces in 4-dimensional space. See Fig. 2. In this way, the identity upon the exchanger is the exchanger upon the identity. The correspondence involves re-parametrizations, and other instances will occur in which exchangers can be re-interpreted in terms of existing glyphs under analogous re-parametrizations as the discussion continues.

The chopstick example generalizes to one for which an arbitrary triple arrow, T, that is a composition of I, X, $\overline{\mathsf{X}}$, $\cap$, and $\cup$ is arranged as a wire assemblage in a box on the right side of the table. Again lift the red chopstick that is on the left with your right hand, and slide the box containing the assemblage to the left with your left hand. Subsequently, place the red chopstick down on the table and to the right.

To fully understand the configuration $\mathsf{X}^{\mathsf{T},\mathsf{I}}_{\mathsf{I},\mathsf{T}}$ it is enough to examine the particular cases $\mathsf{X}^{\cap,\mathsf{I}}_{\mathsf{I},\cap}$ and $\mathsf{X}^{\mathsf{X},\mathsf{I}}_{\mathsf{I},\mathsf{X}}$ since anything else is, up to reflections or rotations, a composition of these. These are depicted in Fig. 3 as immersed surfaces — over/under crossing information is omitted.

Up to re-parametrizations $\mathsf{X}^{\cap,\mathsf{I}}_{\mathsf{I},\cap}$ resembles the ψ-move quadruple arrow , and $\mathsf{X}^{\mathsf{X},\mathsf{I}}_{\mathsf{I},\mathsf{X}}$ resembles the type III move quadruple arrow that is abbreviated as .

---

[2]A gratuitous comment about the capital of Louisiana is obligatory.

[3]According to convention, the two dots should be arranged vertically rather than horizontally. However, the physical metaphor suggests the horizontal arrangement.

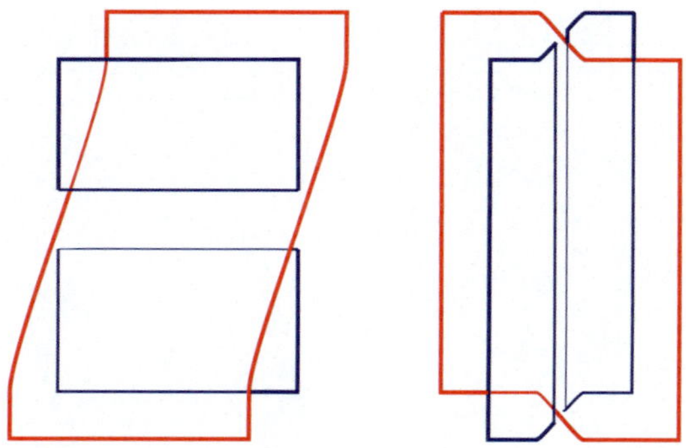

FIGURE 2. The identity upon the exchanger is the exchanger upon the identity

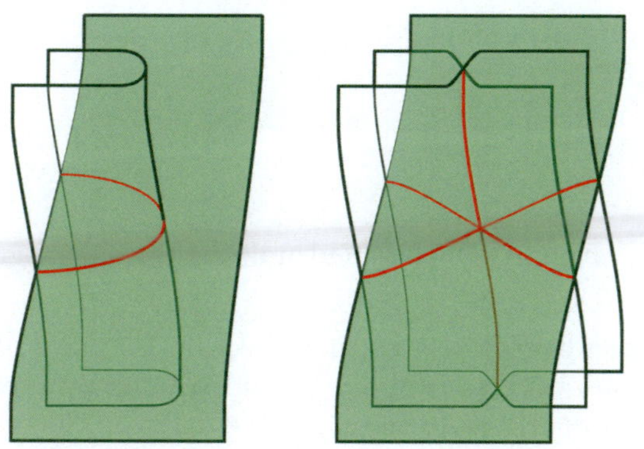

FIGURE 3. Generating figures for the quadruple arrow exchanger

Now suppose that the tangle in the box, T, is either $\begin{bmatrix} \mathsf{X\,I} \\ \mathsf{I\,X} \\ \mathsf{X\,I} \end{bmatrix}$. or $\begin{bmatrix} \cap \\ \mathsf{X} \end{bmatrix}$. These are connected by quadruple arrows

$$\begin{bmatrix} \mathsf{I\,X} \\ \mathsf{X\,I} \\ \mathsf{I\,X} \end{bmatrix} \xleftarrow{\ast} \begin{bmatrix} \mathsf{X\,I} \\ \mathsf{I\,X} \\ \mathsf{X\,I} \end{bmatrix}$$

and

$$\begin{bmatrix} \cap \\ \mathsf{X} \end{bmatrix} \xleftarrow{\curlyvee} [\,\cap\,],$$

respectively. Then naturality of the exchanger with respect to these 4-arrows is illustrated in Fig. 4. Upon reparametrizing, these categorical axioms correspond to the remaining two Roseman moves: respectively, the quadruple point move, and moving a branch point through a transverse sheet.

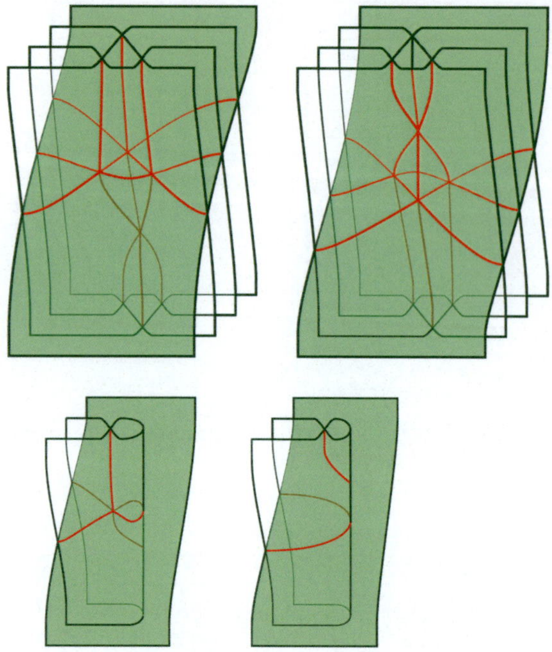

FIGURE 4. Naturality of the quadruple arrow exchanger, part 1

The generating quadruple arrows that describe isotopies of knotted surfaces include the Reidemeister moves of type I, II, and III, the ψ-move, and cusps.
Exercise. Interpret naturality of the quadruple arrow exchanger with respect to the type II move. In the cases of the ψ-move and the cusp, the naturality axioms are drawn in Fig. 5. Naturality with respect to the ψ-move gives the movie move that corresponds to moving a triple point over a fold. Naturality with respect to a cusp can be used to move a double curve through a cusp.

These moves are illustrated as chart moves in Fig.12, page 53 of [**15**]. A glyphographic proof that moving a double arc through a cusp follows is given in Fig. 6.

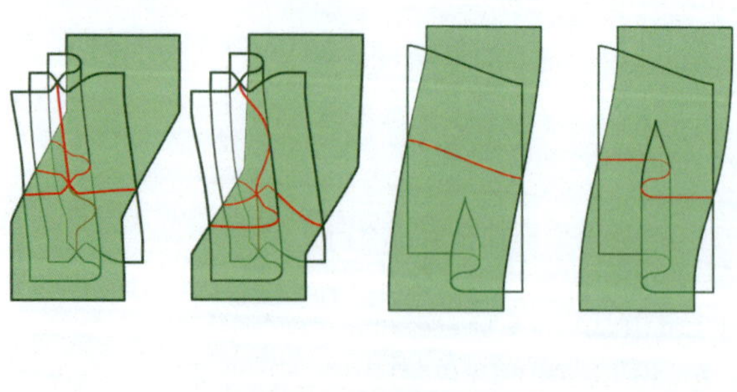

FIGURE 5. Naturality of the quadruple arrow exchanger, part 2

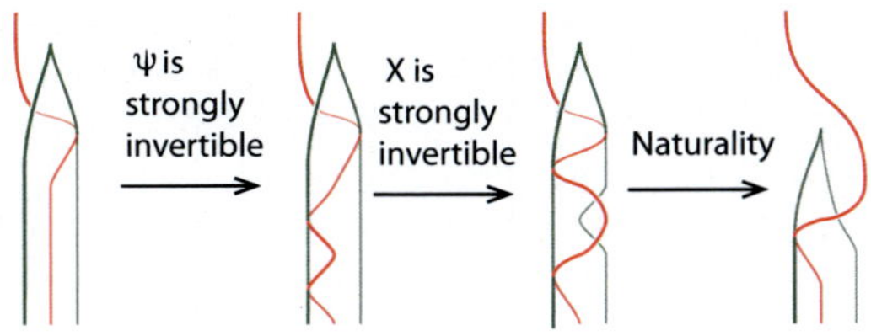

FIGURE 6. Pulling a double arc through a cusp via naturality of the exchanger

1.3. Naturality of exchangers continued. There are other forms of the quadruple arrow exchanger. Let triple arrows F and G be chosen among X, $\overline{\mathsf{X}}$, ⋂, and U. Then consider the quadruple arrow

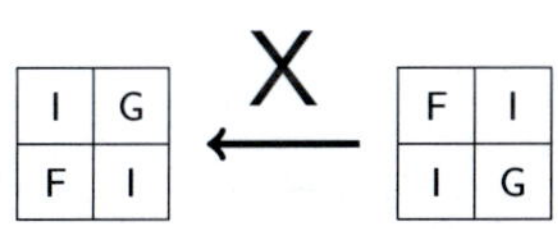

which is analogous to that introduced in Chapter 3, Sections 2.3 and 2.4, and in particular 1.2 where it was introduced to mitigate horizontal composition. The family of naturality 5-isomorphisms that are schematized as indicated below

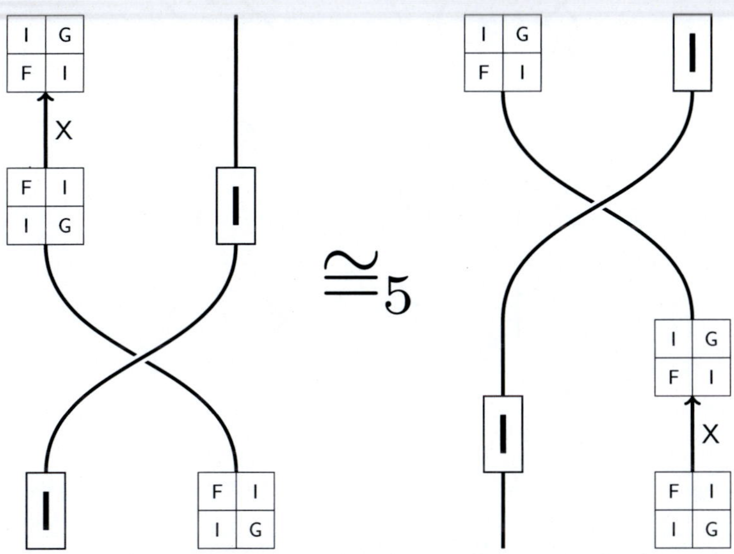

are illustrated in Fig. 7. Therein the cylinders represent the identities upon F and G.

These naturality conditions follow from analogues of the naturality that was illustrated immediately preceding Chapter 6, Section 4.7. Those conditions are among the items illustrated in Fig. 8 in this chapter which are facsimilies of the chart moves contained in Fig. 13, page 53 of [**15**]. Here a thick black line indicates the identity upon X, $\overline{\mathsf{X}}$, ⋂, or U. Serifs on glyphs are omitted, and representative

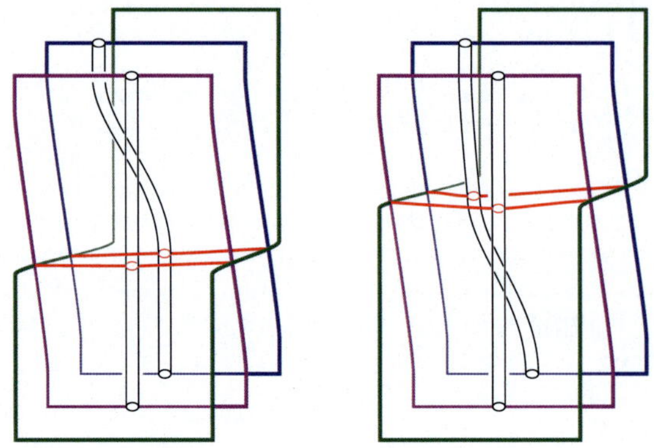

FIGURE 7. More naturality conditions

glyph choices or crossings are indicated. That is to say in these and subsequent figures, not all variations on a given move will be specified.

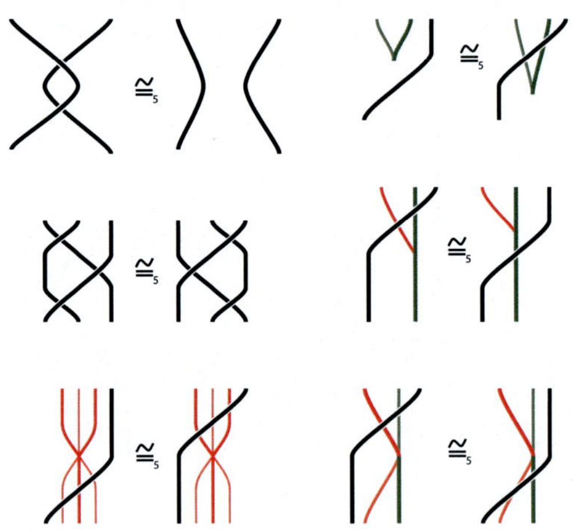

FIGURE 8. Conditions satisfied by exchangers

1.4. The remaining quintuple isomorphisms. Here is a short review of the 5-isomorphisms given from the previous sections.

(1) Critical cancelations, lips, beak-to-beak, swallow-tail, and horizontal cusps are among the 5-isomorphisms because $\cup$ and $\cap$ are weakly invertible triple arrows.

(2) Five of the Roseman moves (Fig. 1, this chapter) are expressions that the type I, II, and III quadruple arrows are strongly invertible.
(3) A zig-zag or critical cancelation occurs along a double arc because of strong invertibility of type II moves.
(4) The ψ-moves are also strongly invertible. So there are 5-isomorphisms that indicate they can be cancelled.
(5) Naturality of a quadruple arrow exchanger gives the quadruple point move, pushing a branch point through a sheet, pushing a type III move over a fold, and pushing a double arc through a cusp (Fig. 1, this chapter and compare with Fig. 13 of [**15**]).
(6) Naturalilty of another exchanger allows for the moves depicted in Fig. 8 in this chapter.

There is some sense in which a type I move is an analogue of a unit or co-unit. Thus there is a method of turning a type I move over in a candy-cane or fish hook fashion. By pushing the branch point through a cusp, the cusp reverses direction. And a branch point can be pushed over an optimal point of the fold set (be it a saddle, fork, birth, or death). In these processes the branch point changes from lying on the left of the optimal point on a fold line to lying on the right. These moves are illustrated in this chapter, Fig. 9.

The bottom two moves that are illustrated in Fig. 9, the top two moves and the bottom left move in Fig. 10 share some properties that we explore. (Cf. the discussion in Chapter 7, Section 3.3 on fish hooks.) First, these have been expressed in terms of saddles or births, but the horizontal reflections that convert saddles to forks and births to deaths produce corresponding moves that also hold. Second, in the presence of birth/fork and saddle/death critical cancelations moving a type I move or a ψ-move past a saddle follows from the ability to move it past a death. The idea of the proof of that claim is contained within Ch. 2, Fig. 4. Suppose, for example, that a type I glyph lies to the left of a saddle point on the fold set. To the left of the type I move, insert a saddle/death canceling pair of critical points. Then the type I move slides to its left over the death, and the death/saddle pair on the right cancels. Third, the collection of fish hook moves in which a glyph slides over an optimal point on a fold are explored in general.

The algebraic analogue is found within the context of adjoint operators in linear algebra. Let us digress. Suppose that $\mathbb{C} \stackrel{\langle \cdot,\cdot \rangle}{\longleftarrow} V \times V$ is a hermitian inner product[4]. Let $V \stackrel{A}{\longleftarrow} V$ denote a linear operator. Then the *adjoint* of A, A^*, is defined via the relationship $\langle A\vec{x}, \vec{y} \rangle = \langle \vec{x}, A^*\vec{y} \rangle$, and in matrix representations, the adjoint is the conjugate transpose of the representative matrix for A.

Now this relationship is schematized in a diagrammatic fashion. Weak inverse k-arrows, t, and $\mathsf{ɟ}$ are paired. A $(k+1)$-arrow G is composed with the $(k+1)$-arrow $\mathsf{F(t)}$ that pairs t and $\mathsf{ɟ}$. The source of G is a weakly invertible k-arrow t', and it is paired with $\mathsf{ɟ}'$. Then the weak or strong inverse $\mathsf{Ↄ}$ is composed on the right.

[4] linear in the first variable and conjugate linear in the second.

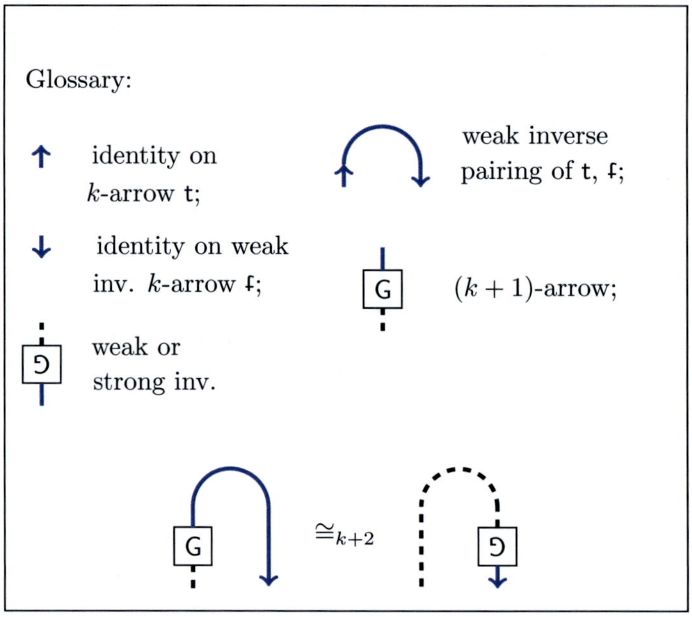

For example, the ψ-move is of this form. In Section 2 (this chapter), the weak inverse pair $\{\mathsf{t}, \text{ʇ}\}$ often is the pair $\{\mathsf{Y}, \curlywedge\}$ and the pairing is one of the zippers. We call these 5-isomorphisms "sliding a 4-arrow over an optimum," and also refer to it as an *adjoint relationship.*

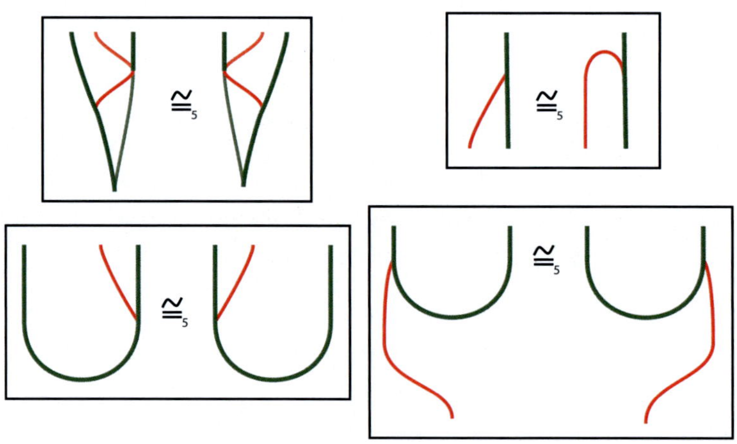

FIGURE 9. 5-isomorphisms that involve type I moves

In the same way, ψ-moves can also move to the left or right of optima on the fold set (birth, death, saddle, or fork). A type III move can be passed over a type II move as indicated in the top right of Fig. 10 which also indicates the remaining moves including the horizontal type II move on the lower right. This 5-isomorphism has a curious appearance that may be related to naturality. We do not now know how to interpret this.

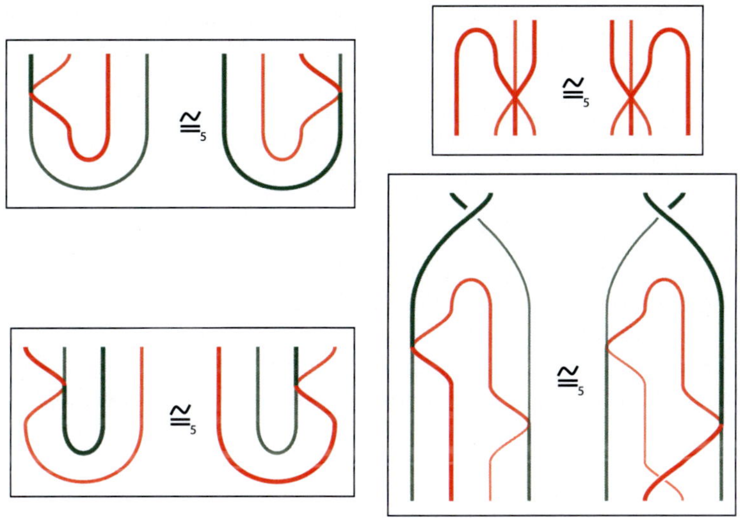

FIGURE 10. Remaining 5-isomorphisms

Extended Exercise. Determine if the horizontal type II move that is located in the lower right corner of Fig. 10 follows from naturality of the exchanger. Determine explicitly which moves are consequences of the other and thereby find a generating set of 5-isomorphisms. Formulate all these moves in categorical terms. See [**46**] and [**7**].

2. Foams in 4-space

Classical knots and links are to knotted trivalent graphs as knotted surfaces are to knotted foams in 4-dimensional space. In this section, the multi-category $\mathcal{T}$ that represents embedded surfaces in 4-dimensional space is extended to the multi-category $\mathcal{F}$ that represents embedded foams in 4-space. First, the double arrows $\curlywedge$ and Y become triple arrows

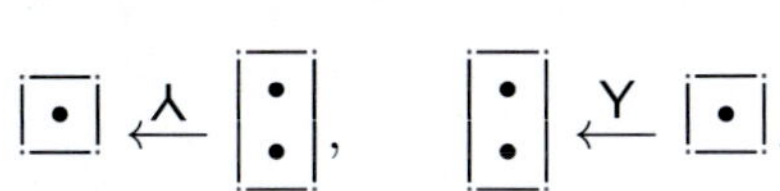

Next include a plethora of new quadruple arrows among the birth, death, saddle, fork, cusp, type I, type II, type III, and ψ-move quadruple arrows. These will be compiled from various places in the text above. For example, the quadruple arrows that were used to define equivalences between diagrams of knotted trivalent graphs in Chapter 9, Section 7 and those from Chapter 9, Section 7.3 are included in the following list.

(1) Strongly invertible associative pairing and co-associative co-pairings (Chapter 4, Table 9) with the obvious strong inverse (doing and undoing is the same as not doing) are added.

(2) The strongly invertible arrows (see Ch. 4, Fig. 13) defining multiplication and co-multiplication in terms of each other (Tables 6 and 7, also in Chapter 4) are considered to be quadruple arrows. These are also described as a fold ending upon a seam.
(3) The zippers (Ch. 4, Table 12) are weakly invertible quadruple arrows. So doing and undoing is not the same as not doing, but there are cuspidal 5-isomorphisms that hold between skew compositions of them.
(4) Naturality of X and $\overline{\mathsf{X}}$ with respect to Y and $\curlywedge$ that are represented by the IY and YI moves of Ch. 9, Fig. 13 and Fig. 14 are strongly invertible quadruple arrows.
(5) The twist move (T-move) that is illustrated in Ch. 9, Fig. 15 is a strongly invertible quadruple arrow.
(6) The associators and co-associators that are depicted in Table 8 (Chapter 4) are among the quadruple arrows. Their degrees of invertibility, weak or strong, will depend upon context.

2.1. Generalities. 1. In settings such as these, authors are allowed to declare the differences that might be identified. So while different things are not equal, sometimes $1 + 1$ really is 2. The main examples that we may consider are the associativity/co-associativity 4-arrows and their possible inverses:  , . In examining foams or embedded foams the 5-arrow

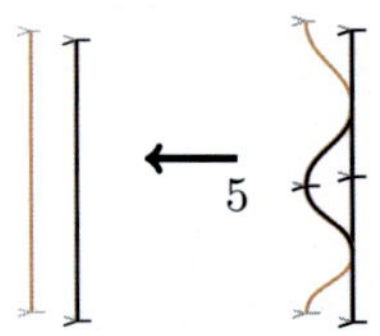

is, at best, a weak isomorphism since on the right there are two vertices and on the left there are none. But some algebraic invariants that will be defined are insensitive to the pair of vertices on the right: one will be measured to be positive, and the other will be measured to be negative. That is to say, indeed, the arrow $a(bc) \leftarrow (ab)c$ is directed, and in the context that we measure, the doing/undoing process $(ab)c \leftarrow a(bc) \leftarrow (ab)c$ is indistinguishable from the identity. The measurement tool is at fault.

In addition, the pentagon relation as expressed as the $(3, 2)$-relation of Chapter 8, Section 2.1 can be used to define a 5-arrow that we may declare to be, or not to be[5], a 5-isomorphism. For example, as in Chapter 9, Section 7.3, the intrinsic topological type of a foam or a regular neighborhood of an embedding thereof may be an object of study. The $(3, 2)$-move preserves a regular neighborhood of the underlying foam.

2. The $(3, 2)$-move is one of eight quintuple arrows that include the quadruple point move. The family of these moves will assist in the definition of a homology

[5]Apologies to W.S.

theory that encompasses homology of associative structures and homology of self-distributive structures. As the section progresses, it will return to the idea of abstract tensors and associate to these eight relationships a system of abstract tensor equations to which the homology theory will provide a family of solutions.

3. Twist vertices are peculiar. Some 5-isomorphisms will be presented that depends upon different ways of twisting a trivalent vertex. However, we do not claim to have a complete set of moves that describes all embedded foam isotopies. Instead, we have a categorical foundation in which additional 5-isomorphisms can be deposited as they become apparent. That situation is unfortunate and is contrast to a claim made by the first named author in [**12**].

In the current context, embedded foams are examined up to an equivalence relation that is generated by a given set of 5-isomorphisms. This context is sufficiently interesting since it is very close to the topological context. But more work needs to be done: either from the singularity point of view, or from the categorical. The algebraic arena was left behind several chapters ago, but it would not be surprising if other algebraic considerations reveal more about the topological.

4. The vertical symmetry between associators and co-associators and the horizontal symmetry between Y and ⅄ are mitigated by the quadruple arrow in which a fold ends along a seam as in Tables 6 and 7 (Chapter 4). The quintuple arrows listed below will be given only in terms of Y and co-associators , or . Perhaps the corresponding relations for associators or for items that involve ⅄ can be obtained by folding the diagrams over by using the given collection of 5-isomorphisms. Instead, the reader is requested to list the corresponding variations.

When shalgebra and qualgebra structures are discussed in Section 3 (this chapter), Y will depict multiplication and ⅄ will depict co-multiplication.

2.2. More quintuple arrows. 1. Knotted Surfaces. All of the 5-isomorphisms that were used to describe isotopies of knotted surfaces are among the 5-isomorphisms for knotted foams. These include the invertibility of the Reidemeister moves and the ψ-move. The moves that involve only folds such as birth/fork and death/saddle critical cancelations, the lips, beak-to-beak, swallow-tail, and horizontal cusp moves from Chapter 6, Section 4 are included. The adjoint (fish hook) 5-isomorphisms in which a type I 4-arrow or a ψ-move 4-arrow are slid over optima are included and also discussed below. Naturality conditions for the quadruple arrow exchangers (Section 1.2, above in this chapter) are included. See also the Figs. 1, 9, and 10 in this chapter.

2. Invertibilities. The YI and IY quadruple arrows, 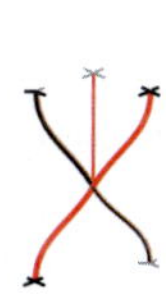 and 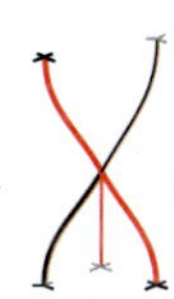, are strongly invertible. Thus there are the 5-isomorphisms:

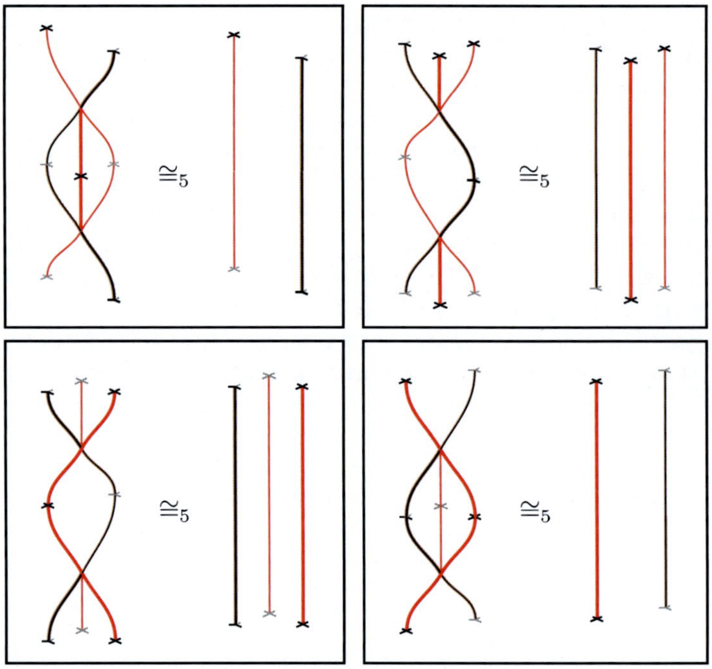

The twist quadruple arrows are strongly invertible. Therefore there are, for example, the following two (of eight) 5-isomorphisms. The remaining 5-isomorphisms are obtained by reflecting in the vertical axis, and changing the serifs from X to $\overline{\mathsf{X}}$.

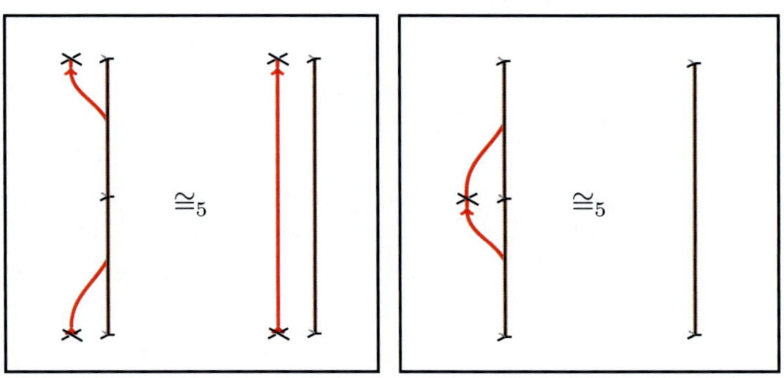

The quadruple arrows in which a fold ends upon a seam (Ch. 4, Table 7) or a seam passes over a fold (Ch. 4, Table 9) are strongly invertible. See Fig. 13 and Fig. 15, respectively (also in Chapter 9).

The associativity and co-associativity quadruple arrow are *assumed to be* strongly invertible. See Ch. 4, Fig. 18. Here, only the 5-isomorphisms for associativity are indicated. Those for co-associativity are obtained by reflecting these through a vertical axis.

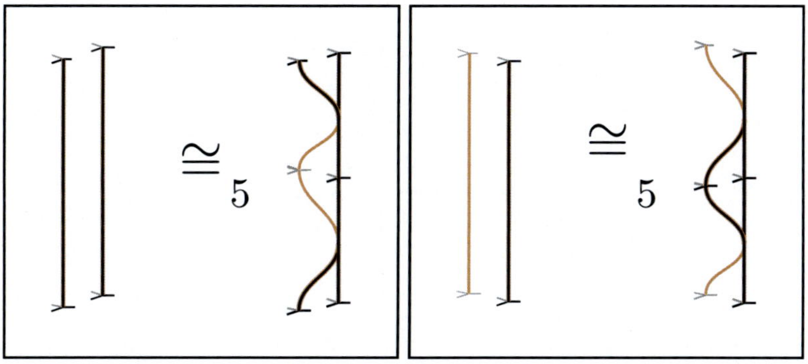

The zippers (Chapter 4, Table 12) are weakly invertible. See Chapter 7, Section 1.2 for the interpretations. The saddle/fork and birth/death arrows describe quintuple arrows, and there are cuspidal 5-isomorphisms that hold. However, quantities that are derived from shalgebra homology, whose definition is outlined below, are invariant under the doing/undoing compositions (births/deaths and saddles/forks for zippers).

---

3. Sliding over optima. There are 5-isomorphisms of the form below.

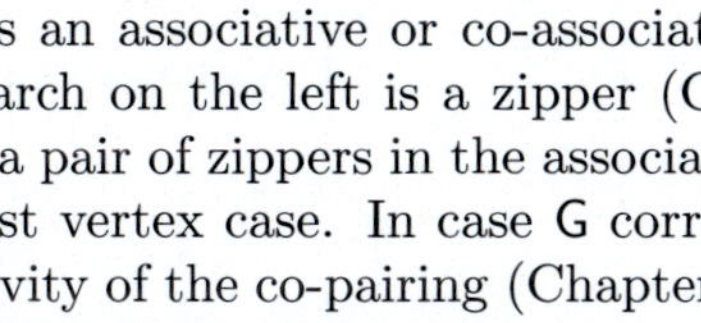

In the cases that G is an associative or co-associative 4-arrow, or that G is a twist vertex, then the arch on the left is a zipper (Chapter 4, Table 12). The dotted arch is doubled as a pair of zippers in the associativity case, or a zipper and a double curve in the twist vertex case. In case G corresponds to associativity of the pairing or co-associativity of the co-pairing (Chapter 4, Table 9), then the arch on the left is a saddle, fork, birth, or death. The horizontal reflection of this move holds as well.

4. Relations that involve twist vertices. Twist quadruple arrows, T, e.g. 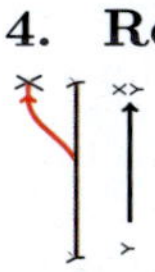 present peculiar problems with respect to determining 5-isomorphisms. In particular, their interactions with the associativity, YI, and IY moves make for interesting contemplation. The first family of 5-isomorphisms is exemplified by the figure given below.

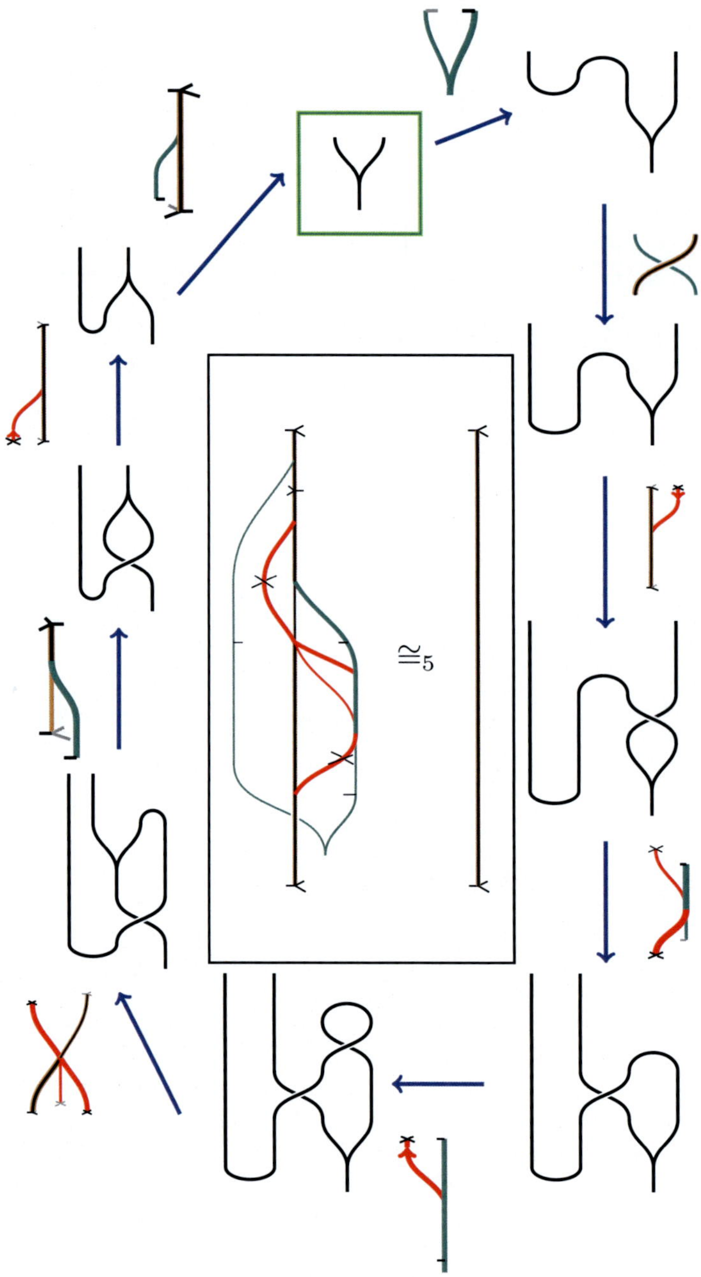

Herein two branches of a Y are twisted from the top going down. Then through a series of other strongly invertible 4-arrows, a twist is affected between another two branches. For lack of space the identity upon Y has been omitted in the movie description, and instead $\boxed{\mathsf{Y}}$ serves to indicate this identity.

The situation is analogous to the following. If the labels $(1, 2, 3)$ are cyclically ordered on a unit circle, a reflection between any pair, $(1, 2)$, $(2, 3)$, or $(1, 3)$ reverses the cyclic order. The 5-isomorphism illustrated above uses two adjacent twists to affect no twists.

Exercise. Determine the intermediate sources and targets and design glyphographic manifestations of the alternate cases.

In the second 5-isomorphism that involves twist vertices, the intermediate stages are described as compositions of X, $\overline{\mathsf{X}}$, Y, ⅄, and I. A twist vertex is reversed by altering between an associativity and a co-associativity.

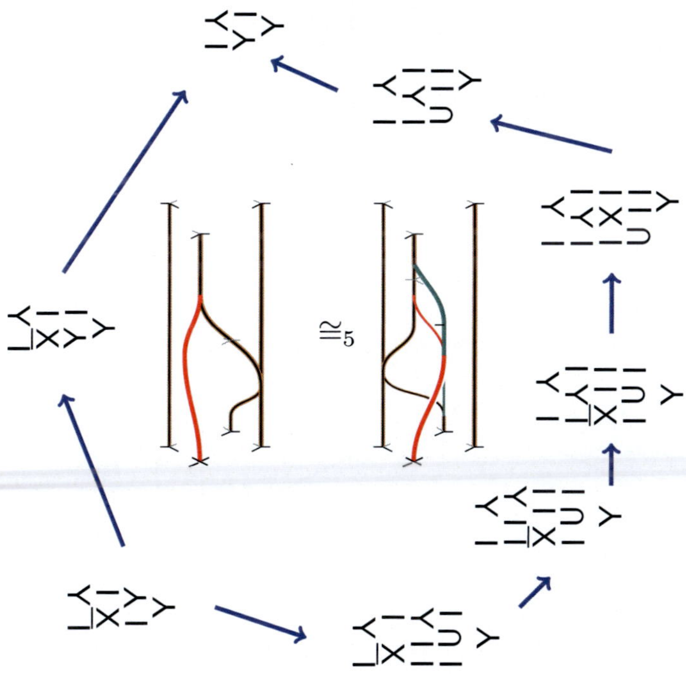

The depictions of the triple arrows in these two illustrations are given in differing styles to point out that sometimes typography is sufficient even though the glyphography is more suggestive.

Both of these quintuple arrows were proposed to the authors in private conversations and correspondences with Masahico Saito. We are grateful for his calling attention to them.

5. The eight most important 5-isomorphisms. When these moves were originally proposed to us by Masahico Saito they were formulated in terms of Y rather than in terms of ⅄. That is how they appear in our mind's eyes. Both of the authors are experienced in mentally transforming figures. So we could have written everything in terms of ⅄. For example, we could have drawn a figure upon a piece of paper, rotated the drawing 180 degrees, and reproduced that drawing in the software. Alternatively, we could draw them in the software as we imagine them, and then ask the software to perform the rotation. Yet the aesthetics in the transformed world seem too foreign, and we would, in writing, tend to slip up in our description confusing left/right and up/down. So to begin, we transform the illustrations from Chapter 8, Section 3.1 into the "Y-up" configuration that feels more comfortable to us.

(1234):

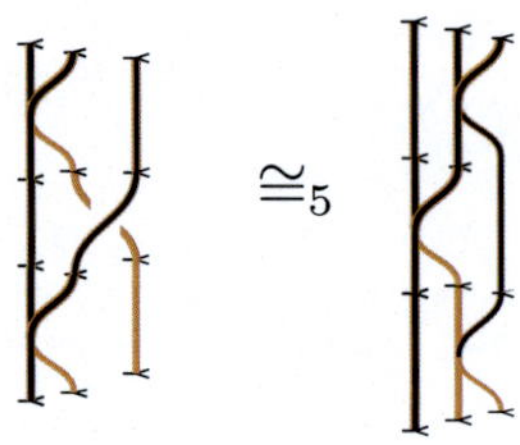

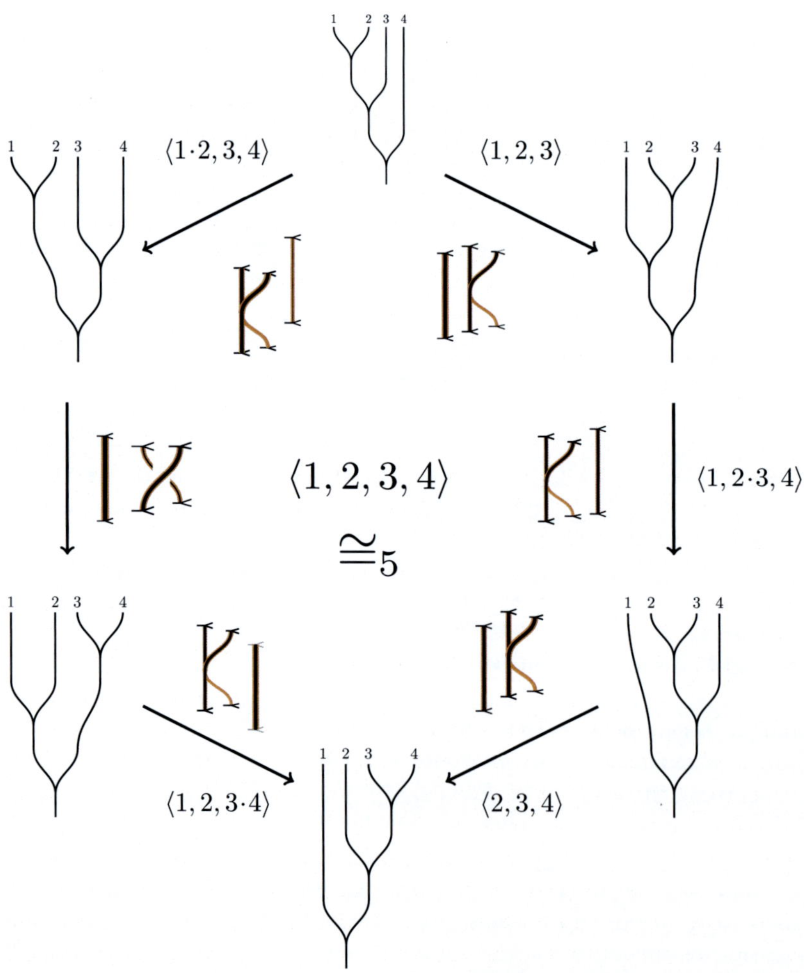

The numeric indices that adorn the trees and the 4-arrows will be explained in Section 4 in this chapter.

This 5-isomorphism is the first of the eight to be exhibited. They are indexed by partitions of the set $(1, 2, 3, 4)$. The Stasheff pentagon (hexagon here since an exchanger is included) corresponds to the unpartitioned quantity 1234. Here the numerals $1, 2, 3, 4$ are simply indicators, and commas, are not included as separators. The remaining seven partitions are 123|4, 12|34, 12|3|4 1|234, 1|23|4, 1|2|34, and 1|2|3|4. Subsequent to the enumeration of these 5-isomorphisms, a further explanation of their *raison d'être* will be given.

(123|4)

The naturality of the quadruple arrow exchanger with respect to the associator is depicted as the 5-isomorphism that is indicated below. The 5-isomorphism is presented glyphographically and as a commutative diagram. Such a diagram is, as above, a method of collapsing the arrows. In that collapse, the things thought of as triple arrows become objects, quadruple arrows become arrows, and the 5-isomorphism is presented as commutative diagram. Within it, the identity arrows that flank the given quadruple arrows are omitted. They are, as they must be, included within the glyphographic representation. Again, the numerical labels will be explained in Section 4 later in this chapter.

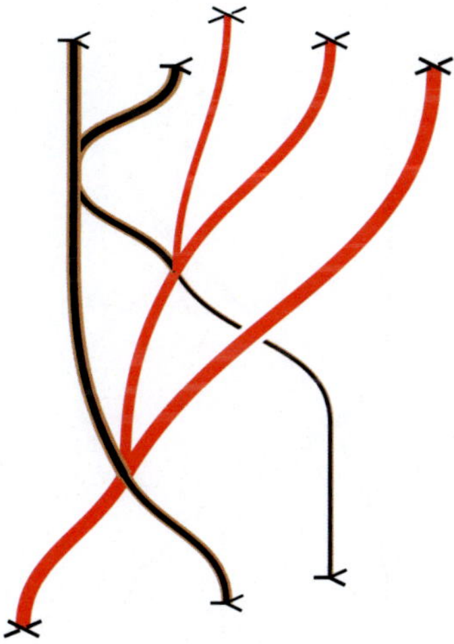

$\cong_5$

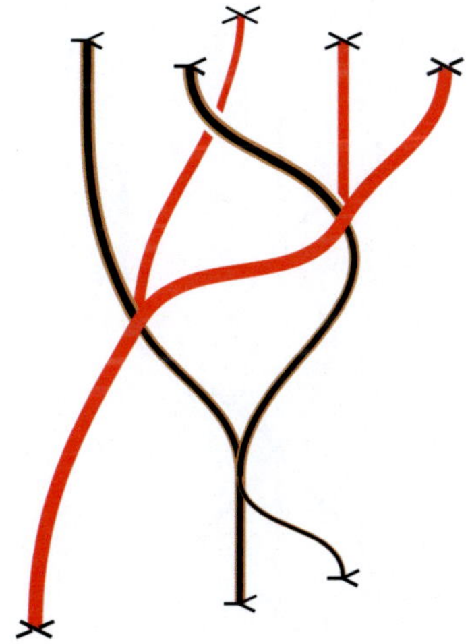

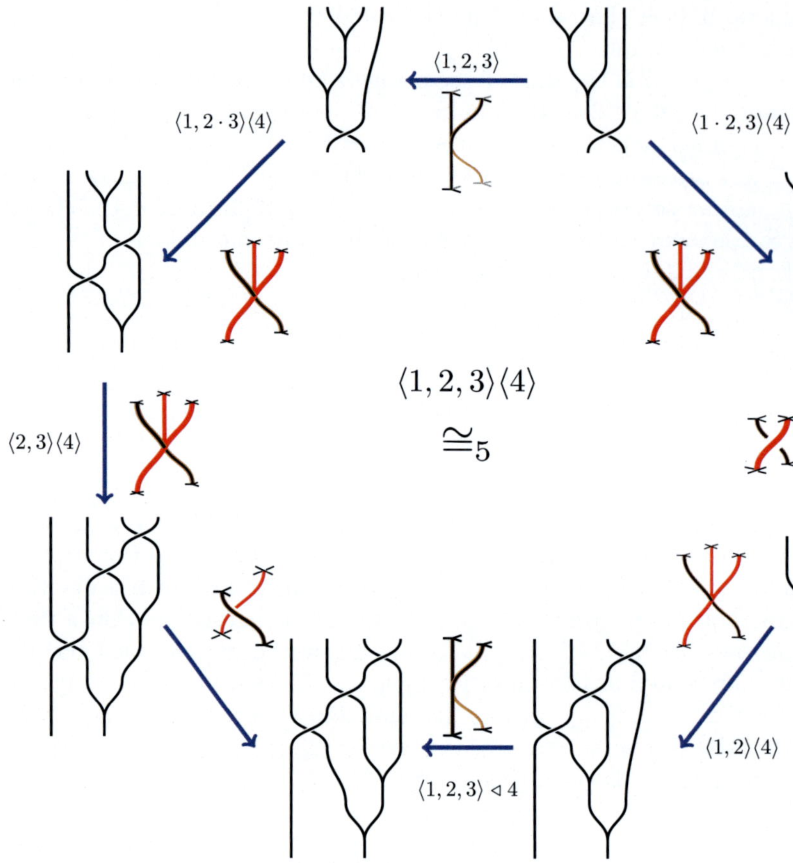
$\langle 1,2,3\rangle$
$\langle 1,2\cdot 3\rangle\langle 4\rangle$
$\langle 1\cdot 2,3\rangle\langle 4\rangle$
$\langle 1,2,3\rangle\langle 4\rangle$
$\cong_5$
$\langle 2,3\rangle\langle 4\rangle$
$\langle 1,2\rangle\langle 4\rangle$
$\langle 1,2,3\rangle \triangleleft 4$

(12|34):

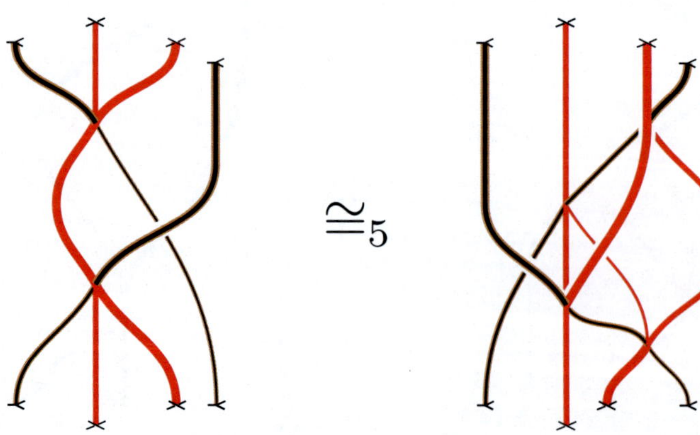
$\cong_5$

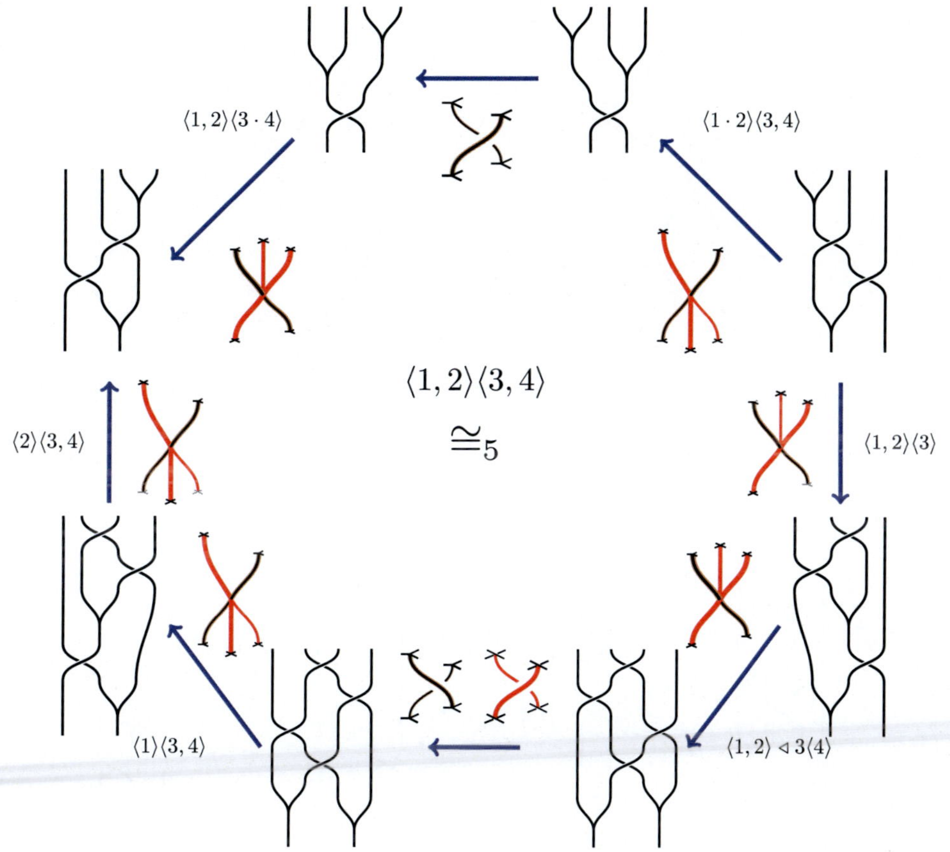

The 5-isomorphism that corresponds to the partition 12|34 is presented above. At the top of the diagram, the compositions $\begin{matrix}\mathsf{Y}\,|\,| \\ |\,\mathsf{Y} \\ \mathsf{X}\end{matrix}$ appears. That is, the vertices at the top have been partitioned into two sets that correspond to $(1,2)$ and $(3,4)$ and the set on right, after having been fused by Y, braids over the one on the left.

(12|3|4):

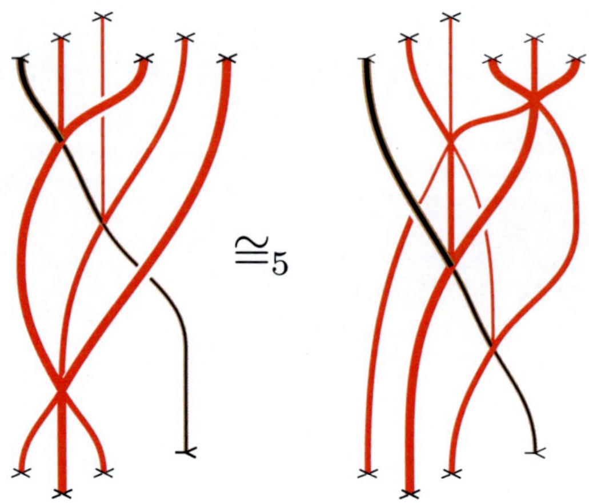
$\cong_5$

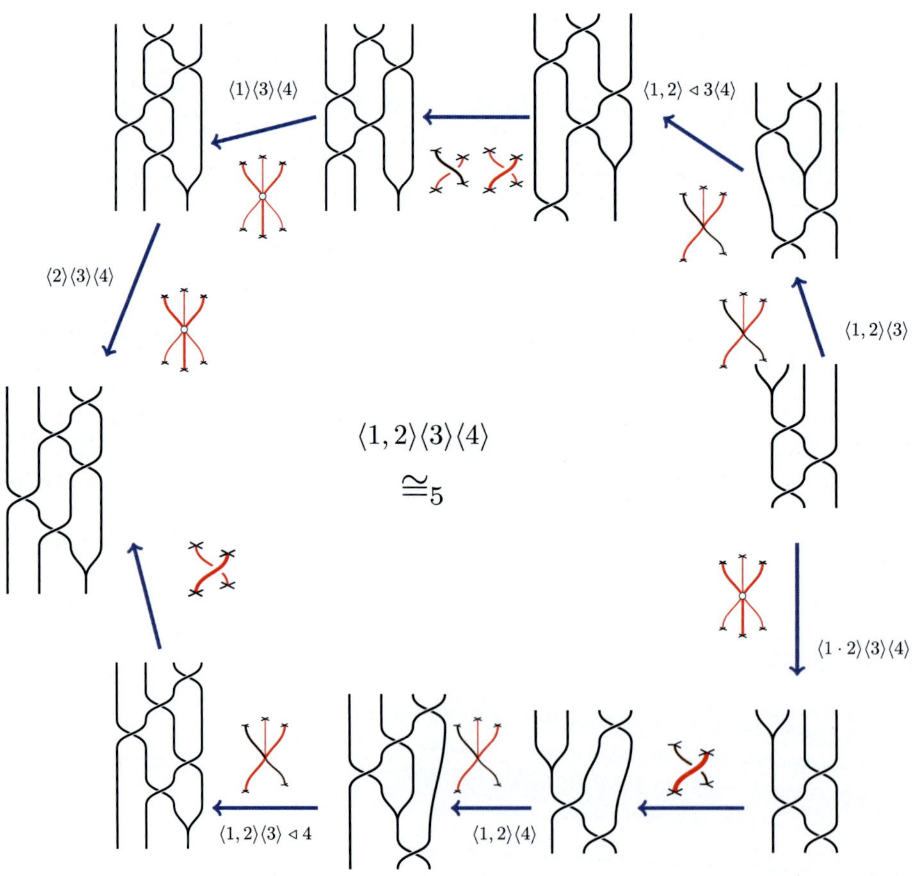
$\langle 1\rangle\langle 3\rangle\langle 4\rangle$
$\langle 1,2\rangle \triangleleft 3\langle 4\rangle$
$\langle 2\rangle\langle 3\rangle\langle 4\rangle$
$\langle 1,2\rangle\langle 3\rangle$
$\langle 1,2\rangle\langle 3\rangle\langle 4\rangle$
$\cong_5$
$\langle 1\cdot 2\rangle\langle 3\rangle\langle 4\rangle$
$\langle 1,2\rangle\langle 3\rangle \triangleleft 4$
$\langle 1,2\rangle\langle 4\rangle$

(1|234):

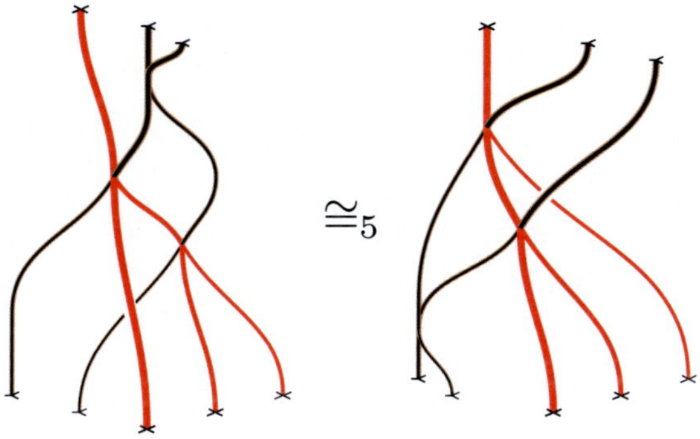

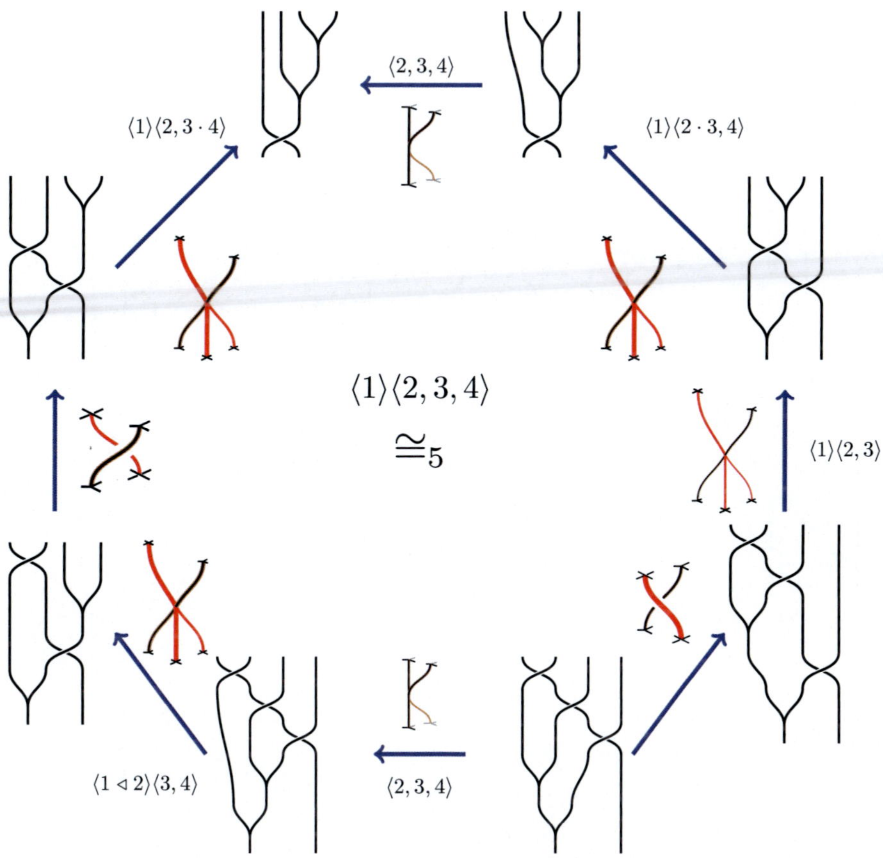

(1|23|4):

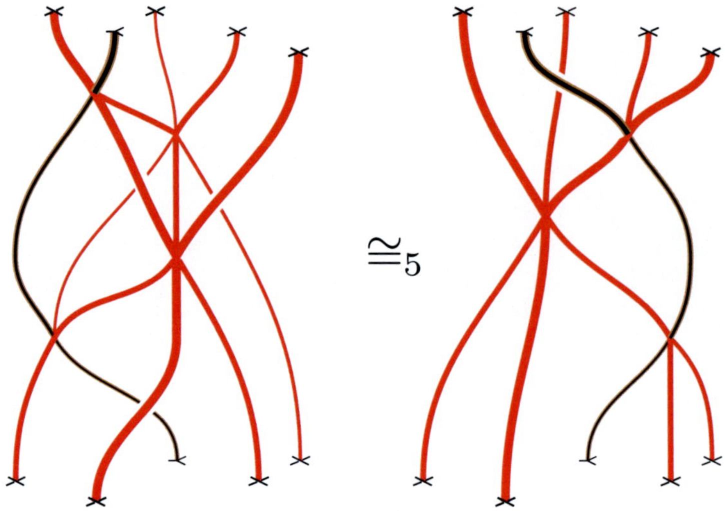

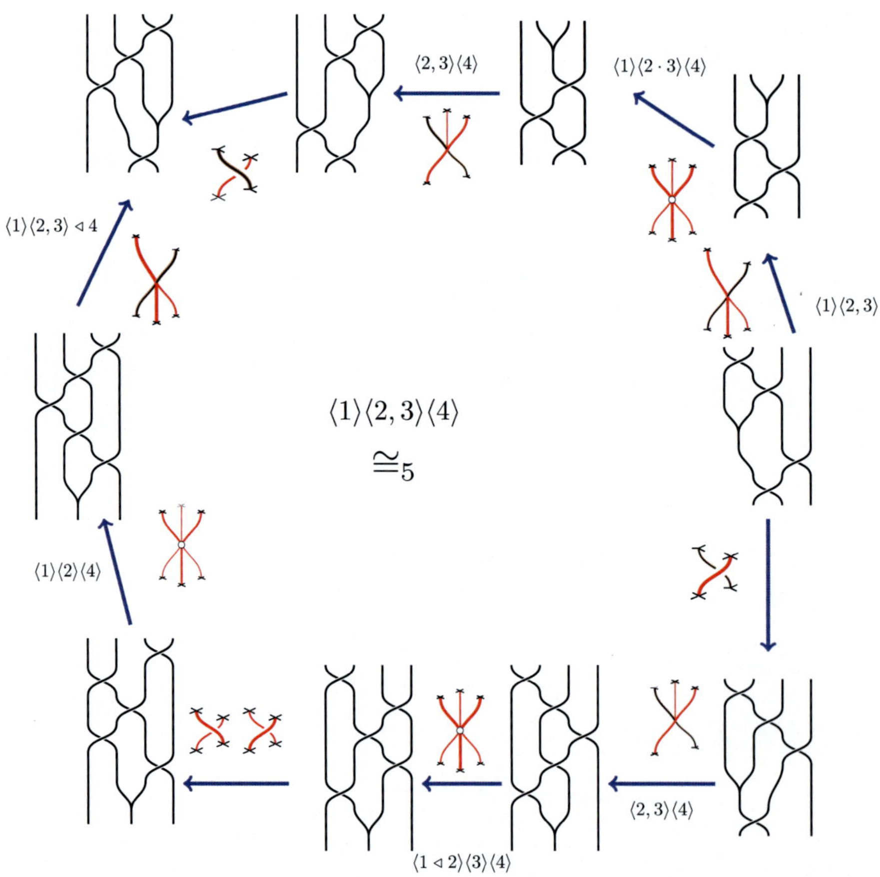

$(1|2|34):$

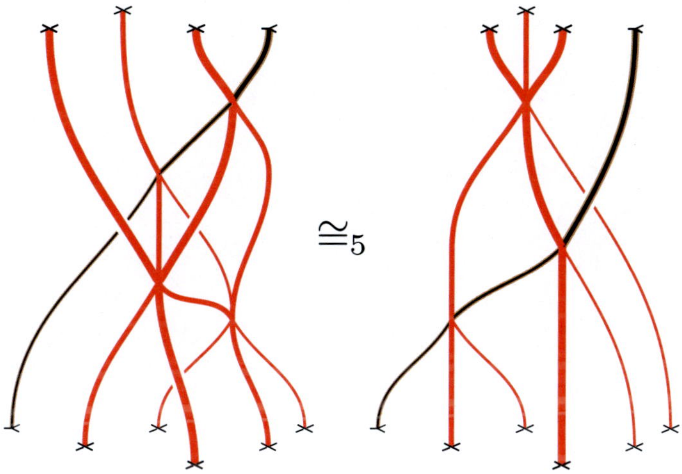

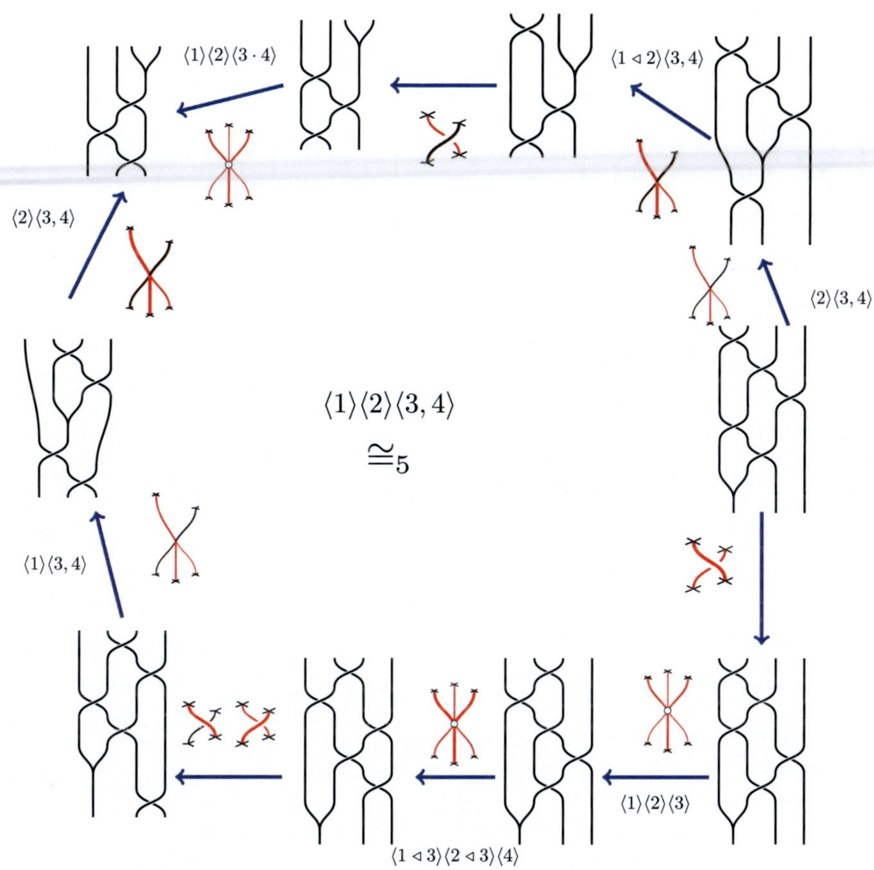

(1|2|3|4):

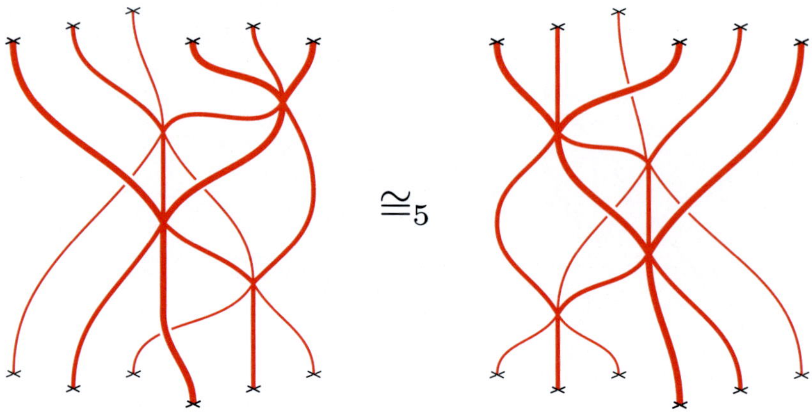

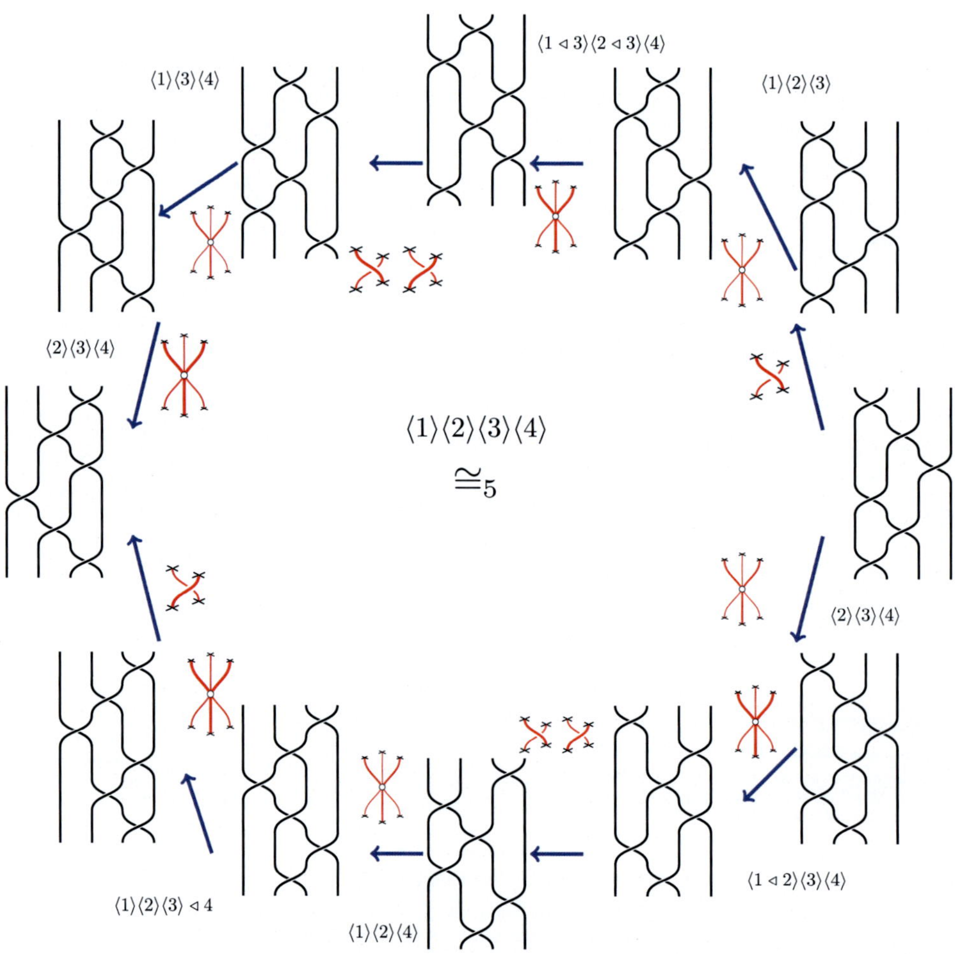

This concludes the catalogue of 5-isomorphisms. The structural aspects of shalgebras and qualgebras (Section 3 in this chapter) provide a contextual framework for these 5-isomorphisms.

3. Shalgebras and qualgebras

In this section, diagrams and relations between diagrams inform a more traditional algebraic structure. The focus is upon relations among Y, X, and $\overline{\mathsf{X}}$. Each font represents a vertex with incident edges that are oriented downward. The end points of each are labeled with elements in a set Q that has two binary operations $(\cdot, \triangleleft)$ defined upon it. Write ${}^{a}\mathsf{Y}_{c}^{b}$ where $c = a \cdot b$. As is customary, $a \cdot b$ will be abbreviated to ab. Write, ${}^{a}_{b}\mathsf{X}^{b}_{d}$ with $d = a \triangleleft b$.

The axioms that these operations will satisfy are informed by the four generating 4-arrows. (H) corresponds to associativity[6], (YI) corresponds to distributivity of $\triangleleft$ over $\cdot$, (YI) corresponds to an exponential rule, and (Ж) corresponds to self-distributivity of $\triangleleft$. The operations $\cdot$ and $\triangleleft$ satisfy the relations:

H: $(ab)c = a(bc)$;
YI: ; $(ab) \triangleleft c = (a \triangleleft c)(b \triangleleft c)$;
IY: $(a \triangleleft b) \triangleleft c = a \triangleleft (bc)$;
Ж: $(a \triangleleft b) \triangleleft c = (a \triangleleft c) \triangleleft (b \triangleleft c)$

that hold for all values $a, b, c \in Q$. Compare these axioms to the labels at the bottom of the trees in the illustrations below.

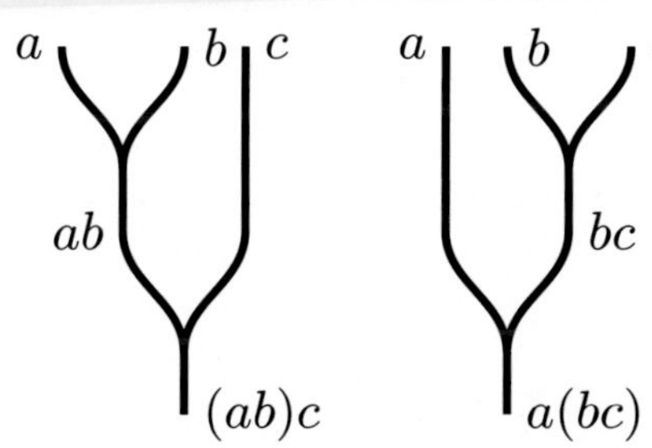

$\langle a, b, c\rangle$

Axiom H:
$(ab)c = a(bc)$

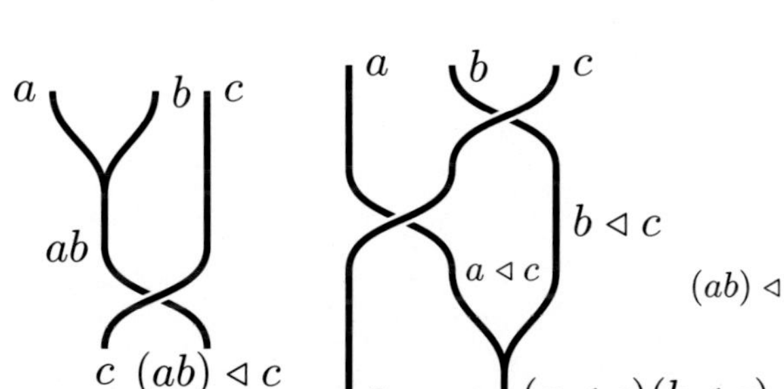

$\langle a, b\rangle\langle c\rangle$

Axiom YI:
$(ab) \triangleleft c = (a \triangleleft c)(b \triangleleft c)$

[6]Since arcs are oriented downward, the relation is read as an associativity axiom rather than a co-associativity axiom.

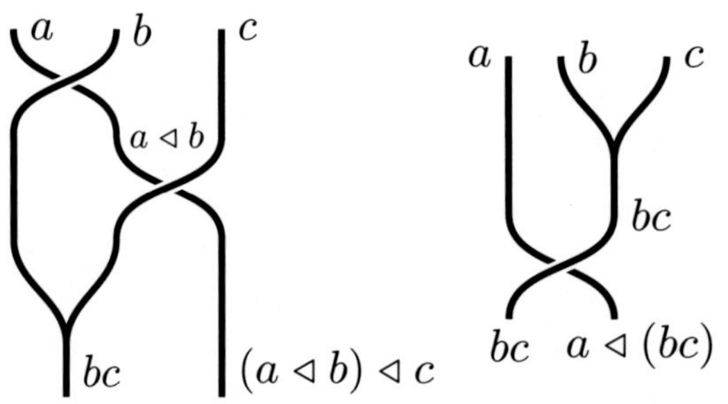

$\langle a \rangle \langle b, c \rangle$

Axiom IY:
$(a \lhd b) \lhd c = a \lhd (bc)$

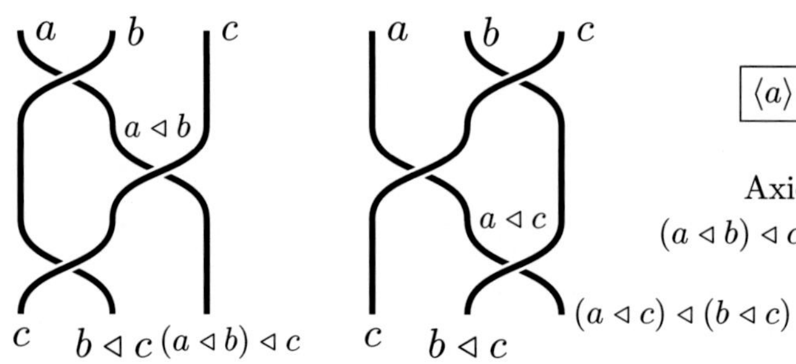

$\langle a \rangle \langle b \rangle \langle c \rangle$

Axiom Ӿ:
$(a \lhd b) \lhd c = (a \lhd c) \lhd (b \lhd c)$

The text will further explicate the items within the framed boxes, e.g. $\langle a, b \rangle \langle c \rangle$ shortly. The font H that indicates associativity is chosen (as suggested to us by Victoria Lebed) because the associativity move is sometimes called the IH move: upon a particular projection, a serifed I transforms into an un-serifed H.

A *shalgebra* is a triple $(Q, \cdot, \lhd)$ that satisfies H, IY, YI, and Ӿ. A *qualgebra* is a shalgebra that also satisfies the axioms I, II, and T that are listed below.

I: $a \lhd a = a$;
II: for all a, b there is a unique c such that $c \lhd b = a$;
T: $b(a \lhd b) = ab$.

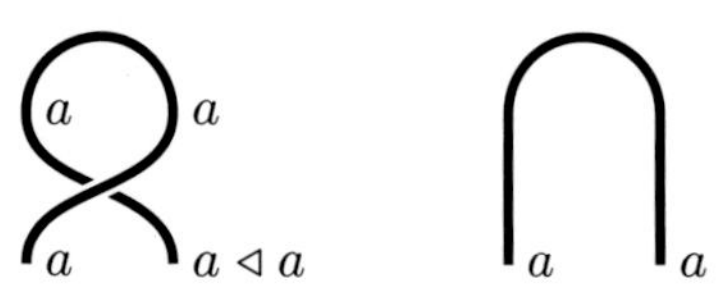

Axiom I:
$a \lhd a = a$

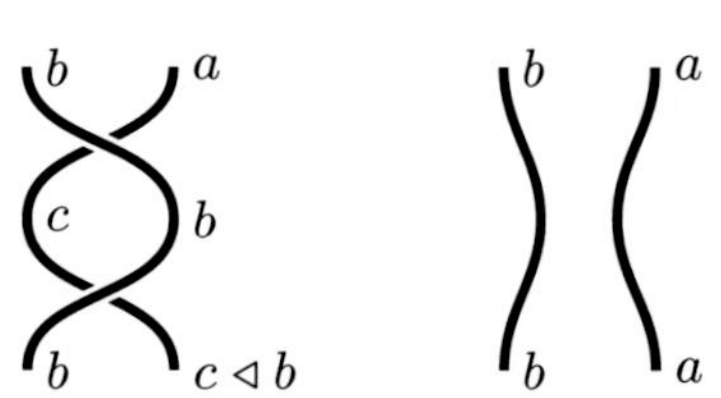

Axiom II:
$\forall a, b, \exists! c : c \lhd b = a$

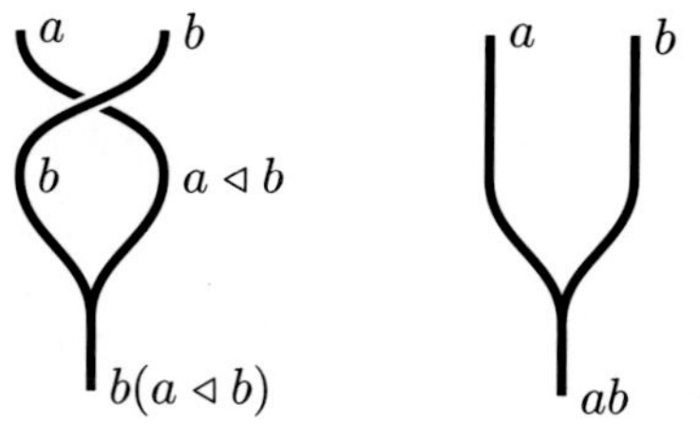

Axiom T:
$ab = b \cdot (a \triangleleft b)$

For example, if Q is a group, $\cdot$ is the group product and $a \triangleleft b = b^{-1}ab$ is conjugation, then $(Q, \cdot, \triangleleft)$ is a qualgebra.

The eight important 5-isomorphisms are placed into a different geometric context. Recall (Chapter 8, Section 2.1), that the *n-dimensional simplex* is the set

$$\Delta^n = \{(x_1, \ldots, x_{n+1}) \in \mathbb{R}^{n+1} : \sum_{j=1}^{n+1} x_j = 1 \ \& \ 0 \leq x_j \text{ for } j = 1, \ldots, n+1\}.$$

Thus, the standard 2-dimensional simplex is an equilateral triangle Δ^2, and the standard 1-dimensional simplex is a line segment Δ^1. A planar embedding of Y is dual to a 2 simplex Δ^2 and X is dual to a square . An *abstract* 2-*chain* $\langle a, b\rangle$ is associated to a Q labeled trivalent vertex Y, and the chain $\langle a\rangle\langle b\rangle$ is associated to a Q labeled X as indicated below. The chains are drawn within the dual structures: triangle or square.

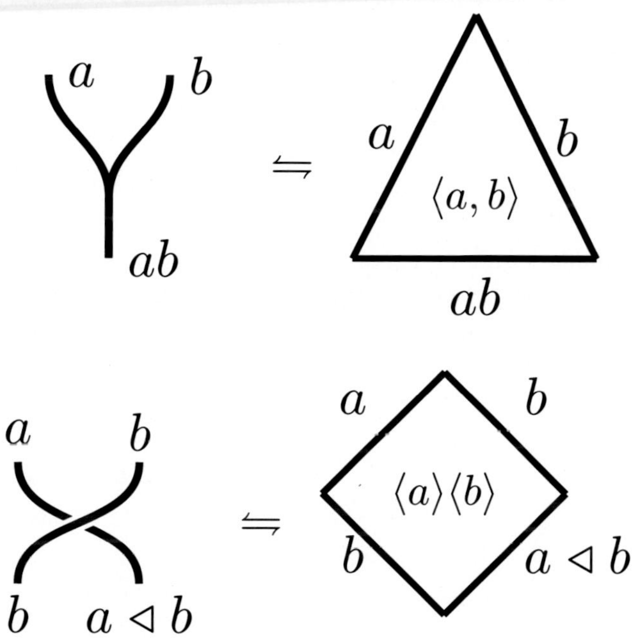

Please notice that the square is 2-dimensional cell that has the structure of $\Delta^1 \times \Delta^1$.

Under these correspondences, the two sides of the H, YI, IY, and Ж moves can be thought of as corresponding to the boundaries of 3-dimensional cells that are respectively decomposed as Δ^3, $\Delta^2 \times \Delta^1$, $\Delta^1 \times \Delta^2$, and $\Delta^1 \times \Delta^1 \times \Delta^1$. But the crossings in the  occurs in a very thin direction above the plane of the paper. So while these are all 4-arrows, the geometric schematic is 3-dimensional. Thus the broken surface diagrams as indicated in Ch. 9, Figs. 5, 11, 14,and 13 fit inside the products of simplices. The associativity move fits into Δ^3, the YI move fits into the prism $\Delta^2 \times \Delta^1$, the IY move fits into the prism $\Delta^1 \times \Delta^2$, and the X move fits into the cube $\Delta^1 \times \Delta^1 \times \Delta^1$.

The correspondences among the algebraic axioms, products of simplices, the expressions of each as a formal chain, and the glyphs representing the triple and quadruple arrows are tabulated.

Y	Δ^2	$\langle a, b\rangle$	$a \cdot b$
X	$\Delta^1 \times \Delta^1$	$\langle a\rangle\langle b\rangle$	$a \triangleleft b$
	Δ^3	$\langle a, b, c\rangle$	$(ab)c = a(bc)$
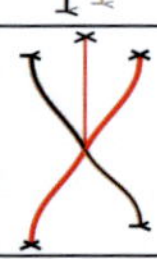	$\Delta^2 \times \Delta^1$	$\langle a, b\rangle\langle c\rangle$	$(ab) \triangleleft c = (a \triangleleft c)(b \triangleleft c)$
	$\Delta^1 \times \Delta^2$	$\langle a\rangle\langle b, c\rangle$	$(a \triangleleft b) \triangleleft c = a \triangleleft (bc)$
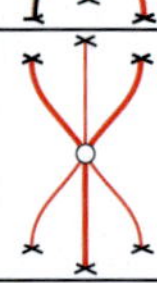	$\Delta^1 \times \Delta^1 \times \Delta^1$	$\langle a\rangle\langle b\rangle\langle c\rangle$	$(a \triangleleft b) \triangleleft c = (a \triangleleft c) \triangleleft (b \triangleleft c)$

The table above has an analogue that describes the eight important 5-isomorphisms of Section 2.2 (this chapter). The correspondences among partitions, products of simplices, and the formal 4-chains that are used to label the 5-isomorphisms are tabulated below. The order in which the entries of the table appear coincides with the order in which the eight important 5-isomorphisms were presented.

(1234)	Δ^4	$\langle a,b,c,d\rangle$
(123|4)	$\Delta^3 \times \Delta^1$	$\langle a,b,c\rangle\langle d\rangle$
(12|34)	$\Delta^2 \times \Delta^2$	$\langle a,b\rangle\langle c,d\rangle$
(12|3|4)	$\Delta^2 \times \Delta^1 \times \Delta^1$	$\langle a,b\rangle\langle c\rangle\langle d\rangle$
(1|234)	$\Delta^1 \times \Delta^3$	$\langle a\rangle\langle b,c,d\rangle$
(1|23|4)	$\Delta^1 \times \Delta^2 \times \Delta^1$	$\langle a\rangle\langle b,c\rangle\langle d\rangle$
(1|2|34)	$\Delta^1 \times \Delta^1 \times \Delta^2$	$\langle a\rangle\langle b\rangle\langle c,d\rangle$
(1|2|3|4)	$\Delta^1 \times \Delta^1 \times \Delta^1 \times \Delta^1$	$\langle a\rangle\langle b\rangle\langle c\rangle\langle d\rangle$

The formal definition of homology of shalgebras will appear in the next section. One motivation is that the boundaries of products of simplices are products of lower dimensional simplices. The correspondences of the H, YI, IY, and Ӿ moves with 3-dimensional prisms are respected among the boundaries of the 4-dimensional hyper-solids that are indicated in the table. This is to say that the eight important 5-isomorphisms arise in a context that naturally reflects some higher dimensional geometry.

4. Homology

The side comment that appears in Chapter 9, Section 6.2 will be summarized and recalled.

Homology measures the "in-exactness" of a chain complex. There are holes that correspond to cycles that do not bound.

A *chain complex* $\mathcal{C}$ consists of a sequence C_n of abelian groups and group homomorphisms $C_{n-1} \stackrel{d_n}{\longleftarrow} C_n$, that satisfy $d_n \circ d_{n+1} = 0$ for all $n = 0, 1, 2, \ldots$. The *kernel* of the homomorphism d_n is the set

$$Z_n = \{v \in C_n : d_n(v) = 0\}.$$

The elements of Z_n are called *n-cycles*. The elements of the set

$$B_n = \{w \in C_n : w = d_{n+1}(u) \text{ for some } u \in C_{n+1}\}$$

are called boundaries. The homology group $H_n = Z_n/B_n$ is the quotient group of the set of cycles modulo the set of boundaries.

Suppose that a set A upon which a binary product $\cdot$ exists is given. For $n = 1, 2, \ldots$, let C_n denote the free abelian group that is generated by n-tuples of elements $(a_1, a_2, \ldots, a_n)$, with $a_j \in A$. And let $C_0 = 0$ denote the trivial group.

For $1 < n$, define

$$\begin{aligned} & d_n(a_1, a_2, \ldots, a_n) \\ = \quad & (a_2, \ldots, a_n) \\ & + \sum_{k=1}^{n-1} (-1)^k (a_1, \ldots, a_k \cdot a_{k+1}, \ldots, a_n) \\ & + (-1)^n (a_1, a_2, \ldots, a_{n-1}), \end{aligned}$$

and define $d_1 = 0$. One can show that $d_n \circ d_{n+1} = 0$ if and only if the product $\cdot$ is associative so $(ab)c = a(bc)$. Indeed, in the case that A is a group, this method gives the definition of group homology [**10**]. The purpose here is to modify this homology theory so that it encompasses both products, $\cdot$ and $\triangleleft$, in a shalgebra.

To that end, the notation is slightly modified. First, a generating k-chain $(a_1, a_2, \ldots, a_k)$ will be offset by an index ℓ. Second, instead of writing $(a_{\ell+1}, a_{\ell+2}, \ldots, a_{\ell+k})$, write $\langle \ell+1, \ell+2, \ldots, \ell+k \rangle$ to indicate such a chain. The subscripts, then, become the labels of the elements. Third, such an offset chain may include an action, via $\triangleleft$, upon all the generating chains to its left. Then

$$\begin{aligned} & d_n\langle \ell+1, \ell+2, \ldots, \ell+k \rangle \\ = \quad & \triangleleft\, (\ell+1)\langle \ell+2 \ldots, \ell+k \rangle \\ & + \sum_{j=1}^{n-1} (-1)^j \langle \ell+1, \ldots, (\ell+j) \cdot (\ell+j+1), \ldots, \ell+k \rangle \\ & + (-1)^k \langle \ell+1, \ell+2, \ldots, \ell+k-1 \rangle. \end{aligned}$$

Fourth, a Leibniz rule (or product rule) is used to compute the boundary of the juxtaposition of two chains:

$$\partial(PQ) = (\partial P)Q + (-1)^{\deg(P)} P(\partial Q).$$

Finally, a rule to interpret some ambiguities is implemented. For example, $\partial\langle a_\ell \rangle = \llcorner\!\lrcorner \triangleleft a_\ell - \llcorner\!\lrcorner$. Notationally, the boundary operator ∂ is used when computing the boundary of the juxtaposition of two chains. A priori, either P or Q may also be the juxtaposition of chains, so the Leibniz rule is used recursively. When the chain is homogeneous, d is applied[7].

It is possible to demonstrate that this process yields a chain complex if and only if $(Q, \cdot, \triangleleft)$ is a shalgebra. That is, when the square of a boundary operator is computed, $\partial^2 = 0$ if and only if $\triangleleft$ and $\cdot$ satisfy axioms H, YI, IY, and X.

How to compute boundaries may not yet clear, so several low dimensional examples will exemplify the process.

$$\partial\langle a, b \rangle = (\triangleleft\, a)\langle b \rangle - \langle ab \rangle + \langle a \rangle.$$

Since there is nothing to the left for the first a to act upon,

$$\partial\langle a, b \rangle = \langle b \rangle - \langle ab \rangle + \langle a \rangle.$$

Thus the boundary of ${}^a\mathsf{Y}^b_{ab}$ is the signed sum of the 1-chains that form the literal boundary of the labeled vertex Y. Or the boundary of the edge colored triangle is the signed sum of the colors of the edges.

[7]Within the specific computations, all boundary maps will be denoted by ∂ and the degree will not be indicated.

Similarly,

$$\begin{aligned}\partial\langle a\rangle\langle b\rangle &= (\partial\langle a\rangle)\langle b\rangle - \langle a\rangle(\partial\langle b\rangle) \\ &= (\partial\langle a\rangle)\langle b\rangle - \langle a\rangle(\_\lhd b - \_) \\ &= \langle b\rangle - \langle b\rangle - \langle a\lhd b\rangle + \langle a\rangle.\end{aligned}$$

So the boundary of the chain ${}^{a}_{b}\mathsf{X}^{b}_{a\lhd b}$ is the difference between the labels on the top and the labels on the bottom of the X. Compare with the illustration of $\langle a\rangle\langle b\rangle$ as a square.

In the next four cases, the boundaries will be computed and compared with the quadruple arrows that correspond to H, YI, IY, and X. In these calculations, notice that the negative terms in the expressions of the boundaries correspond to the Ys and the Xs on the left side of the quadruple arrow while the positive terms correspond to those vertices on the right side.

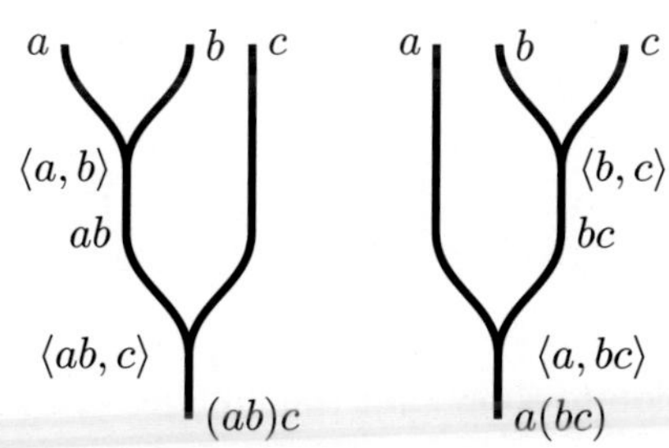

$$\begin{aligned}&d\langle a,b,c\rangle \\ &= \langle b,c\rangle - \langle ab,c\rangle \\ &\quad +\langle a,bc\rangle - \langle a,b\rangle\end{aligned}$$

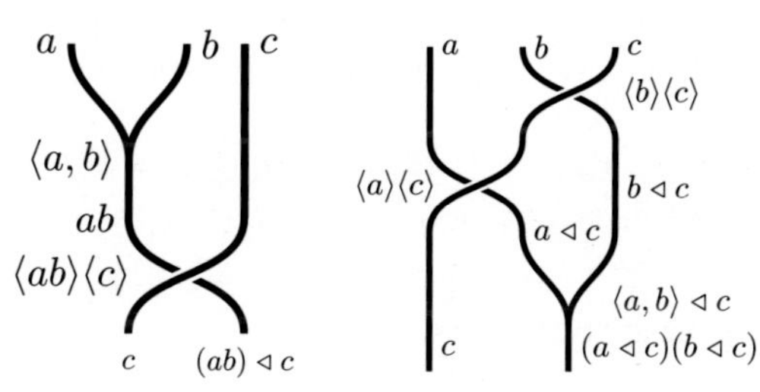

$$\begin{aligned}&\partial\langle a,b\rangle\langle c\rangle \\ &= \langle b\rangle\langle c\rangle - \langle ab\rangle\langle c\rangle \\ &\quad +\langle a\rangle\langle c\rangle \\ &\quad\quad +\langle a,b\rangle\lhd c - \langle a,b\rangle\end{aligned}$$

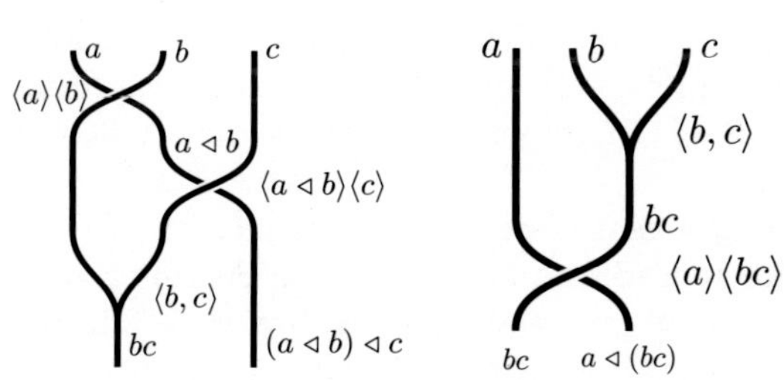

$$\begin{aligned}&\partial\langle a\rangle\langle b,c\rangle \\ &= \langle b,c\rangle - \langle b,c\rangle \\ &\quad -\langle a\lhd b\rangle\langle c\rangle + \langle a\rangle\langle bc\rangle \\ &-\langle a\rangle\langle b\rangle\end{aligned}$$

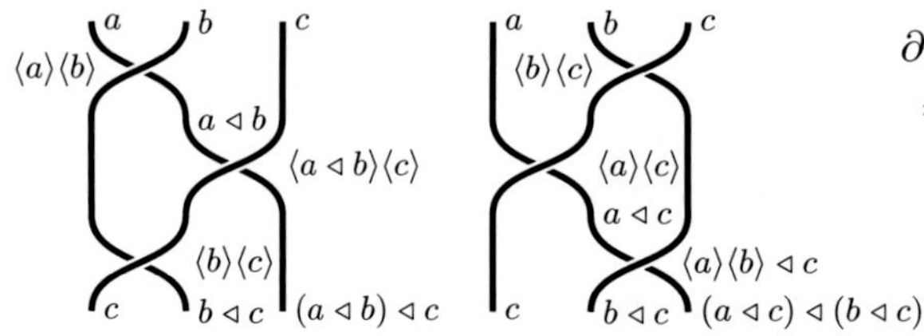

$$\begin{aligned}&\partial\left(\langle a\rangle\langle b\rangle\langle c\rangle\right)\\ &\quad= \langle b\rangle\langle c\rangle - \langle b\rangle\langle c\rangle\\ &\qquad -\langle a \lhd b\rangle\langle c\rangle + \langle a\rangle\langle c\rangle\\ &\qquad +\langle a\rangle\langle b\rangle \lhd c - \langle a\rangle\langle b\rangle\end{aligned}$$

Exercise. Compute the boundaries $\partial\langle a,b,c,d\rangle$ through $\partial\left(\langle a\rangle\langle b\rangle\langle c\rangle\langle d\rangle\right)$ and compare the terms with the labels upon the quadruple arrows in the eight important 5-isomorphisms. For our own notational convenience, we used $1,2,3,4$ as labels rather than a,b,c,d, or x,y,z,w. Notice, in particular, that the positive or negative terms comprise one of the two possible paths from the source quadruple arrow to the target.

In general, it is possible to write down an explicit expression for all the chains of degree n. A generating chain corresponds to a partition

$$1,2,\ldots,\ell_1|\ell_1+1,\ldots,\ell_1+\ell_2|\ldots|\left[\sum_{j=1}^{k-1}\ell_j\right]+1,\ldots,\sum_{j=1}^{k}\ell_j$$

where $\sum_{j=1}^{k}\ell_j = n$.

Exercise. Use mathematical induction to show that the number of such partitions is 2^{n-1}.

The form of the Leibniz rule given suggests the tensor product of two chain complexes is taken. These both define homology for an associative structure, but within this tensor product, later chains act upon earlier ones via $\lhd$. Within the geometric context, trees that express generating chains, and that appear to the right, cross over trees on the left. This over-crossing is an expression of the $\lhd$-action.

An amusing aspect of the homology theory is that it simultaneously encompasses homology of associative and self distributive structures.

5. More abstract tensors

In this section, the glyphographic expressions of the eight 5-isomorphisms that correspond to the partitions $(1,2,3,4)$ through $(1|2|3|4)$ are examined. The glyphs representing the H, YI, IY, and X moves are abstract tensors that have indices at each edge of the glyph. Then a system of eight tensor equations is formulated naively. One of these is a version of the Zamolodchikov tetrahedral equation [**81**]. Finally, cocycles, associated to the homology theory above, are used to formulate solutions. In this manner, the ideas of diagrammatic algebra have traveled full circle.

As a warm-up, the corresponding tensor equations for the H, YI, IY, and X moves are formulated. Compare this material to (1) Chapter 2, Section 3 in which relations among the structure constants of an associative algebra are given, and (2) to the expression of the Yang-Baxter equation as given in Chapter 9, Section 4.1.

5.1. Adorning X and Y with indices. Let Q denote a finite shalgebra. For definiteness you may assume that Q is a group $\cdot$ is ordinary multiplication, and $\triangleleft$ is conjugation. Consider a vector space $V = \mathbb{C}[Q]$ with its basis consisting of the elements of Q; this generalizes the group algebra. One can, of course, only consider a module $\mathbb{Z}[Q]$ depending upon one's specific needs. Think of Y and X as vector space maps $V \otimes V \xleftarrow{\mathsf{Y}} V$ and $V \otimes V \xleftarrow{\mathsf{X}} V \otimes V$. Often these will be written as abstract tensors. So the diagrammatic versions of the shalgebra axioms as abstract tensor equations are formulated. In these equations, and those that follow, summation conventions hold. So a repeated index that appears once as a subscript and once as a super-script is summed over for all (finitely many) values of that index.

For example equation (1) that appears below can be written in terms of structure constants as

$$\sum_{d,e \in Q} \left({}^a\mathsf{Y}^b_d \; \mathsf{I}^c_e \right) {}^d\mathsf{Y}^e_f = \sum_{d,e \in Q} \left(\mathsf{I}^a_d \; {}^b\mathsf{Y}^c_e \right) {}^d\mathsf{Y}^e_f .$$

Herein all the indices range over the finitely many values a, b etc. $\in Q$ where Q is a finite shalgebra.

(1)

$$\left({}^a\mathsf{Y}^b_d \; \mathsf{I}^c_e \right) {}^d\mathsf{Y}^e_f = \left(\mathsf{I}^a_d \; {}^b\mathsf{Y}^c_e \right) {}^d\mathsf{Y}^e_f ,$$

(2)

$$\left({}^a\mathsf{Y}^b_d \; \mathsf{I}^c_e \right) {}^d_f\mathsf{X}^e_g = \left(\mathsf{I}^a_d \; {}^b_e\mathsf{X}^c_h \right) \left({}^d_k\mathsf{X}^e_i \; \mathsf{I}^h_j \right) \left(\mathsf{I}^k_f \; {}^i\mathsf{Y}^j_g \right)$$

(3)

$$\left({}^a_d\mathsf{X}^b_e \; \mathsf{I}^c_h \right) \left(\mathsf{I}^d_i \; {}^e_j\mathsf{X}^h_k \right) \left({}^i\mathsf{Y}^j_f \; \mathsf{I}^k_g \right) = \left(\mathsf{I}^a_d \; {}^b\mathsf{Y}^c_e \right) {}^d_f\mathsf{X}^e_g$$

(4)

$$\left({}^a_d\mathsf{X}^b_e \; \mathsf{I}^c_f \right) \left(\mathsf{I}^d_g \; {}^e_h\mathsf{X}^f_i \right) \left({}^g_j\mathsf{X}^h_k \; \mathsf{I}^i_\ell \right) = \left(\mathsf{I}^a_d \; {}^b_e\mathsf{X}^c_f \right) \left({}^d_g\mathsf{X}^e_h \; \mathsf{I}^f_i \right) \left(\mathsf{I}^g_j \; {}^h_k\mathsf{X}^i_\ell \right)$$

Then for $a, b, c, d \in Q$, define

$$\mathsf{I}^a_b = \delta^a_b = \begin{cases} 1 & \text{if } a = b \\ 0 & \text{otherwise,} \end{cases}$$

$${}^a\mathsf{Y}^b_c = \begin{cases} 1 & \text{if } c = ab \\ 0 & \text{otherwise,} \end{cases}$$

and

$${}^a_c\mathsf{X}^b_d = \begin{cases} 1 & \text{if } c = b \;\&\; d = a \triangleleft b \\ 0 & \text{otherwise.} \end{cases}$$

THEOREM 10.1. *If V has a shalgebra basis, then the above values for I^a_b, ${}^a\mathsf{Y}^b_c$, and ${}^a_c\mathsf{X}^b_d$ satisfy the conditions (1), (2), (3), and (4) that are listed above.*

PROOF. For system of tensor equations, compute the image of a basis element or tensor product of basis elements that appears at the bottom of the expression. Each formal tensor I, Y, or X returns a value of 1 only in case the indices at the bottom depend unequivocally upon the indices at the top.

For example, within the expression, $\left({}^{a}\mathsf{Y}^{b}_{d}\ \mathsf{I}^{c}_{e}\right){}^{d}\mathsf{Y}^{e}_{f}$, the image of the basis element f is the sum

$$\sum_{(ab)c=f} a\otimes b\otimes c.$$

Meanwhile the right-hand side of equation (1), is

$$\sum_{a(bc)=f} a\otimes b\otimes c$$

Since the associative law holds, the image of any basis element f is the same for either side of the equation.

In a similar fashion, the image of $f\otimes g$ in the expression $\left({}^{a}\mathsf{Y}^{b}_{d}\ \mathsf{I}^{c}_{e}\right){}^{d}_{f}\mathsf{X}^{e}_{g}$ is

$$\sum_{\substack{(ab)\triangleleft c=g\\ f=c}} a\otimes b\otimes c.$$

On the other hand, the image of $f\otimes g$ in the expression $\left(\mathsf{I}^{a}_{d}\ {}^{b}_{e}\mathsf{X}^{c}_{h}\right)\left({}^{d}_{k}\mathsf{X}^{e}_{i}\ \mathsf{I}^{h}_{j}\right)\left(\mathsf{I}^{k}_{f}\ {}^{i}\mathsf{Y}^{j}_{g}\right)$ is

$$\sum_{\substack{(a\triangleleft c)(b\triangleleft c)=g,\\ f=c}} a\otimes b\otimes c$$

because I, Y, and X return a value of 1 if and only if the tensor indices at the bottom satisfy "coloring rules" determined by the indices at the top: ${}^{a}\mathsf{Y}^{b}_{c}$ indicates that $c=ab$ while ${}^{a}_{c}\mathsf{X}^{b}_{d}$ indicates that $c=b$ and $d=a\triangleleft b$, and I^{a}_{b} indicates that $a=b$.
Exercise. Show that the remaining two equations follow similarly. □

5.2. Assigning tensor operators to the 4-arrows. Move up one dimension, and assign abstract tensors to the relations H, YI, IY, and Ж. Let Q denote a finite shalgebra, and let $U=V\otimes V$ denote a vector space where V has Q as a basis. The basis element $a=a_1\otimes a_2$ for U will be written as an ordered pair $a=(a_1,a_2)\in Q\times Q$.

In this section, replace the glyphs , 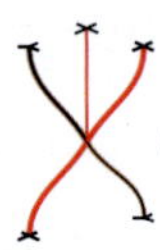, , and  with operators K, W, M, and Ж, respectively. Thus,

$$U\otimes U\xleftarrow{K}U\otimes U,$$

$$U\otimes U\otimes U\xleftarrow{W}U\otimes U,$$

$$U\otimes U\xleftarrow{M}U\otimes U\otimes U,$$

and

$$U\otimes U\otimes U\xleftarrow{\text{Ж}}U\otimes U\otimes U.$$

They correspond, in order, to the H, YI, IY, and Ж quadruple arrows. These operators will be represented as structure constants in an abstract tensor notation. Thus write K^{ab}_{cd}, W^{abc}_{de}, M^{ab}_{cde}, and Ж^{abc}_{def} where the indices $a,b,c,d,e,f\in Q\times Q$

range over all values in the basis for U. Following the development of the desired tensor equations, write $a = (a_1, a_2)$, $b = (b_1, b_2)$, and so on, so that each index corresponds to an ordered pair of indices taken from the shalgebra Q.

For each of the eight important 5-isomorphisms, the source and target edges of the glyphs that comprise either side of the isomorphism will be adorned with indices taken from the set of lower-case italics fonts; these represent elements in $Q \times Q$. Any index that is bounded by a pair of vertices of the glyph is to be summed over. At the top of each illustration, the corresponding abstract tensor equation is formulated.

$$K_{ab}^{dg} K_{dc}^{ef} = K_{bc}^{hj} K_{ah}^{e\ell} K_{\ell j}^{fg}$$

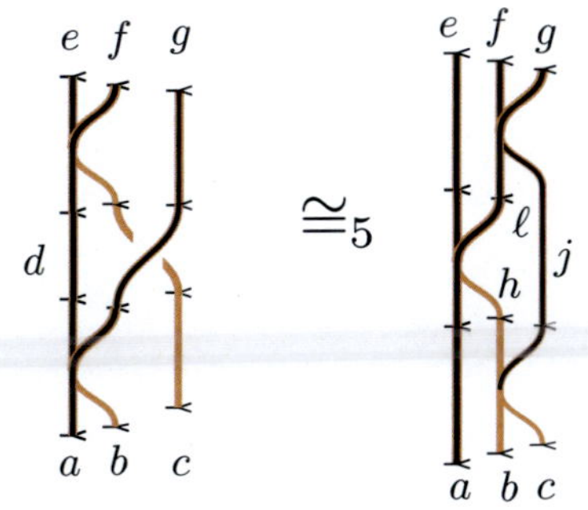

$$W_{ab}^{def} W_{ec}^{g\ell n} K_{dg}^{hj} = K_{bc}^{de} W_{ad}^{h\ell g} W_{ge}^{jnf}$$

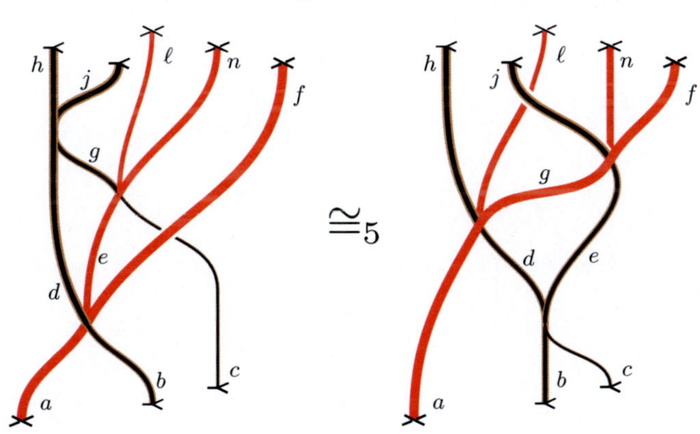

$$M_{abc}^{ef} W_{ed}^{ghj} = W_{cd}^{p\ell n} W_{bp}^{geq} M_{ae\ell}^{hr} M_{rqn}^{jf}$$

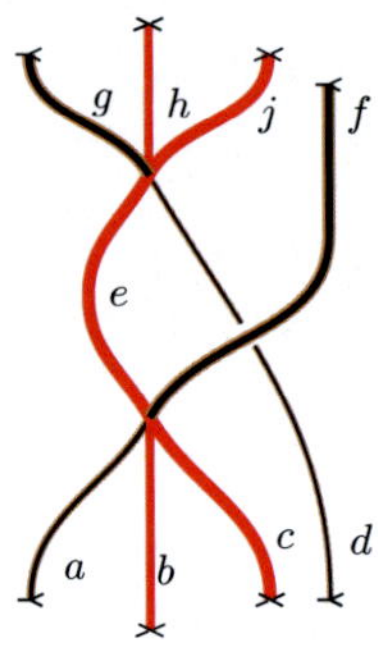

$\cong_5$

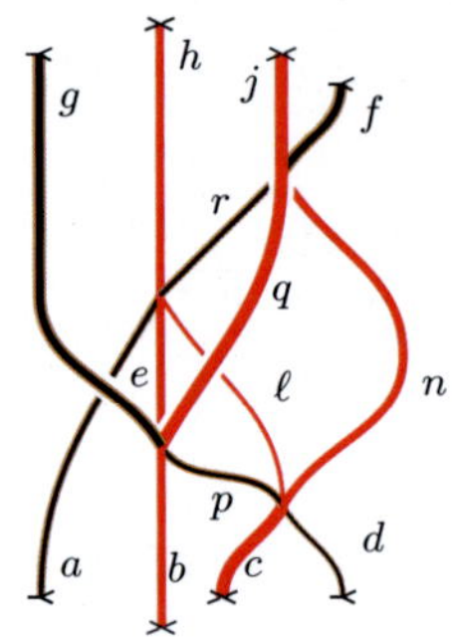

$$Ж_{abc}^{qpn} W_{pd}^{rjf} W_{qr}^{ghe} = W_{cd}^{pq\ell} W_{bp}^{grs} Ж_{arq}^{hjt} Ж_{ts\ell}^{efn}$$

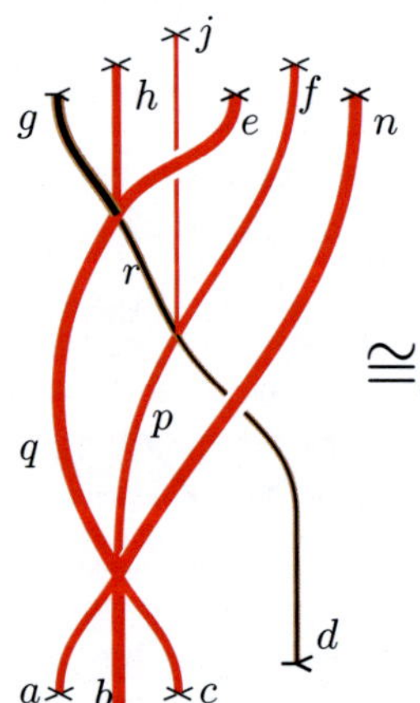

$\cong_5$

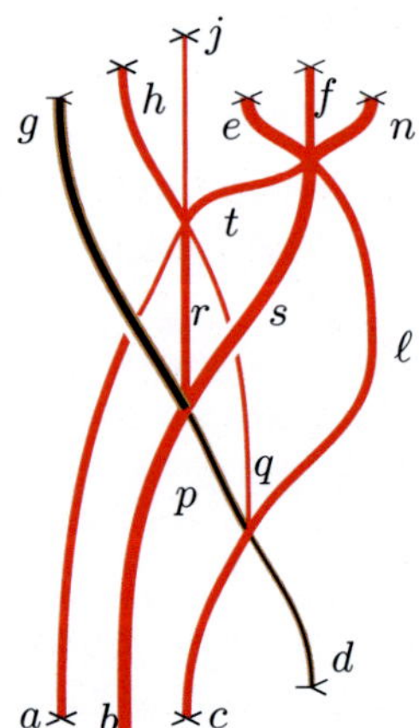

$$M^{g\ell}_{bde}M^{hf}_{acg}K^{jn}_{f\ell} = K^{f\ell}_{ab}M^{gn}_{\ell cd}M^{hj}_{fge}$$

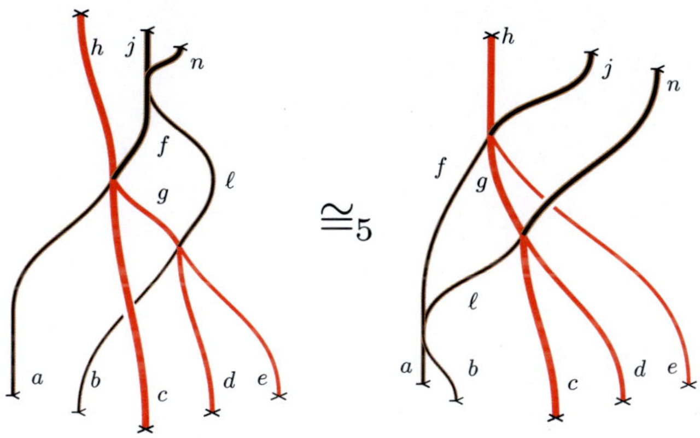

$$W^{qrp}_{ac}\boldsymbol{\ast}^{s\ell n}_{pbd}\boldsymbol{\ast}^{tjf}_{r\ell e}M^{gh}_{qst} = M^{p\ell}_{cde}\boldsymbol{\ast}^{gjr}_{abp}W^{hfn}_{r\ell}$$

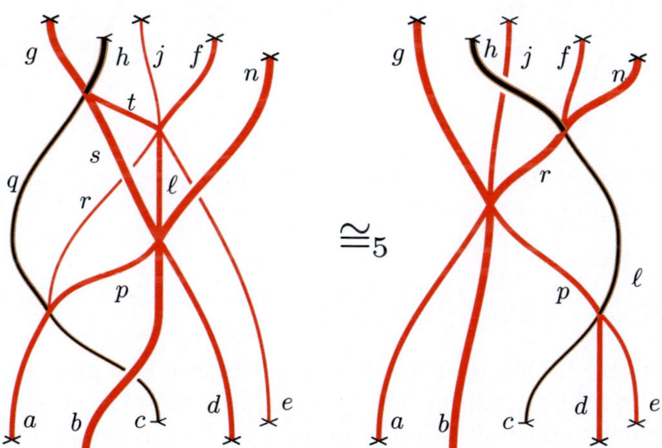

$$\text{Ж}^{p\ell s}_{cef}\text{Ж}^{gqr}_{bdp}M^{ht}_{aq\ell}M^{jn}_{trs} = M^{qp}_{abc}M^{\ell n}_{pde}\text{Ж}^{ghj}_{q\ell f}$$

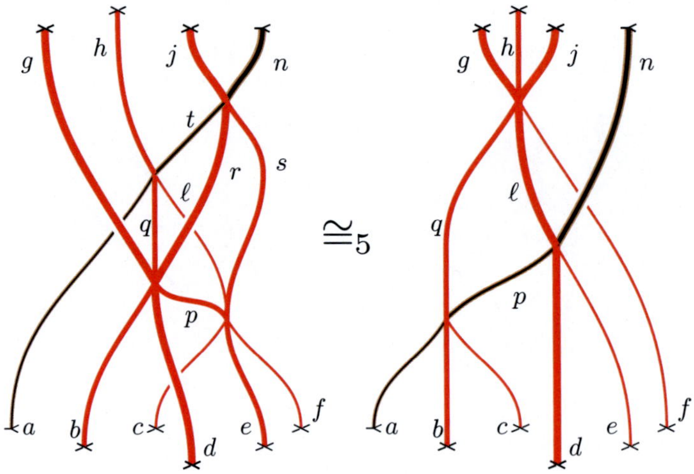

$$\text{Ж}^{ruv}_{cef}\text{Ж}^{g\ell s}_{bdr}\text{Ж}^{hjt}_{a\ell u}\text{Ж}^{npq}_{tsv} = \text{Ж}^{vur}_{abc}\text{Ж}^{s\ell q}_{rde}\text{Ж}^{tjp}_{u\ell f}\text{Ж}^{ghn}_{vst}$$

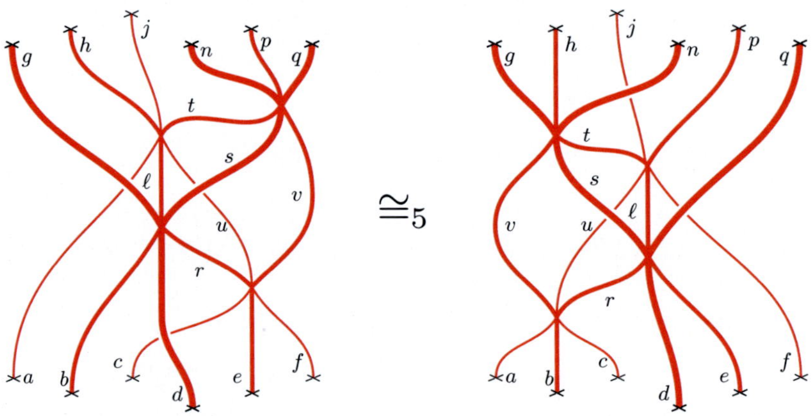

Next four conditions are formulated that are called *coloring conditions* for the indices along the edges of the glyphs. There is a condition for each of the glyphs , , , and . The edge labels a, b, etc. run from left to right, top and then left to right, bottom. Each label is determined by a pair of labels in the shalgebra Q. The coloring conditions are induced from the diagrams of embedded foams that the glyphs represent. So the ordering a, b, c, in comparison with the ordering of the broken surfaces in the foam, is a bit strange. In each of the coloring conditions, three of the foam labels determine the rest. These *independent labels* are specified via tautological equations, e.g. $a_1 = a_1$ while the dependent labels are written in terms of the independent labels in each of the coloring conditions. These are listed.

Coloring condition $\star_H$:

$$\begin{array}{rll} a = (a_1, a_2): & a_1 = a_1; & a_2 = b_1 b_2; \\ b = (b_1, b_2): & b_1 = b_1; & b_2 = b_2; \\ c = (c_1, c_2): & c_1 = a_1 b_1; & c_2 = b_2; \\ d = (d_1, d_2): & d_1 = a_1; & d_2 = b_1. \end{array}$$

Since Q is a shalgebra, $(a_1 b_1) b_2 = a_1 (b_1 b_2)$.

Coloring condition $\star_{YI}$:

$$\begin{array}{rll} a = (a_1, a_2): & a_1 = b_1 \lhd c_2; & a_2 = c_1 \lhd c_2; \\ b = (b_1, b_2): & b_1 = b_1; & b_2 = c_2; \\ c = (c_1, c_2): & c_1 = c_1; & c_2 = c_2; \\ d = (d_1, d_2): & d_1 = b_1 c_1; & d_2 = c_2; \\ e = (e_1, e_2): & e_1 = b_1; & e_2 = c_1. \end{array}$$

Since Q is a shalgebra, $(b_1 c_1) \lhd c_2 = (b_2 \lhd c_2)(c_1 \lhd c_2)$.

Coloring condition $\star_{IY}$:

$$\begin{array}{rll} a = (a_1, a_2): & a_1 = a_1; & a_2 = b_1 b_2; \\ b = (b_1, b_2): & b_1 = b_1; & b_2 = b_2; \\ c = (c_1, c_2): & c_1 = b_1; & c_2 = b_2; \\ d = (d_1, d_2): & d_1 = a_1 \lhd b_1; & d_2 = b_2; \\ e = (e_1, e_2): & e_1 = a_1; & e_2 = b_1. \end{array}$$

Since Q is a shalgebra, $a_1 \lhd (b_1 b_2) = (a_1 \lhd b_1) \lhd b_2$.

Coloring condition $\star_{\text{Ж}}$:

$$\begin{array}{rll} a = (a_1, a_2): & a_1 = b_1 \lhd b_2; & a_2 = c_1 \lhd b_2; \\ b = (b_1, b_2): & b_1 = b_1; & b_2 = b_2; \\ c = (c_1, c_2): & c_1 = c_1; & c_2 = b_2; \\ d = (d_1, d_2): & d_1 = c_1; & d_2 = b_2; \\ e = (e_1, e_2): & e_1 = b_1 \lhd c_1; & e_2 = b_2 \\ f = (f_1, f_2); & f_1 = b_1; & f_2 = c_1. \end{array}$$

Since Q is a shalgebra, $(b_1 \lhd c_1) \lhd c_2 = (b_1 \lhd c_2) \lhd (c_1 \lhd c_2)$.

THEOREM 10.2. *The system of equations:*

$$K^{dg}_{ab} K^{ef}_{dc} = K^{hj}_{bc} K^{e\ell}_{ah} K^{fg}_{\ell j} \tag{5.1}$$

$$W^{def}_{ab} W^{g\ell n}_{ec} K^{hj}_{dg} = K^{de}_{bc} W^{h\ell g}_{ad} W^{jnf}_{ge} \tag{5.2}$$

$$M^{ef}_{abc} W^{ghj}_{ed} = W^{p\ell n}_{cd} W^{geq}_{bp} M^{hr}_{ae\ell} M^{jf}_{rqn} \tag{5.3}$$

$$\text{Ж}^{qpn}_{abc} W^{rjf}_{pd} W^{ghe}_{qr} = W^{pq\ell}_{cd} W^{grs}_{bp} \text{Ж}^{hjt}_{arq} \text{Ж}^{efn}_{ts\ell} \tag{5.4}$$

$$M^{g\ell}_{bde} M^{hf}_{acg} K^{jn}_{f\ell} = K^{f\ell}_{ab} M^{gn}_{\ell cd} M^{hj}_{fge} \tag{5.5}$$

$$W^{qrp}_{ac} \text{Ж}^{s\ell n}_{pbd} \text{Ж}^{tjf}_{r\ell e} M^{gh}_{qst} = M^{p\ell}_{cde} \text{Ж}^{gjr}_{abp} W^{hfn}_{r\ell} \tag{5.6}$$

$$\text{Ж}^{p\ell s}_{cef} \text{Ж}^{gqr}_{bdp} M^{ht}_{aq\ell} M^{jn}_{trs} = M^{qp}_{abc} M^{\ell n}_{pde} \text{Ж}^{ghj}_{q\ell f} \tag{5.7}$$

$$\text{Ж}^{ruv}_{cef} \text{Ж}^{g\ell s}_{bdr} \text{Ж}^{hjt}_{a\ell u} \text{Ж}^{npq}_{tsv} = \text{Ж}^{vur}_{abc} \text{Ж}^{s\ell q}_{rde} \text{Ж}^{tjp}_{u\ell f} \text{Ж}^{ghn}_{vst} \tag{5.8}$$

has a solution that is given by

$$K^{ab}_{cd} = \begin{cases} 1 & \text{if condition } \star_H \textit{ holds;} \\ 0 & \text{otherwise.} \end{cases}$$

$$W^{abc}_{de} = \begin{cases} 1 & \text{if condition } \star_{YI} \text{ holds;} \\ 0 & \text{otherwise.} \end{cases}$$

$$M^{ab}_{cde} = \begin{cases} 1 & \text{if condition } \star_{IY} \text{ holds;} \\ 0 & \text{otherwise.} \end{cases}$$

$$\mathbf{X}^{abc}_{def} = \begin{cases} 1 & \text{if condition } \star_{\mathbf{X}} \text{ holds;} \\ 0 & \text{otherwise.} \end{cases}$$

Before embarking upon the proof, the nature of the tensors K, M, W, and $\mathbf{X}$ is discussed. Because these tensors are chosen as stand-ins for the corresponding glyphs, some information is lost among the indices a, b etc. The dependencies upon initial colors is arcane, and which indices depend upon seams and which depend upon double curves is not apparent in the alphabetic notation. In each case, these indices depend on three input colors. To ease the notational burden, the input variables in Q will be labeled x, y, z and the coloring conditions will be rewritten in terms of these.

Case H.

$$\begin{array}{lll} a = (a_1, a_2): & a_1 = x; & a_2 = yz; \\ b = (b_1, b_2): & b_1 = y; & b_2 = z; \\ c = (c_1, c_2): & c_1 = xy; & c_2 = z; \\ d = (d_1, d_2): & d_1 = x; & d_2 = y. \end{array}$$

Consider the 3-chain $\langle x, y, z\rangle \in C_3 = \mathbb{Z}[Q^3]$. Its boundary is

$$\partial\langle x, y, z\rangle = b - c + a - d$$

which coincides with the tensor indices at the top of K^{ab}_{cd} minus those at the bottom. The tensor K checks whether or not $b-c+a-d$ is the boundary of the chain $\langle x, y, z\rangle$.

Case YI.

$$\begin{array}{lll} a = (a_1, a_2): & a_1 = x \lhd z; & a_2 = y \lhd z; \\ b = (b_1, b_2): & b_1 = x; & b_2 = z; \\ c = (c_1, c_2): & c_1 = y; & c_2 = z; \\ d = (d_1, d_2): & d_1 = xy; & d_2 = z \\ e = (e_1, e_2): & e_1 = x; & e_2 = y. \end{array}$$

Consider the 3-chain $\langle x, y\rangle\langle z\rangle \in C_3 = \mathbb{Z}[Q^2 \times Q] = Z[Q \times Q] \otimes Z[Q]$. Its boundary is

$$\partial\left(\langle x, y\rangle\langle z\rangle\right) = c - d + b + a - e$$

which coincides with the tensor indices at the top of W^{abc}_{de} minus those at the bottom. The indices b, c, and d correspond to double arcs. So, for example, b should be written as $\langle b_1\rangle\langle b_2\rangle = \langle x\rangle\langle z\rangle$. Similar rewrites for c and d apply. The tensor W checks whether or not $c - d + b + a - e$ is the boundary of the chain $\langle x, y\rangle\langle z\rangle$.

Case IY.

$$\begin{array}{lll} a = (a_1, a_2): & a_1 = x; & a_2 = yz; \\ b = (b_1, b_2): & b_1 = y; & b_2 = z; \\ c = (c_1, c_2): & c_1 = y; & c_2 = z; \\ d = (d_1, d_2): & d_1 = x \lhd y; & d_2 = z \\ e = (e_1, e_2): & e_1 = x; & e_2 = y. \end{array}$$

Consider the 3-chain $\langle x\rangle\langle y,z\rangle \in C_3 = \mathbb{Z}[Q\times Q^2] = Z[Q]\otimes Z[Q\times Q]$. Its boundary is

$$\partial\left(\langle x\rangle\langle y,z\rangle\right) = b - c - d + a - e$$

which coincides with the tensor indices at the top of M^{ab}_{cde} minus those at the bottom. The indices a, d, and e correspond to double arcs. So, for example, a should be written as $\langle a_1\rangle\langle a_2\rangle = \langle x\rangle\langle yz\rangle$. Similar rewrites for d and e apply. The tensor M checks whether or not $b-c-d+a-e$ is the boundary of the chain $\langle x\rangle\langle y,z\rangle$.

Case $\mathbf{X}$.

$$\begin{array}{lll} a=(a_1,a_2): & a_1 = x \triangleleft z; & a_2 = y\triangleleft z; \\ b=(b_1,b_2): & b_1 = x; & b_2 = z; \\ c=(c_1,c_2): & c_1 = y; & c_2 = z; \\ d=(d_1,d_2): & d_1 = y; & d_2 = z \\ e=(e_1,e_2): & e_1 = x\triangleleft y; & e_2 = z \\ f=(f_1,f_2): & f_1 = x; & f_2 = y. \end{array}$$

Consider the 3-chain $\langle x\rangle\langle y\rangle\langle z\rangle \in C_3 = \mathbb{Z}[Q\times Q\times Q] = Z[Q]\otimes Z[Q]\otimes Z[Q]$. Its boundary is

$$\partial\left(\langle x\rangle\langle y\rangle\langle z\rangle\right) = c - d - e + b + a - f$$

which coincides with the tensor indices at the top of $\mathbf{X}^{abc}_{def}$ minus those at the bottom. The indices a through f correspond to double arcs. So, for example, a should be written as $\langle a_1\rangle\langle a_2\rangle = \langle x\triangleleft z\rangle\langle y\triangleleft z\rangle$. Similar rewrites for the remaining tensor indices apply. The tensor $\mathbf{X}$ checks whether or not $c-d-e+b+a-f$ is the boundary of the chain $\langle x\rangle\langle y\rangle\langle z\rangle$.

PROOF OF THEOREM 10.2. For each of the eight 5-isomorphisms, there is a source quadruple arrow that will be called an *initial half-twisted forest.* In each of these, labels $x,y,z,w\in Q$ are inserted from left to right on the top of the trees. The vertex Y is a multiplication operation $(\cdot)$, and X induces a self-distributive operation $(\triangleleft)$ on the lower-right arc. In this way, a qualgebra labeling at the top of the initial forest induces a qualgebra labeling of each branch in the forest. These labelings are presented below.

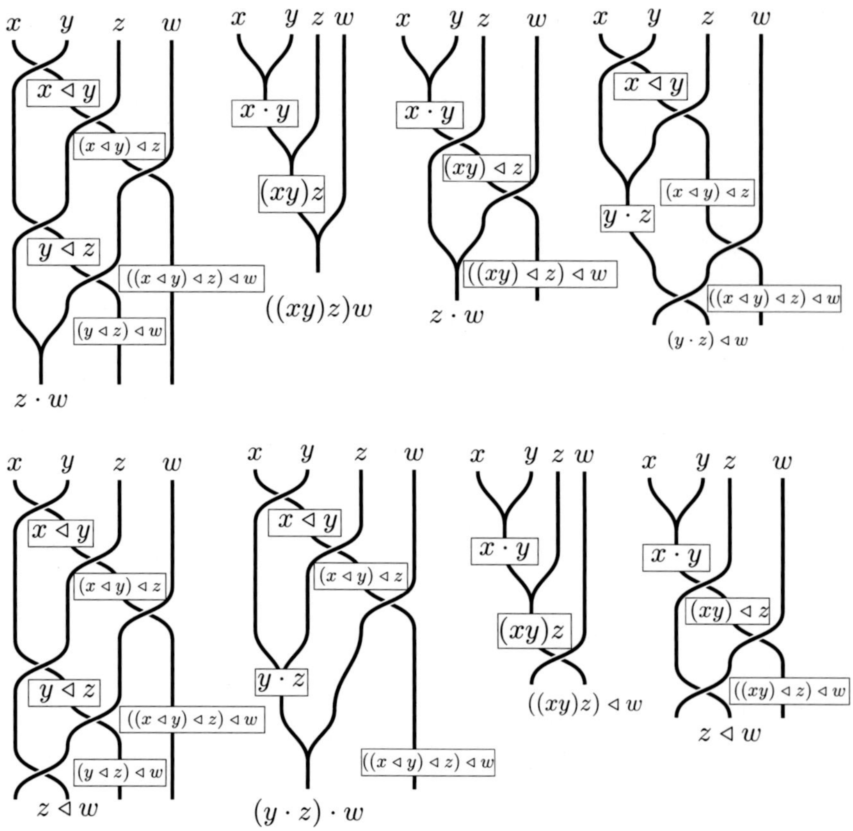

Consider any one of the eight important 5-isomorphisms. The source or target of any of the constituent quadruple arrows will receive a unique and consistent coloring by Q once labels x, y, z, w are chosen at the top of the trees. That is to say that each vertex is labeled as ${}^{a}\mathsf{Y}^{b}_{ab}$ or ${}^{a}_{b}\mathsf{X}^{b}_{a\triangleleft b}$. Moreover, the colorings are consistent from source to target if and only if the connecting tensor, K, M, W, or X takes a value of 1. So for any of the tensor equations 5.1 through 5.8, both the left-hand and the right-hand sides return a value of 1 if and only if the initial half-twisted forest is colored consistently, and that coloring induces a consistent coloring on the subsequent nodes. This completes the proof $\square$.

5.3. Shalgebra cocycles. In this section, the tensor equations 5.1 through 5.8 will be interpreted as cocycle conditions. So a collection of functions K, M, W, and X that satisfy cocycle conditions provide a solution to these equations.

Here is how we will approach the rest of this section. First, the notion of cocycles will be defined in general and exemplified in the lower dimensional case. Then the equations

(1)

$$\left({}^{a}\mathsf{Y}^{b}_{d}\ \mathsf{I}^{c}_{e}\right){}^{d}\mathsf{Y}^{e}_{f} = \left(\mathsf{I}^{a}_{d}\ {}^{b}\mathsf{Y}^{c}_{e}\right){}^{d}\mathsf{Y}^{e}_{f},$$

(2)

$$\left({}^{a}\mathsf{Y}^{b}_{d}\ \mathsf{I}^{c}_{e}\right){}^{d}_{f}\mathsf{X}^{e}_{g} = \left(\mathsf{I}^{a}_{d}\ {}^{b}_{e}\mathsf{X}^{c}_{h}\right)\left({}^{d}_{k}\mathsf{X}^{e}_{i}\ \mathsf{I}^{h}_{j}\right)\left(\mathsf{I}^{k}_{f}\ {}^{i}\mathsf{Y}^{j}_{g}\right)$$

(3)

$$\left({}^{a}_{d}\mathsf{X}^{b}_{e}\ \mathsf{I}^{c}_{h}\right)\left(\mathsf{I}^{d}_{i}\ {}^{e}_{j}\mathsf{X}^{h}_{k}\right)\left({}^{i}\mathsf{Y}^{j}_{f}\ \mathsf{I}^{k}_{g}\right) = \left(\mathsf{I}^{a}_{d}\ {}^{b}\mathsf{Y}^{c}_{e}\right){}^{d}_{f}\mathsf{X}^{e}_{g}$$

(4)

$$\left({}^{a}_{d}\mathsf{X}^{b}_{e}\ \mathsf{I}^{c}_{f}\right)\left(\mathsf{I}^{d}_{g}\ {}^{e}_{h}\mathsf{X}^{f}_{i}\right)\left({}^{g}_{j}\mathsf{X}^{h}_{k}\ \mathsf{I}^{i}_{\ell}\right) = \left(\mathsf{I}^{a}_{d}\ {}^{b}_{e}\mathsf{X}^{c}_{f}\right)\left({}^{d}_{g}\mathsf{X}^{e}_{h}\ \mathsf{I}^{f}_{i}\right)\left(\mathsf{I}^{g}_{j}\ {}^{h}_{k}\mathsf{X}^{i}_{\ell}\right)$$

will be reinterpreted as 2-cocycle conditions. Next, equations 5.1 through 5.8 will be shown to be 3-cocycle conditions.

Let $\mathcal{C}$ denote a chain complex. Recall from Section 4 (this chapter) that each C_n is an abelian group, and $C_{n-1} \xleftarrow{\partial_n} C_n$ is a group homomorphisms such that $\partial_n \circ \partial_{n+1} = 0$. Loosely, we write $\partial^2 = 0$; the maps ∂_n are called boundary operators. The homology of the chain complex is defined as $H_n = \ker(\partial_n)/\mathrm{Im}(\partial_{n+1})$. Let A denote an abelian group; A is fixed and it will not usually enter into the notation. An *A-valued n-cochain* is a homomorphism $A \xleftarrow{\phi} C_n$. The set of n-cochains is denoted as C^n. A *coboundary* is a homomorphism $C^{n+1} \xleftarrow{\delta^n} C^n$. In this case, the coboundary is defined by the remarkably elegant formula

$$\delta^n(\phi)(c) = \phi(\partial_n(c))$$

where ϕ is an n-cochain, and $c \in C_{n+1}$ is an n-chain. An n-cocycle is a cochain ϕ such that $\delta^n(\phi) = 0$.

The behavior of a shalgebra 2-cocycle is to be examined. Let Q denote a shalgebra. The group of 2-chains is the direct sum $\mathbb{Z}[Q^2] \oplus (\mathbb{Z}[Q] \otimes \mathbb{Z}[Q])$. A 2-chain in the first summand is written as $\langle x, y\rangle$ and one in the second is written as $\langle x\rangle\langle y\rangle$. Let $A \xleftarrow{Y} \mathbb{Z}[Q^2]$ and $A \xleftarrow{X} \mathbb{Z}[Q] \otimes \mathbb{Z}[Q]$. Then the ordered pair (X, Y) is a 2-cochain, and there are four possible coboundaries to be computed.

$$\begin{aligned}
&\delta(X,Y)\langle x,y,z\rangle \\
&\quad= (X,Y)\left[\langle y,z\rangle - \langle xy,z\rangle + \langle x,yz\rangle - \langle x,y\rangle\right] \\
&\quad= Y\langle y,z\rangle - Y\langle xy,z\rangle + Y\langle x,yz\rangle - Y\langle x,y\rangle.
\end{aligned}$$

$$\begin{aligned}
&\delta(X,Y)\langle x,y\rangle\langle z\rangle \\
&\quad= (X,Y)\left[\langle y\rangle\langle z\rangle - \langle xy\rangle\langle z\rangle + \langle x\rangle\langle z\rangle + \langle x,y\rangle \lhd z - \langle x,y\rangle\right] \\
&\quad= X\langle y\rangle\langle z\rangle - X\langle xy\rangle\langle z\rangle + X\langle x\rangle\langle z\rangle + Y\langle x,y\rangle \lhd z - Y\langle x,y\rangle.
\end{aligned}$$

$$\begin{aligned}
&\delta(X,Y)\langle x\rangle\langle y,z\rangle \\
&\quad= (X,Y)\left[\langle y,z\rangle - \langle y,z\rangle - \langle x \lhd y\rangle\langle z\rangle + \langle x\rangle\langle yz\rangle - \langle x\rangle\langle y\rangle\right] \\
&\quad= Y\langle y,z\rangle - Y\langle y,z\rangle - X\langle x \lhd y\rangle\langle z\rangle + X\langle x\rangle\langle yz\rangle - X\langle x\rangle\langle y\rangle.
\end{aligned}$$

$$\begin{aligned}
&\delta(X,Y)\langle x\rangle\langle y\rangle\langle z\rangle \\
&\quad = (X,Y)\left[\langle y\rangle\langle z\rangle - \langle y\rangle\langle z\rangle - \langle x \lhd y\rangle\langle z\rangle \right.\\
&\qquad \left. + \langle x\rangle\langle z\rangle + \langle x\rangle\langle y\rangle \lhd z - \langle x\rangle\langle y\rangle\right] \\
&\quad = X\langle y\rangle\langle z\rangle - X\langle y\rangle\langle z\rangle - X\langle x \lhd y\rangle\langle z\rangle \\
&\qquad + X\langle x\rangle\langle z\rangle + X\langle x \lhd z\rangle\langle y \lhd z\rangle - X\langle x\rangle\langle y\rangle.
\end{aligned}$$

If the pair (X,Y) is a cocycle, then

$$Y\langle xy, z\rangle + Y\langle x,y\rangle = Y\langle y,z\rangle + Y\langle x, yz\rangle,$$

$$X\langle xy\rangle\langle z\rangle + Y\langle x,y\rangle = X\langle y\rangle\langle z\rangle + X\langle x\rangle\langle z\rangle + Y\langle x \lhd z, y \lhd z\rangle,$$

$$Y\langle y,z\rangle + X\langle x \lhd y\rangle\langle z\rangle + X\langle x\rangle\langle y\rangle = Y\langle y,z\rangle + X\langle x\rangle\langle yz\rangle,$$

and

$$X\langle y\rangle\langle z\rangle + X\langle x \lhd y\rangle\langle z\rangle + X\langle x\rangle\langle y\rangle = X\langle y\rangle\langle z\rangle + X\langle x\rangle\langle z\rangle + X\langle x \lhd z\rangle\langle y \lhd z\rangle.$$

Now assign the cocycle value $X\langle a\rangle\langle b\rangle$ to the shalgebra colored crossing ${}^a_b\mathsf{X}^b_{a\lhd b}$, and assign $Y\langle a,b\rangle$ to the trivalent vertex ${}^a\mathsf{Y}^b_{ab}$. The value $-X\langle a\rangle\langle b\rangle$ is assigned to the negative crossing ${}^b_a\overline{\mathsf{X}}^{a\lhd b}_b$. The four equations above are written with these restricted colors as follows:

(1)

$${}^a\mathsf{Y}^b_{ab}\ {}^{ab}\mathsf{Y}^c_{(ab)c} = {}^b\mathsf{Y}^c_{bc}\ {}^a\mathsf{Y}^{bc}_{a(bc)},$$

(2)

$${}^a\mathsf{Y}^b_{ab}\ {}^{ab}_{c}\mathsf{X}^c_{(ab)\lhd c} = {}^b_c\mathsf{X}^c_{b\lhd c}\ {}^a_c\mathsf{X}^c_{a\lhd c}\ {}^{a\lhd c}\mathsf{Y}^{b\lhd c}_{(a\lhd c)(b\lhd c)},$$

(3)

$${}^a_b\mathsf{X}^b_{a\lhd b}\ {}^{a\lhd b}_{c}\mathsf{X}^c_{(a\lhd b)\lhd c}\ {}^b\mathsf{Y}^c_{bc} = {}^b\mathsf{Y}^c_{bc}\ {}^{a}_{bc}\mathsf{X}^{bc}_{a\lhd(bc)},$$

and

(4)

$${}^a_b\mathsf{X}^b_{a\lhd b}\ {}^{a\lhd b}_{c}\mathsf{X}^c_{(a\lhd b)\lhd c}\ {}^b_c\mathsf{X}^c_{b\lhd c} = {}^b_c\mathsf{X}^c_{b\lhd c}\ {}^a_c\mathsf{X}^c_{a\lhd c}\ {}^{a\lhd c}_{b\lhd c}\mathsf{X}^{b\lhd c}_{(a\lhd c)\lhd(b\lhd c)}.$$

The four coboundary conditions correspond to these equations which in turn were derived from moves to handlebody knots as illustrated in Section 4 (this chapter).

THEOREM 10.3. *A shalgebra 2-cocycle that is assigned to a colored trivalent graph embedded in 3-space via* $X\langle a\rangle\langle b\rangle \leftrightharpoons {}^a_b\mathsf{X}^b_{a\lhd b}$, *and* $Y\langle a,b\rangle \leftrightharpoons {}^a\mathsf{Y}^b_{ab}$ *remains unchanged under four of the moves to handlebody knots.*

It is possible to create invariants for handlebody knots via this approach. See [**32**] for further details.

The previous calculation will be imitated to interpret equations 5.1 through 5.8 as cocycle conditions. To begin assume, then, that Q is a shalgebra and that the tensor indices satisfy conditions $\star_H$, $\star_{YI}$, $\star_{IY}$, and $\star_{\text{Ж}}$. Let A denote an abelian group. Then let

$$A \xleftarrow{K} \mathbb{Z}[Q^3], \quad A \xleftarrow{M} \mathbb{Z}[Q^2] \otimes \mathbb{Z}[Q],$$

$$A \xleftarrow{W} \mathbb{Z}[Q] \otimes \mathbb{Z}[Q^2], \quad A \xleftarrow{\text{Ж}} \mathbb{Z}[Q] \otimes \mathbb{Z}[Q] \otimes \mathbb{Z}[Q].$$

The quadruple $(K, M, W, \text{Ж})$ represents the direct sum of these homomorphisms, and the coboundary of this is computed upon each of the cochains $\langle x, y, z, w\rangle$ through $\langle x\rangle\langle y\rangle\langle z\rangle\langle w\rangle$. The eight cocycle conditions are presented below, and paired with the corresponding tensor equations.

$$\langle x, y, z, w\rangle :$$

$$K\langle x, y, z\rangle + K\langle x, y \cdot z, w\rangle + K\langle y, z, w\rangle$$
$$= K\langle x \cdot y, z, w\rangle + K\langle x, y, z \cdot w\rangle;$$

$$K^{dg}_{ab} K^{ef}_{dc} \quad = \quad K^{hj}_{bc} K^{e\ell}_{ah} K^{fg}_{\ell j}$$

$$\langle x, y, z\rangle\langle w\rangle :$$

$$W\langle x \cdot y, z\rangle\langle w\rangle + W\langle x, y\rangle\langle w\rangle + K\langle x \lhd w, y \lhd w, z \lhd w\rangle$$
$$= K\langle x, y, z\rangle + W\langle x, y \cdot z\rangle\langle w\rangle + W\langle y, z\rangle\langle w\rangle;$$

$$W^{def}_{ab} W^{g\ell n}_{ec} K^{hj}_{dg} \quad = \quad K^{de}_{bc} W^{h\ell g}_{ad} W^{jnf}_{ge}$$

$$\langle x, y\rangle\langle z, w\rangle :$$

$$M\langle x \cdot y\rangle\langle z, w\rangle + W\langle x, y\rangle\langle z \cdot w\rangle$$
$$= W\langle x, y\rangle\langle z\rangle + W\langle x \lhd z, y \lhd z\rangle\langle w\rangle + M\langle x\rangle\langle z, w\rangle + M\langle y\rangle\langle z, w\rangle;$$

$$M^{ef}_{abc} W^{ghj}_{ed} \quad = \quad W^{p\ell n}_{cd} W^{geq}_{bp} M^{hr}_{ae\ell} M^{jf}_{rqn}$$

$$\langle x, y\rangle\langle z\rangle\langle w\rangle :$$

$$\text{Ж}\langle x \cdot y\rangle\langle z\rangle\langle w\rangle + W\langle x, y\rangle\langle w\rangle + W\langle x \lhd w, y \lhd w\rangle\langle z \lhd w\rangle$$
$$= W\langle x, y\rangle\langle z\rangle + W\langle x \lhd z, y \lhd z\rangle\langle w\rangle + \text{Ж}\langle x\rangle\langle z\rangle\langle w\rangle + \text{Ж}\langle y\rangle\langle z\rangle\langle w\rangle;$$

$$\text{Ж}^{qpn}_{abc} W^{rjf}_{pd} W^{ghe}_{qr} \quad = \quad W^{pq\ell}_{cd} W^{grs}_{bp} \text{Ж}^{hjt}_{arq} \text{Ж}^{efn}_{ts\ell}$$

$$\langle x\rangle\langle y, z, w\rangle :$$

$$K\langle y, z, w\rangle + M\langle x \lhd y\rangle\langle z, w\rangle + M\langle x\rangle\langle y, z \cdot w\rangle$$
$$= M\langle x\rangle\langle y, z\rangle + M\langle x\rangle\langle y \cdot z, w\rangle + K\langle y, z, w\rangle;$$

$$M^{g\ell}_{bde} M^{hf}_{acg} K^{jn}_{f\ell} \quad = \quad K^{f\ell}_{ab} M^{gn}_{\ell cd} M^{hj}_{fge}$$

$$\langle x\rangle\langle y,z\rangle\langle w\rangle :$$

$$\begin{aligned}\text{Ж}\langle x\rangle\langle y\rangle\langle w\rangle + \text{Ж}\langle x \triangleleft y\rangle\langle z\rangle\langle w\rangle + W\langle y,z\rangle\langle w\rangle + M\langle x\rangle\langle y,z\rangle \triangleleft w \\ = M\langle x\rangle\langle y,z\rangle + W\langle y,z\rangle\langle w\rangle + \text{Ж}\langle x\rangle\langle y\cdot z\rangle\langle w\rangle;\end{aligned}$$

$$W^{qrp}_{ac}\text{Ж}^{s\ell n}_{pbd}\text{Ж}^{tjf}_{r\ell e}M^{gh}_{qst} \quad = \quad M^{p\ell}_{cde}\text{Ж}^{gjr}_{abp}W^{hfn}_{r\ell}$$

$$\langle x\rangle\langle y\rangle\langle z,w\rangle :$$

$$\begin{aligned}M\langle y\rangle\langle z,w\rangle + M\langle x \triangleleft y\rangle\langle z,w\rangle + \text{Ж}\langle x\rangle\langle y\rangle\langle z\cdot w\rangle \\ = \text{Ж}\langle x\rangle\langle y\rangle\langle z\rangle + \text{Ж}\langle x \triangleleft z\rangle\langle y \triangleleft z\rangle\langle w\rangle + M\langle x\rangle\langle z,w\rangle + M\langle y\rangle\langle z,w\rangle;\end{aligned}$$

$$\text{Ж}^{p\ell s}_{cef}\text{Ж}^{gqr}_{bdp}M^{ht}_{aq\ell}M^{jn}_{trs} \quad = \quad M^{qp}_{abc}M^{\ell n}_{pde}\text{Ж}^{ghj}_{q\ell f}$$

$$\langle x\rangle\langle y\rangle\langle z\rangle\langle w\rangle :$$

$$\begin{aligned}\text{Ж}\langle x\rangle\langle y\rangle\langle z\rangle + \text{Ж}\langle x \triangleleft z\rangle\langle y \triangleleft z\rangle\langle w\rangle + \text{Ж}\langle x\rangle\langle z\rangle\langle w\rangle + \text{Ж}\langle y\rangle\langle z\rangle\langle w\rangle \\ = \text{Ж}\langle y\rangle\langle z\rangle\langle w\rangle + \text{Ж}\langle x \triangleleft y\rangle\langle z\rangle\langle w\rangle + \text{Ж}\langle x\rangle\langle y\rangle\langle w\rangle + \text{Ж}\langle x\rangle\langle y\rangle\langle z\rangle \triangleleft w.\end{aligned}$$

$$\text{Ж}^{ruv}_{cef}\text{Ж}^{g\ell s}_{bdr}\text{Ж}^{hjt}_{a\ell u}\text{Ж}^{npq}_{tsv} \quad = \quad \text{Ж}^{vur}_{abc}\text{Ж}^{s\ell q}_{rde}\text{Ж}^{tjp}_{u\ell f}\text{Ж}^{ghn}_{vst}$$

In order to write the tensor indices in terms of x, y, z, and w, take the boundaries of the arguments of the cochains, K, M, W, and Ж, and recapitulate them at the top and bottom of the tensors. For example,

$$\partial\langle x,y,z\rangle = \langle y,z\rangle - \langle xy,z\rangle + \langle x,yz\rangle - \langle x,y\rangle,$$

so in the tensor K^{hj}_{bc}, the values of the indices are $h = \langle x,yz\rangle$, $j = \langle y,z\rangle$, $b = \langle xy,z\rangle$, and $c = \langle x,y\rangle$.

6. Conclusion

The eight important 5-isomorphisms are consequences of the coherence axiom for associativity and the naturality conditions of a quadruple arrow exchanger. In this sense, they arise in common categorical contexts. In turn, they correspond to moves to embedded foams in 4-space, and with the exception of the coherence relation for associativity, they represent isotopies of these embedded foams. Moreover, other moves to foams have nice categorical interpretations: weak and strong invertibility of a given quadruple arrow, naturality of exchangers, and adjoint relations (sliding arrows over optima).

By coloring foams with shalgebra (or qualgebra) elements, the strongly invertible quadruple arrows give birth to 3-cocycle conditions, and the eight important 5-isomorphisms engender 4-cocycle conditions.

The 4-cocycle conditions are related to half-twisted forests in which four branches appear at the top. The corresponding 5-cocycle conditions correspond to half-twisted forests with five branches on the top, and these correspond to codimension 3 singularities for planar trivalent graphs. In particular, we encourage research into the associated higher categorical structures that are formulated by studying 3-foams in 5-space along the lines of Chapter 7.

CHAPTER 11

Higher dimensional braids

There are four aspects of braid theory that will be outlined within this chapter: (1) geometric, (2) algebraic, (3) surface braids and their higher dimensional analogues, and (4) the categorical/diagrammatic or glyphographic representation of these.

The geometric aspect of braids is expressed (Section 1) by thinking of them as strings in space that flow monotonically downwards. From the algebraic perspective, the n-string braid group, B_n, is given by the generators $\sigma_1, \ldots, \sigma_{n-1}$ and the relations stated in Chapter 9, Section 5. In Section 2, consequences of the relations will be discussed and described glyphographically. A closed geometric braid, embedded in a solid torus $D^2 \times S^1$ is a (non-regular) covering of the core $\{0\} \times S^1$ of the torus (See [**55**] for the definition of a covering space). In an analogous manner, a closed orientable surface that is embedded in $D^2 \times S^2$ in such a way that the composition $F \subset D^2 \times S^2 \stackrel{p_2}{\to} S^2$ is a (simple) branched covering is called a *closed (simple) surface braid*. The definition of a (simple) branched cover will be given in Section 3. Simple surface braids can be described by means of their charts which are glyphographic representations of quadruple arrows in a multi-category that is closely related to the multi-category of embedded surfaces in 4-space. The relationship is presented in Section 3.4. Further categorical and geometric descriptions of simple branched coverings will be presented by means of several examples in Section 4.

1. Geometric braids

A *geometric braid* or simply a *braid* of degree n is a compact 1-manifold b in $\mathbb{R}^2 \times [0,1]$ which is the union of n mutually disjoint arcs $a_1, \ldots, a_n$ such that the projection $a_i \to [0,1]$ of a_i to the second factor is an embedding for each $i = 1, \ldots, n$, and the boundary ∂b consist of the points $([n] \times \{0\}) \times \{0,1\}$ where $[n] = \{1, 2, \ldots, n\} \subset \mathbb{N} \subset \mathbb{R}$. The interval $[0, n+1] \subset \mathbb{R}$ is depicted from left to right along the page, the interval $[0,1]$ runs down the page, and the second $\mathbb{R}$ factor in $\mathbb{R}^2 \times [0,1]$ is perpendicular to the reader's line of sight. Imagine this as a thin region that contains the braid. In this way, braids are described by their diagrams in $\mathbb{R} \times [0,1]$ by using the projection $\mathbb{R}^2 \times [0,1] \to \mathbb{R} \times [0,1]; ((x,y),s) \to (x,s)$, and the page that you are reading is contained in this plane. A diagram of a braid is derived, by means of adding crossing information, from an immersed 1-manifold in $\mathbb{R} \times [0,1]$ which is monotone with respect to the $[0,1]$ factor and whose boundary is $[n] \times \{0,1\}$. A diagram of a geometric braid is referred to as a *braid diagram*. Usually in a braid diagram, all multiple points are transverse double points, and they have over/under information indicated as in the case of knot diagrams. The resolved double points are called *crossings* or *crossing points* of the diagram. Also, the heights of the crossings of a braid diagram are usually all distinct.

The *trivial braid* or the *identity braid* is the braid $([n] \times \{0\}) \times [0,1]$.

The most elementary non-trivial braids are braids $\sigma_1, \ldots, \sigma_{n-1}$ and their inverses $\sigma_1^{-1}, \ldots, \sigma_{n-1}^{-1}$. The illustration of these is recalled from Chapter 9, Section 5. Here downward pointing orientations are included.

σ_i :

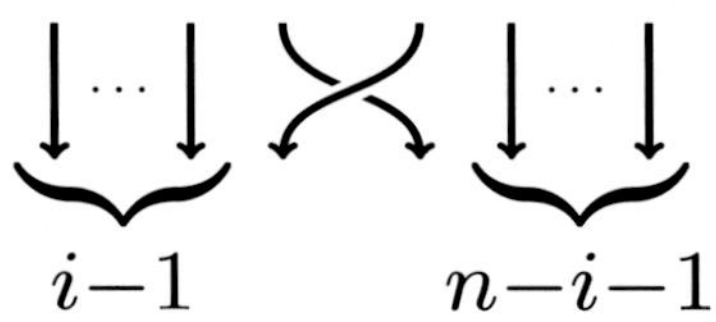

σ_i^{-1} :

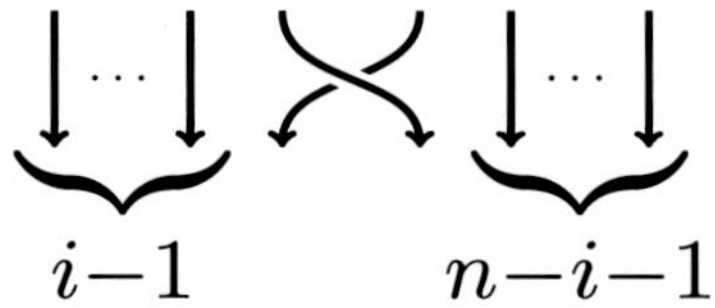

Two (geometric) braids b and b' are regarded as the same or equivalent if there is a one-parameter family of braids $\{b_s\}$ with $b_0 = b$ and $b_1 = b'$. The equivalence class $[b]$ of a geometric braid b is also called a *braid*, and the equivalence class $[b]$ may be denoted by b.

For two braids b and b' of the same degree n, the product bb' is defined by putting b' under b. Under this composition, the trivial braid $([n]\times\{0\})\times[0,1]$ serves as an identity. The product is associative, and inverses exist because the equivalence relation allows for the braid diagrams to be rescaled as needed.

The *braid group*, B_n, of degree n is the group of (the equivalence classes) of braids of degree n. This group is generated by

$$\sigma_1, \ldots, \sigma_{n-1},$$

which are called the *standard generators* or the *Artin's braid generators*. The braid group has the group presentation below.

$$\left\langle \sigma_1, \ldots, \sigma_{n-1} : \begin{array}{ll} \sigma_i\sigma_{i+1}\sigma_i = \sigma_{i+1}\sigma_i\sigma_{i+1}, & \text{for } i = 1, \ldots, n-2, \\ \sigma_i\sigma_j = \sigma_j\sigma_i, & \text{for } |i-j| > 1 \end{array} \right\rangle.$$

The relations above are called *braid relations*.

In the categorical language, there is a multi-category with one object $*$, only the identity 1-arrow defined upon it, and a non-identity double arrow $\boxed{\bullet}$ as presented in Chapter 9, Section 1. In this way, $([n]\times\{0\})$ corresponds

to

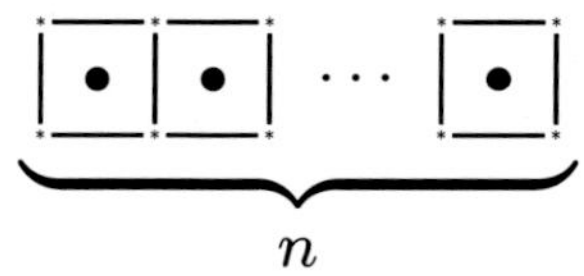

,

and the braid generator corresponds to an exchanger . That the exchanger is strongly invertible allows for the existence of , and its naturality with respect to other triple arrows gives the braid relation $\sigma_i\sigma_{i+1}\sigma_i = \sigma_{i+1}\sigma_i\sigma_{i+1}$. According to the conventions that different things are not equal (Chapter 2, Section 5), this relation and the relations $\sigma_i^{\pm i}\sigma_i^{\mp 1} = 1$ are strongly invertible quadruple arrows. Glyphs will be assigned to them in the next section. The commutativity of distant braid generators is an expression of the (strongly invertible) quadruple arrow exchanger.

Notice that the braid group, B_n, could have been defined abstractly via the presentation above.

2. Glyphographic description of surface braids

The braid relations

$$\begin{aligned} &\sigma_i\sigma_{i+1}\sigma_i = \sigma_{i+1}\sigma_i\sigma_{i+1}, && \text{for } i = 1, \dots, n-2, \\ &\sigma_i\sigma_j = \sigma_j\sigma_i, && \text{for } |i-j| > 1 \end{aligned}$$

with the trivial relations

$$\sigma_i^{-1}\sigma_i = 1, \quad \sigma_i\sigma_i^{-1} = 1, \qquad \text{for } i = 1, \dots, n-1$$

induce the following relations

$$\begin{aligned} &\sigma_i^{-1}\sigma_{i+1}\sigma_i = \sigma_{i+1}\sigma_i\sigma_{i+1}^{-1}, && \text{for } i = 1, \dots, n-2, \\ &\sigma_i\sigma_{i+1}\sigma_i^{-1} = \sigma_{i+1}^{-1}\sigma_i\sigma_{i+1}, && \text{for } i = 1, \dots, n-2, \\ &\sigma_i^{-1}\sigma_{i+1}^{-1}\sigma_i = \sigma_{i+1}\sigma_i^{-1}\sigma_{i+1}^{-1}, && \text{for } i = 1, \dots, n-2, \\ &\sigma_i\sigma_{i+1}^{-1}\sigma_i^{-1} = \sigma_{i+1}^{-1}\sigma_i^{-1}\sigma_{i+1}, && \text{for } i = 1, \dots, n-2, \\ &\sigma_i^{-1}\sigma_{i+1}^{-1}\sigma_i^{-1} = \sigma_{i+1}^{-1}\sigma_i^{-1}\sigma_{i+1}^{-1}, && \text{for } i = 1, \dots, n-2, \\ &\sigma_i^{-1}\sigma_j = \sigma_j\sigma_i^{-1}, && \text{for } |i-j| > 1, \\ &\sigma_i\sigma_j^{-1} = \sigma_j^{-1}\sigma_i, && \text{for } |i-j| > 1, \\ &\sigma_i^{-1}\sigma_j^{-1} = \sigma_j^{-1}\sigma_i^{-1}, && \text{for } |i-j| > 1. \end{aligned}$$

The braid relations and these consequences will be schematized in a glyphographic format because items 1 and 2 from Chapter 2, Section 5 suggest that different things are not equal, but should be compared by arrows. The quadruple arrows that correspond to these relations are all strongly invertible. In particular, there are 5-isomorphisms that are oriented versions of those given in Chapter 10, Section 1.1 for the type II and type III move. Moreover, there are zig-zag moves or cuspidal 5-isomorphisms that involve the cancelations of $\sigma_i^{\pm 1}\sigma_i^{\mp 1}$. In place of a bookmark, indices such as i, $i+1$

or j are placed along the arcs of the glyphs that indicate the corresponding index of the braid generator ($\uparrow$) or its inverse ($\downarrow$).

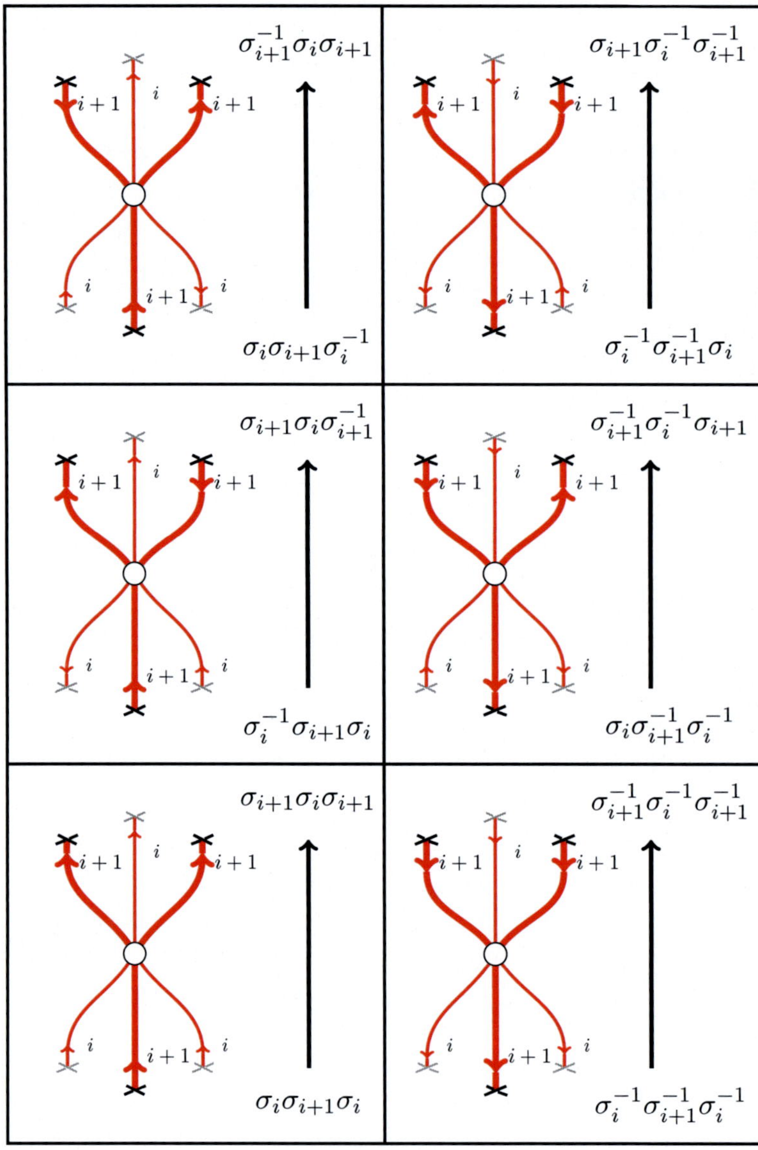

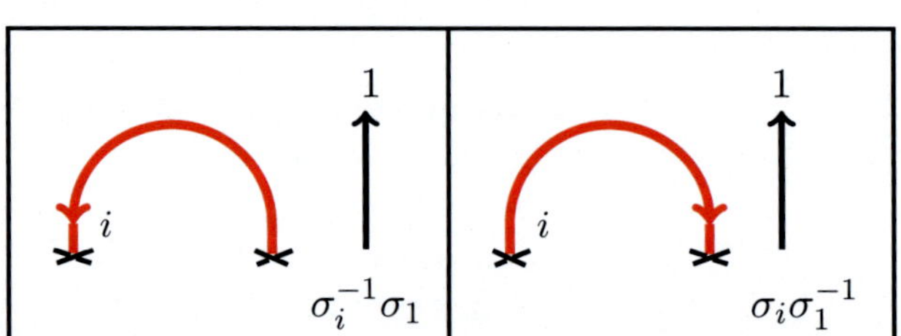

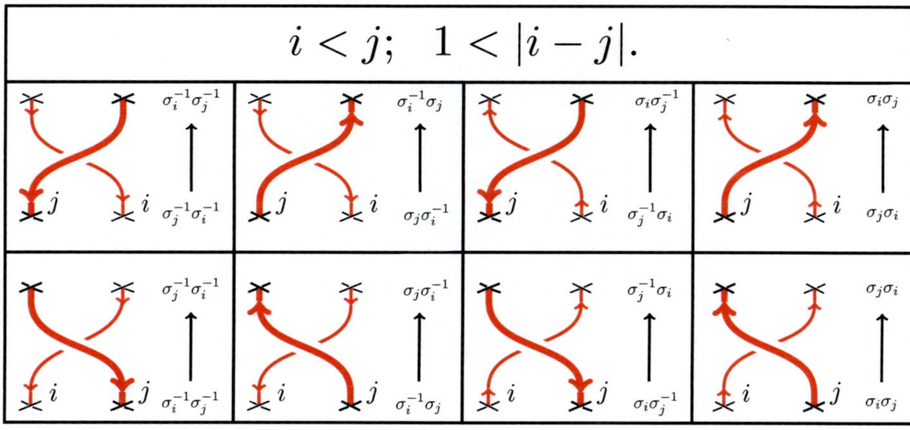

2.1. Additional glyphs. The consequences of the braid relations are strongly invertible quadruple arrows that were presented using glyphographic notation above. These quadruple arrows together with the four that are presented in this section are used to give the categorical description of surface braids. The geometric context for these remaining four quadruple arrows will be explained in Section 3 (this chapter). Any one of the quadruple arrows that is presented in the table below is called a *black vertex*.

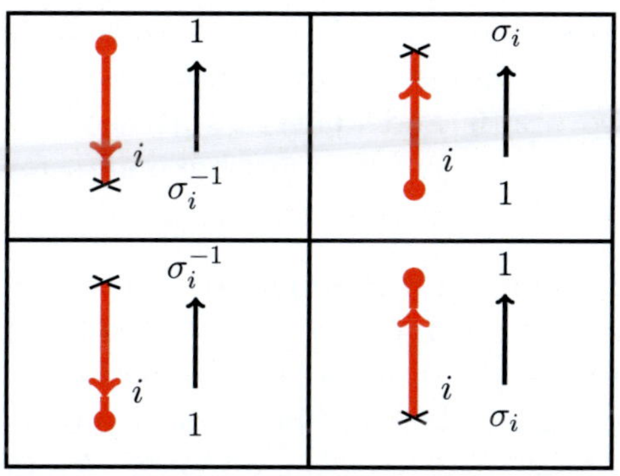

A broken surface diagram for the simple surface braid that is represented by a black vertex is shown in Fig. 1.

Black vertices are weakly invertible in the sense that there are quintuple arrows given in Section 3.5 (this chapter). Those quintuple arrows are not 5-isomorphisms, but there are 6-arrows from their compositions to the appropriate identity 5-arrows. The weak inverse to any one of the black vertices appears in the same column as the given vertex. As weakly invertible quadruple arrows, black vertices share certain categorical similarities with the unit, , and co-unit, , arrows. In particular, additional quintuple arrows that resemble those given in Chapter 4, Tables 1, 2, and 13 will be used to help describe braided 3-dimensional solids. And even further, the schemata of Chapter 7 and in particular 1.1 will apply. See also Chapter 4, Fig. 14 and Chapter 5, Section 3.1. However, in the context of surface braids, the compositions of black vertices never cancel.

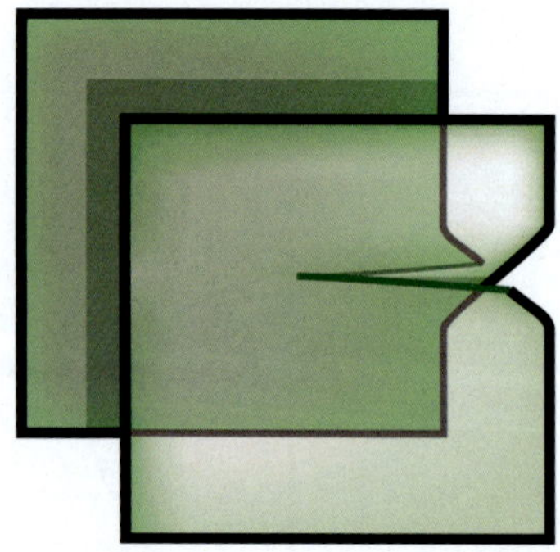

FIGURE 1. The broken surface diagram near a black vertex

Serifs will not be drawn upon the glyphs in the sections that follow.

3. Surface braids

The section begins with a short exposition of branched coverings between surfaces (Section 3.1). Then braided surfaces and surface braids are introduced (Section 3.2). Two examples of surface braids are presented from the point of view of their movies and as quadruple arrows. Braid charts are introduced, and the chart moves are articulated (Section 3.3). The relationships between surface braids and knotted surfaces in 4-dimensional space will be given (Section 3.4). Self-intersection points for surface braids are introduced in Section 3.6 where their axioms are discussed.

3.1. Branched coverings of surfaces. Let $U = \{z \in \mathbb{C} : |z| < 1\}$ denote the unit disk in the complex plane. Consider the function $U \to U$ defined by $z \mapsto z^d$. The origin 0 in the domain is the prototype for a *branch point of degree* d. The case $d = 2$ will be of most interest.

Let $f : M \to N$ be a map between surfaces. A point p of M is called a *branch point* of f with *degree* $d \geq 2$ if there is a neighborhood $\nu(p)$ of p in M and a neighborhood $\nu(f(p))$ in N together with homeomorphisms $\varphi : \nu(p) \to U$ with $\varphi(p) = 0$ and $\psi : \nu(f(p)) \to U$ with $\psi(f(p)) = 0$ such that $\psi \circ f \circ \varphi^{-1} : U \to U$ is $z \mapsto z^d$ for all $z \in U$. The integer d is also called the *local degree* of f at p, and is denoted by $\deg(f;p)$. A point p of M is called a *regular point* of f if there is a neighborhood $\nu(p)$ of p in M and a neighborhood $\nu(f(p))$ of $f(p)$ in N such that $f|\nu(p) : \nu(p) \to \nu(f(p))$ is a homeomorphism. For convenience, for a regular point p, the local degree $\deg(f;p)$ is defined to be 1. A point q of N is called a *branch value* if it is the image of a branch point. A point q of N is called a *regular value* if each point of $f^{-1}(q)$ is a regular point.

A map $f : M \to N$ between surfaces is called a *branched covering map* of degree n if each point of M is a regular point or a branch point and the restriction $f|_{M \setminus f^{-1}(\Sigma)} : M \setminus f^{-1}(\Sigma) \to N \setminus \Sigma$ is a covering map of degree

n, where Σ is the set of branch values. The covering map $f|_{M\setminus f^{-1}(\Sigma)}$ is called the *covering map associated with* the branched covering map f. By definition, a covering map is a branched covering map without branch points.

A branched covering map $f : M \to N$ is called *simple* if for every branch value $q \in N$, there is a unique branch point p in $f^{-1}(q)$ and the local degree at the branch point is 2. By definition, a covering map is a simple branched covering map.

3.2. 2-dimensional braids. Let n denote a positive integer, let $D^2 = \{(x, y) : x \in [0, n+1] \ \& \ y \in [-1, 1]\}$, and let $B^2 = \{(z, w) : z, w \in [0, 1]\}$. Topologically, both D^2 and B^2 are 2-dimensional disks. Let $p_{zw} : D^2 \times B^2 \to B^2$ denote the projection $p_{zw}(x, y, z, w) = (z, w)$.

Let F be a compact oriented surface embedded in $D^2 \times B^2$. The embedded surface F is a *braided surface* of braid index n if the map $f : F \to B^2$, which is the restriction of the projection $p_{zw} : D^2 \times B^2 \to B^2$ to F, is a branched covering map of degree n, and $\partial F = F \cap \partial(D^2 \times B^2) \subset (\mathrm{int} D^2) \times (\partial B^2)$. The square boundary of B^2 is identified with the unit circle S^1 in the usual topological manner. A braided surface is called a *2-dimensional braid* or a *surface braid* if $\partial F = ([n] \times \{0\}) \times (\partial B^2)$, where $[n] = \{1, 2, \ldots, n\} \subset [0, n+1]$. A 2-dimensional braid is called *simple* if its branched covering map is simple.

Let F be a 2-dimensional braid. The coordinates of $D^2 = [0, n+1] \times [-1, 1]$ are given as (x, y) and those for B^2 are given as (z, w). For each $w \in [0, 1]$, the restriction of F to the set $D^2 \times [0, 1] \times \{w\}$ is a (geometric) braid or a "singular braid" of degree n, say b_w. The one-parameter family $\{b_w\}_{w \in [0,1]}$ is called the *movie* or the *braid movie* of F.

Fig. 2 is an example of a movie of a 2-dimensional braid of braid index 3 in which the singular braids occur at levels $w = w_1, w_2, w_3$, and w_4. Fig. 3 contains its glyphographic representation. The braid indices, rather than the bookmarks, are edge labels, and serifs are not present along the glyphs. The initial braid b_0 and the terminal braid b_1 are both trivial 3-braids.

Fig. 4 contains an example of a movie of a 2-dimensional braid of braid index 4. Singular braids b_w occur at $w = w_1, w_2, w_3, w_4$. The initial braid b_0 and the terminal braid b_1 are the trivial 4-braids. The glyphographic representation appears in Fig. 5.

The movie of Fig. 2 can be described a sequence of braid words similar to the sequence below (which represents the movie of Fig. 4). The symbol v_i denotes the singular point between the ith and the $(i+1)$st strings.

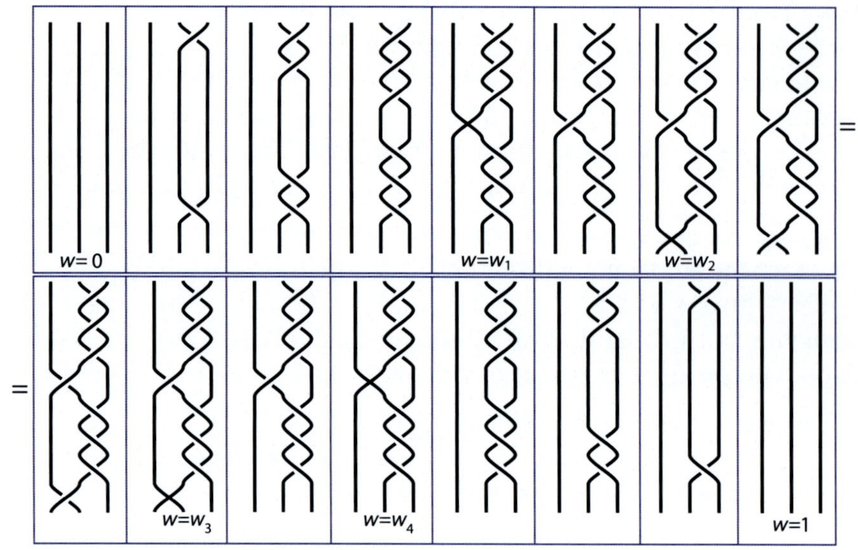

FIGURE 2. A braid movie

$$
\begin{array}{rcl}
b_0 & = & e \\
b_1 & = & \sigma_1^{-1}\sigma_1 \\
(w = w_1) & & \sigma_1^{-1}v_1 \\
b_2 & = & \sigma_1^{-1} \\
(w = w_2) & & v_3\sigma_1^{-1} \\
b_3 & = & \sigma_3\sigma_1^{-1} \\
b_4 & = & \sigma_2^{-1}\sigma_2\sigma_3\sigma_1^{-1} \\
b_5 & = & \sigma_2^{-1}\sigma_1^{-1}\sigma_1\sigma_2\sigma_3\sigma_1^{-1} \\
b_6 & = & \sigma_2^{-1}\sigma_1^{-1}\sigma_1\sigma_2\sigma_1^{-1}\sigma_3 \\
b_7 & = & \sigma_2^{-1}\sigma_1^{-1}\sigma_2^{-1}\sigma_1\sigma_2\sigma_3 \\
(w = w_3) & & \sigma_2^{-1}\sigma_1^{-1}v_2\sigma_1\sigma_2\sigma_3 \\
b_8 & = & \sigma_2^{-1}\sigma_1^{-1}\sigma_1\sigma_2\sigma_3 \\
b_9 & = & \sigma_2^{-1}\sigma_2\sigma_3 \\
b_{10} & = & \sigma_3 \\
(w = w_4) & & v_3 \\
b_{11} & = & e
\end{array}
$$

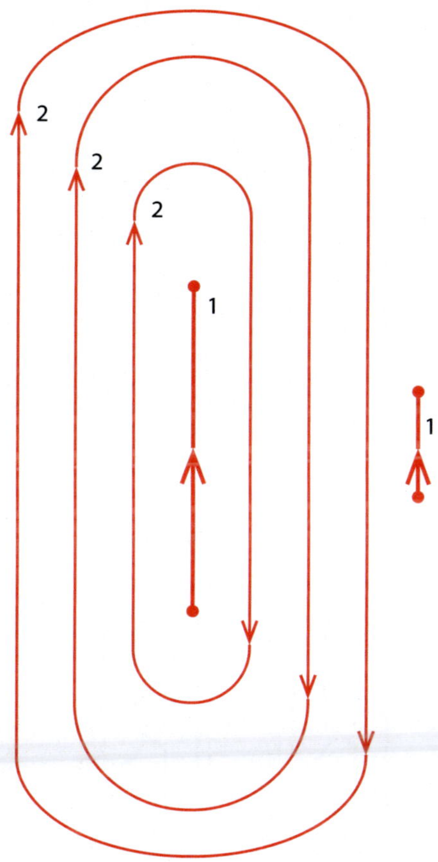

FIGURE 3. The glyphographic representation or chart

3.3. Charts and chart moves. A *chart* of degree, or braid index, n is a finite graph, $\Gamma = \Gamma(n)$, in B^2 whose edges are oriented and labelled with integers in $\{1, \ldots, n-1\}$ satisfying the following conditions.

(i) Each vertex has degree one, four, or six.
(ii) The edge adjacent to a degree-one vertex is either oriented towards or away from the vertex.
(iii) The four edges adjacent to a degree-four vertex are labeled with i and j alternately with $|i-j| > 1$, and diagonally opposite edges are oriented coherently.
(iv) The six edges adjacent to a degree-six vertex are labelled with i and $i \pm 1$ alternately, three consecutive edges are oriented inwardly, and the other three are outwardly.

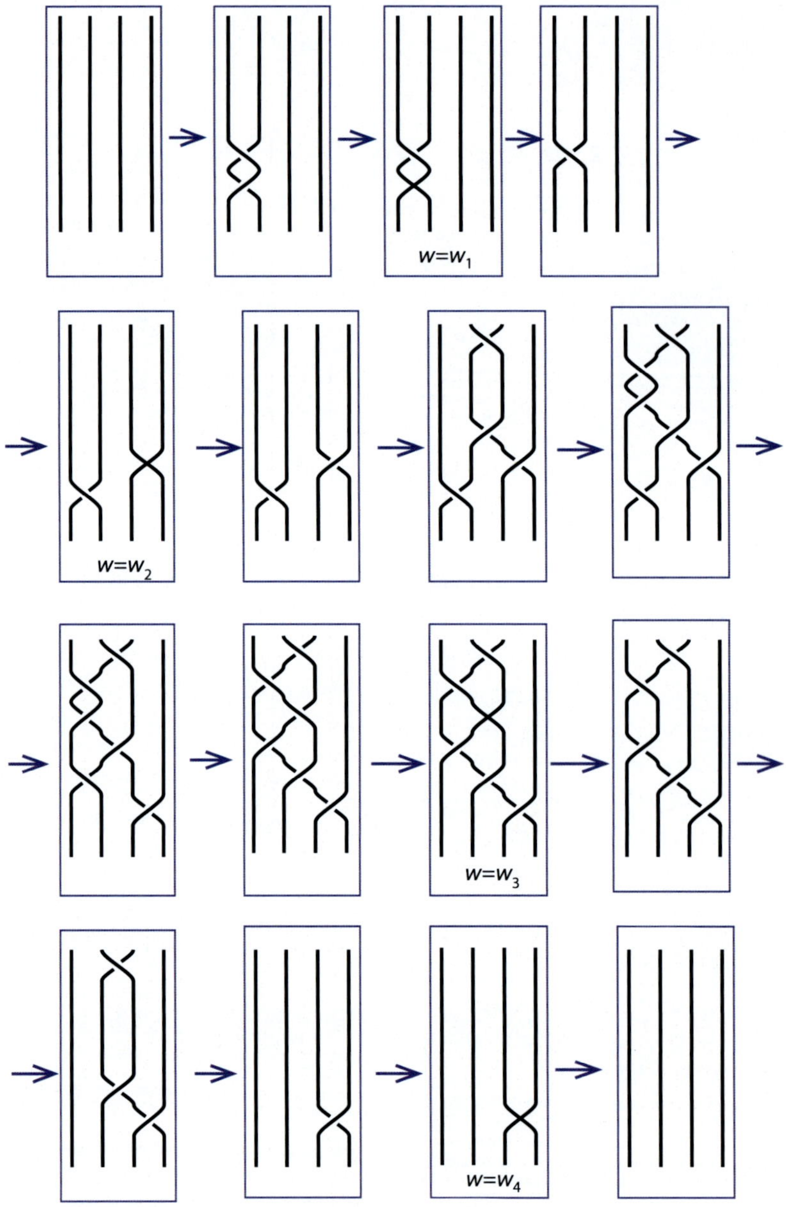

FIGURE 4. A 4-string braid movie

Furthermore, the graph is assumed to be disjoint from the boundary ∂B^2 of B^2. A *black vertex* is a degree-one vertex, a *crossing* is a degree-four vertex, and a *white vertex* is a degree-six vertex.

The glyphs given in Section 2 (this chapter) describe vertices and critical points for a chart. Recall that $B^2 = \{(z, w) : z, w \in [0, 1]\}$. The chart Γ is assumed to be embedded in B^2 in such a way that the last coordinate w

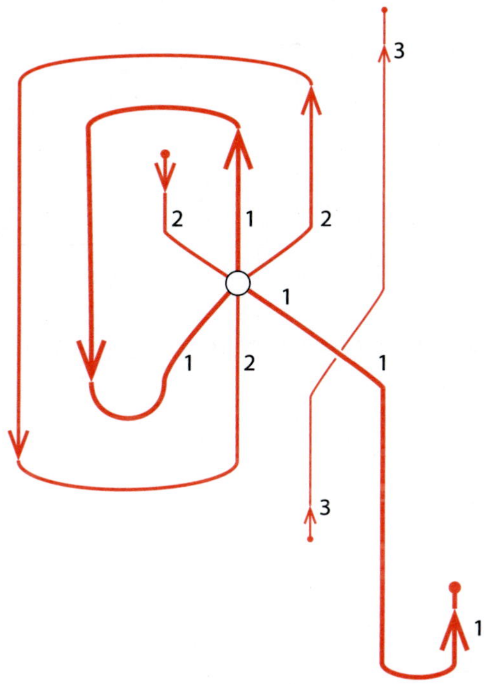

FIGURE 5. Its glyphographic representation or chart

serves as a height function for Γ. Non-degenerate critical points correspond to type II glyphs, white vertices correspond to type III glyphs, crossings correspond to quadruple arrow exchangers, and black vertices were described in Section 2.1. Because simultaneity is illusory (Chapter 2, Section 5, item 4) critical points and vertices usually appear at distinct heights; these are values of w. Figs. 3 and 5 contain two examples of charts.

The black vertices in a chart correspond to branch values of a simple surface braid. A simple surface braid can be isotoped in the 4-ball $D^2 \times B^2$ so (1) it can be described in movie form, (2) the projection p_{zw}, when restricted to the surface F, is a simple branched covering, and the chart in B^2 describes the double points, triple points, and branch points of the restriction to F of the projection p_{xzw}. Moreover, there is an unambiguous reconstruction of the surface braid from its corresponding chart.

Two charts Γ_1 and Γ_2 represent simple surface braids that are isotopic through simple surface braids if and only if they are related by a finite sequence of moves taken from the lists that appear below. The lists will repeat much of the information from Chapter 10, Section 1 since in many cases, the same moves apply in the knotted surface or categorical context. There are fourteen kinds of moves. Within the illustrations, $i+3 < j+1 < k$. Unoriented arcs indicate that the move is valid with any orientation.

Item 1. The quadruple arrow exchanger is strongly invertible.

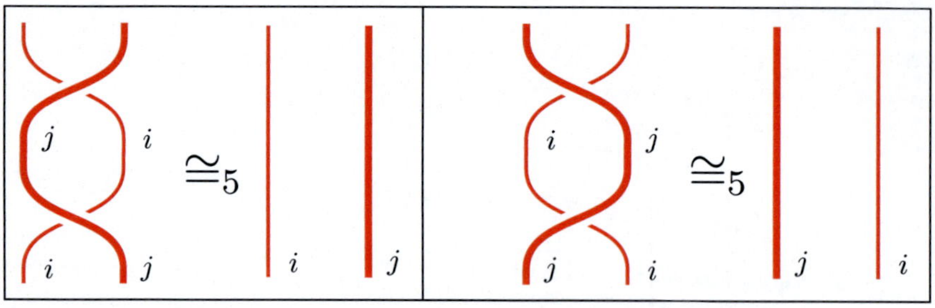

Item 2. The family of quadruple arrow exchangers are natural with respect to other quadruple arrows.

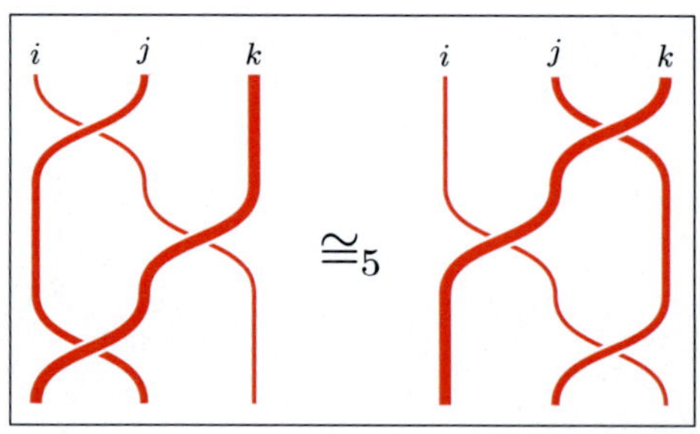

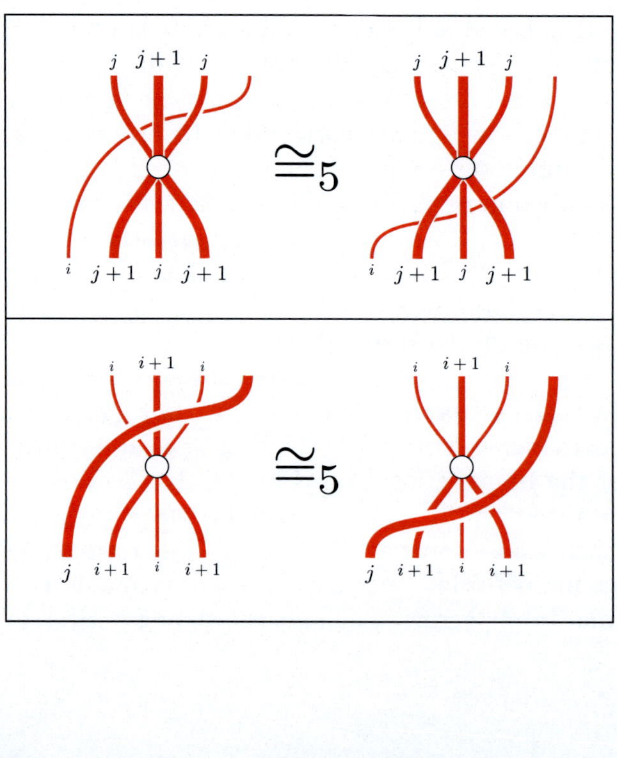

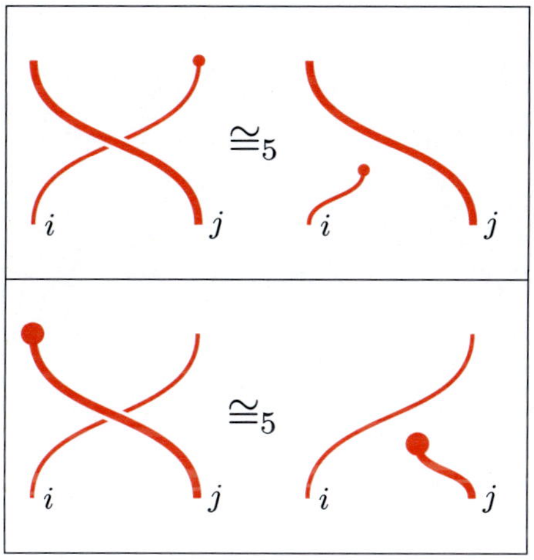

Item 3. As a consequence of the naturality of the quadruple arrow exchanger with respect to type II quadruple arrows, there are ψ-type moves.

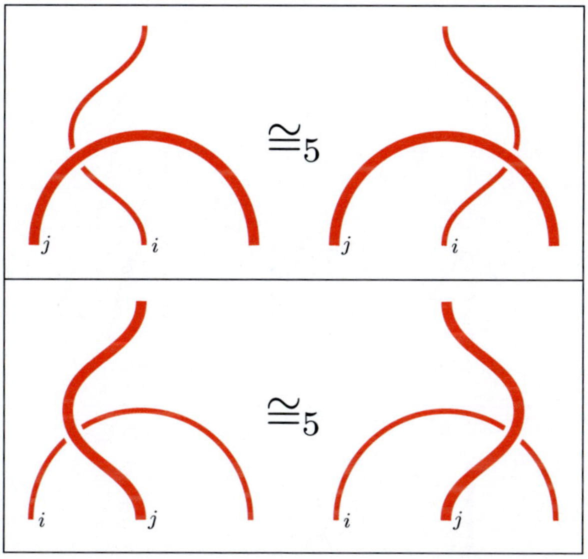

Item 4. As in Chapter 10, Section 1.3, naturality of the quadruple point exchanger implies the tetrahedral move, and the ability to pull black vertices

through transverse sheets.

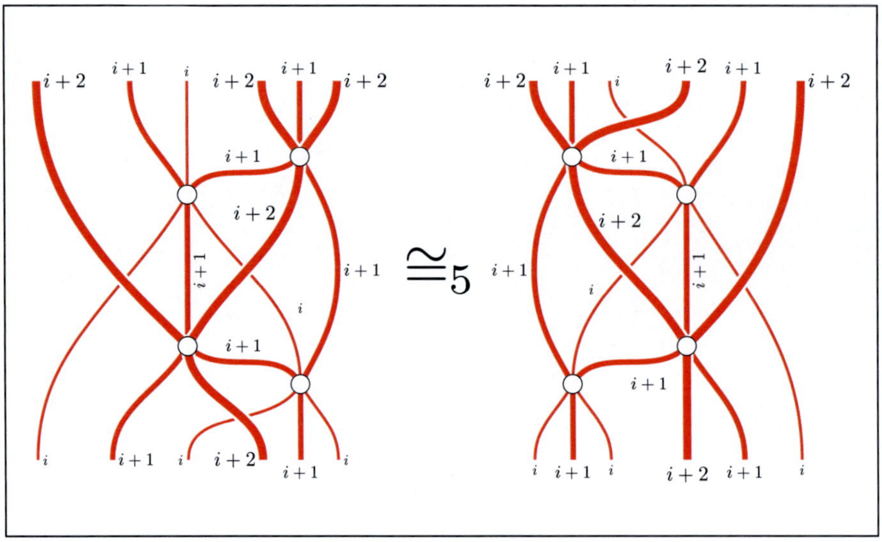

In the illustrations below, orientations are necessary. The arcs are oriented from $-i$ to $+i$. When several arcs are to be oriented the signs are respective, and they indicate that orientations among the arcs must be consistent, but can be changed uniformly. The notation $\pm i$ is a short-hand notation for a pair $(\pm, i)$. The second index is the braid index i.

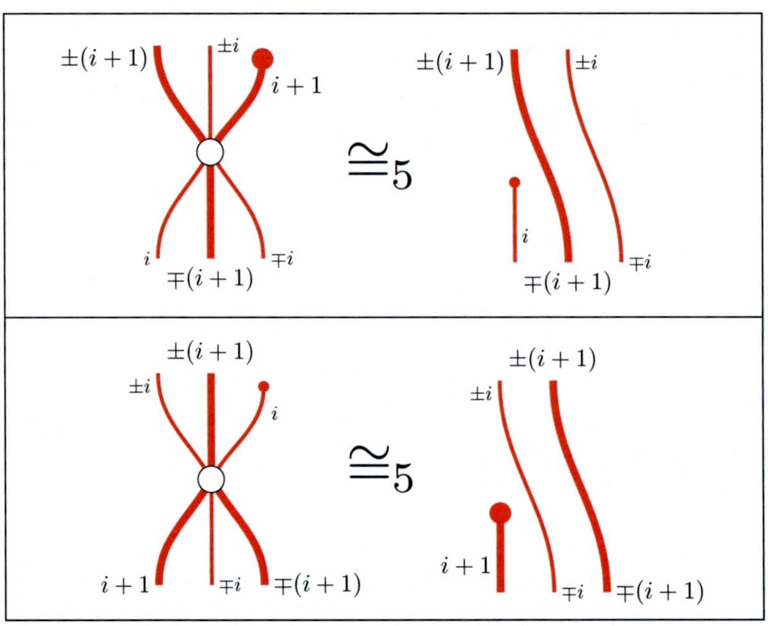

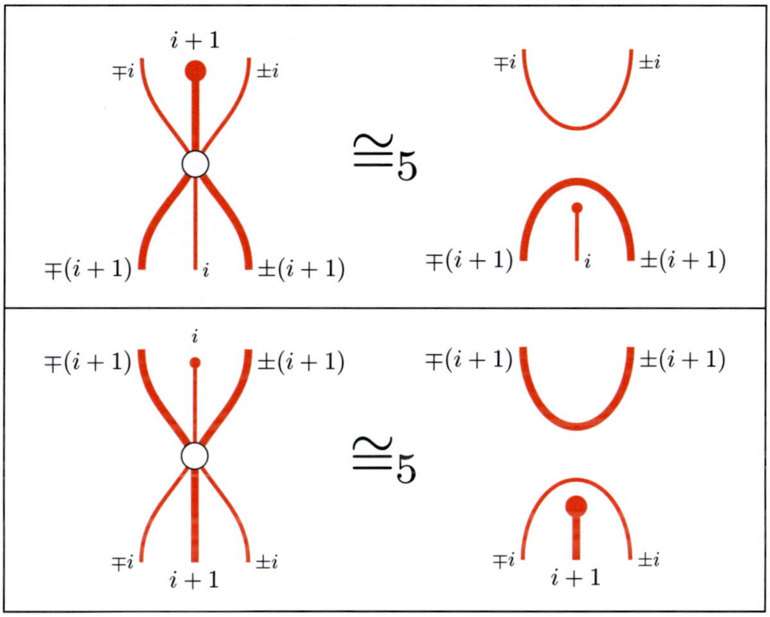

Item 5. The type II and type III moves (white vertices) are strongly invertible. In particular, there are cuspidal 5-isomorphisms for the skew compositions of a type II birth/death pair.

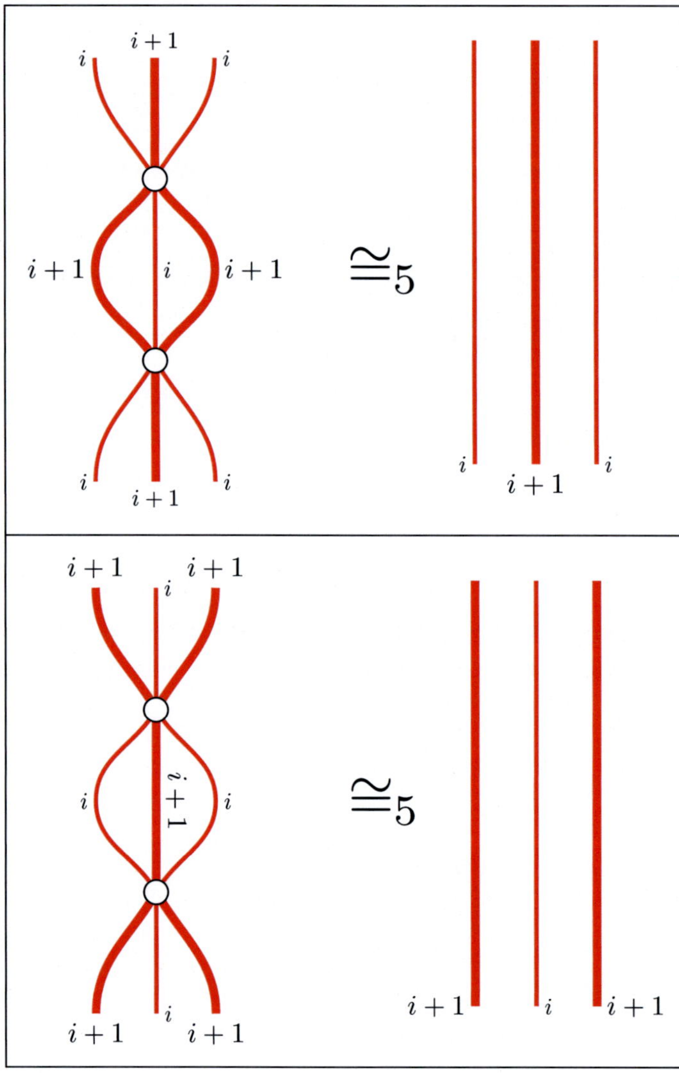
i
i + 1
i
i + 1
i
i + 1
≅5
i
i + 1
i
i
i + 1
i
i + 1
i + 1
i
i
i + 1
i
≅5
i + 1
i + 1
i
i + 1
i + 1
i
i + 1

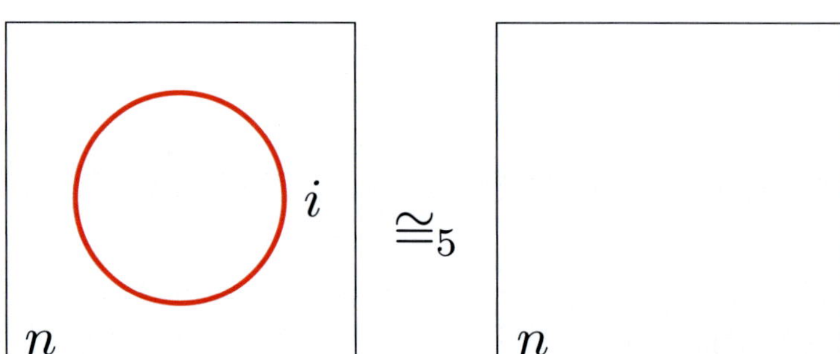
i
n
≅5
n

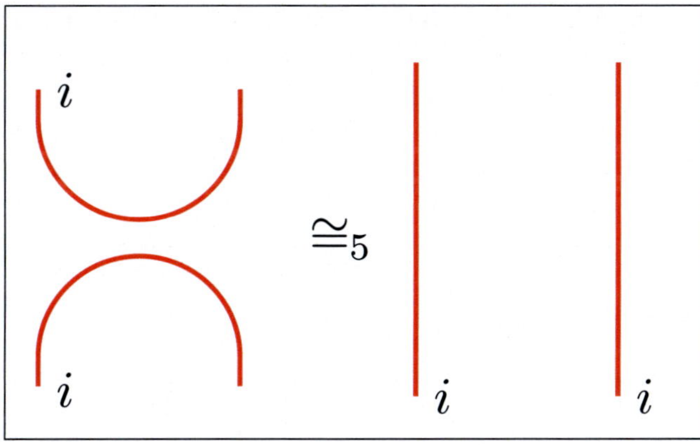
i
$\cong_5$
i
i
i

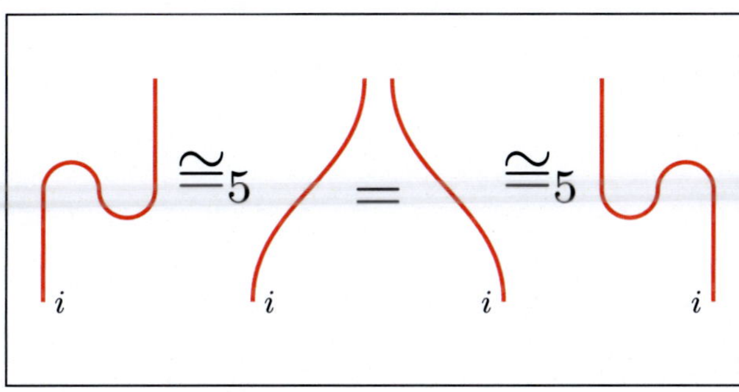
$\cong_5$
$=$
$\cong_5$
i
i
i
i

Item 6. White vertices and black vertices can be pushed over optima. In other words, the type II optima have the adjoint or fish hook property.

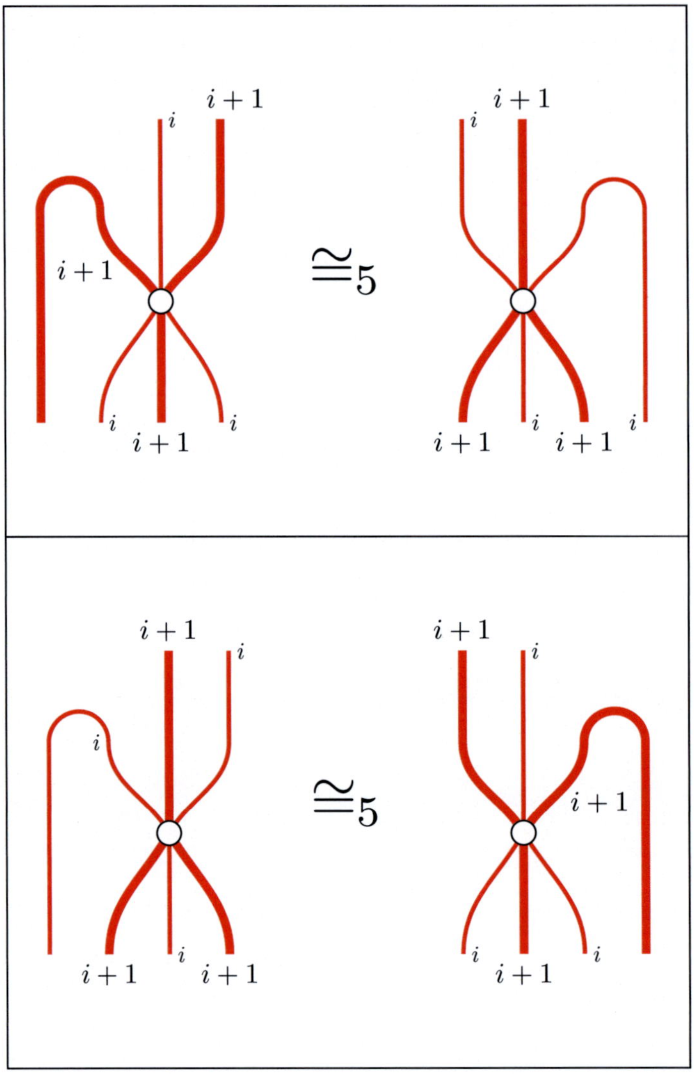

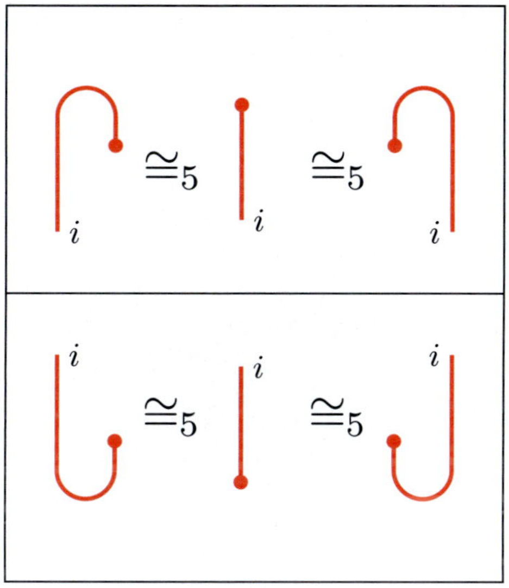

3.4. Relations with embedded surfaces. Consider the example of an n-string braid b contained in $([0,n+1]\times[-1,1])\times[0,1]$. Its closure lies in a solid torus $D^2\times S^1$, where the disk D^2 is identified with $[0,n+1]\times[-1,1]$, and the interval factor $[0,1]$ is considered as a subinterval on the circle: $[0,1]\subset S^1$. The closure, $\hat{b}$ consists of the union $b\cup\big(([n]\times\{0\})\times(S^1\setminus[0,1])\big)$. The projection $p: D^2\times S^1\to S^1$ when restricted to $\hat{b}$ is a (possibly non-regular) covering map. In this section, an analogous construction with surface braids is presented.

1. Given a chart $\Gamma=\Gamma_n$ of degree n, there is a canonical simple surface braid $F\subset D^2\times B^2$ whose projection has black vertices as its branch values. The edges and the white vertices are the images of the double point arcs and triple points, respectively, of the restriction of the projection p_{xzw} : $([0,n+1]\times\mathbb{R})\times([0,1]\times[0,1])\to[0,n+1]\times([0,1]\times[0,1])$.

Since the restriction of the embedding of F in $D^2\times B^2$ to the boundary is a trivial braid, the boundary of F consists of n circles. A closed surface $\hat{F}$ is obtained as the union $\hat{F}=F\cup(\cup_{i=1}^n D_i^2)$ where the union occurs when each boundary circle of F is identified with the boundary of exactly one 2-dimensional disk D_i^2. The surface braid $F\subset D^2\times B^2$ is extended to a branched covering of $\hat{F}$ onto the 2-sphere S^2.

The 2-sphere can be thought of as the union of two 2-dimensional disks, for example, as the northern and southern hemispheres. Or better yet, using geographical coordinates, pick the “square” region bounded by the two latitudes $\pm35°$, and the longitudes 15° west and 45°W. On our home planet, this square contains parts of South America and parts of West Africa. The complementary disk on the sphere contains the north and south poles, Asia, Australia, Europe, and North America. The lines of longitude 60°E and 120°W form a great circle upon which the complementary disk is folded.

The union of disks $\cup_{i=1}^{n} D_i^2$ are stacked and trivially cover the rest of the sphere. Fig. 6 contains a schematic depiction. The branched covering of F whose chart is indicated fits into this square region.

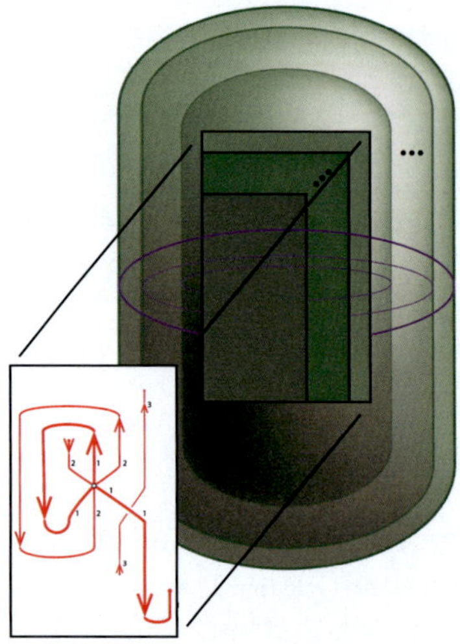

FIGURE 6. Closing a surface braid

2. A black vertex can be perturbed in 4-space to present locally a general position embedded surface in 4-space. This perturbation is described via movies and glyphs, so that a categorical 5-isomorphism exist. One such perturbation is indicated in Fig. 7. This perturbation is called the *chevron move*.

Using chevron moves any simple surface braid can be perturbed into a general position embedded surface in 4-space that, in turn, is described via birth, death, saddle, fork, cusp, ψ-move, and type I, II, and III glyphs. Equivalent surface braids represent isotopic knotted surfaces in 4-space. In [**38**], the second named author demonstrated that a given orientable embedded surface could be put into the form of a closed simple surface braid thereby developing an analogue of the classical Alexander isotopy. There are inequivalent braid charts that represent isotopic knotted surfaces.

3. In [**39**] an analogue of the Markov Theorem that indicates when two surface braids represent the same surface in 4-space was given. Here this idea is sketched. Let $\Gamma = \Gamma_n$ indicate a chart of degree n that indicates an n sheeted surface braid in the disk. Let $v = v(B)$ denote a braid word that represents the n-string braid B. Then the move that is indicated below is called a *chart conjugation*. Write $\Gamma \leftrightharpoons \Gamma^v$.

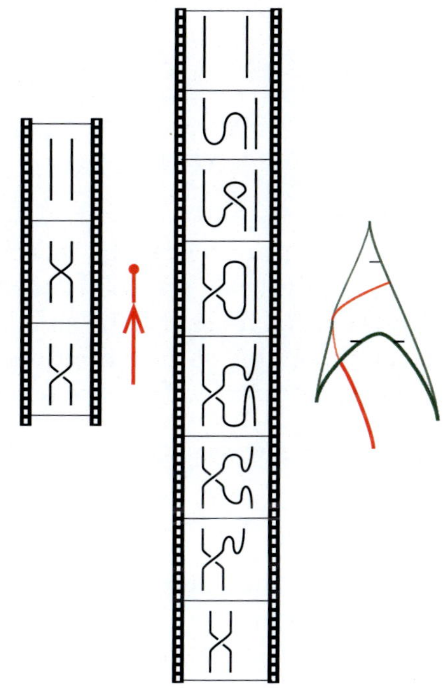

FIGURE 7. Perturbing a black vertex

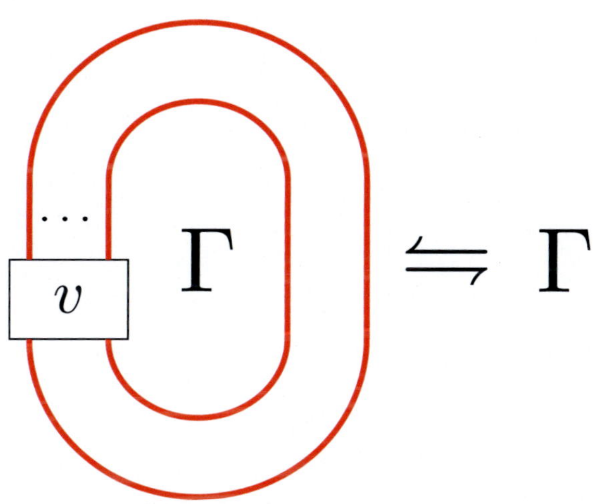

A surface braid stabilization is achieved by taking the chart Γ_n for a braid index n surface braid, and replacing it with a chart of degree $(n+1)$ that contains a disjoint edge bounded by a pair of black vertices and that has label $(n+1)$. Its orientation does not matter. The result is denoted Γ'_{n+1}.

$$\Gamma_n \leftrightharpoons \Gamma_n$$
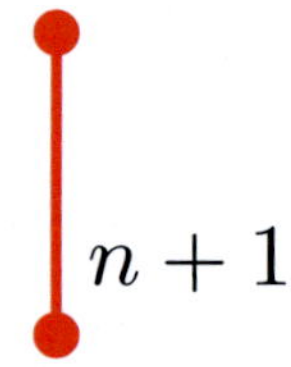

$$= \Gamma'_{n+1}$$

Suppose two charts Γ_0 and Γ_1 are related by means of a finite sequence of applications of moves (1) equivalence up to 5-isomorphisms, (2) $\Gamma \leftrightharpoons \Gamma^v$, or (3) $\Gamma_n \leftrightharpoons \Gamma'_{n+1}$. Then the charts are said to be simple Markov equivalent. Each chart represents a corresponding surface braid, and so the represented surface braids are also said to be simple Markov equivalent.

THEOREM 11.1. *The closed surface braids $\widehat{F}(\Gamma)$ and $\widehat{F}(\Gamma^v)$ represent isotopic surfaces in 4-space. The closed surface braids $\widehat{F}(\Gamma_n)$ and $\widehat{F}(\Gamma'_{n+1})$ represent isotopic surfaces in 4-space.*

PROOF. The proof is left to the reader as an exercise in applying the 5-isomorphisms for embedded surfaces. When considering the braid closure $\hat{F}$, compare to Fig. 6; nested birth, death, and fold glyphs surround the representative chart $\square$.

Any oriented surface-link in $\mathbb{R}^4$ can be described by a surface braid, namely, there is a surface braid whose closure is ambient isotopic to the surface-link. One can take such a surface braid to be simple. It is not known if such a simple surface braid is unique up to simple Markov equivalence in the above sense. However, it is unique up to Markov equivalence. Here Markov equivalence is an equivalence relation among all (simple or non-simple) surface braids generated by ambient isotopies of $D^2 \times B^2$, conjugations and (de)stablizations (See [**40**]).

As a trivial example, consider the closed surface braid — as schematized in Fig. 6 — of the chart

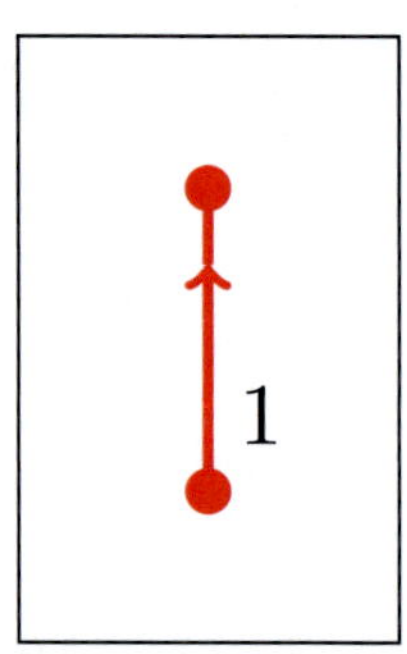

which is the 2-fold branched cover of the 2-sphere S^2 that has two branch points. It coincides with the graph of the extension of $z \mapsto z^2$ to $\mathbb{C} \cup \{\infty\} = S^2$ in which both 0 and ∞ are simple branch points. See Fig. 8. This chart represents a braid movie that can be described as the sequence $(1, \sigma_1, 1)$ of braid words in B_2. It is Markov equivalent to the degree 0 empty chart. The closed surface braid of such a chart is the standardly embedded 2-sphere, $S^2 \subset \mathbb{R}^3$. An explicit isotopy can be constructed using the chevron moves that are defined via Fig. 7. See also Fig. 9.

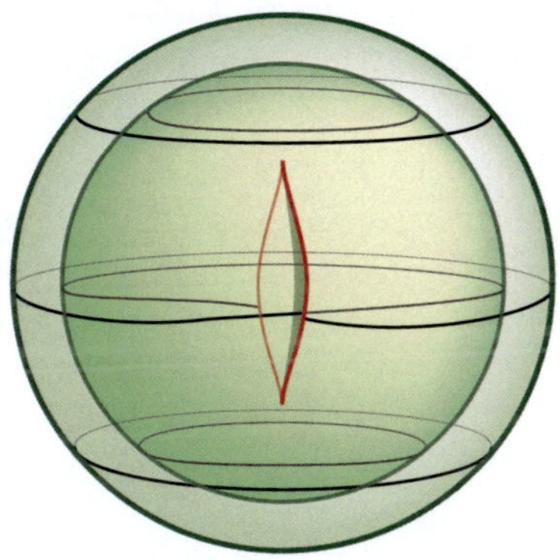

FIGURE 8. The broken surface diagram of an edge in a chart

3.5. Weak invertibility of black vertices. In the following, the quadruple arrow  is assumed to be weakly invertible. In particular, there are four (unnamed) quintuple arrows as indicated below. These are also weakly invertible.

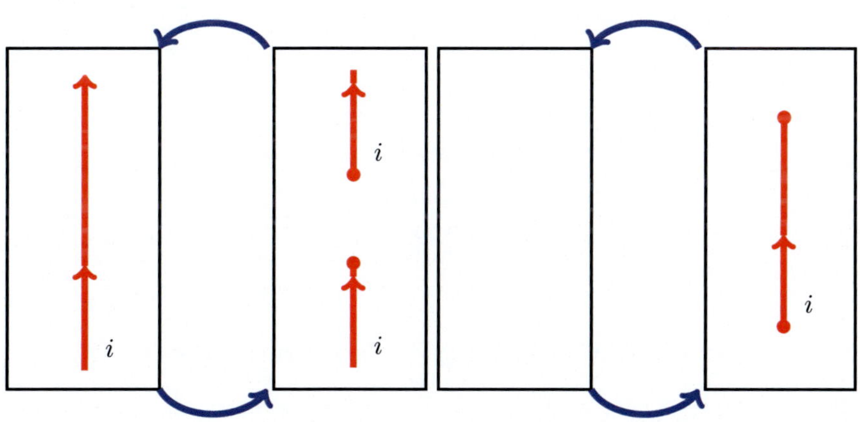

Figs. 9 and 10 demonstrate, via the chevron map, the relationship between these arrows that are paired as (arrow, its reverse) and the (saddle, fork) reverse pairs. After chevron maps are applied, the diagrams are manipulated through a series of 5-isomorphisms in which strong inverses cancel. Then the weakly invertible saddle and fork are paired. These pairing correspond to attaching a 2-handle (from left to right in the illustrations) to the surface. Subsequently, additional strong inverse 5-isomorphisms are applied. *Therefore, pairing weakly invertible black vertices corresponds to a handle attachment between surfaces that are embedded in* 4*-space.*

In Section 20.1 of [**40**], the author, without the use of the chevron map, demonstrates that inserting an edge with label i, bounded by two black vertices, corresponds to the attachment of a 1-handle to the surface braid. Cutting an arc of a chart into two and inserting a pair of black vertices also is realized by the insertion of an edge as indicated below. So that operation also corresponds to attaching a 1-handle.

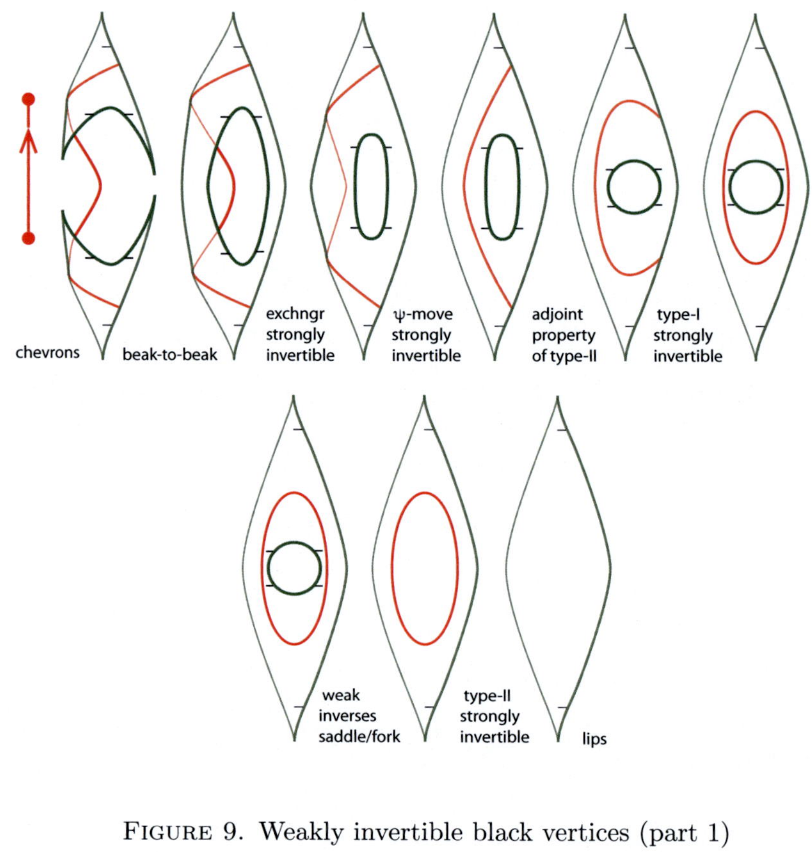

FIGURE 9. Weakly invertible black vertices (part 1)

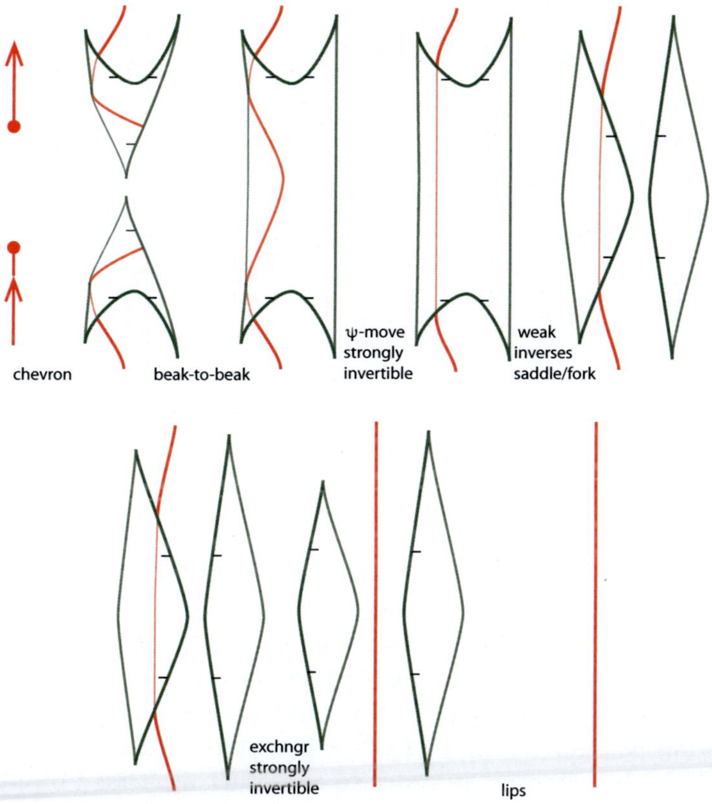

FIGURE 10. Weakly invertible black vertices (part 2)

At the end of Chapter 7, Section 4.3, an illustration indicates that the orientation of an arc in the plane can be affected in the presence of fish hook moves (Chapter 7, Section 3.3) that have since been codified as adjoint properties (Chapter 10, Section 1.4). Since the type II move has the adjoint property with respect to black vertices, the weak invertibility of 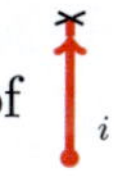 implies that of .

Furthermore, in the presence of the type II adjoint 5-isomorphism (fish hook), one of the weak isomorphisms between a composition of black vertices yields the other. This further indicates why both saddle/fork cancelations correspond to handle attachment.

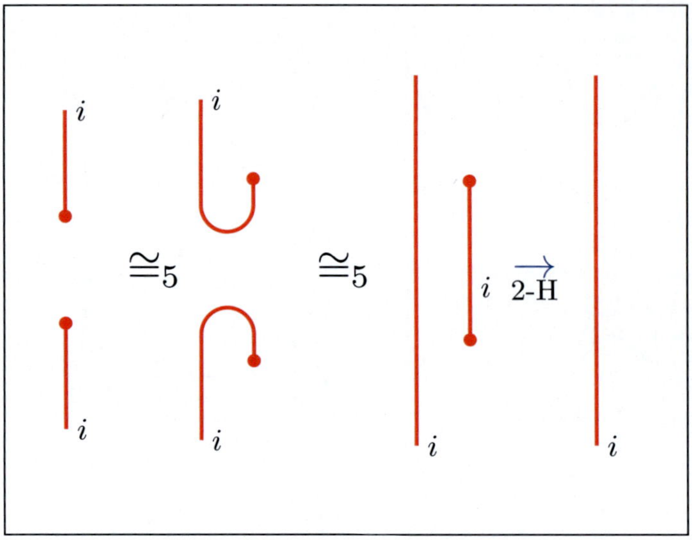

3.6. Self-interesction points for surface braids. A *syllepsis* or *self-interesection point* is introduced into the braid theory by means of the glyphs below. The glyph on the left that is indicated by a $(-)$ is called a *sink*; that on the right, $(+)$, is a *source*. These quadruple arrows are not included when studying surface braids up to isotopy, but they are strongly invertible when considering singular surface braids. The strong invertibility will be used in Section 4 (this chapter) in order to construct immersed braidings of 3- and 4-dimensional manifolds. In particular, all of the higher dimensional birth/death and saddle/fork pairs of multiple arrows that involve syllepses are assumed to be strongly invertible.

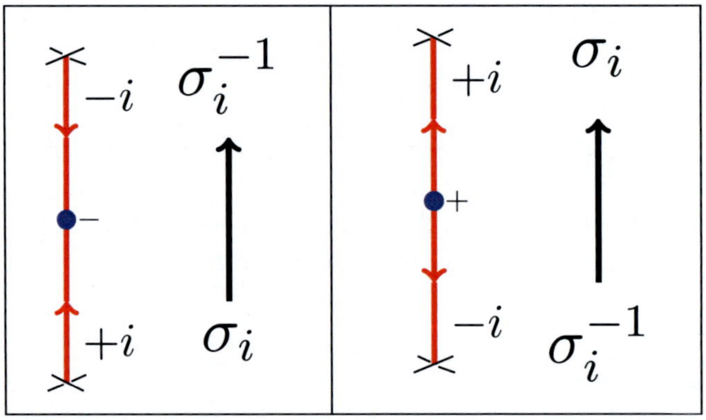

The behavior of syllepses with respect to black vertices, white vertices crossings, and type II moves can be summarized as follows.

(1) A syllepsis can be created or cancelled at a black vertex. Here only the sink is drawn.

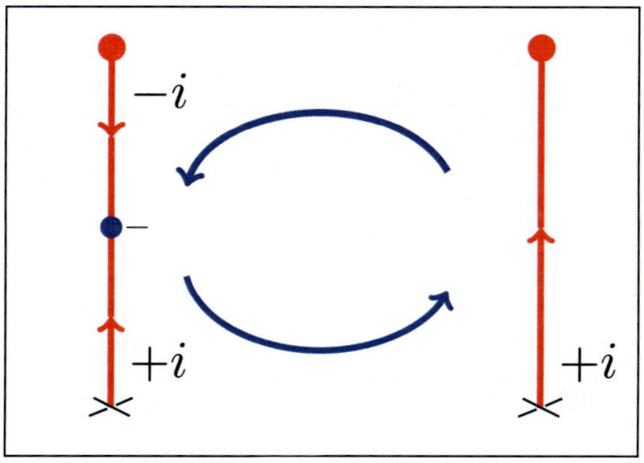

(2) A pair of oppositely signed syllepses that are adjacent along an edge can be created or annihilated.

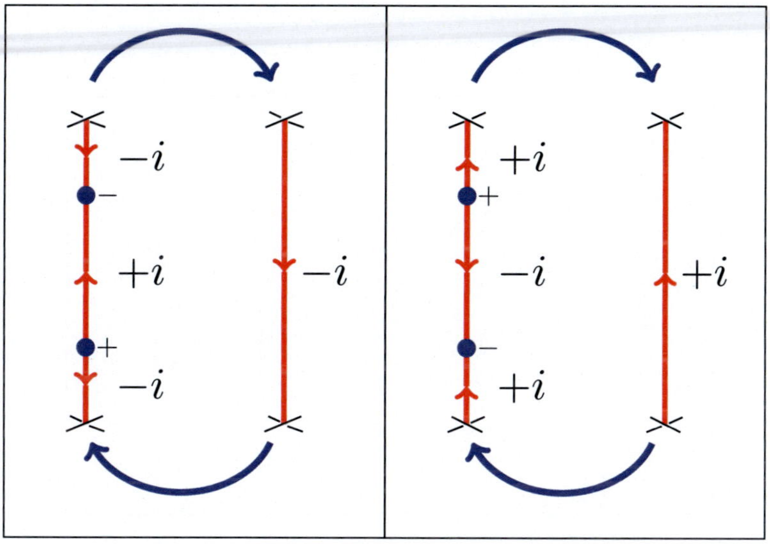

(3) A syllepsis can travel across a crossing because the crossing expresses a natural exchanger quadruple arrow. For example:

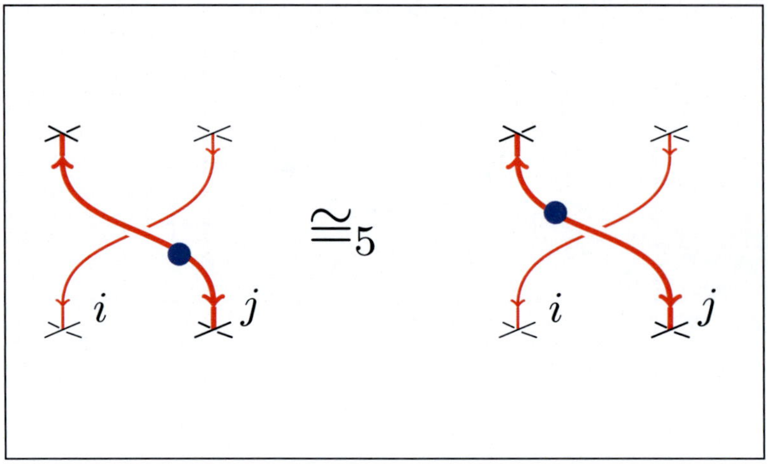

(4) A syllepsis can travel across a type II move since the type II move has the adjoint property.

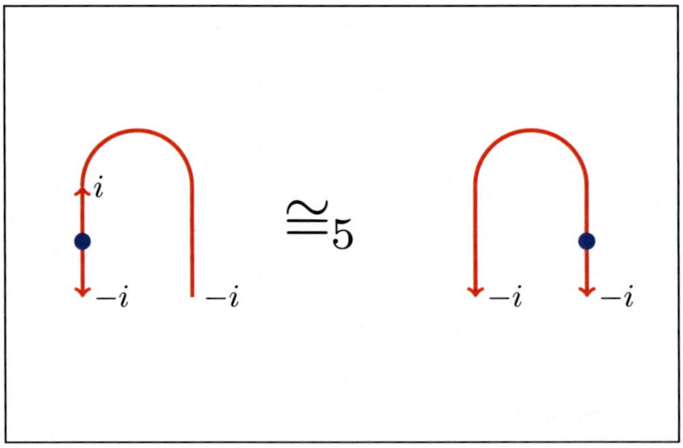

(5) A syllepsis can travel through white vertices along certain edges. These motions imitate those of a black vertex through a white vertex, and they follow because a quadruple arrow exchanger is natural with respect to other quadruple arrows.

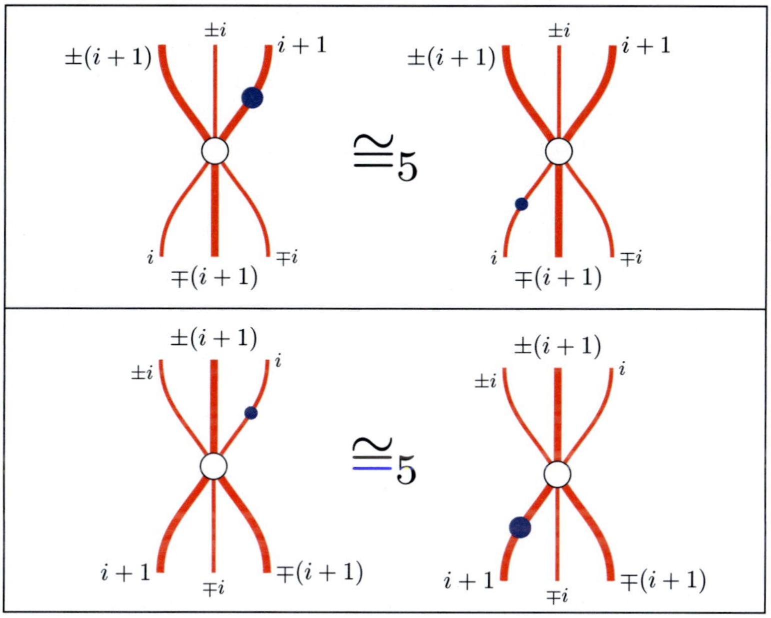

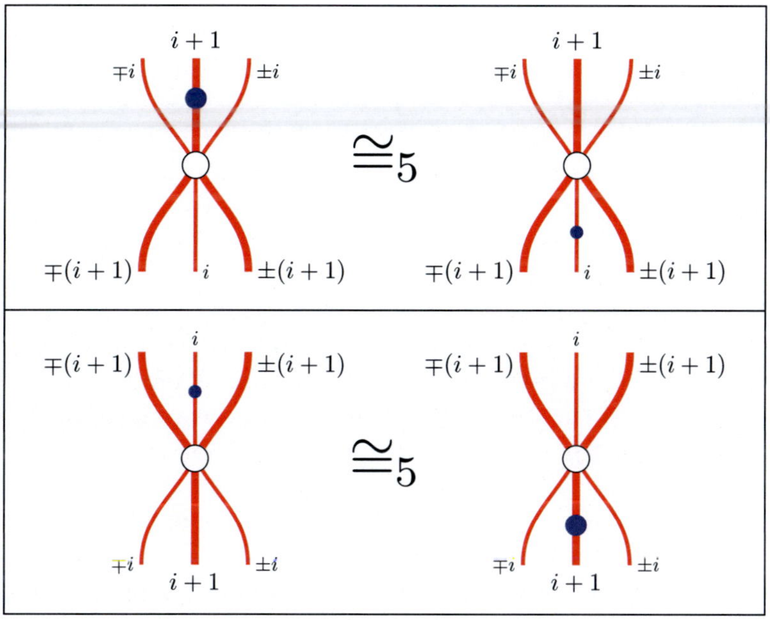

There is a notational distinction indicated between the first two items above and the remaining items. Suppose either that a syllepsis is introduced at a black vertex and there are no other syllepses, or that a pair of oppositely signed syllepses are born. Then an embedded surface braid has been transformed into an immersed surface braid. A syllepsis corresponds to a double point of the represented surface in 4-space. The source and target quadruple arrows represent things in different worlds: without syllepses, the braidings are embedded surfaces; when a syllepses is present, the braiding is immersed. Thus these changes are indicated by 5-arrows rather than 5-isomorphisms. Therefore, the first two items change the *isotopy* classes of immersed surface braids, but do not change the *homotopy* classes. The remaining items do not change the isotopy classes.

In Section 4 (this chapter), some of the higher dimensional braids of 3- and 4-manifolds that are exemplified are projections of immersed braided manifolds: these are 3- and 4-dimensional spaces that are placed in 5- and 6-dimensional spaces with some self-intersections which lie along the set of syllepses, but that project nicely onto spheres so that the projection induces branched coverings.

4. Charts in 3- and 4-dimensions

A 3-dimensional manifold will be called a *solid*, and a 4-dimensional manifold will be called a *hyper-solid*. Recall from Chapter 5, Section 3.4 that a manifold is a nice space for which each point has a neighborhood that is homeomorphic to an open subset of $\mathbb{R}^m_+ = \{(x_1, x_2, \ldots, x_m) \in \mathbb{R}^m : 0 \leq x_m\}$. Here $m = 3$ or 4. The language of Section 3.1 (this chapter) is modified and imitated to define simple branched covers in higher dimensions.

Let $U = \{z \in \mathbb{C} : |z| < 1\}$. Let $f : M^m \to N^m$ denote a map between solids ($m = 3$) or hyper-solids ($m = 4$). A point p of M is called a *branch point* of f of *degree* d if there is a neighborhood $\nu(p)$ of p in M and a neighborhood $\nu(f(p))$ in N together with homeomorphisms $\varphi : \nu(p) \to U \times D^{m-2}$ with $\varphi(p) = (0,0)$ and $\psi : \nu(f(p)) \to U \times D^{m-2}$ with $\psi(f(p)) = (0,0)$ such that $\psi \circ f \circ \varphi^{-1} : U \times D^{m-2} \to U \times D^{m-2}$ is $(z, x) \mapsto (z^d, x)$ for all $(z, x) \in U \times D^{m-2}$. A point p of M is called a *regular point* of f if there is a neighborhood $\nu(p)$ of p in M and a neighborhood $\nu(f(p))$ of $f(p)$ in N such that $f|\nu(p) : \nu(p) \to \nu(f(p))$ is a homeomorphism. A point q of N is called a *branch value* if it is the image of a branch point. A point q of N is called a *regular value* if each point of $f^{-1}(q)$ is a regular point.

A map $f : M^m \to N^m$ between manifolds is called a *branched covering map* of degree n if each point of M is a regular point or a branch point and the restriction $f|_{M \setminus f^{-1}(\Sigma)} : M \setminus f^{-1}(\Sigma) \to N \setminus \Sigma$ is a covering map of degree n, where Σ is the set of branch values. The covering map $f|_{M \setminus f^{-1}(\Sigma)}$ is called the *covering map associated with* the branched covering map f. By definition, the set of branch values Σ is a codimension 2 submanifold of N. (We do not consider the more general definition of a branched covering in which Σ is a codimension 2 subcomplex).

Only simple branched coverings will be considered. So $d = 2$ and the branch set is a submanifold of dimension $m - 2$ in M. In dimension $m = 3$, the branch set is a knot or link, and in dimension $m = 4$ it is a (possibly non-orientable) surface.

It is known classically [**67**], that the 2-fold branched cover of S^3 which is branched along a knot or link is constructed as follows. Create two copies of the 3-sphere S^3, say S_1 and S_2. Split each along an orientable surface that is bounded by the knot. After the cutting, a pair of pita-bread like surfaces exist in each of S_1 and S_2; these are the doubled surfaces along which the solids were cut; the edges are the original knot. Each surface has a positive and negative side. Glue S_1 at the positive side to S_2 along the negative side of its pita surface, and glue the positive side of the pita surface in S_2 to the negative side of the pita surface in S_1. Several examples of 2-fold branched covers of S^3 are among the examples of braided solids that are given below.

4.1. Chart movies and curtains. In this section, movies of charts are used to describe branched coverings of solids. A *chart movie* consists of a finite sequence of charts $(C_0, C_1, \ldots C_k, C_{k+1})$ each of braid index n such that C_0 and C_{k+1} are both empty, and C_{i+1} is obtained from C_i by exactly one of the following:

(1) a chart move which is described in Section 3.3 as a 5-isomorphism;
(2) the creation or cancelation of a pair of branch points as described in Section 3.5;
(3) an exchange of heights between vertices or critical points in a chart (this is a 5-isomorphism exchanger);
(4) the creation of a syllepsis at a black vertex that changes the sign of the vertex (Section 3.6);
(5) the creation or cancelation of an oppositely signed pair of syllepses (Section 3.6); or
(6) a syllepsis moves through a vertex of a chart as described in Section 3.6.

In particular, C_1 is obtained from C_0 either by the insertion of an arc (oriented up or down) or a circle (oriented clockwise or anti-clockwise) with edge label i and C_{k+1} is obtained from C_k by the deletion of such an edge or circle. However, the labels on the inserted and deleted edges or circle do not have to agree.

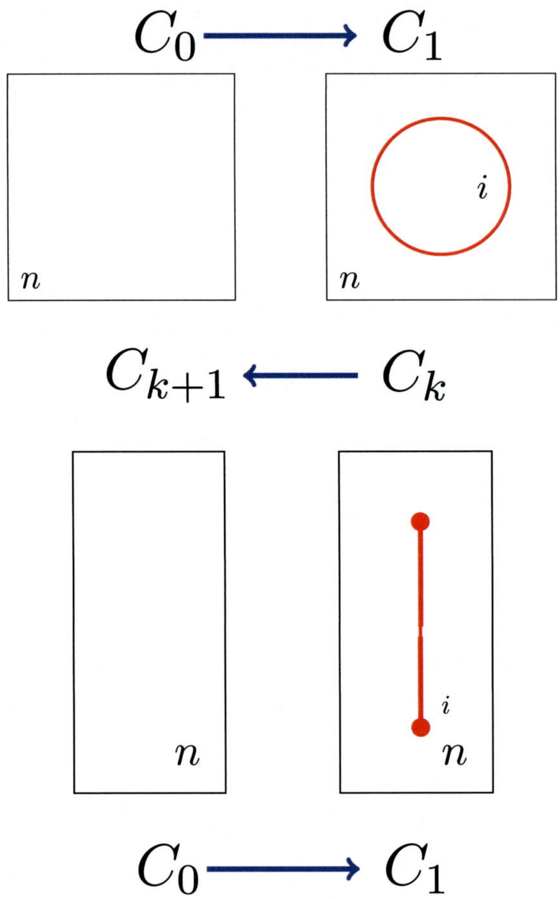

The chart C_j is called the *jth still* in the movie.

Given a chart movie, a degree n, simple, branched cover of the 3-dimensional sphere S^3 can be obtained by forming the closure of each chart as indicated in Fig. 6, interpolating between successive stills, and interpolating by identities on the closures. Here are more details.

The charts C_0 and C_{k+1} represent the identity on the trivial n-string braids. These charts, then, each represents a trivial covering of the rectangular disk B^2 that stacks like pages in a book. The closure thereof is a collection of nested 2-dimensional spheres. Closing towards an empty C_{-1} and an empty[1] C_{k+2} makes each sphere successively bound a ball. Assume that C_1 is obtained from C_0 by the insertion of an edge, and C_{k+1} is obtained from C_k by the deletion of such an edge. As discussed in Section 3.5, the transition from C_0 to C_1 is tantamount to adding a 1-handle between the ith and $(i+1)$st sphere. Similarly, the transition from C_k to C_{k+1} is a 2-handle attachment that splits a connection between levels i and $(i+1)$

[1] These are empty in a manner that differs from the emptiness of C_0 and C_{k+1}. Those trivial charts represent a trivial braiding since they are charts with braid index n. The charts C_{-1} and C_{k+2} simply do not exist, and therefore represent the empty covering of the empty manifold.

into concentric spheres. The introduction or deletion of any such a pair of branch points between stills C_j and C_{j+1} that occurs in a manner analogous to the cancelation of unit/co-unit, arch, or canal (see Chapter 4, Table 13 and Chapter 5, Section 3) also corresponds to a 1- or 2-handle attachment. Furthermore, a pair of surfaces that differ from each other by means of a handle attachment cobound a solid. It is this solid that braids over the interpolation.

With the exceptions of the moves that involve syllepses, the remaining changes in charts are 5-isomorphisms that have the property that the differing charts represent isotopic surface braids (or singular surface braids) in 4-space. The isotopies that relate them interpolate between them.

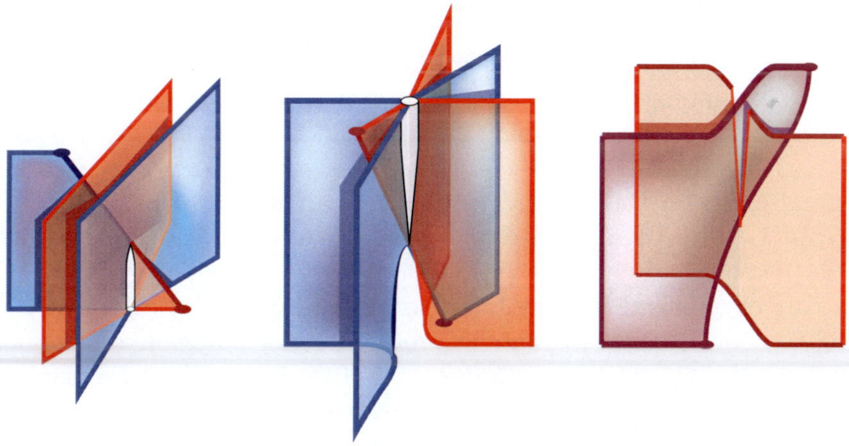

FIGURE 11. Naturality of exchangers with respect to black vertices

Suppose that C_j and C_{j+1} differ by a 5-isomorphism from Section 3.3. A curtain-like structure relates them. When a black vertex pushes though a crossing or a white vertex, then examples of the curtains are illustrated in Fig. 11. A white vertex in a chart also represents a type III move. That white vertex is the cross-section of a "white-dowel" found, for example, within the curtain as shown in Fig. 11. Along a white dowel six sheets of curtain meet. Meanwhile, a crossing point (4-valent vertex in a chart) evolves in a curtain to form another sort of dowel along which four sheets of curtains meet. This dowel can be depicted as a broken surface, and in that sense its invertibility is analogous to a type II move, and its naturality is analogous to a type III move.

A curtain that interpolates between the two sides of the quadruple point move is less apparent, but imagine that the four white vertices converge to a common point, and then diverge. Each edge of the chart describes a surface in the interpolation, and the white vertices describe edges at which three sheets converge. None of the examples below include a quadruple point move. Within some of the illustrations hyper-glyphs that indicate transitions between stills of the chart movies are indicated. In others, these hyper-glyphs are critical regions for the curtains. They are the analogues of those described in the previous paragraph.

Thus a *curtain* is a singular surface that interpolates between stills of a chart movie. It is a composition of quintuple arrows that is used to describe a braided solid in 5-dimensional space. That solid may be immersed, and its double point set is 1-dimensional and formed from the syllepses. The branch set of such a braiding is a knot or link in $D^3 \subset S^3$. This branch set forms the boundary of the curtain. The braid closure of a curtain provides an immersed or embedded (when there are no syllepses present) solid in $D^2 \times S^3 \subset S^5$ such that the projection onto S^3 is a branched covering.

The first few examples illustrate braidings of the 2-fold branched covers of S^3 that are branched over knots and links. Take any 2-sphere cross-section of the 3-sphere that transversely intersects a surface which is bounded by the knot. The intersection is either empty or a collect of closed curves and arcs. In the latter case, consider the curves and arcs to be oriented by way of the normal orientation to the surface. Assign an index 1 to each arc or curve, and interpret that intersection as a braid chart.

Example 1. The branched cover of S^3 branched along an unknot.

The (incomplete) chart movies that are given immediately below represent a pair of quintuple arrows that are weak inverses of each other (Section 3.5). On the left, the movie represents the addition of a 1-handle between a pair of concentric spheres, and that on the right represents the addition of a 2-handle. The 2-handle cuts out the double arc and branch points to return the configuration to a pair of concentric spheres (Fig. 9).

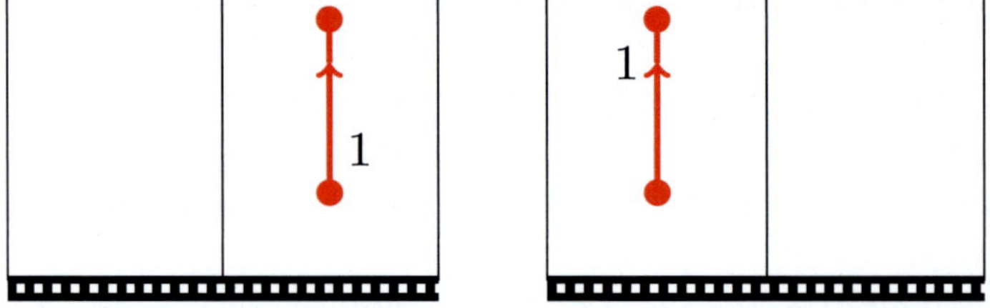

By composing these quintuple arrows, the chart movie below is obtained.

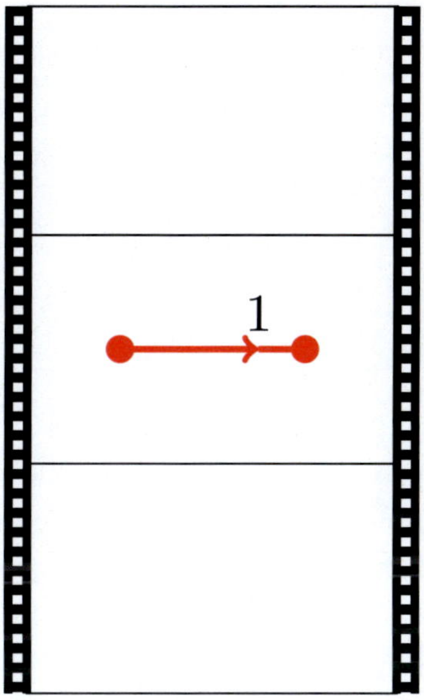

The corresponding curtain is a disk.

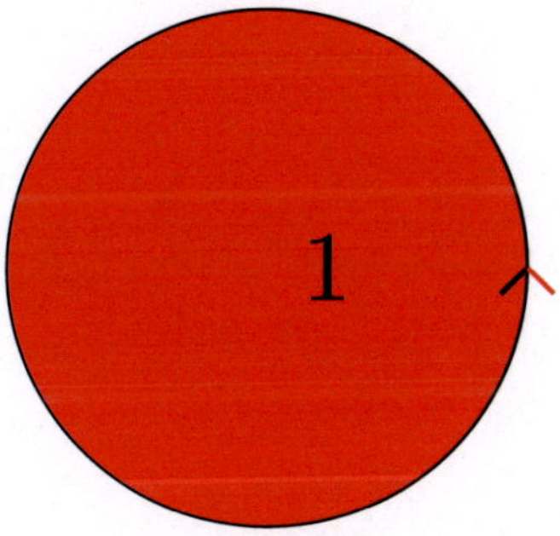

The 2-fold branched cover of the 3-sphere, S^3, that is branched over an unknotted circle also is a 3-dimensional sphere. The curtain that consists of a disk represents a braiding in 5-space thereof.

The 2-fold branched cover of S^3 that is branched along an unlink.
Consider the chart movie below.

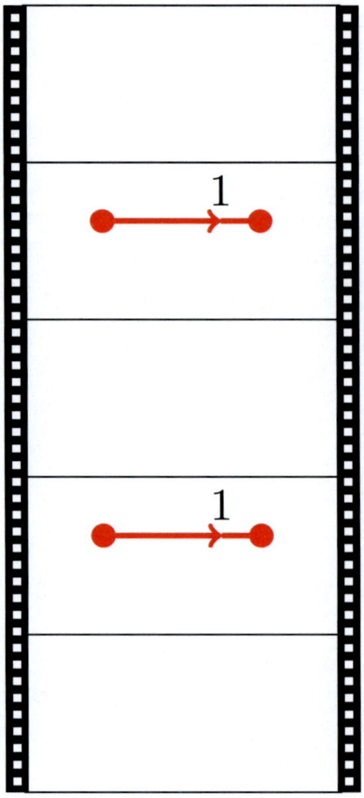

The corresponding curtain is a pair of disks:

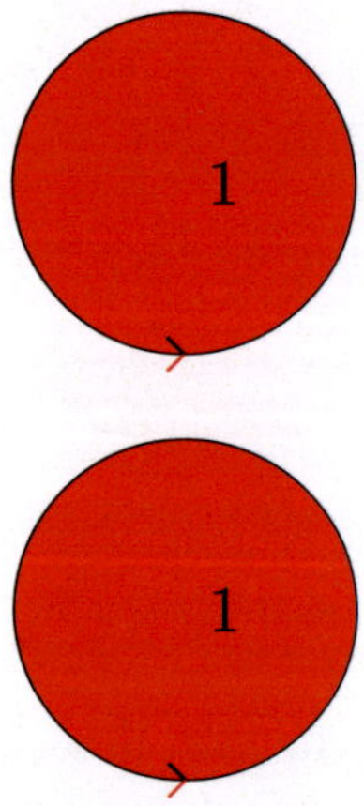

And the 2-fold branched cover of S^3 that is branched along a pair of unknotted unlinked circles is the product space $S^2 \times S^1$. The top disk creates a 3-sphere, and the bottom disk attaches a 1-handle to it. The curtain that consists of these two labeled disks represents a braiding of this manifold.

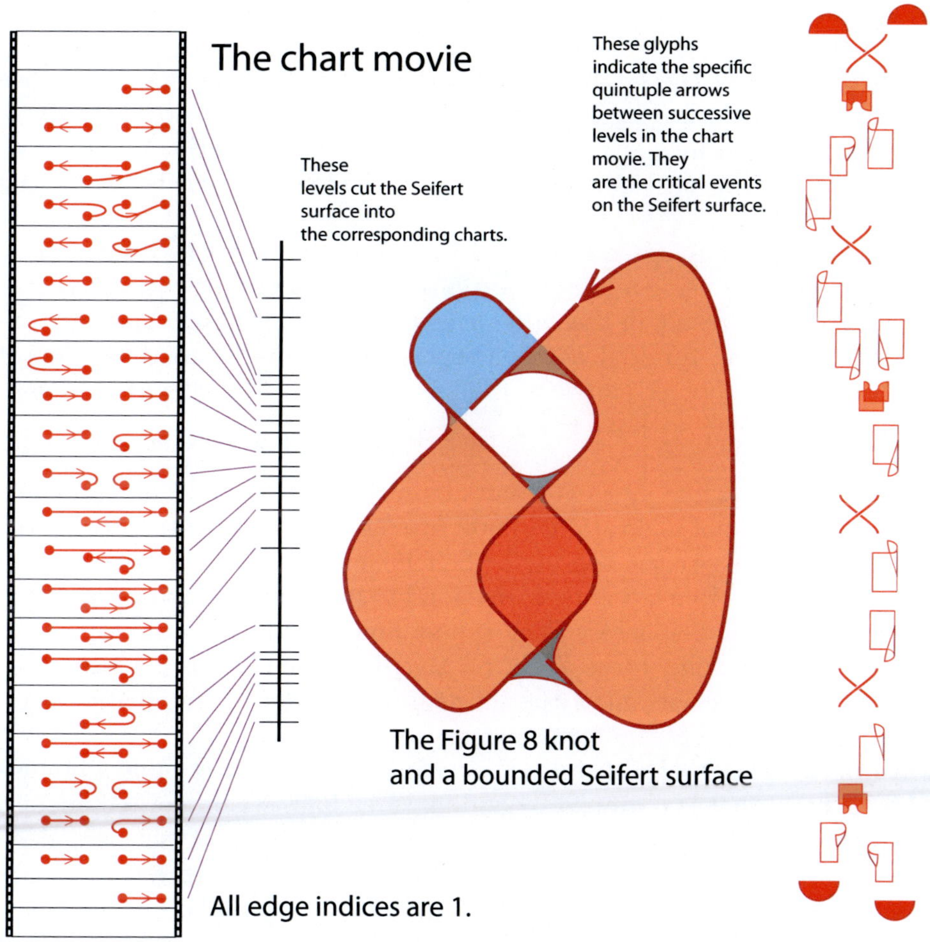

FIGURE 12. A chart movie for the Figure 8 knot

Example 3: The 2-fold branched cover of S^3 that is branched along a Figure-8 knot. Fig. 12 indicates the 2-fold branched cover of the figure-8 knot (4_1 in the tables).

A finite sequence of latitudinal transverse 2-spheres gives rise to a chart movie when the latitudes are chosen to lie successively with a unique critical event between them.

In Fig. 12, a Seifert surface that is bounded by the figure-8 knot is indicated as the central object of the figure. The Seifert surface is color coded so that red portions are positively oriented (outwards normals point towards the reader) and blue portions are negatively oriented. The surface was obtained by applying the Seifert algorithm. An axis to the left of the knot diagram has tally marks that indicate non-critical slices. Extend the tally marks horizontally to the right to slice the illustration of the Seifert surface. On the left of the figure, the corresponding chart movie of braid index 2 charts are drawn. The stills in the chart movie are the transverse slices at the tally levels.

On the right of the figure, glyphs indicate the sequence of transitions between the critical levels. These can also be read from the illustration of the Seifert surface which is interpreted, now, as the composition of quintuple arrows. The resulting branched cover is embedded in the 5-dimensional sphere S^5.

In general, a lens space is formed by gluing two solid tori along their common boundary using a homeomorphism of the bounding torus. Isotopic homeomorphisms result in the same lens space. The isotopy classes of orientation preserving toroidal homeomorphisms correspond to the group of (2×2) invertible integer matrices, $\mathrm{SL}_2(\mathbb{Z})$. (Consider the effect on a longitude/meridian pair). If $\begin{bmatrix} a & b \\ c & d \end{bmatrix}$ is such a matrix, then $ac - bd = 1$, and a pair of relatively prime integers results by way of the Euclidean algorithm.

The figure-8 knot is a rational knot, and its 2-fold branched cover is the lens space $L(5, 2)$. At the top of the Seifert surface, two pairs of births of oppositely signed branch points, correspond to two 1-handles being attached between concentric 2-spheres. Each sphere bounds a ball, so the top of the figure indicates a solid torus $D^2 \times S^1$ that is embedded in 5-space. The same is true at the bottom of the illustration. In the intermediate stages, the homeomorphism of the torus occurs.

Example 2: The trefoil knot. The trefoil knot (3_1 in the tables) bounds a Möbius band that is obtained by adding three half twists to a rectangular strip before attaching opposite edges. By inserting a loop of syllepses along the core of the Möbius band, a curtain is obtained that represents an immersed lens space $L(3, 1)$ in S^5. Fig. 13 indicates the details. On the left of the illustration, a chart movie is drawn in which each still represents a chart of degree 2. Most of these contain an oppositely signed pair of syllepses at which the triple arrow interchanger X converts to $\overline{\mathsf{X}}$, or vice versa in the associated braid movie. Thus the closure of each of these stills represents an immersed braided surface in S^4. The first few and the last few charts represent immersed solid tori in five dimensional balls. The intermediate charts each represents a 2-fold cover of the 2-sphere that has four simple branch points. Each intermediate still represents a torus immersed in S^4. Here the three-fold twisting is apparent from the Mobius band, and so the homeomorphism of the intermediate torus is achieved by a $(3, 1)$ curve. A version of this immersed branched covering was first given in [**14**].

Example 3: The Hopf link. Fig. 14 is organized differently from the previous two illustrations. It indicates a 4-fold branched cover of the 3-sphere that is branched along a Hopf link. The curtain that is associated to the chart movie appears on the left of the illustration as a pair of disks that intersect. The color coding does not represent front and back of a Seifert surface, but instead is used to emphasize that the two sheets in the curtains have differing braid generators assigned.

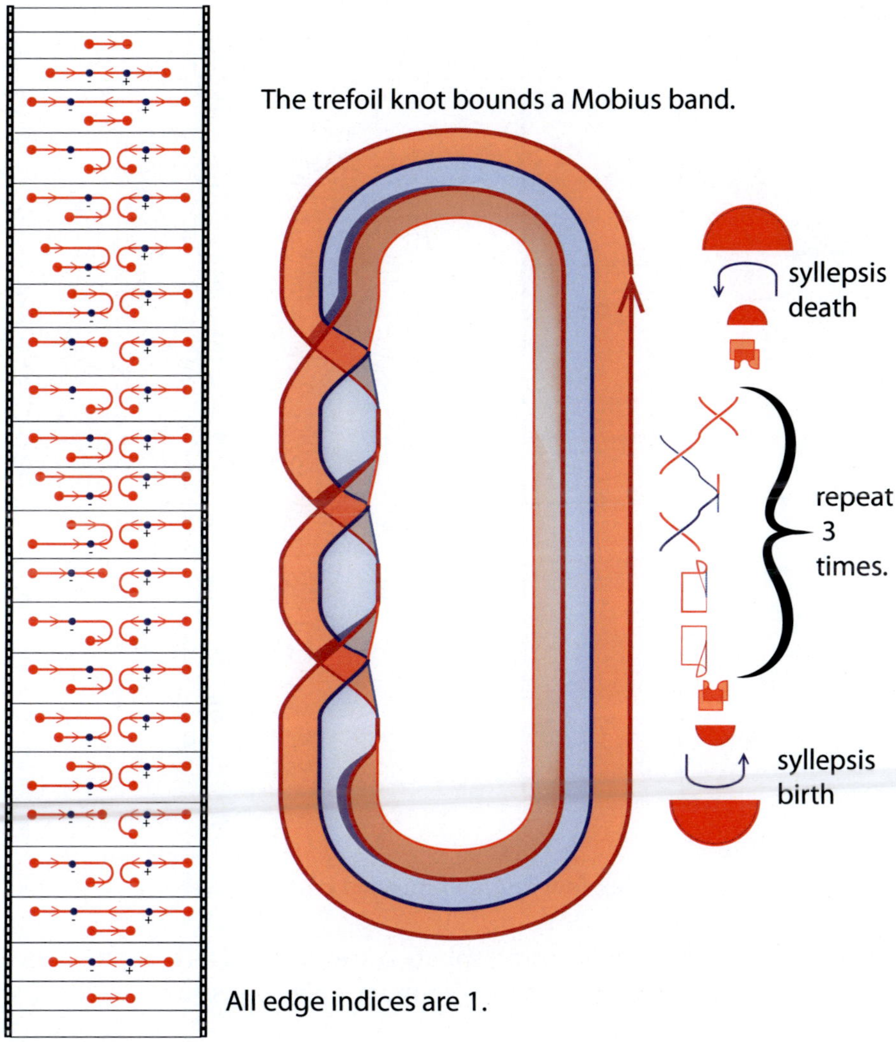

FIGURE 13. A chart movie for the trefoil that bounds a Möbius band

The fundamental group of the complement of the Hopf link in S^3 is the free abelian group of rank two, $\mathbb{Z} \times \mathbb{Z}$. A simple branched covering of S^3 with a given knot or link as its branch set is determined by a representation of the fundamental group into a permutation group in which the generators of $\mathbb{Z} \times \mathbb{Z}$ are assigned to distant transpositions: $(1, 2)$ and $(3, 4)$. In the case at hand, the group $\mathbb{Z}_2 \times \mathbb{Z}_2$ is lifted to the braid group such that one component of the link has the braid generator σ_1 assigned and the other has σ_3 assigned.

Within the illustration, each still in the chart movie has an associated braid movie indicated. At the left and right ends of the braid movies, the concentric circles are successively capped with disks, and at the top and

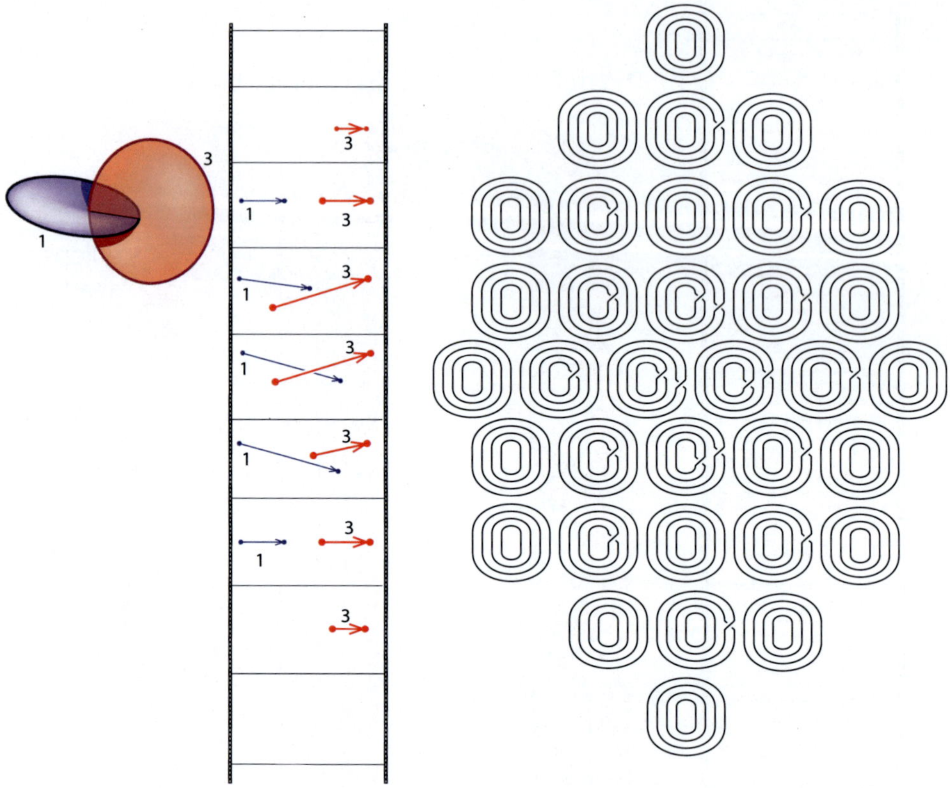

FIGURE 14. An index four simple branched covering over the Hopf link

bottom the resulting concentric spheres are successively capped with 3-dimensional balls. However, those capping operations are not illustrated. Cf. Fig. 6.

Example 4: The 3-fold branched cover of the 3-sphere branched along a trefoil. This example was first given in [**14**]. The fundamental group of the complement in S^3 of the trefoil is given by the presentation

$$\pi_1(K(3_1)) = \langle x, y : xyx = yxy \rangle.$$

A representation into the permutation group Σ_3 is used to construct a 3-fold branched covering. The representation we have in mind is $x \mapsto (1,2)$ and $y \mapsto (2,3)$. This representation lifts to a braid representation where x is assigned to σ_1, and y is assigned to σ_2. After all, this fundamental group is, in fact, the three string braid group B_3. From these representation the chart movie, braid colorings, and curtain structures in Fig. 15 are presented.

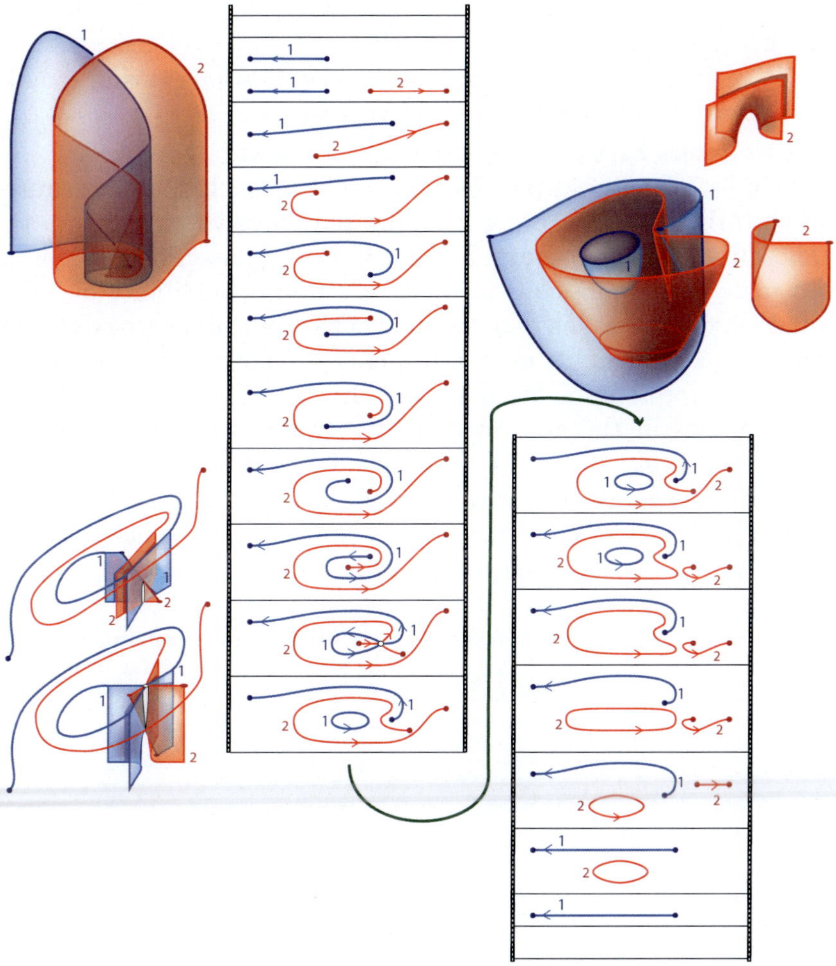

FIGURE 15. A braiding of a 3-sphere branched over the trefoil

4.2. Curtain movies and 4-dimensional spaces in 6-space. Following the process that is outlined in Chapter 7, the movie metaphor will be pushed up by one more dimension. Specifically, transitions between chart movies will be thought of as 6-arrows, and these will be composed to produce embedded and immersed braidings of 4-dimensional manifolds that are simple branched covers of the 4-sphere in which the branching set is an embedded surface in 4-dimensional space.

Fix a braid index n. Most of the ingredients needed to give an example of *curtain movie* are in place. This is a sequence of curtains, $(U_0, U_1, \ldots, U_{\ell+1})$ such that

(1) U_0 and $U_{\ell+1}$ are empty curtains.
(2) U_1 and U_ℓ each is a disk with an anti-clockwise oriented boundary or a 2-sphere with labels in $\{1, 2, \ldots, i-1\}$. As in the case of chart

movies, the label on the disk or sphere U_1 does not have to coincide with the label on U_ℓ.

(3) U_{j+1} differs from U_j by an application of 6-isomorphism, some other 6-arrow, or an exchange of critical levels.

In the examples that are presented here, U_1 and U_ℓ are disks.

The long list of 6-isomorphisms that might occur is not enumerated, in part, because the examples presented here only involve a few of these, and in part, because we do not know all of the 6-isomorphisms that might apply. However, many are expressions of weak or strong invertibility of 5-arrows. Many others are expressions of naturality of the quintuple arrow exchanger.

Each still in a curtain movie represents a chart movie. Each still in a chart movie represents a braid movie. Each still in a braid movie is a sequence of words in the generators of the braid group. Successive stills in a braid movie differ by equivalences of braid words, the introduction or deletion of a braid generator, or a syllepsis that converts a generator to its inverse or vice versa. A braid movie is a composition of quadruple arrows. A chart movie is a composition of quintuple arrows. And a curtain movie is the composition of 6-arrows. In any case, successive stills are related by some geometric transformation to the underlying strings, surfaces, or solids.

The 2-fold branched cover of S^4 branched over a torus. Consider the curtain movie:

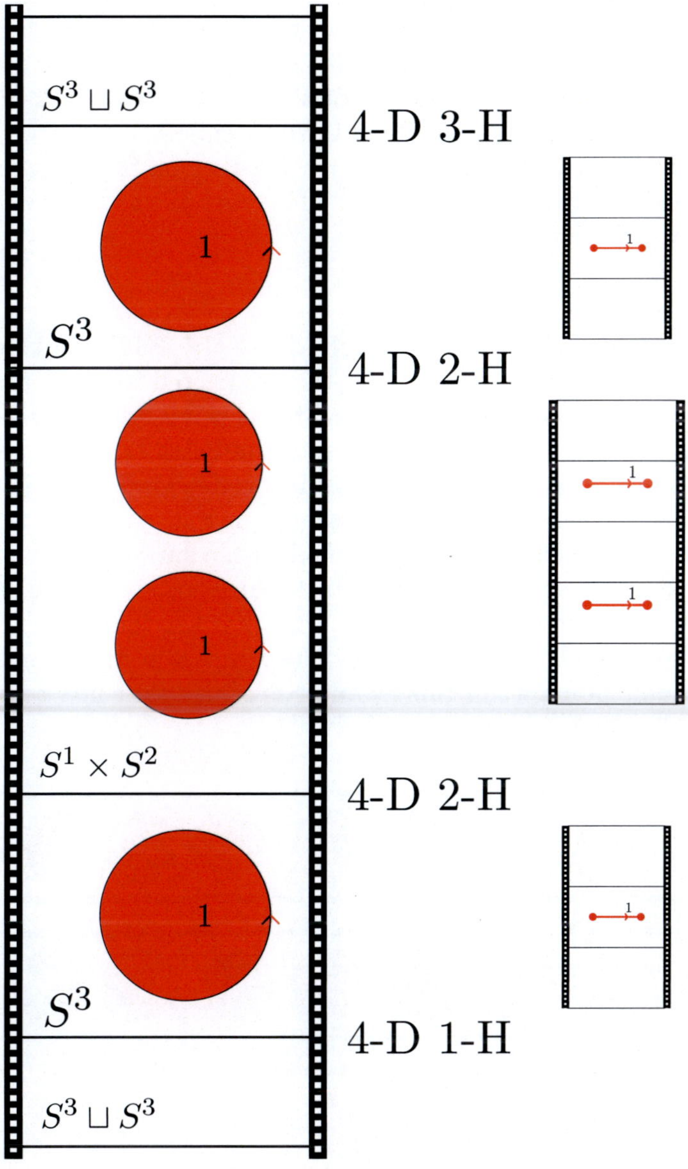

The transitions are as follows. The empty chart at the bottom represents, upon the braid closure, a pair of concentric 3-dimensional spheres embedded in $S^3 \times B^2 \subset S^5$. These are obtained as a pair of 4-dimensional 0-handles. The transition to the 2-fold branched cover of S^3 branched along the unknotted circle is achieved by a 4-dimensional 1-handle being attached. The transition to the braided solid $S^1 \times S^2$ as the 2-fold branched covering of S^3 branched along a pair of unlinked circles is achieved by the attachment of a 4-dimensional 2-handle. The pair of disks in the middle still of the movie

are obtained by splitting the disk in the previous still into 2 by a saddle type move analogous to that in Chapter 5, Section 3.1. The next transition is achieved by a 4-dimensional 2-handle. A 4-dimensional 3-handle is attached to transition from a simple closed curve to an empty chart. Finally, each sphere in $S^3 \sqcup S^3$ is capped of by a 4-dimensional 4-handle. The branch surface represented by the evolution of the boundaries is a 2-dimensional torus ($S^1 \times S^1$) standardly embedded in the 3-sphere, $S^3 \subset S^4$.
Exercise. Prove that the 2-fold branched cover of S^4 branched along a standardly embedded torus is $S^2 \times S^2$. The curtain movie gives a braided embedding of $S^2 \times S^2$ into S^6.

An immersed braiding with branched set the Klein bottle. A space that is closely related to $S^2 \times S^2$ is the connected sum of the complex projective space with its complex conjugate. It is a 2-fold branched cover of the 4-sphere, S^4, where the branched set is the Klein bottle.

To construct the immersed braiding that is depicted in Fig. 16, a movie of the Klein bottle (for example as presented in Fig. 34 of [**4**]) is developed for the Klein bottle. Then each still is filled with a surface. In this case, each still is a circle or a pair of circles, and so the surfaces are all disks. Then the orientations on the boundary are chosen to be consistent from still to still. Of course this cannot be done, so a syllepsis arc is born and subsequently dies. The double point set of this example is a disk that lifts to a sphere in the 4-dimensional space.

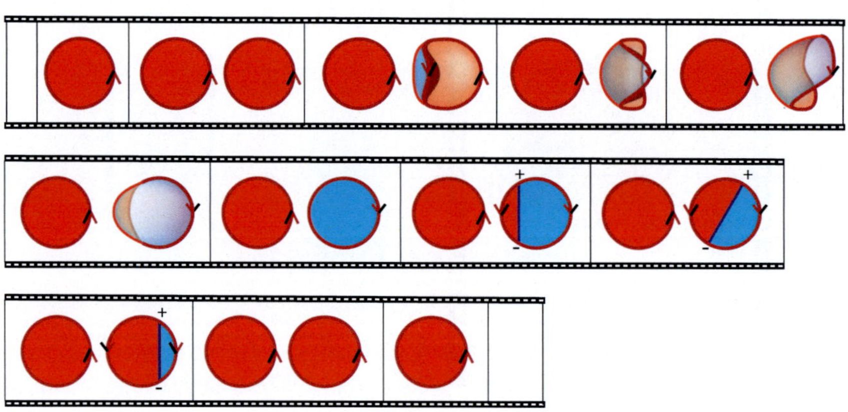

FIGURE 16. An immersed braiding of $\mathbb{C}P^2 \sharp \overline{\mathbb{C}P^2}$

A braiding of the 3-fold branched cover of S^4 branched over the spun trefoil. In the last example of this section, a curtain movie is presented as a sequence of chart movies in Fig. 17. The actual size of the original drawing of this figure is about 25 square meters, but it has been reduced to fit the page. The branch set is the spun trefoil which is among the most simple knotted surfaces. The fundamental group of the exterior of the surface coincides with that of the trefoil — the group is B_3, the three string braid group. The surface's movie is palindromic as is the curtain movie that is presented here.

From step 0 to step 1, a blue disk with label 1 is born. From step 1 to step 2, a red disk with label 2 is born. Between steps 2 and step 9 these two disks fold and interwine. From step 9 to step 10, an arc of white vertices is created. From step 10 to step 11 it is split into two arcs. From 11 to 12 a 1-handle is attached to the curtain surface. From 12 to 13 a canceling pair of fish hooks are introduced, and from 13 to 14 the red surface is split using a weak invertibility condition of black vertices. Close ups of the curtain transformations are indicated in Fig. 18.

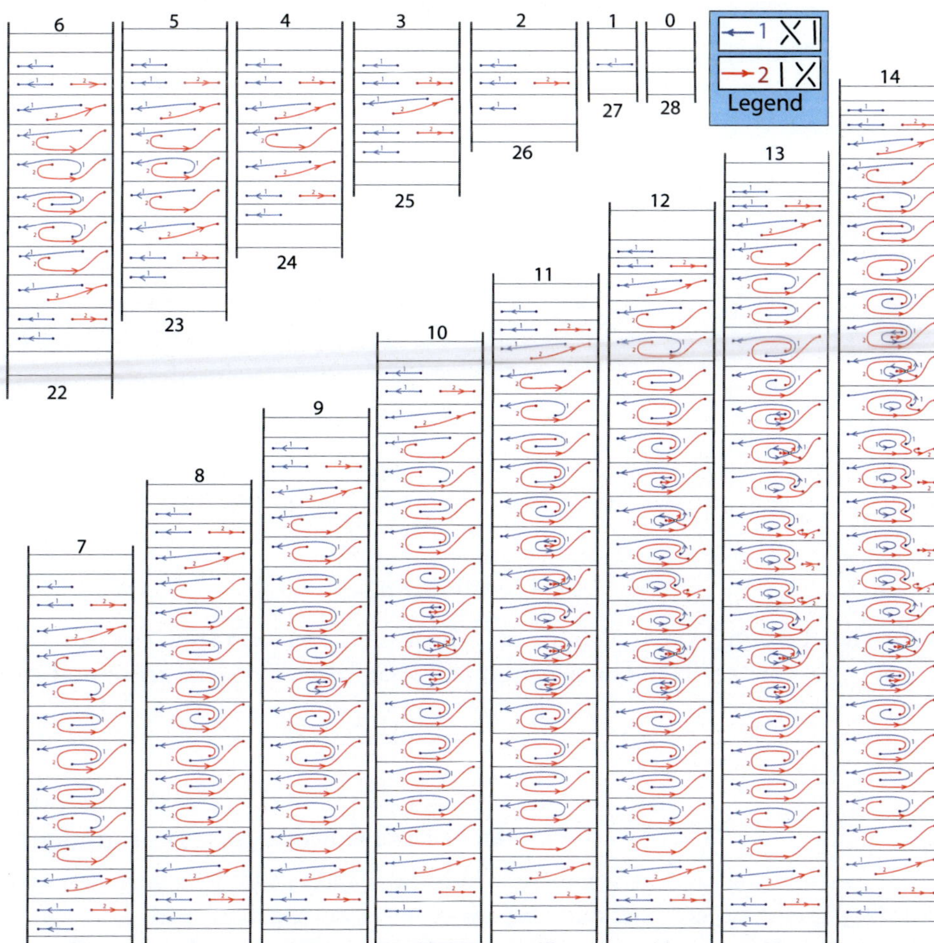

FIGURE 17. A braiding of the 3-fold branched cover of S^4 branched over the spun trefoil

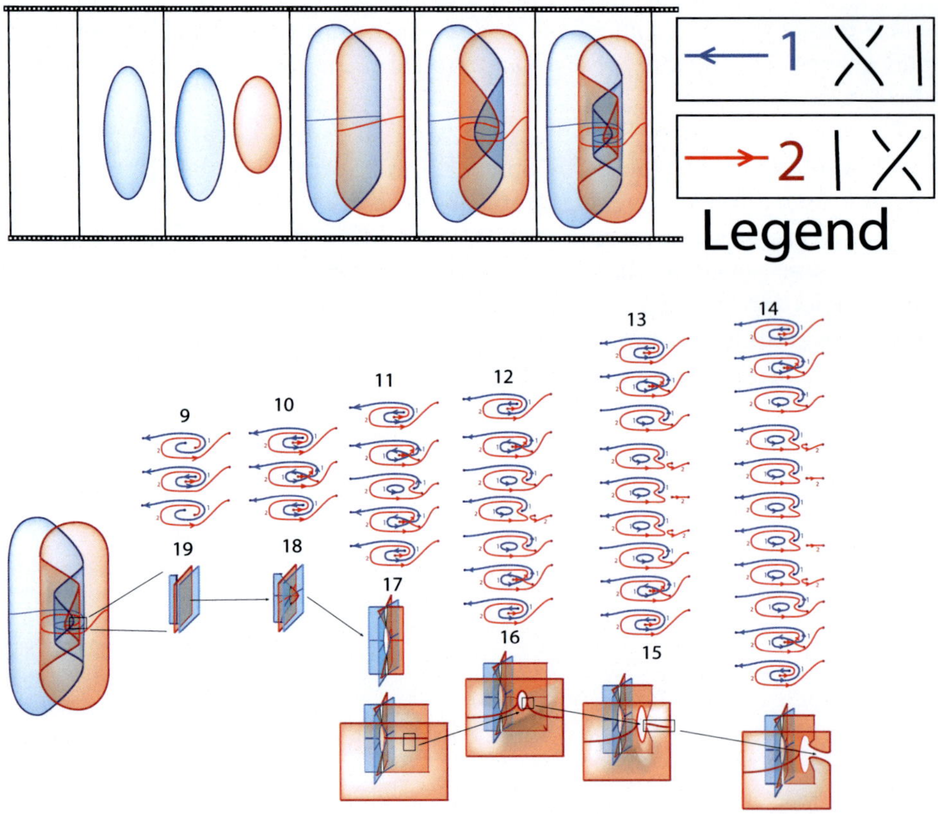

FIGURE 18. Details of the changes between curtains

5. Conclusion

In the introduction to Chapter 3, Section 1, we asserted that mathematics is abstracted from the arithmetical structure of the natural numbers, $\mathbb{N} = \{0, 1, 2, \ldots\}$. For example, the (multiple) arrow $\bullet \overset{\perp}{\longleftarrow} \emptyset$ or more generally $(n+1) \overset{\perp}{\longleftarrow} n$ has the effect of adding 1, and the fact that every number has a successor, is among one of the Peano axioms. By considering the operation $\top$ as a weak inverse, Chapter 7 indicated how surfaces with boundary could be constructed. So starting from the ability to add 1, the concept of subtracting 1 was weakened, and considerations of the higher order structures lead to the study of critical points and critical interactions.

Black vertices generalize $\perp$ and $\top$. A label and a direction are added to an incident edge, and relations among edges (type II, white vertices and crossings) give rise to the theory of surface braids. The calculus of surface braids can be summarized as chart diagrams that are considered modulo 5-isomorphisms or that are related by quintuple arrows and that give rise to higher order braiding. Relations among these give higher dimensional braidings still.

The purpose, then, of diagrammatic algebra might be stated as giving a method of calculating with and/or visualizing higher dimensional objects in a rigorous manner. Knottings and braidings are among the most common things that can be represented using the diagrammatic calculus that has been developed here. These calculi also can be found and illustrated in multiple contexts, and analogies among the contexts abound.

CHAPTER 12

Globular multi-categories

GLOBULAR is a web-based program that runs most effectively on Google Chrome. It requires a three-button mouse and an initial connection to the internet. One can access it by typing the URL GLOBULAR.SCIENCE. Additional web-based resources include the YouTube video

{https://youtu.be/QXZJvXgsOxk}

and the GLOBULAR page

{https://ncatlab.org/nlab/show/Globular}

at the n-lab.

The program was developed by Krzysztof Bar, Katherine Casey, Aleks Kissinger, Jamie Vicary and Caspar Wylie. Indeed, this book has been an attempt by the authors to reverse engineer some of the workings of GLOBULAR since we have found it to be useful in illustrating and manipulating knotted surfaces and other higher dimensional phenomena.

It is both unfortunate and fortunate that computer-based resources change rapidly. For example, it is fortunate that we no longer are using punch cards or dial-up modems to communicate to the "main-frame" resource. It is unfortunate that items and processes towards which we have grown familiar have vanished or changed beyond recognition. And we regret that a favorite book is obsolete since it was written with version 3.0 of a widely used program in mind. The program at this writing is at version 12.0.

Although the GLOBULAR.SCIENCE software has been produced by reliable colleagues, we are obliged to write about the program in an indirect fashion, because, in general, a book is a more lasting medium than a computer program.

An early version of the current chapter contained step-by-step instructions for constructing the multi-categories of the prior chapters within the constructs of the specific software. The authors decided to deposit that early version at

http://www.southalabama.edu/mathstat/
personal_pages/carter/LastChapter.pdf

while revising the chapter to its current form. In addition, GLOBULAR worksheets are provided

http://globular.science/2102.004v1
http://globular.science/2102.006v1

`http://globular.science/2102.005v1`

In particular, the worksheet

`http://globular.science/2102.003v1`

is available to construct explicit embedded foams, and braided immersions and embeddings in codimension 2. Beware, all the moves that are needed to generate curtain movies may not be included in the signature. The interested reader can view these as analogues of the "director's cut." But more precisely, they are analogous to scenes from a screenplay that never made it to the film.

Beyond the explicit movie moves, braid chart moves, and curtain moves that are presented within the signature of these GLOBULAR worksheets, implicit moves exist at the axiomatic level of the program itself. In particular, the naturality axiom for the exchanger produce, by fiat, seven of the eight important moves given as item 5 in Chapter 10, Section 2.2. On the other hand the strong invertibility of cuspidal compositions have to be entered by the user.

In this chapter, a guide to construct worksheets that are similar in spirit to those that function in GLOBULAR are developed. Subtle aspects of invertible arrows are discussed. The constructions that induce exchangers are discussed at a fundamental level.

1. Arrows and cells

Multi-categories that have a small finite number of objects will be our primary interest. An object is also called a 0-cell. It is thought of as a point. For any 0-cell A, there is an identity 1-arrow I_A that has the 0-cell A as both its source and target. More generally for any n-cell N, there is an $(n+1)$-cell I_N that has N as its source and object. To be sure, the ideas of source and targets that are associated to n-cells need to be explained. Once a 0-cell has been declared, identities upon identities upon it can easily be manufactured. One imagines the identity upon a 0-cell as a homogeneous 1-cell, the identity upon the identity as a homogeneous 2-cell, etc.

In general, suppose that $1 \le n$, and that n-cells f_0 and f_1 have been specified. Associated to each are a source, $s(f_j) \in \{(n-1)\text{–cells}\}$, and a target $t(f_j) \in \{(n-1)\text{–cells}\}$, for $j = 0, 1$. Assume further that $s(f_0) = s(f_1)$ and $t(f_0) = t(f_1)$. Then an $(n+1)$-cell F can be constructed with source $S(F) = f_0$ and target $T(F) = f_1$. To reiterate, there are two stages of source and targets that can be applied. We require that $s(S(F)) = s(T(F))$ and that $t(T(F)) = t(S(F))$ — the source of the source is the same as the source of the target while the target of the target is the target of the source.

Let 0-cells a and b be given. These may or may not be the same objects. Declare[1] a to be the *source* of a 1-cell, f_0, and declare b to be its *target.*

[1]The declarations are usually done in the context of the given software.

Schematically, $a \,-\!\boxed{f_0}\!\to b$. There can be more than one 1-cell that have the same source and target. So also declare a to be the source of the 1-cell, f_1, and declare b to be its target. Such a 1-cell looks similar $a \,-\!\boxed{f_1}\!\to b$.

These can be compared by means of a 2-cell with a source declared to be f_0, and a target declared to be f_1. Name the 2-cell F. The source of the source $s(S(\mathsf{F})) = s(f_0) = a$, the source of the target $s(T(\mathsf{F})) = s(f_1) = a$, The target of the source $t(S(\mathsf{F})) = t(f_0) = b$, and the target of the target $t(T(\mathsf{F})) = t(f_1) = b$. Those conditions are necessary and sufficient in order to declare the existence of such a 2-cell.

After a discussion about composition, the 2-cell F will be declared to be invertible. Its inverse will be denoted in the text by Ⅎ. *But in the context of the current chapter, invertibility is a different notion than has been presented in prior sections.* Here are the schematics for such 2-cells:

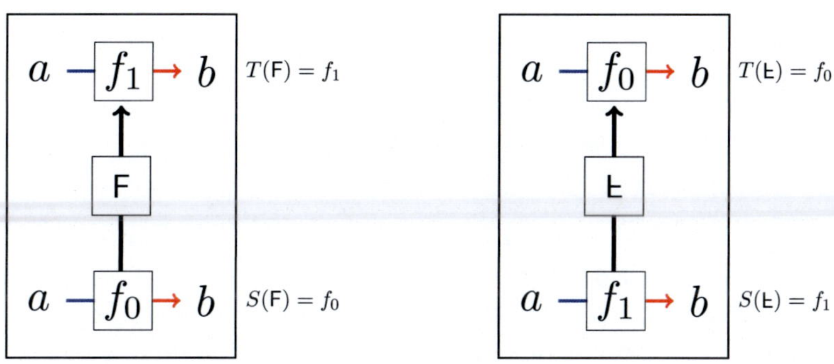

Compositions of 1-cells. For full generality, suppose that there are 0-cells, $a_0, \ldots, a_{n+1}$, together with 1-cells $a_i \,-\!\boxed{f_i}\!\to a_{i+1}$ so that the source $s(f_i) = a_i$ and that $t(f_i) = a_{i+1}$. Then there is an unambiguously defined composition of 1-cells $f = f_0 \circ f_1 \circ \cdots f_n$ that has source $s(f) = a_0$ and target $t(f) = a_{n+1}$:

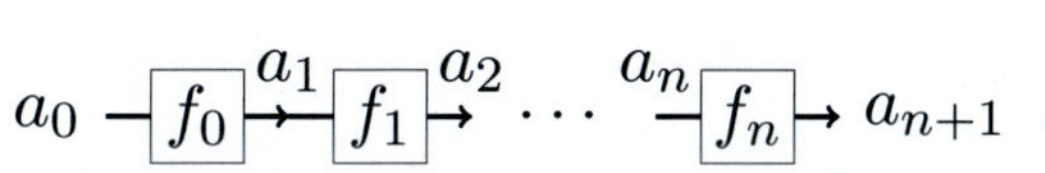

The composition is unambiguous since it is associative. The nodes f_0, f_1, $\ldots$, f_n are equally dispersed along the segment that represents the composition. For the purpose of further schematization, labels upon the 1-cells will be removed:

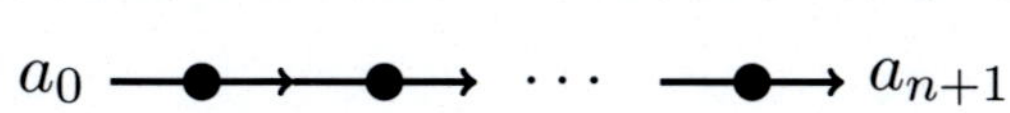

Comparison between 1-cells. Let us consider two exemplary compositions for which their sources and targets agree.

and

These are different things. Label one f and the other g. A comparison between them is a 2-cell H that is drawn in schematic fashion. To the right of the illustration is a "decatigorification" or more specifically, a projection of the 2-cell to a 1-cell so that source and target are apparent.

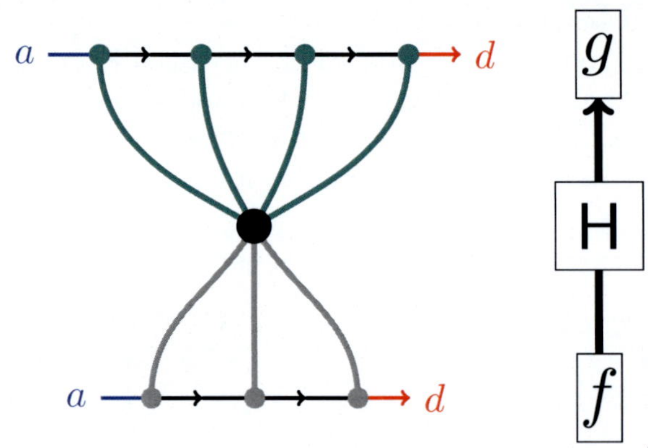

A necessary and sufficient condition for the existence of such a 2-cell, H, is that $s(f) = s(g)$ and that $t(f) = t(g)$. That is, $s(S(\mathsf{H})) = s(T(\mathsf{H}))$ and $t(S(\mathsf{H})) = t(T(\mathsf{H}))$. The intermediate sources and targets for the composite arrows do not affect the existence, or non-existence of such a 2-cell.

Invertibility of 1-cells in the current context. Herein, to assert that the 1-cell 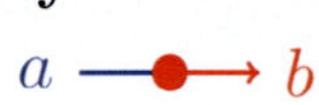 is invertible is the assertion that there is a 1-cell

 together with invertible 2-cells

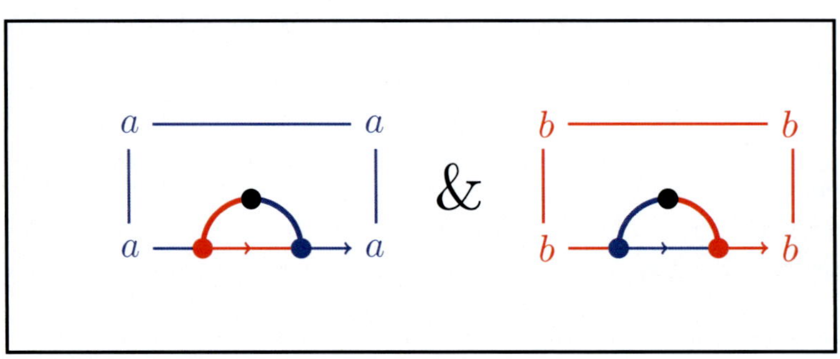

and their reverses. Furthermore, there are 3-cells and their reverses such that there are 4-cells (and reverses) that compare the composition of reversible pairs with the appropriate identities. Higher and higher compositions of cells with their reverses are comparable to identity cells. The differences in the current situation and the prior situations: Chapter 5 Section 3.2, Chapter 6, and Chapter 7 Section 2 are as follows.

The assertion that a given n-cell $a \dashv \boxed{f} \mapsto b$ is invertible is the assertion that there a reversing n-cell $b \dashv \boxed{g} \mapsto a$ together with all the invertible higher dimensional cells formed by comparing the compositions $f \circ g$ and $g \circ f$ to the identity upon the appropriate source or target. **We are not asserting that cuspidal lips, beak-to-beak, horizontal cusps, and swallow-tail isomorphisms exist.** These all have to be constructed at each level.

So even though the declaration that the 2-cell F is invertible includes the assertion that there is a 3-cell such as that depicted below and three additional 3-cells that are not drawn, cuspidal 4-cells and relations among them remain to be constructed.

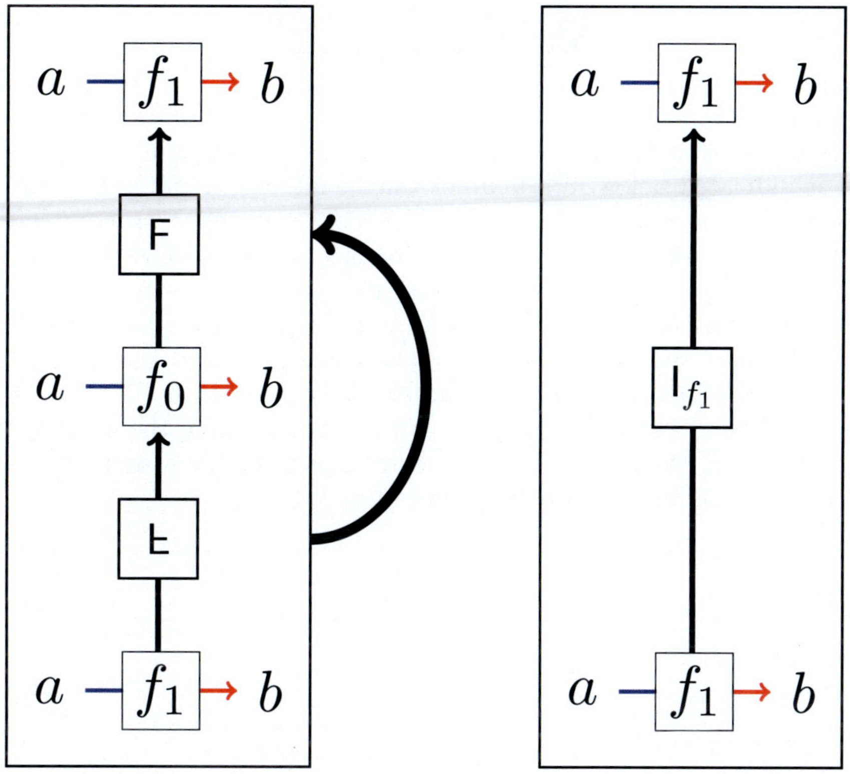

The mythos of n-cells. As a data structure, an n-cell f has $(n-1)$-cells a and b as its source and target respectively. The requirement that $s(a) = s(b)$ and $t(a) = t(b)$ is required. A 0-cell is a point. A 1-cell is depicted as an interval or an arrow, $a \longrightarrow b$, but as a data structure, it exists as a name of the 1-cell with a specific source and target to be found among the 0-cells. In a similar fashion, a 2-cell has a specific source, f_0, and target, f_1, that are found among the collection of 1-cells, with the conditions $s(f_0) = s(f_1)$ and $t(f_0) = t(f_1)$ required.

Nevertheless, we imagine a 2-cell to appear as a rectangle

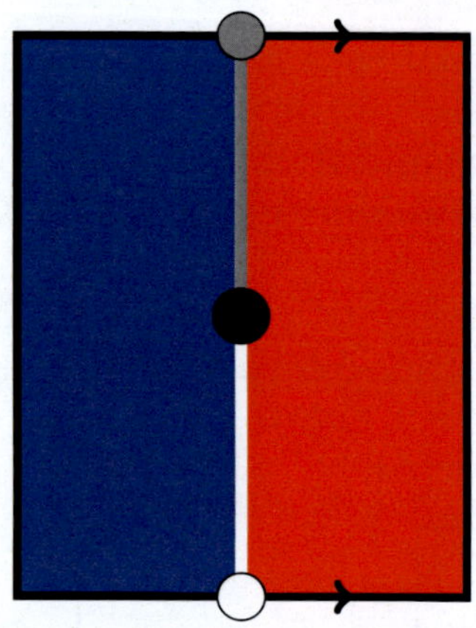

that has a name, say, H_0.

A 3-cell H that has the 2-cell H_0 as its source and the 2-cell H_1 as its target is indicated below in a manner that combines its data structure and the graphical conceptualization of its source and target. To complete the depiction, H is conceptualized as a 3-dimensional cube. The faces $z = 0, 1$ correspond, respectively, to the source, H_0, and target, H_1, 2-cells. The $y = 0$ face corresponds to the identity I_{f_0} upon f_0 — the identity upon the source $S(\mathsf{H}_0) = S(\mathsf{H}_1)$. The $y = 1$ face corresponds to the identity, I_{f_1}, upon f_1 — the identity upon the target $T(\mathsf{H}_0) = T(\mathsf{H}_1)$. The $x = 0$ face corresponds to the identity I_{I_a} which is the identity upon the identity upon the source of f_0. And the $x = 1$ face corresponds to the identity I_{I_b} — the identity upon the identity of the target of f_1.

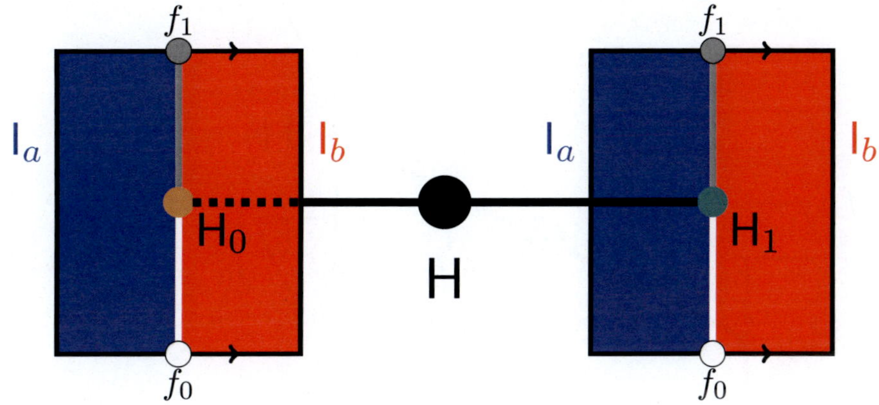

As a final schematic, the illustration below indicates a 4-cell H that has a 3-cell H_0 as its source and a 3-cell H_1 as its target. The 4-cell H is conceptualized as occupying the 4-dimensional cube

$$\{(x_1, x_2, x_3, x_4) : 0 \le x_j \le 1 \text{ for } j = 1, 2, 3, 4\}.$$

In that conceptualization, the various 3-cells that occupy the boundary 3-dimensional cubes will be listed.

- The face $x_4 = 0$ contains the source H_0. The face $x_4 = 1$ contains the target H_1.
- The face $x_3 = 0$ contains the identity upon F_0 which is the source of H_0 as well as the source of H_1. The face $x_3 = 1$ contains the identity upon F_1 which is the target of H_0 as well as the target of H_1.
- The face $x_2 = 0$ contains the identity upon the identity upon f_0 which is the source of the 2-cell F_0. The 1-cell f_0 is also the source of the 2-cell F_1. The face $x_2 = 1$ contains the identity upon the identity upon f_1 which is the target of the 2-cell F_0. The 1-cell f_1 is also the target of the 2-cell F_1.
- The face $x_1 = 0$ contains the 3-stage identity upon the source $s(f_0) = s(f_1)$. The face $x_1 = 1$ contains the 3-stage identity upon the target $t(f_0) = t(f_1)$. For example, the source of f_0 is a point. The identity upon it is the source of F_0. The 2-stage identity upon the source of f_0 is the identity upon the source of F_0. The 3-stage identity upon f_0, is the identity upon the source of H_0.

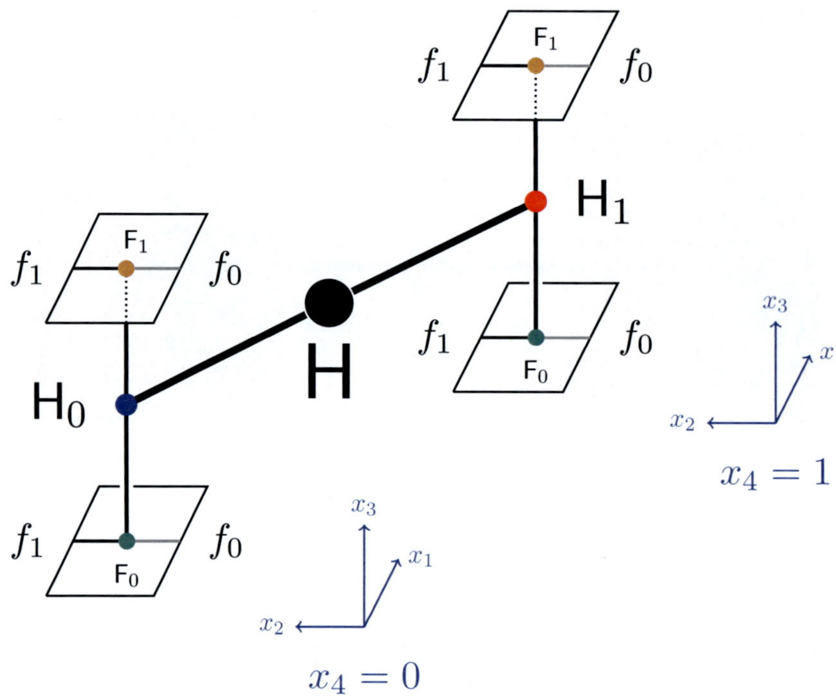

Within the GLOBULAR software, there is a projection into the 2-dimensional plane wherein the source of the cell appears at the bottom and the target appears at the top, for any n-cell with $2 \leq n$. In moving from the schematic above, to an n-dimensional cube, a similar projection can be defined. The identity upon the common source appears on the left, and the identity upon the common target appears on the right. For larger n, both the source and target may be sliced. Each may be a composite, and a sequence of slices describe the intermediate cells that are composed.

1.1. Axiomatics of n-cells. An inductive process for declaring n-cells will be presented. At the earliest level, suppose that a (finite) set of 0-cells has been declared and named. For $0 \leq n$, there is an identity $(n+1)$-cell for any given n-cell. In particular, for any 0-cell, there are inductively defined identity n-cells for any dimension n.

A 1-cell H can be declared by specifying a source and target $s(\mathsf{H}) = \mathsf{H}_0, t(\mathsf{H}) = \mathsf{H}_1 \in \{0\text{–cells}\}$. Two 1-cells F and H can be composed to form a 1-cell $\mathsf{G} = \mathsf{F} \circ \mathsf{H}$ provided that $\mathsf{F}_1 = \mathsf{H}_0$ — the source of H is the target of F. For the record, $\mathsf{G}_0 = s(\mathsf{G}) = s(\mathsf{F}) = \mathsf{F}_0$ while $\mathsf{G}_1 = t(\mathsf{G}) = t(\mathsf{H}) = \mathsf{H}_1$. In the composition $\mathsf{G} = \mathsf{F} \circ \mathsf{H}$, the 1-cell F is called *the lower cell*, and the 1-cell H the *higher cell*. When a 1-cell H is declared to be invertible, the inverse of H is automatically considered to be an allowable 1-cell.

The set of 1-cells consist of those that have been declared and virtual 1-cells that are formed as possible compositions of 1-cells, and allowable inverses. Composition of 1-cells is strictly associative and identity 1-cells serve as identities under composition. They may be inserted or deleted as needed.

Suppose that a generating set of n-cells has been declared. Given two n-cells[2] H_0 and H_1 that have coincident sources and coincident targets ($s(\mathsf{H}_0) = s(\mathsf{H}_1) = \mathsf{H}_{00} = \mathsf{H}_{01}$ and $t(\mathsf{H}_0) = t(\mathsf{H}_1) = \mathsf{H}_{10} = \mathsf{H}_{11}$), an $(n+1)$-cell H may be declared with $s(\mathsf{H}) = \mathsf{H}_0$ and $t(\mathsf{H}) = \mathsf{H}_1$.

For $1 \leq n$, the (direct) composition of $(n+1)$-cells is defined as in the case of the composition of 1-cells. The target of the lower $(n+1)$-cell must agree with the source of the higher $(n+1)$-cell.

A *globular signature* is the collection of cells that have been declared.

From the schema that are outlined, skew compositions of n-cells for $1 < n$ can be constructed. Moreover, the schematics of exchangers (or interchangers) that were given in Chapter 3, Section 2.3 can be implemented. At the software level of GLOBULAR.SCIENCE exchanger n-cells are implicitly given within the program. Moreover, the naturality axioms of exchangers are higher order exchangers.

2. Group presentations

Recall that a group is a category that has one object and for which every morphism is invertible. In this section, the eight element dihedral group

$$D = \langle R, T : R^4 = T^2 = (RT)^2 = 1 \rangle$$

will be presented as a globular 3-category. The object is $*$. The generating 1-cells are 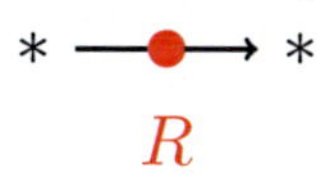

and 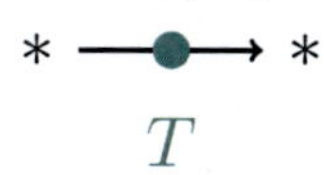

. In general, a group element is represented as a word in T and R. Such a word is a composition of 1-cells, of a form similar to this:

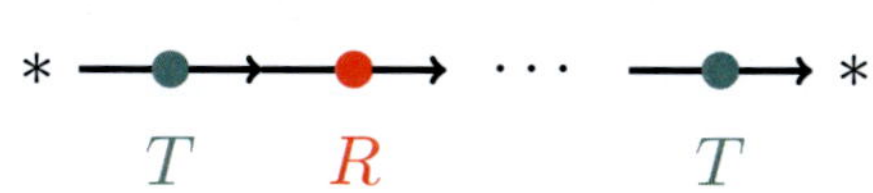

.

The relations in the group can be expressed as 2-cells for which the source (bottom edge) is the given word, e.g. R^4, T^2, or $(RT)^2$, and the target 1-cell is the identity $* \longrightarrow *$. These 2-cells should be declared to be invertible. For example, there are 2-cells that are obtained from the

[2]These n-cells may be found among those that have been declared, or they may be constructed by means of composition.

declaration that R is invertible:

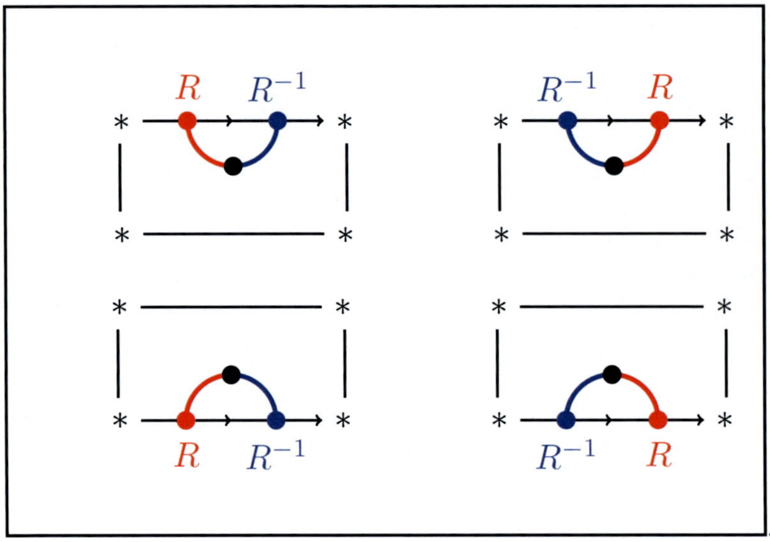

Since $T^2 = 1$, an inverse is not explicitly drawn, but this relation gives rise to the 2-cells below.

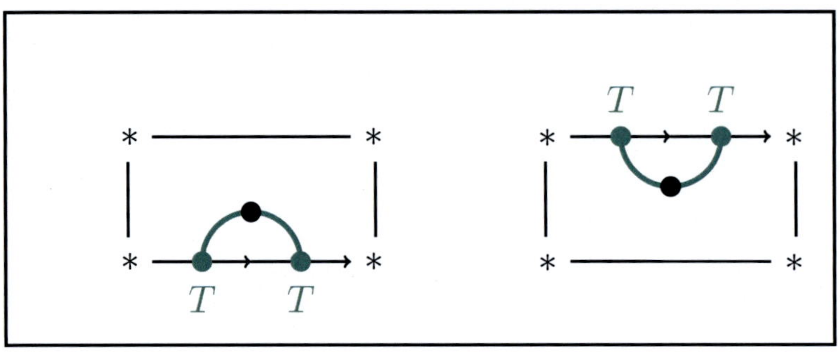

The remaining relational 2-cells are depicted.

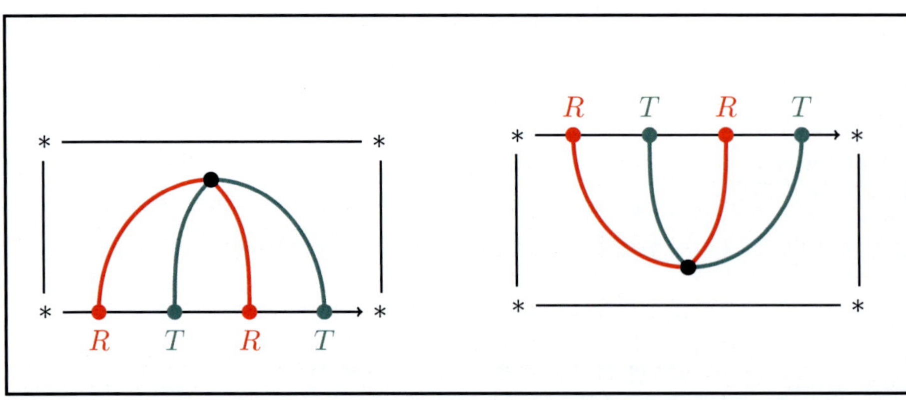

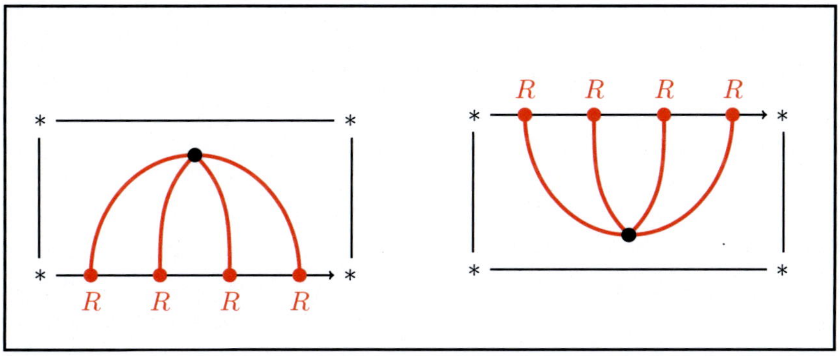

It can be shown that the elements in this group are found among the representative words $\{1, R, R^2, R^3, T, TR, TR^2, TR^3\}$. Key facts that are often used to prove this are the following consequences of the relations: $R^{-1} = R^3$, $R^{-2} = R^2$, and $RT = TR^3$. In order to demonstrate these consequences, several 2-cells that are of the form $R^{-1} \Rightarrow R^3$, $R^{-2} \Rightarrow R^2$, and $RT \Rightarrow TR^3$ will be constructed from the generating set of 2-cells. These also should be considered to be invertible.

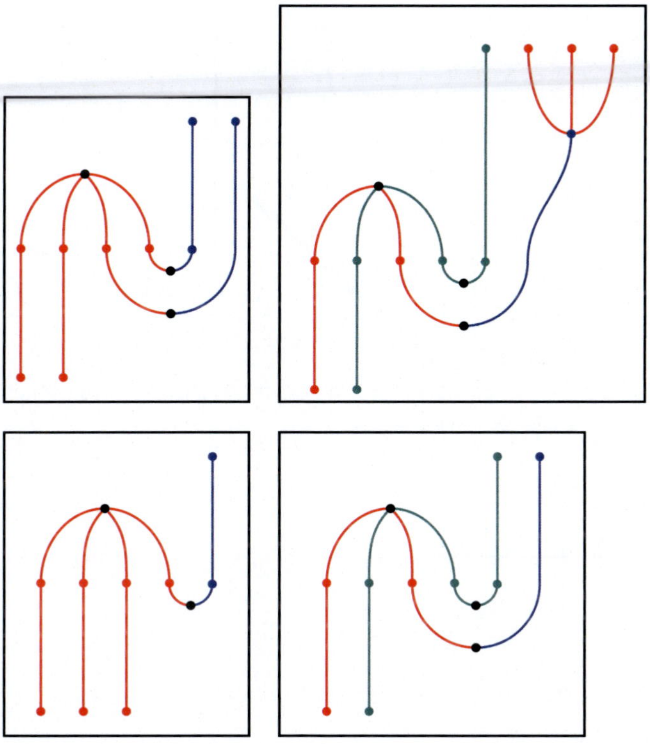

To say these cells are of that form is to assert that the source, at the bottom edge of the cell, coincides with the left-hand expression, while the target, at the top, coincides with the right-hand-side. The cells are constructed in a graphical fashion. In the illustration that is above, the 2-cells imply $R^2 = R^{-2}$, $RT = TR^3$, $R^3 = R^{-1}$, and $RT = TR^{-1}$. The 2-cell in the upper right corner is a consequence of the inverse of the 2-cell that is depicted in the lower left corner.

At the last stage of the process, Theorems that are 3-cells are asserted. These will have as their source the composition of such 2-cells and as their targets the purported equations. For completeness of thought, glyphs label these 3-cells. At their bottom, sparse schematics indicate compositions of 2-cells that coincide with the black vertices on the source 2-cell. At their top, a single black vertex emerges. It indicates that the top 2-cell is the composition of those at the bottom.

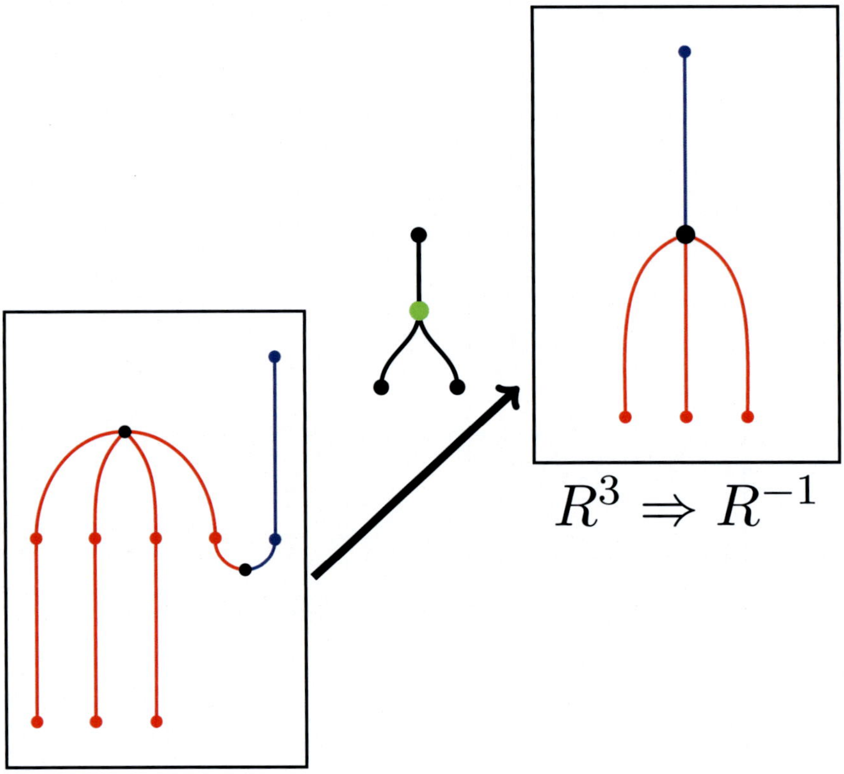

$R^3 \Rightarrow R^{-1}$

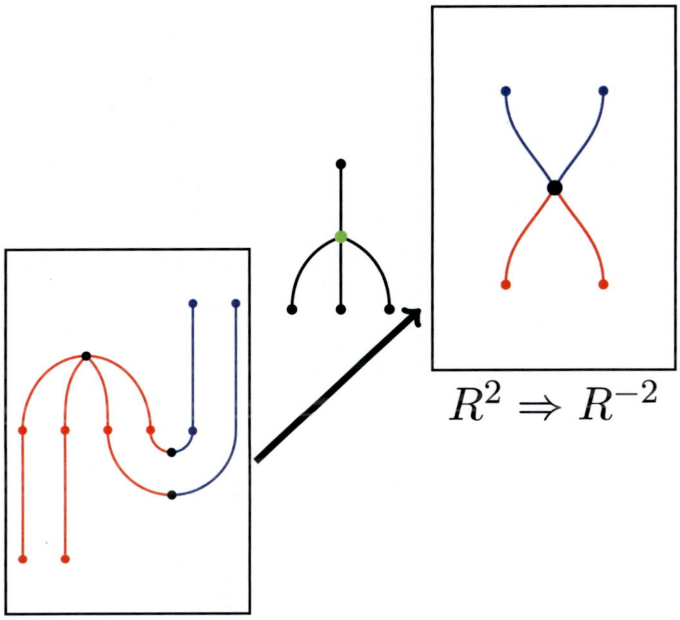

$R^2 \Rightarrow R^{-2}$

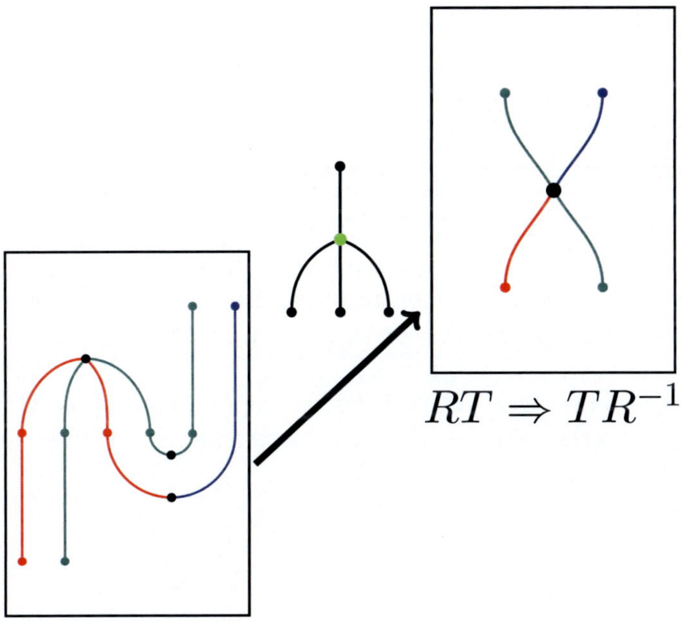

$RT \Rightarrow TR^{-1}$

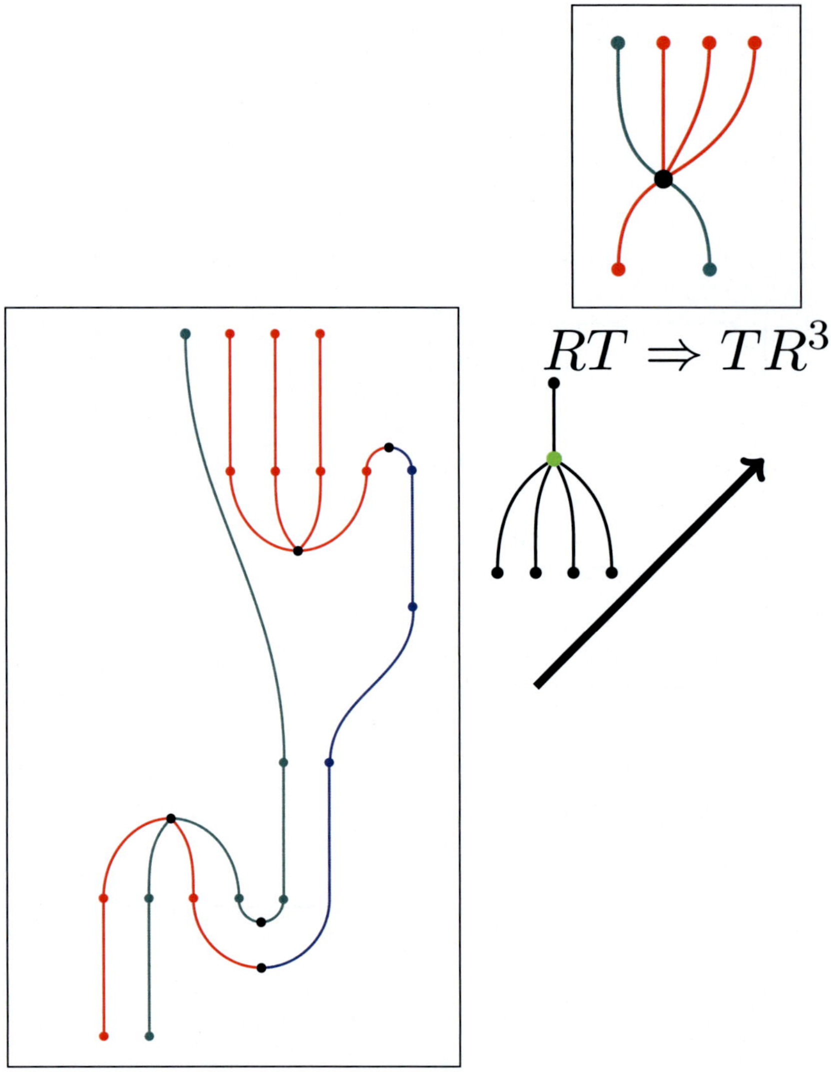

Exercise. Choose a group presentation of your favorite finite group. Construct 1-cells that represent generators and 2-cells that represent relators. Determine a set of consequences for these relators and express them via the determination that each generator is invertible. In case that you know identities among the relations construct the corresponding 3-cells. Compare the resulting glyphs to the vertices of braid charts that are given in Chapter 11, Section 3.3.

3. Conclusion

The five principles framing this book are recalled:

(1) Different things are not the same. They may be naturally isomorphic.
(2) Arrows in a category are used to compare things. In particular, arrows quantify differences between things.
(3) When an action (or arrow) is applied and the process is reversed, the result may, or may not, be the same as inaction. So an arrow followed by its reverse can be compared (by an arrow of larger degree) to staying still. Therefore, "doing" followed by "undoing" may be, or may not be, the same as not doing.
(4) Simultaneity is illusory. This leads to the idea of exchange.
(5) Change followed by exchange is comparable to exchange followed by change via an arrow of higher degree.

The things t and ʇ are not the same. The comparisons between them are $\mathsf{t} \xrightarrow{\mathsf{p}}$ ʇ and ʇ $\xrightarrow{\mathsf{b}} \mathsf{t}$. These might be thought of as venturing out and returning, and a round trip in either direction can be compared to remaining in place. When these and other higher arrows (or cells) are considered (or constructed) such a small amount of information leads to the study of properly embedded surfaces in space.

In case, there is but one thing, $*$ and an identity arrow $* — *$, then a 2-cell, $\boxed{\bullet}$ that begins at and ends at that identity can be declared to exist. Such a generating double arrow allows for counting and positional exchange to take place. The naturality of the exchanger with respect to other triple arrows leads to the theory of knots and braids. An example of naturality is the naturality with respect to another exchange. "Change followed by exchange is exchange followed by change" expresses, in particular, the type III move.

The maxim "simultaneity is illusory" is usually stated in physical terms that two particles cannot occupy the space at the same time. That physical principle ignores the natures of space, time, and particle. In our approach, time is a metaphor for the ultimate spacial dimension, and particle is the 2-cell or double arrow $\boxed{\bullet}$. Space emerges by constructing more and more comparisons.

Two particles can exchange place: X. They can merge into one ⅄. One can split into two: Y. They can annihilate: ∩. And they can be created from the nothingness that is the identity upon the identity upon $*$: U. Apply a process and its reverse. The comparisons of doing followed by undoing to not doing create the atomic pieces of surface and foam. Specific interactions among these comparisions become the substance of the next particulate level to form the vertices of the glyphs. Edges of glyphs are the identities of the interactions.

Within the braid context, the things to be compared are crossing and not crossing. The comparison is a black vertex. Also, positive and negative crossings (exchanges) are compared, and in that case the Fs and Ⅎs are type II Reidemeister moves.

Syllepses compare X and $\overline{\mathsf{X}}$. Set t = X and ʇ = $\overline{\mathsf{X}}$ with source and sink syllepses as b and p. Within the context of braids, lines and surface intersections emerge.

More can occur once syllepses are introduced. Braids reduce to permutations. In Section 2 (this chapter), an example outlined how to conceive of relations in a group as 2-cells, and consequences of relations as 3-cells. Whereas finite groups have faithful permutation representations, finite groups might also be considered within the context of symmetric globular multi-categories that are formed by means of exchanging 3-cells and syllepses. Structural similarities between this situation and that of braid movies, curtains, and curtain movies deserve exploration.

We have encoded the glyphs in the group context in a cartoonish fashion because globular multi-categories only see an arrow (or cell) as a source and a target, and compare how these are similar and how they differ. Our hope is that the reader can enhance such glyphs with bookmarks and serifs in order to create functional languages that encapsulate the algebraic system that is more easily expressed diagrammatically.

A final statement about the scope of this treatise is appropriate. The diagrammatic representation of algebraic and categorical structures is far more encompassing than a text such as this can cover. Our language choices may, or may not be, standard with respect to the remaining literature. Occasionally, these choices are made out of ignorance. At other times, they are made in hopes of establishing our tropes as standards. Our hope is that by molding the ideas that have been exemplified here, the reader can create additional and deeper understanding of diagrammatic algebra. Thereby you can create your own methods. We wish you the best of luck in these endeavors!

Bibliography

[1] Samson Abramsky and Bob Coecke. Categorical quantum mechanics. In *Handbook of quantum logic and quantum structures—quantum logic*, pages 261–323. Elsevier/North-Holland, Amsterdam, 2009.

[2] Colin C. Adams. *The knot book*. American Mathematical Society, Providence, RI, 2004. An elementary introduction to the mathematical theory of knots, Revised reprint of the 1994 original.

[3] J. W. Alexander and G. B. Briggs. On types of knotted curves. *Ann. of Math. (2)*, 28(1-4):562–586, 1926/27.

[4] Yongju Bae, J. Scott Carter, Seonmi Choi, and Sera Kim. Non-orientable surfaces in 4-dimensional space. *J. Knot Theory Ramifications*, 23(11):1430002, 52, 2014.

[5] John C. Baez, Brandon Coya, and Franciscus Rebro. Props in network theory. *Theory Appl. Categ.*, 33:Paper No. 25, 727–783, 2018.

[6] John C. Baez and James Dolan. Higher-dimensional algebra and topological quantum field theory. *J. Math. Phys.*, 36(11):6073–6105, 1995.

[7] John C. Baez and Laurel Langford. Higher-dimensional algebra. IV. 2-tangles. *Adv. Math.*, 180(2):705–764, 2003.

[8] Dror Bar-Natan. On Khovanov's categorification of the Jones polynomial. *Algebr. Geom. Topol.*, 2:337–370, 2002.

[9] Dror Bar-Natan. Khovanov's homology for tangles and cobordisms. *Geom. Topol.*, 9:1443–1499, 2005.

[10] Kenneth S. Brown. *Cohomology of groups*, volume 87 of *Graduate Texts in Mathematics*. Springer-Verlag, New York, 1994. Corrected reprint of the 1982 original.

[11] J. W. Bruce and P. J. Giblin. *Curves and singularities*. Cambridge University Press, Cambridge, second edition, 1992. A geometrical introduction to singularity theory.

[12] J. Scott Carter. Reidemeister/Roseman-type moves to embedded foams in 4-dimensional space. In *New ideas in low dimensional topology*, volume 56 of *Ser. Knots Everything*, pages 1–30. World Sci. Publ., Hackensack, NJ, 2015.

[13] J. Scott Carter, Daniel E. Flath, and Masahico Saito. *The classical and quantum 6j-symbols*, volume 43 of *Mathematical Notes*. Princeton University Press, Princeton, NJ, 1995.

[14] J. Scott Carter and Seiichi Kamada. How to fold a manifold. In *New ideas in low dimensional topology*, volume 56 of *Ser. Knots Everything*, pages 31–77. World Sci. Publ., Hackensack, NJ, 2015.

[15] J. Scott Carter and Masahico Saito. *Knotted surfaces and their diagrams*, volume 55 of *Mathematical Surveys and Monographs*. American Mathematical Society, Providence, RI, 1998.

[16] Jean Cerf. La stratification naturelle des espaces de fonctions différentiables réelles et le théorème de la pseudo-isotopie. *Inst. Hautes Études Sci. Publ. Math.*, (39):5–173, 1970.

[17] Bob Coecke, Fabrizio Genovese, Martha Lewis, Dan Marsden, and Alex Toumi. Generalized relations in linguistics & cognition. *Theoret. Comput. Sci.*, 752:104–115, 2018.

[18] Bob Coecke and Aleks Kissinger. Categorical quantum mechanics I: causal quantum processes. In *Categories for the working philosopher*, pages 286–328. Oxford Univ. Press, Oxford, 2017.

[19] J. H. Conway. An enumeration of knots and links, and some of their algebraic properties. In *Computational Problems in Abstract Algebra (Proc. Conf., Oxford, 1967)*, pages 329–358. Pergamon, Oxford, 1970.

[20] Louis Crane and Igor B. Frenkel. Four-dimensional topological quantum field theory, Hopf categories, and the canonical bases. volume 35, pages 5136–5154. 1994. Topology and physics.

[21] *et al.* eds. C. Adams. *Encyclopedia of Knot Theory.* CRC.

[22] P. Freyd, D. Yetter, J. Hoste, W. B. R. Lickorish, K. Millett, and A. Ocneanu. A new polynomial invariant of knots and links. *Bull. Amer. Math. Soc. (N.S.)*, 12(2):239–246, 1985.

[23] Cole A. Giller. Towards a classical knot theory for surfaces in $\mathbf{R}^4$. *Illinois J. Math.*, 26(4):591–631, 1982.

[24] M. Golubitsky and V. Guillemin. *Stable mappings and their singularities.* Springer-Verlag, New York-Heidelberg, 1973. Graduate Texts in Mathematics, Vol. 14.

[25] Amar Hadzihasanovic. A diagrammatic axiomatisation for qubit entanglement. In *2015 30th Annual ACM/IEEE Symposium on Logic in Computer Science (LICS 2015)*, pages 573–584. IEEE Computer Soc., Los Alamitos, CA, 2015.

[26] Amar Hadzihasanovic. A combinatorial-topological shape category for polygraphs. *Appl. Categ. Structures*, 28(3):419–476, 2020.

[27] Allen Hatcher. *Algebraic topology.* Cambridge University Press, Cambridge, 2002.

[28] M. A. Hennings. On solutions to the braid equation identified by Woronowicz. *Lett. Math. Phys.*, 27(1):13–17, 1993.

[29] Morris W. Hirsch. *Differential topology.* Springer-Verlag, New York-Heidelberg, 1976. Graduate Texts in Mathematics, No. 33.

[30] Jim Hoste, Morwen Thistlethwaite, and Jeff Weeks. The first 1,701,936 knots. *Math. Intelligencer*, 20(4):33–48, 1998.

[31] Atsushi Ishii. Moves and invariants for knotted handlebodies. *Algebr. Geom. Topol.*, 8(3):1403–1418, 2008.

[32] Atsushi Ishii, Masahide Iwakiri, Yeonhee Jang, and Kanako Oshiro. A G-family of quandles and handlebody-knots. *Illinois J. Math.*, 57(3):817–838, 2013.

[33] Shyuichi Izumiya, Maria del Carmen Romero Fuster, Maria Aparecida Soares Ruas, and Farid Tari. *Differential geometry from a singularity theory viewpoint.* World Scientific Publishing Co. Pte. Ltd., Hackensack, NJ, 2016.

[34] V. F. R. Jones. Hecke algebra representations of braid groups and link polynomials. *Ann. of Math. (2)*, 126(2):335–388, 1987.

[35] Vaughan F. R. Jones. A polynomial invariant for knots via von Neumann algebras. *Bull. Amer. Math. Soc. (N.S.)*, 12(1):103–111, 1985.

[36] André Joyal and Ross Street. Braided tensor categories. *Adv. Math.*, 102(1):20–78, 1993.

[37] David Joyce. A classifying invariant of knots, the knot quandle. *J. Pure Appl. Algebra*, 23(1):37–65, 1982.

[38] Seiichi Kamada. Alexander's and Markov's theorems in dimension four. *Bull. Amer. Math. Soc. (N.S.)*, 31(1):64–67, 1994.

[39] Seiichi Kamada. Arrangement of Markov moves for 2-dimensional braids. In *Low-dimensional topology (Funchal, 1998)*, volume 233 of *Contemp. Math.*, pages 197–213. Amer. Math. Soc., Providence, RI, 1999.

[40] Seiichi Kamada. *Braid and knot theory in dimension four*, volume 95 of *Mathematical Surveys and Monographs.* American Mathematical Society, Providence, RI, 2002.

[41] Louis H. Kauffman. *On knots*, volume 115 of *Annals of Mathematics Studies.* Princeton University Press, Princeton, NJ, 1987.

[42] Louis H. Kauffman. State models and the Jones polynomial. *Topology*, 26(3):395–407, 1987.

[43] Louis H. Kauffman. *Knots and physics*, volume 53 of *Series on Knots and Everything.* World Scientific Publishing Co. Pte. Ltd., Hackensack, NJ, fourth edition, 2013.

[44] Louis H. Kauffman and Sóstenes L. Lins. *Temperley-Lieb recoupling theory and invariants of 3-manifolds*, volume 134 of *Annals of Mathematics Studies.* Princeton University Press, Princeton, NJ, 1994.

[45] Louis H. Kauffman and David E. Radford. Invariants of 3-manifolds derived from finite-dimensional Hopf algebras. *J. Knot Theory Ramifications*, 4(1):131–162, 1995.

[46] Kengo Kawamura, Kanako Oshiro, and Kokoro Tanaka. Independence of Roseman moves including triple points. *Algebr. Geom. Topol.*, 16(4):2443–2458, 2016.

[47] Akio Kawauchi. *A survey of knot theory.* Birkhäuser Verlag, Basel, 1996. Translated and revised from the 1990 Japanese original by the author.

[48] Mikhail Khovanov. A categorification of the Jones polynomial. *Duke Math. J.*, 101(3):359–426, 2000.

[49] Mikhail Khovanov. A categorification of the Jones polynomial. *Duke Math. J.*, 101(3):359–426, 2000.

[50] Robion Kirby. A calculus for framed links in S^3. *Invent. Math.*, 45(1):35–56, 1978.

[51] Greg Kuperberg. Involutory Hopf algebras and 3-manifold invariants. *Internat. J. Math.*, 2(1):41–66, 1991.

[52] R. Lawrence. Homological representations of the Hecke algebra. *Comm. Math. Phys.*, 135:141–192, 1990.

[53] Charles Livingston. *Knot theory*, volume 24 of *Carus Mathematical Monographs*. Mathematical Association of America, Washington, DC, 1993.

[54] Shahn Majid. *Foundations of quantum group theory*. Cambridge University Press, Cambridge, 1995.

[55] William S. Massey. *A basic course in algebraic topology*, volume 127 of *Graduate Texts in Mathematics*. Springer-Verlag, New York, 1991.

[56] S. V. Matveev. Transformations of special spines, and the Zeeman conjecture. *Izv. Akad. Nauk SSSR Ser. Mat.*, 51(5):1104–1116, 1119, 1987.

[57] Sergei Matveev. Distributive groupoids in knot theory. *Mat. Sb. (N.S.)*, 119(161)(1):78–88, 160, 1982.

[58] Jenelle McAtee, Daniel S. Silver, and Susan G. Williams. Coloring spatial graphs. *J. Knot Theory Ramifications*, 10(1):109–120, 2001.

[59] Marston Morse and Stewart S. Cairns. *Critical point theory in global analysis and differential topology: An introduction*. Pure and Applied Mathematics, Vol. 33. Academic Press, New York-London, 1969.

[60] Kunio Murasugi. *Knot theory and its applications*. Birkhäuser Boston, Inc., Boston, MA, 1996. Translated from the 1993 Japanese original by Bohdan Kurpita.

[61] Udo Pachner. P.L. homeomorphic manifolds are equivalent by elementary shellings. *European J. Combin.*, 12(2):129–145, 1991.

[62] Roger Penrose. Applications of negative dimensional tensors. In *Combinatorial Mathematics and its Applications (Proc. Conf., Oxford, 1969)*, pages 221–244. Academic Press, London, 1971.

[63] Riccardo Piergallini. Standard moves for standard polyhedra and spines. *Rend. Circ. Mat. Palermo (2) Suppl.*, (18):391–414, 1988. Third National Conference on Topology (Italian) (Trieste, 1986).

[64] Kurt Reidemeister. Knoten und Gruppen. *Abh. Math. Sem. Univ. Hamburg*, 5(1):7–23, 1927.

[65] N. Reshetikhin and V. G. Turaev. Invariants of 3-manifolds via link polynomials and quantum groups. *Invent. Math.*, 103(3):547–597, 1991.

[66] J. H. Rieger. Families of maps from the plane to the plane. *J. London Math. Soc. (2)*, 36(2):351–369, 1987.

[67] Dale Rolfsen. *Knots and links*, volume 7 of *Mathematics Lecture Series*. Publish or Perish, Inc., Houston, TX, 1990. Corrected reprint of the 1976 original.

[68] Dennis Roseman. Reidemeister-type moves for surfaces in four-dimensional space. In *Knot theory (Warsaw, 1995)*, volume 42 of *Banach Center Publ.*, pages 347–380. Polish Acad. Sci. Inst. Math., Warsaw, 1998.

[69] Colin Patrick Rourke and Brian Joseph Sanderson. *Introduction to piecewise-linear topology*. Springer Study Edition. Springer-Verlag, Berlin-New York, 1982. Reprint.

[70] Kentaro Saji. Criteria for singularities of smooth maps from the plane into the plane and their applications. *Hiroshima Math. J.*, 40(2):229–239, 2010.

[71] P. Selinger. A survey of graphical languages for monoidal categories. In *New structures for physics*, volume 813 of *Lecture Notes in Phys.*, pages 289–355. Springer, Heidelberg, 2011.

[72] Mei Chee Shum. Tortile tensor categories. *J. Pure Appl. Algebra*, 93(1):57–110, 1994.

[73] Richard P. Stanley. *Catalan numbers*. Cambridge University Press, New York, 2015.

[74] James Dillon Stasheff. Homotopy associativity of H-spaces. I, II. *Trans. Amer. Math. Soc. 108 (1963), 275-292; ibid.*, 108:293–312, 1963.

[75] V. G. Turaev. The Yang-Baxter equation and invariants of links. *Invent. Math.*, 92(3):527–553, 1988.

[76] V. G. Turaev and O. Ya. Viro. State sum invariants of 3-manifolds and quantum $6j$-symbols. *Topology*, 31(4):865–902, 1992.

[77] Hassler Whitney. On singularities of mappings of euclidean spaces. I. Mappings of the plane into the plane. *Ann. of Math. (2)*, 62:374–410, 1955.

[78] Edward Witten. Quantum field theory and the Jones polynomial. *Comm. Math. Phys.*, 121(3):351–399, 1989.

[79] Takeshi Yajima. On simply knotted spheres in R^4. *Osaka Math. J.*, 1:133–152, 1964.

[80] David N. Yetter. Quantum groups and representations of monoidal categories. *Math. Proc. Cambridge Philos. Soc.*, 108(2):261–290, 1990.

[81] A. B. Zamolodchikov. Tetrahedron equations and the relativistic S-matrix of straight-strings in $2 + 1$-dimensions. *Comm. Math. Phys.*, 79(4):489–505, 1981.

Index

Selected Published Titles in This Series

264 **J. Scott Carter and Seiichi Kamada,** Diagrammatic Algebra, 2021

259 **William Heinzer, Christel Rotthaus, and Sylvia Wiegand,** Integral Domains Inside Noetherian Power Series Rings, 2021

258 **Pramod N. Achar,** Perverse Sheaves and Applications to Representation Theory, 2021

257 **Juha Kinnunen, Juha Lehrbäck, and Antti Vähäkangas,** Maximal Function Methods for Sobolev Spaces, 2021

256 **Michio Jimbo, Tetsuji Miwa, and Fedor Smirnov,** Local Operators in Integrable Models I, 2021

255 **Alexandre Boritchev and Sergei Kuksin,** One-Dimensional Turbulence and the Stochastic Burgers Equation, 2021

254 **Karim Belabas and Henri Cohen,** Numerical Algorithms for Number Theory, 2021

253 **Robert R. Bruner and John Rognes,** The Adams Spectral Sequence for Topological Modular Forms, 2021

252 **Julie Déserti,** The Cremona Group and Its Subgroups, 2021

251 **David Hoff,** Linear and Quasilinear Parabolic Systems, 2020

250 **Bachir Bekka and Pierre de la Harpe,** Unitary Representations of Groups, Duals, and Characters, 2020

249 **Nikolai M. Adrianov, Fedor Pakovich, and Alexander K. Zvonkin,** Davenport–Zannier Polynomials and Dessins d'Enfants, 2020

248 **Paul B. Larson and Jindrich Zapletal,** Geometric Set Theory, 2020

247 **István Heckenberger and Hans-Jürgen Schneider,** Hopf Algebras and Root Systems, 2020

246 **Matheus C. Bortolan, Alexandre N. Carvalho, and José A. Langa,** Attractors Under Autonomous and Non-autonomous Perturbations, 2020

245 **Aiping Wang and Anton Zettl,** Ordinary Differential Operators, 2019

244 **Nabile Boussaïd and Andrew Comech,** Nonlinear Dirac Equation, 2019

243 **José M. Isidro,** Jordan Triple Systems in Complex and Functional Analysis, 2019

242 **Bhargav Bhatt, Ana Caraiani, Kiran S. Kedlaya, Peter Scholze, and Jared Weinstein,** Perfectoid Spaces, 2019

241 **Dana P. Williams,** A Tool Kit for Groupoid C^*-Algebras, 2019

240 **Antonio Fernández López,** Jordan Structures in Lie Algebras, 2019

239 **Nicola Arcozzi, Richard Rochberg, Eric T. Sawyer, and Brett D. Wick,** The Dirichlet Space and Related Function Spaces, 2019

238 **Michael Tsfasman, Serge Vlăduţ, and Dmitry Nogin,** Algebraic Geometry Codes: Advanced Chapters, 2019

237 **Dusa McDuff, Mohammad Tehrani, Kenji Fukaya, and Dominic Joyce,** Virtual Fundamental Cycles in Symplectic Topology, 2019

236 **Bernard Host and Bryna Kra,** Nilpotent Structures in Ergodic Theory, 2018

235 **Habib Ammari, Brian Fitzpatrick, Hyeonbae Kang, Matias Ruiz, Sanghyeon Yu, and Hai Zhang,** Mathematical and Computational Methods in Photonics and Phononics, 2018

234 **Vladimir I. Bogachev,** Weak Convergence of Measures, 2018

233 **N. V. Krylov,** Sobolev and Viscosity Solutions for Fully Nonlinear Elliptic and Parabolic Equations, 2018

232 **Dmitry Khavinson and Erik Lundberg,** Linear Holomorphic Partial Differential Equations and Classical Potential Theory, 2018

For a complete list of titles in this series, visit the
AMS Bookstore at **www.ams.org/bookstore/survseries/**.